Tomorrow's Chemistry Today

Edited by
Bruno Pignataro

Related Titles

Mathias Christmann, Sefan Bräse (eds.)

Asymmetric Synthesis – The Essentials

2nd, completely revised edition

2007
ISBN: 978-3-527-32093-6

Tomas Hudlicky and Josephine W. Reed

The Way of Synthesis

Evolution of Design and Methods for Natural Products

2007
ISBN: 978-3-527-32077-6

Carsten Schmuck, Helma Wennemers (eds.)

Highlights in Bioorganic Chemistry

Methods and Applications

2004
ISBN: 978-3-527-30656-5

Kyriacos C. Nicolaou, Scott A. Snyder

Classics in Total Synthesis II

More Targets, Strategies, Methods

2003
ISBN: 978-3-527-30685-5

Guo-Qiang Lin, Yue-Ming Li, Albert S. C. Chan

Principles and Applications of Asymmetric Synthesis

2001
ISBN: 978-0-471-40027-1

Tomorrow's Chemistry Today

Concepts in Nanoscience, Organic Materials
and Environmental Chemistry

Edited by
Bruno Pignataro

WILEY-VCH Verlag GmbH & Co. KGaA

The Editor

Professor Bruno Pignataro
Department of Physical Chemistry
University of Palermo
Valle delle Scienze
90128 Palermo
Italy

■ All books published by Wiley-VCH are carefully produced. Nevertheless, authors, editors, and publisher do not warrant the information contained in these books, including this book, to be free of errors. Readers are advised to keep in mind that statements, data, illustrations, procedural details or other items may inadvertently be inaccurate.

Library of Congress Card No.:
applied for

British Library Cataloguing-in-Publication Data
A catalogue record for this book is available from the British Library.

Bibliographic information published by the Deutsche Nationalbibliothek
The Deutsche Nationalbibliothek lists this publication in the Deutsche Nationalbibliografie; detailed bibliographic data are available in the Internet at <http://dnb.d-nb.de>.

© 2008 WILEY-VCH Verlag GmbH & Co. KGaA, Weinheim

All rights reserved (including those of translation into other languages). No part of this book may be reproduced in any form – by photoprinting, microfilm, or any other means – nor transmitted or translated into a machine language without written permission from the publishers. Registered names, trademarks, etc. used in this book, even when not specifically marked as such, are not to be considered unprotected by law.

Typesetting SNP Best-set Typesetter Ltd., Hong Kong
Printing betz-druck GmbH, Darmstadt
Binding Litges & Dopf Buchbinderei GmbH, Heppenheim
Cover Adam Design, Weinheim

Printed in the Federal Republic of Germany
Printed on acid-free paper

ISBN: 978-3-527-31918-3

Contents

Preface *XV*
Author List *XXI*
Member Societies *XXV*

Part One Self-Organization, Nanoscience and Nanotechnology

1 Subcomponent Self-Assembly as a Route to New Structures and Materials *3*
   Jonathan R. Nitschke
1.1 Introduction *3*
1.2 Aqueous Cu(I) *5*
1.3 Chirality *7*
1.4 Construction *8*
1.4.1 Dicopper Helicates *8*
1.4.2 Tricopper Helicates *10*
1.4.3 Catenanes and Macrocycles *11*
1.4.4 [2 × 2] Tetracopper(I) Grid *12*
1.5 Sorting *13*
1.5.1 Sorting Ligand Structures with Cu(I) *13*
1.5.2 Simultaneous Syntheses of Helicates *13*
1.5.3 Sorting within a Structure *14*
1.5.4 Cooperative Selection by Iron and Copper *17*
1.6 Substitution/Reconfiguration *20*
1.6.1 New Cascade Reaction *20*
1.6.2 Hammett Effects *22*
1.6.3 Helicate Reconfigurations *23*
1.6.4 Substitution as a Route to Polymeric Helicates *24*
1.7 Conclusion and Outlook *27*
1.8 Acknowledgments *27*

2	**Molecular Metal Oxides and Clusters as Building Blocks for Functional Nanoscale Architectures and Potential Nanosystems** *31*	
	Leroy Cronin	
2.1	Introduction *31*	
2.2	From POM Building Blocks to Nanoscale Superclusters *33*	
2.3	From Building Blocks to Functional POM Clusters *37*	
2.3.1	Host–Guest Chemistry of POM-based Superclusters *38*	
2.3.2	Magnetic and Conducting POMs *39*	
2.3.3	Thermochromic and Thermally Switchable POM Clusters *40*	
2.4	Bringing the Components Together – Towards Prototype Polyoxometalate-based Functional Nanosystems *42*	
2.5	Acknowledgments *44*	
3	**Nanostructured Porous Materials: Building Matter from the Bottom Up** *47*	
	Javier García-Martínez	
3.1	Introduction *47*	
3.2	Synthesis by Organic Molecule Templates *48*	
3.3	Synthesis by Molecular Self-Assembly: Liquid Crystals and Cooperative Assembly *50*	
3.4	Spatially Constrained Synthesis: Foams, Microemulsions, and Molds *57*	
3.4.1	Microemulsions *57*	
3.4.2	Capping Agents *57*	
3.4.3	Foams *58*	
3.4.4	Molds *59*	
3.5	Multiscale Self-Assembly *59*	
3.6	Biomimetic Synthesis: Toward a Multidisciplinary Approach *61*	
3.7	Acknowledgments *69*	
4	**Strategies Toward Hierarchically Structured Optoelectronically Active Polymers** *73*	
	Eike Jahnke and Holger Frauenrath	
4.1	Hierarchically Structured Organic Optoelectronic Materials via Self-Assembly *73*	
4.2	Toward Hierarchically Structured Conjugated Polymers via the Foldamer Approach *74*	
4.3	*"Self-Assemble, then Polymerize"* – A Complementary Approach and Its Requirements *78*	
4.3.1	Topochemical Polymerization Using Self-Assembled Scaffolds *79*	
4.3.2	Self-Assembly of β-Sheet Forming Oligopeptides and Their Polymer Conjugates *80*	
4.4	Macromonomer Design and Preparation *82*	
4.5	Hierarchical Self-Organization in Organic Solvents *85*	

4.6	A General Model for the Hierarchical Self-Organization of Oligopeptide–Polymer Conjugates *89*
4.7	Conversion to Conjugated Polymers by UV Irradiation *92*
4.8	Conclusions and Perspectives *95*
4.9	Acknowledgments *95*

5 Mimicking Nature: Bio-inspired Models of Copper Proteins *101*
Iryna A. Koval, Patrick Gamez and Jan Reedijk

5.1	Environmental Pollution: How Can "Green" Chemistry Help? *101*
5.2	Copper in Living Organisms *102*
5.2.1	Type 1 Active Site *102*
5.2.2	Type 2 Active Site *103*
5.2.3	Type 3 Active Site *104*
5.2.4	Type 4 Active Site *104*
5.2.5	The Cu_A Active Site *104*
5.2.6	The Cu_B Active Site *105*
5.2.7	The Cu_Z Active Site *105*
5.3	Catechol Oxidase: Structure and Function *105*
5.3.1	Catalytic Reaction Mechanism *107*
5.4	Model Systems of Catechol Oxidase: Historic Overview *108*
5.5	Our Research on Catechol Oxidase Models and Mechanistic Studies *114*
5.5.1	Ligand Design *114*
5.5.2	Copper(I) and Copper(II) Complexes with [22]py4pz: Structural Properties and Mechanism of the Catalytic Reaction *114*
5.5.3	Copper(I) and Copper(II) Complexes with [22]pr4pz: Unraveling Catalytic Mechanisms *118*
5.6	Concluding Remarks *124*
5.7	Acknowledgments *125*

6 From the Past to the Future of Rotaxanes *129*
Andreea R. Schmitzer

6.1	Introduction *129*
6.2	Synthesis of Rotaxanes *131*
6.2.1	Van der Waals Interactions in the Synthesis of Rotaxanes *132*
6.2.2	Hydrophobic Interactions in the Synthesis of Rotaxanes *133*
6.2.3	Hydrogen Bonding in Rotaxane Synthesis *134*
6.2.4	Donor–Acceptor Interactions in the Synthesis of Rotaxanes *135*
6.2.5	Transition-Metal Coordination in the Synthesis of Rotaxanes *136*
6.3	Applications of Rotaxanes *137*
6.3.1	Rotaxanes as Molecular Shuttles *137*
6.3.1.1	Acid–Base-controlled Molecular Shuttle *139*
6.3.1.2	A Light-driven Molecular Shuttle *140*
6.3.2	Molecular Lifts *142*

6.3.3	Artificial Molecular Muscles	143
6.3.4	Redox-activated Switches for Dynamic Memory Storage	144
6.3.5	Bioelectronics	147
6.3.6	Membrane Transport	149
6.3.7	Catalytically Active Rotaxanes as Processive Enzyme Mimics	151
6.4	Conclusion and Perspectives	152

7 Multiphoton Processes and Nonlinear Harmonic Generations in Lanthanide Complexes 161
Ga-Lai Law

7.1	Introduction	161
7.2	Types of Nonlinear Processes	162
7.3	Selection Rules for Multiphoton Absorption	164
7.4	Multiphoton Absorption Induced Emission	165
7.5	Nonlinear Harmonic Generation	176
7.6	Conclusion and Future Perspectives	181
7.7	Acknowledgments	181

8 Light-emitting Organic Nanoaggregates from Functionalized *para*-Quaterphenylenes 185
Manuela Schiek

8.1	Introduction to *para*-Phenylene Organic Nanofibers	185
8.2	General Aspects of Nanofiber Growth	187
8.3	Synthesis of Functionalized *para*-Quaterphenylenes	189
8.4	Variety of Organic Nanoaggregates from Functionalized *para*-Quaterphenylenes	193
8.5	Symmetrically Functionalized *p*-Quaterphenylenes	194
8.6	Differently Di-functionalized *p*-Quaterphenylenes	197
8.7	Monofunctionalized *p*-Quaterphenylenes	199
8.8	Tailoring Morphology: Nanoshaping	200
8.9	Tailoring Optical Properties: Linear Optics	201
8.10	Creating New Properties: Nonlinear Optics	203
8.11	Summary	205
8.12	Acknowledgments	205

9 Plant Viral Capsids as Programmable Nanobuilding Blocks 215
Nicole F. Steinmetz

9.1	Nanobiotechnology – A Definition	215
9.2	Viral Particles as Tools for Nanobiotechnology	216
9.3	General Introduction to CPMV	216
9.4	Advantages of Plant Viral Particles as Nanoscaffolds	219
9.5	Addressable Viral Nanobuilding Block	220
9.6	From Labeling Studies to Applications	222
9.7	Immobilization of Viral Particles and the Construction of Arrays on Solid Supports	229

9.8	Outlook	*231*
9.9	Acknowledgments	*232*

10 New Calorimetric Approaches to the Study of Soft Matter 3D Organization *237*
J.M. Nedelec and M. Baba

10.1	Introduction *237*	
10.2	Transitions in Confined Geometries *238*	
10.2.1	Theoretical Basis *239*	
10.2.1.1	Confinement Effect on Triple-point Temperature *239*	
10.2.2	Porosity Measurements via Determination of the Gibbs–Thomson Relation *240*	
10.2.2.1	Thermoporosimetry *241*	
10.2.2.2	NMR Cryoporometry *241*	
10.2.2.3	Surface Force Apparatus *241*	
10.2.3	Thermoporosimetry and Pore Size Distribution Measurement *242*	
10.3	Application of Thermoporosimetry to Soft Materials *243*	
10.3.1	Analogy and Limitations *243*	
10.3.2	Examples of Use of TPM with Solvent Confined by Polymers and Networks *244*	
10.3.2.1	Elastomers *244*	
10.3.2.2	Hydrogels *246*	
10.3.2.3	Polymeric Membranes *246*	
10.3.2.4	Crosslinking of Polyolefins *246*	
10.4	Study of the Kinetics of Photo-initiated Reactions by PhotoDSC *247*	
10.4.1	The PhotoDSC Device *247*	
10.4.2	Photocuring and Photopolymerization Investigations *247*	
10.5	Accelerated Aging of Polymer Materials *251*	
10.5.1	Study of Crosslinking of Polycyclooctene *251*	
10.5.1.1	Correlation between Oxidation and Crystallinity *251*	
10.5.1.2	Crosslinking and Crystallizability *253*	
10.5.1.3	Photo-aging Study by Macroperoxide Concentration Monitoring *254*	
10.5.2	Kinetics of Chain Scissions during Accelerated Aging of Poly(ethylene oxide) *255*	
10.5.2.1	Chain Scission Kinetics from Melting *256*	
10.6	Conclusion *258*	

Part Two Organic Synthesis, Catalysis and Materials

11 Naphthalenediimides as Photoactive and Electroactive Components in Supramolecular Chemistry *265*
Sheshanath Vishwanath Bhosale

11.1	Introduction *265*	
11.2	General Syntheses and Reactivity *266*	

11.2.1	Synthesis of Core-substituted NDIs 268
11.2.2	General Chemical and Physical Properties 268
11.3	Redox and Optical Properties of NDIs 271
11.3.1	NDIs in Host–Guest Chemistry 272
11.3.2	NDI-DAN Foldamers 272
11.3.3	Ion Channels 273
11.3.4	NDIs in Material Chemistry 275
11.4	Catenanes and Rotaxanes 276
11.4.1	NDIs Used as Sensors 277
11.4.2	Nanotubes 279
11.5	NDIs in Supramolecular Chemistry 281
11.5.1	Energy and Electron Transfer 281
11.5.2	Covalent Models 281
11.5.3	Noncovalent Models 284
11.6	Applications of Core-Substituted NDIs 287
11.7	Prospects and Conclusion 290
11.8	Acknowledgment 290
12	**Coordination Chemistry of Phosphole Ligands Substituted with Pyridyl Moieties: From Catalysis to Nonlinear Optics and Supramolecular Assemblies** 295
	Christophe Lescop and Muriel Hissler
12.1	Introduction 295
12.2	π-Conjugated Derivatives Incorporating Phosphole Ring 296
12.2.1	Synthesis and Physical Properties 296
12.2.2	Fine Tuning of the Physical Properties via Chemical Modifications of the Phosphole Ring 298
12.3	Coordination Chemistry of 2-(2-Pyridyl)phosphole Derivatives: Applications in Catalysis and as Nonlinear Optical Molecular Materials 300
12.3.1	Syntheses and Catalytic Tests 300
12.3.2	Isomerization of Coordinated Phosphole Ring into 2-Phospholene Ring 301
12.3.3	Square-Planar Complexes Exhibiting Nonlinear Optical Activity 303
12.3.4	Ruthenium Complexes 304
12.4	Coordination Chemistry of 2,5-(2-Pyridyl)phosphole Derivatives: Complexes Bearing Bridging Phosphane Ligands and Coordination-driven Supramolecular Organization of π-Conjugated Chromophores 305
12.4.1	Bimetallic Coordination Complexes Bearing a Bridging Phosphane Ligand 305
12.4.1.1	Pd(I) and Pt(I) Bimetallic Complexes 306
12.4.1.2	Cu(I) Bimetallic Complexes 307

12.4.2	Supramolecular Organization of π-Conjugated Chromophores via Coordination Chemistry: Synthesis of Analogues of [2.2]-Paracyclophanes *310*
12.5	Conclusions *314*
12.6	Acknowledgments *315*

13 Selective Hydrogen Transfer Reactions over Supported Copper Catalysts Leading to Simple, Safe, and Clean Protocols for Organic Synthesis *321*
Federica Zaccheria and Nicoletta Ravasio

13.1	Chemoselective Reduction of Polyunsaturated Compounds via Hydrogen Transfer *323*
13.2	Alcohol Dehydrogenation *325*
13.3	Racemization of Chiral Secondary Alcohols *331*
13.4	Isomerization of Allylic Alcohols *331*
13.5	Conclusions *333*

14 Selective Oxido-Reductive Processes by Nucleophilic Radical Addition under Mild Conditions *337*
Cristian Gambarotti and Carlo Punta

14.1	Introduction *337*
14.2	Nucleophilic Radical Addition to N-heteroaromatic Bases *338*
14.2.1	Acylation of N-heteroaromatic Bases *338*
14.2.2	Acylation of N-heteroaromatic Bases Catalyzed by N-hydroxyphthalimide *340*
14.2.3	Photoinduced Nucleophilic Radical Substitution in the Presence of TiO_2 *341*
14.2.4	Hydroxymethylation of N-heteroaromatic Bases *343*
14.2.5	Perfluoroalkylation of N-heteroaromatic Bases and Quinones *344*
14.3	Nucleophilic Radical Addition to Aldimines *345*
14.3.1	Nucleophilic Radical Addition Promoted by $TiCl_3/PhN_2^+$ Systems *345*
14.3.2	Nucleophilic Radical Addition Promoted by $TiCl_3$/Pyridine Systems *347*
14.3.3	Nucleophilic Radical Addition Promoted by $TiCl_3$/Hydroperoxide Systems *348*

Part Three Health, Food, and Environment

15 Future Perspectives of Medicinal Chemistry in the View of an Inorganic Chemist *355*
Palanisamy Uma Maheswari

15.1	Introduction *355*
15.1.1	Conventional versus Targeted Therapy *358*
15.2	Ruthenium Anticancer Drugs *359*
15.2.1	Ru–Polypyridyl Complexes *359*

15.2.2	Ru–Polyaminocarboxylate Complexes *361*
15.2.3	Ru-Dimethyl Sulfoxide Complexes *362*
15.2.4	Ru–Arylazopyridine Complexes *363*
15.2.5	Ru–Organometallic Arene Complexes *365*
15.2.6	NAMI-A Type Complexes *366*
15.2.7	The Transferrin Delivery Mechanism *367*
15.2.8	Discerning Estrogen Receptor Modulators Based on Ru *368*
15.2.9	Ru–Ketoconazole Complexes *369*
15.2.10	Protein Kinase Inhibitors Based on Ru *369*
15.2.11	Ru–RAPTA Complexes *370*
15.3	Chemical Nucleases as Anticancer Drugs *373*
15.4	Inorganic Chemotherapy for Cancer: Outlook *378*
15.5	Acknowledgments *381*

16 Speeding Up Discovery Chemistry: New Perspectives in Medicinal Chemistry *389*
Matteo Colombo and Ilaria Peretto

16.1	Solid-phase Extraction *390*
16.2	Polymer-assisted Solution-phase Synthesis *392*
16.3	Microwave-assisted Organic Synthesis [10, 11] *395*
16.4	Flow Chemistry *400*
16.5	Analytical Instrumentation *404*
16.6	Conclusions *405*

17 Overview of Protein-Tannin Interactions *409*
Elisabete Barros de Carvalho, Victor Armando Pereira de Freitas and Nuno Filipe da Cruz Batista Mateus

17.1	Phenolic Compounds *409*
17.2	Tannin Structures *410*
17.2.1	Dietary Burden and Properties of Phenolic Compounds *411*
17.3	Interactions between Proteins and Tannins *412*
17.4	Experimental Studies of the Interactions between Proteins and Tannins *412*
17.4.1	Nephelometric Studies of BSA and Condensed Tannin Aggregation *413*
17.5	Factors That Influence the Interactions between Proteins and Tannins *415*
17.5.1	Structural Features *415*
17.5.2	pH and Ionic Strength *415*
17.5.3	Influence of Polysaccharide on the Interactions between Protein and Tannin *417*
17.6	Flow Nephelometric Analysis of Protein–Tannin Interactions *419*
17.7	Interactions of Tannins with Salivary Proteins – Astringency *421*
17.8	Polysaccharides and Astringency *423*
17.9	Acknowledgments *425*

18	**Photochemical Transformation Processes of Environmental Significance** *429*	
	Davide Vione	
18.1	Introduction and Overview of Environmental Photochemistry *429*	
18.1.1	Photochemical Processes in the Atmosphere *429*	
18.1.2	Photochemical Reactions in Ice and Snow *434*	
18.1.3	Photochemical Reactions in Surface Waters *435*	
18.2	Transformation Reactions Induced by $^{\bullet}OH$, $^{\bullet}NO_2$ and $Cl_2^{\bullet-}$ in Surface Waters *437*	
18.2.1.	Reactions Induced by $^{\bullet}OH$ *437*	
18.2.2	Reactions Induced by $^{\bullet}NO_2$ *445*	
18.2.3	Reactions Induced by $Cl_2^{\bullet-}$ *446*	
18.3	Conclusions *448*	
18.4	Acknowledgments *449*	

Index *455*

Preface

Contrary to the general image that chemistry has in public opinion, chemists are great observers, admirers, and lovers of Nature. Chemists have a relationship with Nature at a molecular level, learn from it, and attempt to copy its perfection and harmony. In their activities, chemists work to find solutions for human health; to widen the range of sustainable processes and materials; to prevent pollution and maintain the quality of climate; to devise clean, renewable energy sources; to preserve and restore the cultural heritage; and to develop new technologies for improving everyday life. Using synthetic processes and discovering and manipulating molecules, chemists are increasingly establishing a primary role within prominent interdisciplinary scientific and technological fields such as those of nanoscience, nanotechnology, and biotechnology. Alluding to precisely this great potential, "Long life to chemistry" said Jean Marie Lehn at the end of his plenary during the 1st European Chemistry Congress held in Budapest on 27–31 August 2006. This sentiment has to be related also to the fact that young chemists are producing new paradigms opening up excellent perspectives for future research.

The plan for this book was originated during the preparation of the European Young Chemists Award that I had the honor to chair and that was held during the First European Chemistry Congress. At that congress a number of young chemists showed the results of their research, presenting fascinating ideas and original conclusions and proposing radically new materials, molecules, supramolecules, and superstructures. About 120 chemists from all over the world, and all less than 34 years old, participated in the Award. According to the supporting letters, there were several excellent candidates. Just to give you an idea of the type and level of assessments contained in those letters, let me cite few of them: "outstanding scientist, who in spite of the young age has already accomplished a lot"; "unusually talented chemist"; "this rapid rise through the academic ranks is almost unprecedented and is testament to extraordinary talent"; "particularly bright and full of original ideas and also hard working"; "totally reliable and highly professional, gives continuous input of original solutions"; "truly outstanding synthetic organic chemist with a glittering future ahead"; "the mobility and international cooperation experience of the candidate are great examples for the future generation of scientists not only in Europe but also outside"; "ambitious, successful young scientist who is goal oriented on challenging scientific topics." About half of the participants were judged top level by the Award jury.

Most of the candidates presented fundamental research issues, although possible applications were almost always also considered. They dealt with a variety of problems in keeping with chemical tradition.

I was then encouraged to collect in a book what I felt to be the most interesting topics by different candidates for the Award. *Tomorrow's Chemistry Today* is therefore a book intended to showcase excellence in chemistry by inviting a selection of young chemists each to write a chapter on their research field, their main results, and the perspectives they envision for the future.

Many of the 18 contributions are interdisciplinary and involve interfaces such as:
- organic-synthesis/polymer science/supramolecular science;
- supramolecular chemistry/material science and nanotechnology/optoelectronics;
- bioorganic chemistry or inorganic chemistry/medicinal chemistry;
- organic synthesis/analytical chemistry/protein biochemistry;
- biology/nanoscience/physical chemistry;
- biology/supramolecular chemistry.

Reading the book, one will find many new ideas and innovations. It is clear that important steps forward, at the forefront of modern chemical science and technology, are made in several area with the contributions of these talented authors. These concern at least the following fields:
- New synthetic procedures, reaction routes, and schemes intended to give supramolecular motifs.
- Development of real bottom-up molecular technology as well as nanotechnology through supramolecular chemistry.
- New chemical products or materials with unusual properties for potential applications in various devices.
- Hybrid nanomaterials involving organic, inorganic, as well as biological systems or assemblies.
- Molecular systems having intense industrial interest in medicine.
- Structure–property relationship and biomimetic chemistry.
- New "green" catalysts for environmentally friendly industrial processes.
- Advanced characterization methods.

The book has been divided into three main parts:
1. Self-organization, Nanoscience, and Nanotechnology
2. Organic Synthesis, Catalysis, and Materials
3. Health, Food, and Environment

In the first part, emphasis is given to the efforts made in the exploitation of improved knowledge of noncovalent interactions to synthesize new molecules having hierarchical structure, possibly to mimic Nature. Molecules are often designed to utilize precisely these noncovalent interactions and molecular recognition processes, particularly those based upon hydrogen bonding, metal–ligand coordination, π–π interactions, hydrophobic interaction, ion pairing, and van der Waals interactions. This is in order to stabilize well-defined conformations and therefore function.

Powerful methods for the synthesis of elaborate and intricate supramolecular systems and the technique of *subcomponent self-assembly* for the creation of increasingly complex structures are presented. Particular strategies of synthesis are described such as *"self assemble, then polymerize, and then fold into hierarchical structures"* or vice versa, as well as successful strategies involving the incorporation of aromatic heterocycles into the backbone of π-conjugated systems for the design and assembly of structures having desired properties. In some cases the parameters controlling the exact nature of the observed hierarchical structures are discussed. Fascinating architectures, or molecular topologies if you like, are demonstrated that have an almost unmatched range of physical properties involving different types of molecules such as polyoxometallates, co-oligomers alternating phosphole and thiophene and/or pyridine rings, catenanes, rotaxane, naphthalenediimides, and so on. Their potentialities in everyday life as catalysts, sensors, molecular machines, switches, photoactive or electroactive components for optoelectronics as well as light-emitting diodes, thin-film transistors, photovoltaic cells, nanodevices, and so on, are discussed. Reading these works it is easily understood that, as one of the contributors says, "Chemists are in an ideal position to develop such a molecular approach to functional nanostructures because they are able to design, synthesize, investigate, and organize molecules– i.e., make them react or bring them together into larger assemblies. And at the end a better understanding of the rules and principles guiding a self-assembly process can allow one to utilize these rules synthetically, creating new structures possessing new functions for engineering at the molecular level."

The book continues with other contributions in the area of materials and catalysis. Important concepts are treated, like that of exploiting nonlinear optical behavior of certain classes of materials which emit in the short wavelength region, such as the visible region, when excited by another region such as the infrared. This property leads to many advantages, especially in biological studies, telecommunications, and three-dimensional optical storage, and it is potentially important for bioimaging. The bottom-up approach is again amply exploited to prepare nanostructured materials with hierarchical organization, leading to properties which can be tuned by judicious modification of their synthesis conditions. New synthetic techniques based again on weak interactions are continually being developed to gain more precise control over the organization of solids. In particular, template-assisted synthesis, self-assembly, and biomimetic methods are highlighted as likely to become widely used in the fabrication of materials with controlled porosity. The important method of spatially constrained synthesis is described. The bottom-up nanoengineering approach is used in another contribution dealing with the preparation of light-emitting aggregates from functionalized *para*-quaterphenylene. This work ends with the question: "which chemically functionalized oligomers would still undergo a similar self-assembly process and allow creation of quantitative amounts of crystalline nanofibers with tailored morphologies and optical, electrical, mechanical and even new properties?"

Moving to other contributions, one can readily appreciate that Nature still has plenty of things to teach us for engineering at molecular level and preparing useful materials. This motif is present, for example, in a contribution reporting the study of bio-inspired models of copper proteins elucidating model compounds of the copper-containing enzyme catechol oxidase and aiming to understand its mechanism of action.

Nature has always been a source of inspiration for chemists and materials scientists. In addition to the inspiration, Nature is also giving us "materials" useful for nanotechnology. This concept is vividly and beautifully presented in a further contribution in which plant viral particles are used as programmable nanobuilding blocks. The focus of this chapter is in the area of nanobiotechnology and the exploitation of biomolecules for technological applications. A new field is emerging, says the author: "a highly interdisciplinary area which involves collaborations between virologists, chemists, physicists, and materials scientists. It is exciting at the virus–chemistry interface."

The book collects contributions in the field of characterization of materials also, and these are reported in various chapters. In addition to this, a particular chapter is dedicated to interesting new calorimetric approaches to the study of soft-matter three-dimensional organization intended to demonstrate methods able to make a contribution to our understanding of hierarchical porous structures in which matter and void are organized in regular and controlled patterns.

Studies in the catalytic-organic chemistry area are enriched here by an elegant contribution on selective hydrogen transfer reactions over supported copper catalysts leading to simple, safe, and clean protocols for organic synthesis.

Contributions to organic synthesis, in some respects more traditional than those previously mentioned and concerning different areas from those potentially important for nanotechnology, materials, or catalysis, are also reported.

In one of these contributions, organic synthetic procedures regarding nucleophilic radical addition under mild conditions is described, underlining and confirming the idea that high reactivity is not necessarily associated with low selectivity.

In the last part of the book some examples are reported on the importance of the contribution of chemical studies to fields that are of increasing concern for the public opinion such as health, food, and the environment. One of these describes investigation of the protein–tannin interaction in order to better understand organoleptic properties of foodstuffs, and in particular those of red wine.

Two other chapters give an overview of the analogues and derivatives of cisplatin and the alternatives for it, the ruthenium-based drugs reported in the last 30 years for tumor biology, and present both future perspectives of medicinal chemistry for speeding up discovery chemistry in the field and future strategies for drug design. Last but not least, a chapter is devoted to the important photochemical transformation processes of environmental significance and their possible influence on climate change.

The contributions reported in this book clearly show that chemistry is not a static science and that this is because it is continuously developing its knowledge base, techniques, and paradigms, adapting its potentialities to the demands of society, implementing its own tradition and collaborating with other scientific areas to open up entirely new fields at the interface with physics or life sciences to generate hybrid systems. It is important to stress that the systems chemists can create may have characteristics or properties that are not even present in Nature. Either exploiting the synthetic arts such as those presented in many chapters of this book or creating hybrid systems with living organisms, chemists are, as stated at the beginning, in the ideal position to contribute to our civil and societal development. The perspective for this science and for the products that it can give to society are therefore excellent, considering especially that a number of talented young researchers are very active in the area.

In conclusion, I hope that such a book, directed to a broad readership, will be a source of new ideas and innovation for the research work of many scientists, the contributions covering many of the frontier issues in chemistry. Our future is undoubtedly on the shoulders of the new scientific generation, but I would like to express the warning that in any case there will be no significant progress if – together with the creativity of young scientists and their will to develop interdisciplinary and collaborative projects – there is not established a constructive political will that takes care of the growth of young scientists and their research.

I cannot finish this preface without acknowledging all the authors and the persons who helped me in the book project. I am very grateful to Professor Natile (President of the European Association for Chemical and Molecular Sciences) and Professor De Angelis (President of the Italian Chemical Society) for their stimulation and suggestions. And of course, I thank all the Societies (see the book cover) that motivated and sponsored the book.

Palermo, August 2007
Bruno Pignataro

Author List

Mohammed Baba
Laboratoire de Thermodynamique des Solutions et des Polymères, CNRS UMR 6003, TransChiMiC, Université Blaise Pascal, Clermont-Ferrand 2 24 Avenue des Landais, Bâtiment Chimie 7, 63 177 Aubière, France
E-mail: mohammed.baba@univ-bpclermont.fr

Sheshanath Vishwanath Bhosale
School of Chemistry, Monash University, Wellington Road, Clayton, Victoria 3800, Australia
E-mail: shehanath.bhosale@sci.monash.edu.au

Elisabete Barros de Carvalho
Centro de Investigação em Química, Universidade do Porto, Faculdade de Ciências, Departamento de Química, Rua do Campo Alegre, 687 4169-007 Porto, Portugal
E-mail: elisabete.carvalho@fc.up.pt

Matteo Colombo
NiKem Research srl, via Zambeletti, 25, 20021 Baranzate (Milan), Italy
E-mail: matteo.colombo@nikemresearch.com

Leroy Cronin
Department of Chemistry, The University of Glasgow, Glasgow, G12 8QQ, UK
E-mail: L.Cronin@chem.gla.ac.uk

Holger Frauenrath
Eidgenössische Technische Hochschule Zürich, Department of Materials, Wolfgang-Pauli-Str. 10, HCI H515, CH-8093 Zürich, Switzerland
E-mail: frauenrath@mat.ethz.ch

Victor Armando Pereira de Freitas
Centro de Investigação em Química, Universidade do Porto, Faculdade de Ciências, Departamento de Química, Rua do Campo Alegre, 687 4169-007 Porto, Portugal
E-mail: vfreitas@fc.up.pt

Cristian Gambarotti
Politecnico di Milano, Materiali ed Ingegneria Chimica "Giulio Natta", Via Mancinelli 7, I-20131 Milano, Italy
E-mail: cristian.gambarotti@polimi.it

Patrick Gamez
Leiden University, Leiden Institute of Chemistry, P.O. Box 9502, 2300 R.A. Leiden, The Netherlands
E-mail: p.gamez@chem.leidenuniv.nl

Tomorrow's Chemistry Today. Concepts in Nanoscience, Organic Materials and Environmental Chemistry.
Edited by Bruno Pignataro
Copyright © 2008 WILEY-VCH Verlag GmbH & Co. KGaA, Weinheim
ISBN: 978-3-527-31918-3

Javier García-Martínez
Molecular Nanotechnology Laboratory, Department of Inorganic Chemistry, University of Alicante, Carretera San Vicente s/n. E-03690, Alicante, Spain
E-mail: j.garcia@ua.es

Muriel Hissler
Université de Rennes I Sciences Chimiques de Rennes UMR 6226 CNRS-IR1 263 avenue du Général Leclerc 35042 Rennes, France
E-mail: muriel.hissler@univ-rennes1.fr

Eike Jahnke
Eidgenössische Technische Hochschule Zürich, Department of Materials, Wolfgang-Pauli-Str. 10, HCI H515, CH-8093 Zürich, Switzerland
E-mail: jahnke@mat.ethz.ch

Iryna A. Koval
TNO Quality of Life, P.O. Box 718, Polarisavenue 151, 2130 A.S. Hoofddorp, The Netherlands
E-mail: i.koval@chem.leidenuniv.nl

Ga-Lai Law
Department of Chemistry, The University of Hong Kong, Chong Yuet Ming Chemistry Building, Pokfulam Road, Hong Kong
E-mail: galai_law@hotmail.com

Christophe Lescop
Université de Rennes I Sciences Chimiques de Rennes UMR 6226 CNRS-IR1 263 avenue du Général Leclerc 35042 Rennes, France
E-mail: christophe.lescop@univ-rennes1.fr

Palanisamy Uma Maheswari
Leiden Institute of Chemistry, Leiden University, P.O. Box, 9502, 2300 R.A. Leiden, The Netherlands
E-mail: p.maheswari@chem.leidenuniv.nl

Nuno Filipe da Cruz Batista Mateus
Centro de Investigação em Química, Universidade do Porto, Faculdade de Ciências, Departamento de Química, Rua do Campo Alegre, 687 4169-007 Porto, Portugal
E-mail: nbmateus@fc.up.pt

J.M. Nedelec
Laboratoire des Matériaux Inorganiques, CNRS UMR 6002, TransChiMiC, Université Blaise Pascal, Clermont-Ferrand 2 & Ecole Nationale Supérieure de Chimie de Clermont-Ferrand, 24 Avenue des Landais, 63 177 Aubière, France
E-mail: j-marie.nedelec@univ-bpclermont.fr

Jonathan R. Nitschke
University of Cambridge, Department of Chemistry, Lensfield Road Cambridge, CB2 1EW, United Kingdom
E-mial: jrn34@cam.ac.uk

Ilaria Peretto
NiKem Research srl, via Zambeletti, 25, 20021 Baranzate (Milan), Italy
E-mail: ilaria.peretto@nikemresearch.com

Carlo Punta
Politecnico di Milano, Dipartimento di Chimica, Materiali ed Ingegneria Chimica "Giulio Natta", Via Mancinelli 7, I-20131 Milano, Italy
E-mail: carlo.punta@polimi.it

Nicoletta Ravasio
Università di Milano, Dipartimento di Chimica Inorganica, Metallorganica e Analitica, Via Venezian, 21, 20133 Milano, Italy
E-mail: n.ravasio@istm.cnr.it

Jan Reedijk
Leiden University, Leiden Institute of Chemistry, P.O. Box 9502, 2300 R.A. Leiden, The Netherlands
E-mail: reedijk@chem.leidenuniv.nl

Manuela Schiek
University of Southern Denmark Mads Clausen Institute, NanoSYD, Alsion 2, 6400 Sonderborg, Denmark
E-mail: Manuela.Schiek@uni-oldenburg.de

Andreea R. Schmitzer
Department of Chemistry, Université de Montréal, 2900 Edouard Montpetit, succursale Centre ville CP 6128, Montréal H3C 3J7, Canada
E-mail: ar.schmitzer@umontreal.ca

Nicole F. Steinmetz
The Scripps Research Institute, Department of Cell Biology, CB262 La Jolla CA 92037, USA
E-mail: nicoles@scripps.edu

Davide Vione
Dipartimento di Chimica Analitica, Università di Torino, Via P. Giuria 5, 10125 Torino, Italy
E-mail: davide.vione@unito.it

Federica Zaccheria
Università di Milano, Dipartimento di Chimica Inorganica, Metallorganica e Analitica, Via Venezian, 21, 20133 Milano, Italy
E-mail: federica.zaccheria@unimi.it

Member Societies

A

Society of Albanian Chemists (SAC)
University of Tirana
Faculty of Natural Sciences
Tirana
Albania

GÖCH – Gesellschaft Österreichischer
Chemiker
Nibelungengasse 11/6
1010 Vienna
Austria

ASAC – Austrian Society for Analytical
Chemistry
Universitaet Linz
Altenbergerstrasse 69
4040 Linz
Austria

B

KVCV – Koninklijke Vlaamse
Chemische Vereniging
Celestijnenlaan 200F
3001 Herverlee
Belgium

Société Royale de Chimie
Université Libre de Bruxelles
CP 160/07
Av F.D. Roosevelt 50
1050 Bruxelles
Belgium

Union of Chemists in Bulgaria (UCB)
108 Rakovski Street
PO Box 431
1000 Sofia
Bulgaria

C

Croatian Chemical Society
Horvatovac 102a
10000 Zagreb
Croatia

Pancyprian Union of Chemists (PUC)
P. O. Box 28361
2093 Nicosia
Cyprus

Czech Chemical Society
Novotného lávka 5
116 68 Praha 1
Czech Republic

D

Danish Chemical Society
Universitetsparken 5
2100 København Ø
Denmark

Kemiingeniorgruppen
Ingeniørforeningen i Danmark
Kalvebod Brygge 31-33
1780 København V
Denmark

E

Estonian Chemical Society
Akadeemia tee 15
12618 Tallinn
Estonian

F

Association of Finnish Chemical
Societies
Urho Kekkosen katu 8 C 31
00100 Helsinki
Finnland

Société Française de Chimie
250, rue Saint-Jacques
75005 Paris
France

G

Dechema
Gesellschaft für Chemische Technik
und Biotechnologie e.V.
Theodor-Heuss-Allee 25
60486 Frankfurt am Main
Germany

Gesellschaft Deutscher Chemiker e.V.
(GDCh)
Postfach 90 04 40
60444 Frankfurt am Main
Germany

Deutsche Bunsen-Gesellschaft (DBG)
für Physikalische Chemie
Postfach 150104
60061 Frankfurt am Main
Germany

VAA
Geschäftstelle Köln
Mohrenstraße 11-17
50670 Köln
Germany

Greece
Association of Greek Chemists
info@eex.gr

H

The Hungarian Chemical Society
(MKE)
Fö utca 63
1027 Budapest
Hungary

I

Institute of Chemistry of Ireland
PO Box 9322
Cardiff Lane
Dublin 2
Republic of Ireland

The Israel Chemical Society (ICS)
P.O. Box 26
76100 Rehovot
Israel

Societa Chimica Italiana
Viale Liegi 48c
00198 Roma
Italy

Consiglio Nazionale dei Chimici (CNC)
Piazza San Bernardo, 106
00187 Roma
Italy

L

Latvian Chemical Society
21 Aizkraukles Street
1006 Riga
Latvia

Lithuanian Chemical Society
G A Gostauto
Vilnius LT 2600
Lithuania

Association des Chimistes
Luxembourgeois (ACHIL)
16 rue J. Theis
9286 Diekirch
Luxembourg

Society of Chemists and Technologists
of Macedonia (SCTM)
Faculty of Technology and Metallurgy
Rudjer Boskovic
1000 Skopje
Macedonia

Chemical Society of Montenegro
(CSM)
Hemijsko Drustvo Crne Gore
Tehnolosko-metalurski fakultet
Cetinjski pu bb
81000 Podgorica,
Republic of Montenegro

N

Royal Netherlands Chemical
Society–KNCV
Vlietweg 16
2266 KA Leidschendam
Netherlands

Norwegian Chemical Society
President: Tor Hemmingsen
Universitetet i Stavanger
4036 Stavanger
Norway

P

Polish Chemical Society (PCS)
ul. Freta 16
00-227 Warszawa
Poland

Portuguese Electrochemical Society
(SPE)
University of Coimbra
Department of Chemistry
3004-535 Coimbra
Portugal

Sociedade Portuguesa de Química
Av. Republica, 37, 4
1050-187 Lisboa
Portugal

R

Romanian Society of Analytical
Chemistry
13 Bulevardul Republicii
70346 Bucharest
Romania

Romanian Chemical Society
c/o The Romanian Academy
Calea Victoriei nr. 125
71102 Bucharest
Romania

Mendeleev Russian Chemical Society
The Federation of Mendeleev Chemical
Societies (FCS)
Prof N. N. Kulov (Vice President)
Department of International Relations
Kurnakov Institute of General and
Inorganic Chemistry of the Russian
Academy of Sciences
31 Leninskii prospekt
119991 Moscow
Russia

Russian Scientific Council on Analytical
Chemistry
Council Secretariat
Institute of General and Inorganic
Chemistry
31 Leninskii prospekt
119991 Moscow
Russia

S

Serbian Chemical Society
Karnegijeva 4
11120 Belgrade
Serbia

Slovak Chemical Society (SCS)
Radlinského 9
81237 Bratislava
Slovakia

Slovenian Chemical Society
Hajdrihova 19
p.p. 660
1000 Ljubljana
Slovenia

National Chemical Association of Spain (ANQUE)
C/ Lagasca, 27
28001 Madrid
Spain

La Real Sociedad Española de Química (RSEQ)
Secretary General
Facultade de Quimica
Universidad Complutense
28040 Madrid
Spain

Spanish Society of Analytical Chemistry (SEQA)
Universidad Complutense de Madrid
Dra. Carmen Cámara
Dpto de Química Analítica
Facultad de Químicas
Universidad Complutense de Madrid
Madrid
Spain

Swedish Chemical Society
Wallingatan 24 3 tr
11124 Stockholm
Sweden

Schweizerische Chemische Gesellschaft (SCG)
Schwarztorstrasse 9
3007 Bern
Switzerland

T

Chemical Society of Turkey
Halaskargazi Gad No. 53 D 8
Harhiye
80230 Istanbul
Turkey

U

Ukranian Chemical Society
ul. Murmanskaya 1
02094 Kiev
Ukraine

Royal Society of Chemistry, Cambridge
Thomas Graham House
Science Park
Milton Road
Cambridge CB4 0WF
United Kingdom

Part One
Self-Organization, Nanoscience and Nanotechnology

1
Subcomponent Self-Assembly as a Route to New Structures and Materials

Jonathan R. Nitschke

1.1
Introduction

The concept of self-organization cuts across many of the sciences [1]. Gas and dust within nebulae coalesce into stars, which in turn arrange into galaxies that adopt a remarkable foamlike large-scale structure [2]. Biomolecules likewise self-organize – and organize *each other* – into the complex structures that compose cells, which may undergo further self-organization to create multicellular organisms. Living creatures aggregate also into herds, populations, communities, and biomes [3].

Human intelligence has proven skilled at examining organized structures and deducing the principles of self-organization that lead to their formation from less complex matter. This kind of deductive reasoning is one of the cornerstones of science, allowing for future predictions to be made based upon the principles uncovered.

The soul of chemistry lies in the art of synthesis. Through understanding chemical self-organization, new synthetic possibilities arise in ways that are not possible in the other sciences. When chemical bonds are formed in well-defined ways under thermodynamic [4, 5], as opposed to kinetic [6], control, one may refer to the resulting self-organization process as "self-assembly." An understanding of the rules and principles guiding a self-assembly process can allow one to utilize these rules synthetically, creating new structures that possess new functions.

As one example, the observation of hydrogen bonding [7] in natural systems such as peptide helices and DNA base pairs led to a theoretical understanding of this phenomenon. This understanding has permitted the use of hydrogen bonding in synthesis, leading to the preparation of such diverse structures as Rebek's capsules [8], Lehn's supramolecular polymers [9], and Whitesides' rosettes [10].

Over the course of the past four years, we have developed and employed the technique of *subcomponent self-assembly* toward the creation of increasingly complex structures. This technique, itself a subset of metallo-organic self-assembly [11–13], involves the simultaneous formation of covalent (carbon–heteroatom) and dative (heteroatom–metal) bonds, bringing both ligand and complex into being at the same time. The roots of subcomponent self-assembly lie in the template synthesis of Busch [14]. Before and after the inception of our research program, other researchers have employed this method to synthesize a wealth of structures, including macrocycles [15], helicates [16, 17], rotaxanes [18], catenanes [19], grids [20, 21], and a Borromean link [22].

The preparation of the Borromean link, in particular, represents a remarkable synthetic accomplishment. This topological *tour de force* was created in one pot from zinc acetate and the two simple subcomponents shown at right in Figure 1.1, one of which is commercially available!

Stoddart *et al.*'s preparation of a Borromean link [22] and, more recently, a Solomon link [23] together with Leigh *et al.*'s preparations of rotaxanes [18] and catenanes [19] demonstrates the power of subcomponent self-assembly to prepare topologically complex structures from simple building blocks. In order to construct the Borromean link of Figure 1.1, for example, one cannot start with pre-formed macrocycles – it is necessary that the subcomponents be linked reversibly (via imine bonds in this case). These reversible linkages allow the macrocycles to open up in order to generate the topological complexity of the final product.

In our own laboratories, initial proof-of-concept experiments established the utility of subcomponent self-assembly based upon copper(I) coordination and imine bond formation, most usefully in aqueous solution [24]. We have subsequently developed our research program along three main lines, seeking responses to a series of questions.

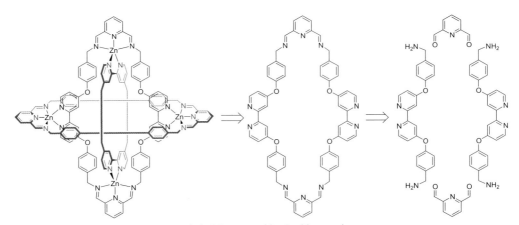

Figure 1.1 Borromean link [22] prepared by Stoddart *et al.* from zinc(II) and the dialdehyde and diamine subcomponents shown at right via the simultaneous, reversible formation of covalent (C=N) and coordinative (N→Zn) bonds. Two of the six corners of the link (at left) have been replaced with straight lines for clarity.

Our first line of research asks how simple subcomponents might be used to create complex structures via self-assembly. How may self-assembly information be encoded into the subcomponents? What other means of encoding self-assembly information into the system might be employed, such as solvent effects and pH? Are there structures that are readily accessible using subcomponent self-assembly that are difficult or impossible to create otherwise? How may this method be used to generate topological complexity?

Our second line of investigation delves into the possibility of utilizing this methodology in *sorting* complex mixtures, using the techniques and ideas of dynamic combinatorial [25, 26] chemistry: Is it possible to direct given subcomponents into specific places within assemblies? Can one observe the clean formation of two distinct structures from a common pool of ligand subcomponents? May the coordinative preferences of two different metal ions be used to induce different sets of ligand subcomponents to assemble around each metal?

Our third line of inquiry deals with the dynamic-evolutionary aspects of the structures we create. The reversibly-formed linkages holding these structures together allow a wide range of *substitution* and *reconfiguration* chemistry, on both dynamic covalent [5] (C=N) and coordinative (N→Metal) levels: What driving forces may be harnessed to effect the transformation of one structure into another, cleanly and in high yield? Can one address the two different levels, coordinative and covalent, independently? Is it possible to preferentially substitute a single subcomponent within a structure or a mixture that contains several different possible sites of attack?

1.2
Aqueous Cu(I)

The first study we undertook [24] validated the use of subcomponent self-assembly using aqueous copper(I), as well as taking initial steps in the directions of *construction*, *sorting*, and *reconfiguration*.

In aqueous solution Cu^I is frequently observed to disproportionate to Cu^{II} and copper metal, and imines are in most cases the minority species when amines and carbonyl compounds are mixed in water [27]. When imines and copper(I) are present in the same solution, however, this pattern of stability reverses. Imines are excellent ligands for Cu^I, stabilizing the metal in this oxidation state, and metal coordination can prevent imines from hydrolyzing. We were thus able to prepare complex **1** from the precursors shown at left in Scheme 1.1 [24].

Conceptually, one may imagine two different spaces within the flask wherein **1** self-assembles: a dynamic covalent [5] space and a supramolecular [11] space (Figure 1.2). The dynamic covalent space consists of all of the different possible ligand structures that could self-assemble from a given set of ligand subcomponents, and the supramolecular space consists of all possible metal complexes of these possible ligands.

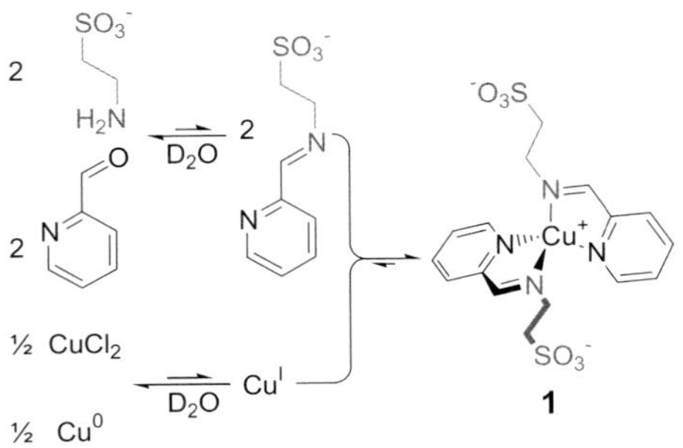

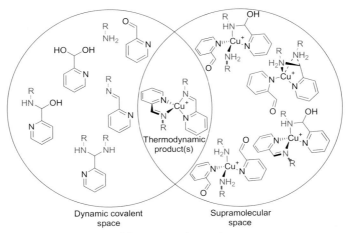

Scheme 1.1 Mutual stabilization of imines and Cu^I in aqueous solution during the formation of **1**.

Figure 1.2 Intersection of dynamic covalent and supramolecular spaces during subcomponent self-assembly.

Certain ligand structures are certain to be favored, and others not present at all ("virtual") [26]. Likewise, certain metal complexes are thermodynamically more stable than others. Since dynamic interconversion is possible on both covalent and supramolecular levels, both ligand and metal preferences may act in concert to amplify a limited subset of structures out of the dynamic library of all possible structures. The preparation of **1** thus represents a sorting of the dynamic combinatorial library of Figure 1.2.

Due to the strong preference of copper(I) for imine ligands, the set of observed structures is often much smaller than the set of possible structures, such as those containing aminal or hemiaminal ligands. Copper(I)/imine systems are thus particularly fruitful for use in subcomponent self-assembly. We are very interested in deciphering the selection rules that dictate the products observed under a given set of conditions, with the goal of being able to understand and exploit the basic

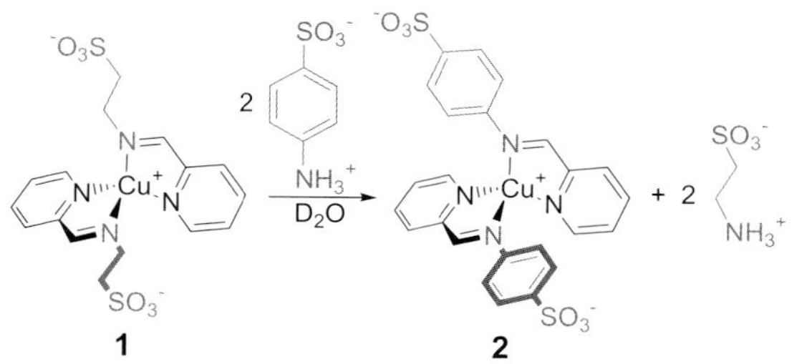

Scheme 1.2 Subcomponent substitution driven by differences in acidity.

"programming language" that might enable the formation of complex structures based on simple starting materials.

Although thermodynamically stable in aqueous solution, complex **1** nonetheless readily underwent covalent imine substitution in the presence of sulfanilic acid to form **2** (Scheme 1.2).

This reaction occurred with greater than 95% selectivity. The driving force behind this imine exchange may be understood in terms of the difference in acidity between sulfanilic acid (pK_a = 3.2) and taurine (pK_a = 9.1), which favors the displacement of the protonated form of the weaker acid (taurine) from **1** and the incorporation of the deprotonated form of the stronger acid (sulfanilic acid) during the formation of **2** [24].

1.3
Chirality

The copper(I) centers of **1** and **2** are chiral. The proximity of another chiral center gives diastereomers, differentiating the energies of the *P* and *M* metal-based stereocenters of the mononuclear complex.

Initial investigations [28] revealed that (*S*)-3-aminopropane-1,2-diol may be used to synthesize a mononuclear complex similar to **1** (Scheme 1.3). In dimethyl

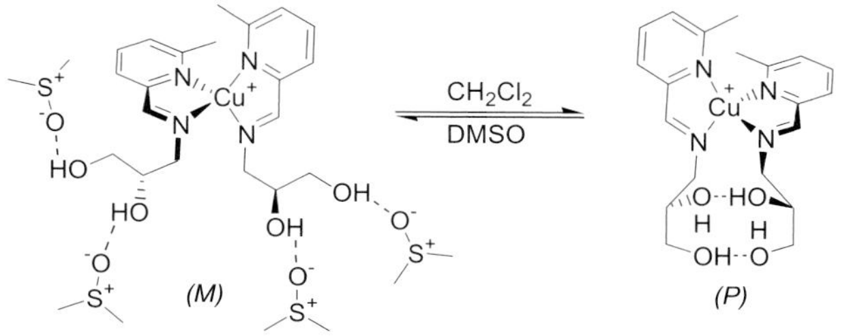

Scheme 1.3 Postulated structures of a mononuclear complex containing a chiral amine subcomponent in DMSO (left, *M* predominating) and CH_2Cl_2 (right, *P* exclusively).

sulfoxide (DMSO) solution, circular dichroism (CD) and NMR spectra indicated that one diastereomer is present in 20% excess over the other. In dichloromethane solution, however, only one diastereomer was observed by NMR. The CD spectrum indicated, however, that it had the opposite chirality at copper than the one favored in DMSO!

In dichloromethane, the hydroxyl groups appeared to be strongly associated with each other, rigidifying the structure and leading to efficient chiral induction. In contrast, DMSO would be expected to interact strongly with the hydroxyl groups, acting as a hydrogen bond acceptor (Scheme 1.3, left). The effect should be to pull the hydroxyl groups out into the solvent medium. One of the two diastereomers should allow for more energetically favorable interactions between the hydroxyl groups and the solvent, leading to the observed diastereoselectivity.

This interpretation is also supported by the results of a study correlating the observed diastereomeric excess with the Kamlet–Taft β parameter [29], a measure of the hydrogen-bond acceptor strength. A linear free energy relationship was found to exist between β and the diastereomeric excess for those solvents having α (hydrogen bond donor strength) = 0 [28].

1.4
Construction

Following our preparation of mononuclear complexes **1** and **2**, we sought to employ subcomponent self-assembly to prepare polynuclear assemblies of greater structural complexity. The use of a copper(I) template allowed the linking of two amine and two aldehyde subcomponents in a well-defined way, in which the two imine ligands lie at a 90° angle about the copper center, as shown in Scheme 1.4. We sought to build more complex structures by using this motif as a tecton [30], or fundamental building block, as described below.

1.4.1
Dicopper Helicates

The reaction of sulfanilic acid with 2,9-diformyl-1,10-phenanthroline, copper(I) oxide, and sodium bicarbonate gave a quantitative yield of the anionic double

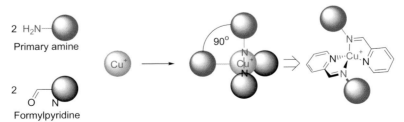

Scheme 1.4 The organization of subcomponents around a metal center.

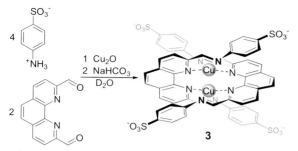

Scheme 1.5 Construction of double helicate **3** from subcomponents.

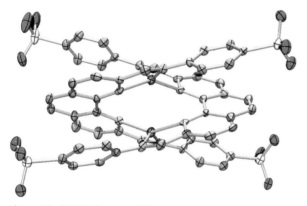

Figure 1.3 ORTEP diagram of dianionic **3**.

helicate **3**, as shown in Scheme 1.5 [17]. Two of the bis-pyridine(imine) building blocks shown in Scheme 1.2 were thus incorporated into a single phenanthroline-bis(imine) subcomponent. The geometry of the phenanthroline molecule prevents all four nitrogen atoms of one of the ligands of **3** from coordinating to a single copper(I) ion, but two copper centers may readily be chelated together to generate the helical structure of **3**.

In the crystal, the copper(I) centers of **3** adopt a flattened tetrahedral geometry (Figure 1.3), in very similar fashion to what has been observed in related structures [31, 32]. The deep green color of such complexes has been noted [31] to be extremely unusual for copper(I), being more frequently associated with copper(II). The color is associated with a local minimum in the UV-Visible spectrum of **3** at 560 nm, between higher-energy absorptions associated with π–π* transitions and a broad abs centered around 690 nm. We suspect this latter feature to be associated with one or more metal-to-ligand charge transfer transitions. The 2.73 Å distance between the copper centers might allow a photoexcited state in which the addi-

Table 1.1 Helicate formation selection rules in water and acetonitrile.

Amine	Helicate in H$_2$O	Helicate in CH$_3$CN
H$_2$N~~~OH	Yes	Yes
H$_2$N~~~O~~~OH	Yes	Yes
H$_2$N—CH(OH)—CH$_2$OH	Yes	No
H$_2$N—C$_6$H$_4$—SO$_3^-$ Na$^+$	Yes	No
H$_2$N~~~SO$_3^-$ Na$^+$	Yes	No
H$_2$N—C(OH)(OH)(OH)	No	No
H$_2$N~~~N(CH$_3$)$_3^+$ CF$_3$SO$_3^-$	No	No
H$_2$N—C$_6$H$_4$—N$^+$ CF$_3$SO$_3^-$	No	No

tional positive charge is delocalized across both copper ions, as has been seen in other dicopper(I) structures [33]. Theoretical investigations are underway.

In addition to sulfanilic acid, numerous other primary amines could be used to construct helicates. The conditions under which different amines were incorporated into these helicates were investigated. Table 1.1 summarizes the selection rules discovered.

Water was preferred to acetonitrile as the solvent, allowing moderately hindered and anionic amines to self-assemble. Acetonitrile is a much better ligand for copper(I) than water, making it more difficult for hindered ligands (such as the one formed from serinol, third entry in Table 1.1) to form complexes in competition with the solvent. More hindered amines as well as cationic amines were not incorporated in either solvent, which we attribute to steric and Coulombic repulsion, respectively.

1.4.2
Tricopper Helicates

Tricopper helicates could also be synthesized using a simple modification of the dicopper helicate preparation [28]. When three equivalents of copper(I) were employed and 8-aminoquinoline was used in place of an aniline, tricopper double-helicate **4** was formed as the unique product (Scheme 1.6).

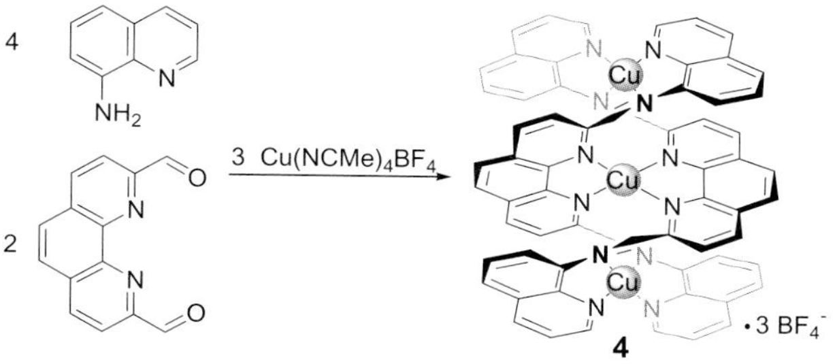

Scheme 1.6 The preparation of tricopper helicate **4**.

1.4.3
Catenanes and Macrocycles

When short, flexible diamine **a** was used as a subcomponent in helicate formation, as shown on the left side of Scheme 1.7, only one topological isomer of product was observed: twisted macrocycle **5**. This diamine is not long enough to loop around the phenanthroline to form a catenated structure [28].

When a longer diamine subcomponent that contained rigid phenylene segments was used, as shown in Scheme 1.7 at right, the formation of such macrocyclic structures became energetically disfavored. The orientation of the rigid phenylene groups readily allowed the flexible chains to bridge across the backs of the phenanthroline groups, giving rise to the catenated structure **6**. This interpenetration of two identical macrocycles was the only observed product [28].

Unlike the original Sauvage catenates [34], catenate **6** is helically chiral in addition to possessing the possibility of becoming topologically chiral through the incorporation of an asymmetrical dianiline. The investigation of both kinds of chirality in catenates similar to **4** is currently under investigation.

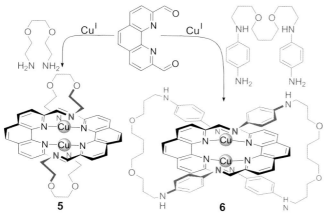

Scheme 1.7 The selection of a macrocyclic (**5**) or catenated (**6**) topology based on the rigidity and length of the subcomponents employed.

1.4.4
[2 × 2] Tetracopper(I) Grid

The aqueous reaction of Cu^I, pyridine-2-carbaldehyde and a water-soluble *m*-phenylenediamine resulted in the quantitative formation of the tetracopper(I) grid complex **7** shown in Scheme 1.8 [21].

The crystal structure of the grid (Figure 1.4) suggested the presence of strain, an unusual feature for a quantitatively self-assembled structure. Intriguingly, no grid was observed to form in any solvent except water. We hypothesize that the hydrophobic effect plays an essential role in the self-assembly process, causing ligands and metal ions to wrap together into a compact structure in which the hydrophobic ligand surfaces are minimally exposed to the aqueous environment.

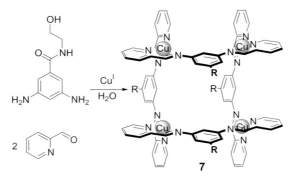

Scheme 1.8 Self-assembly of [2 × 2] grid complex **7** that forms only in water among all solvents tried (R = —CONHCH$_2$CH$_2$OH).

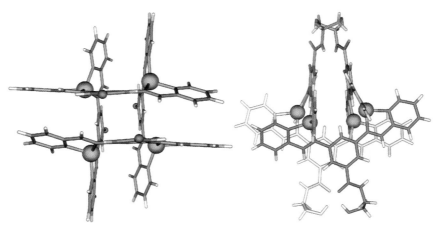

Figure 1.4 Orthogonal views of the crystal structure of the tetracationic grid **7** (the —CONHCH$_2$CH$_2$OH groups of the ligands are replaced by purple spheres at left).

A "diffuse pressure" applied by the hydrophobic effect would compensate the strain thus engendered. Extension of this strategy may permit the use of self-assembly to construct other strained structures, which tend to have unusual and technologically interesting properties [35].

1.5 Sorting

A particular challenge of subcomponent self-assembly lies in the fact that one must employ building blocks that contain proportionally more self-assembly information than is required in the case of presynthesized ligands: "assembly instructions" for both ligands and supramolecular structure must be included. It is therefore worthwhile to investigate ways in which this information might be encoded, such that individual subcomponents might be directed to react with specific partners within mixtures. This idea allows complex dynamic libraries [36] to be sorted into a limited number of structures, or individual subcomponents to be directed to specific locations within larger structures.

1.5.1 Sorting Ligand Structures with Cu(I)

In initial work [37], we demonstrated that complexes containing different imine ligands could be synthesized in each others' presence. When pyridine-2-carbaldehyde and benzaldehyde-2-sulfonate were mixed in aqueous solution with the diamine shown in Scheme 1.9, a library of ligands is created in dynamic equilibrium with the starting materials. The addition of copper(I) eliminated all but two of these ligands, forming complexes **8** and **9** in quantitative yield [37].

The simultaneous formation of **8** and **9** results in a situation in which all copper(I) ions are tetracoordinate and all of the ligands' nitrogen atoms are bound to copper centers. Any other structures formed from this mixture of subcomponents would either contain more than one metal center (entropically disfavored) or have unsatisfied valences at either metal or ligand (enthalpically disfavored).

1.5.2 Simultaneous Syntheses of Helicates

This concept may also be extended to polynuclear helicates [38]. When 2-aminoquinoline and 4-chloroaniline were mixed with the phenanthroline dialdehyde shown in Scheme 1.10, a dynamic library of potential ligands was observed to form. The addition of copper(I) causes this library to collapse, generating only dicopper and tricopper helicates. As in the mononuclear case of Scheme 1.9, the driving force behind this selectivity appeared to be the formation of structures in which all ligand and metal valences are satisfied. The use of supramolecular (coordination) chemistry to drive the covalent reconfiguration of intraligand bonds thus

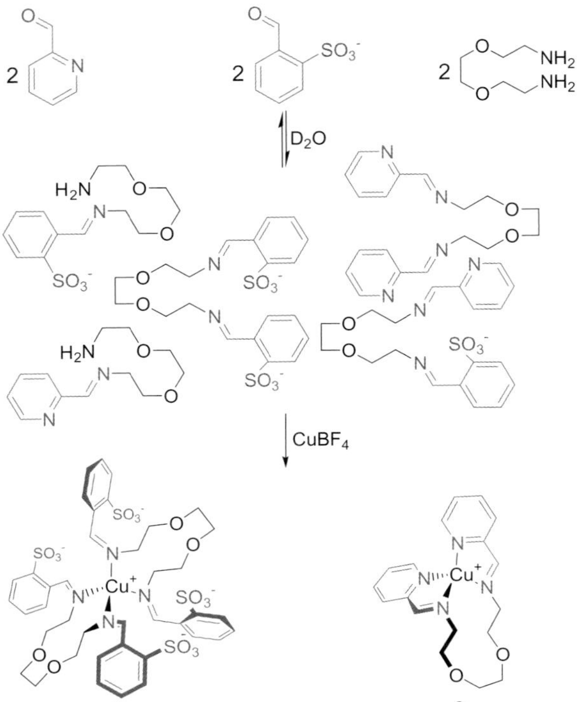

Scheme 1.9 The dynamic reconstitution of a library of imine ligands into a mixture of **8** and **9** following the addition of copper(I).

appears to be a general phenomenon, applicable in polynuclear as well as mononuclear cases.

1.5.3
Sorting within a Structure

The preparation of structure **10**, shown in Scheme 1.11, requires a different kind of selectivity in the choice of ligand subcomponents. Whereas during the simultaneous formation of dicopper and tricopper helicates (Scheme 1.10) all mixed ligands were eliminated from the dynamic library initially formed, in Scheme 1.11 the mixed ligand forms the unique structure selected during equilibration [39].

This differential selectivity results from the differing numbers of donor atoms offered by the two dialdehydes upon which these structures are based. Phenanthroline dicarbaldehyde readily lends itself to the construction of a set of *homo-ligands* bearing a number of donor atoms divisible by 4, matching the coordination preference of copper(I), as seen in the dicopper and tricopper helicate structures discussed earlier.

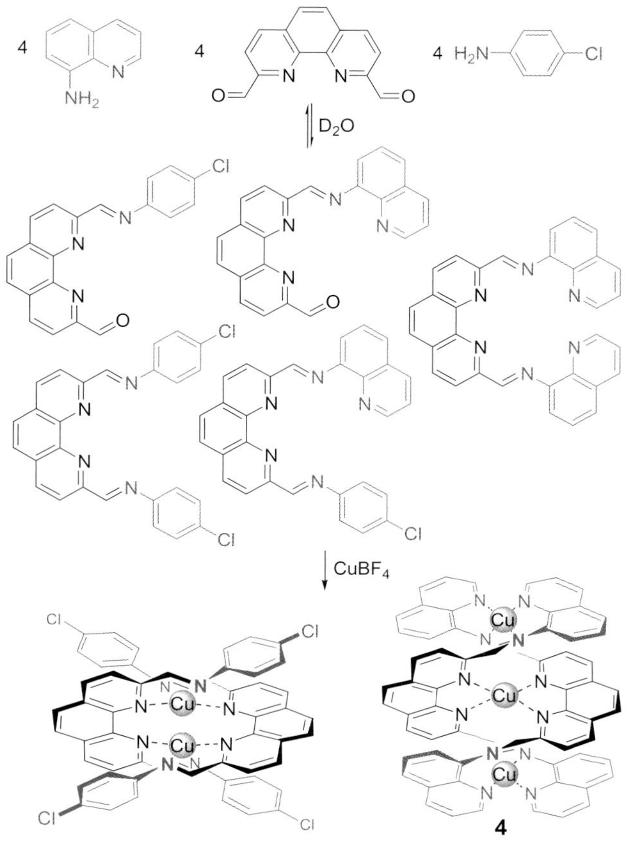

Scheme 1.10 Simultaneous preparation of dicopper and tricopper helicates from a dynamic library of ligands.

In contrast, pyridine dicarbaldehyde must make homo-ligands incorporating an odd number of donor sites. In order to generate ligand sets bearing a number of donor sites divisible by 4, *hetero-ligands* are necessary. In following this principle, the formation of hetero-ligand-containing structure **10** is selected from the components shown in Scheme 1.11.

The special stability of compound **10** was demonstrated by the fact that it could also be generated by mixing together the two homo-ligand-containing complexes **11** and **12** (Scheme 1.11, bottom). Although both of these complexes are thermodynamically stable, **11** contains only three donor atoms per copper, whereas **12** contains five such donors. The possibility of achieving coordinative saturation thus drives an imine metathesis reaction, redistributing the subcomponents to give structure **10** as the uniquely observed product. We are not aware of another such case in which different subcomponents are sorted *within* a single product structure.

In addition to structure **10**, in which the ligand adopts a head-to-head orientation, we were able to prepare structure **13** (Scheme 1.12), in which the ligands

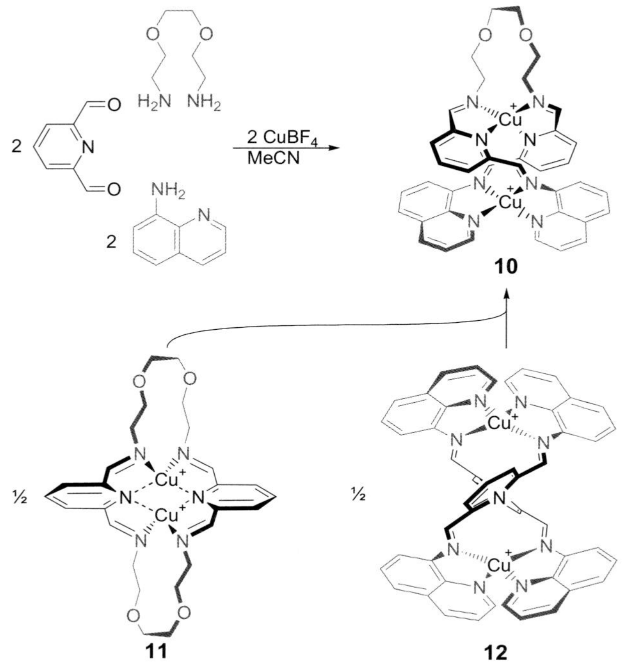

Scheme 1.11 The preparation of structure **10** from a mixture of ligand subcomponents (top) and through the covalent conproportionation of subcomponents from preformed structures **11** and **12** (bottom).

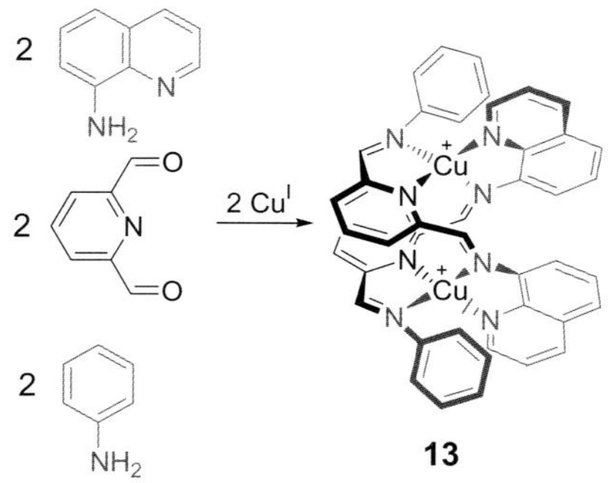

Scheme 1.12 The preparation of **13**, in which the ligands adopt a head-to-tail configuration.

adopt a head-to-tail orientation [39]. This orientation is favored by the antiparallel orientation of the ligands' dipoles, which is not possible in the case of **10** due to the insufficient length of the diamine subcomponent. In addition to the constitutional selectivity engendered through the avoidance of valence frustration, we were able thus to observe an orientational selectivity by employing either two equivalents of aniline or one equivalent of diamine subcomponent.

1.5.4
Cooperative Selection by Iron and Copper

Extending this sorting methodology further, we have examined a larger self-organizing system in which Fe^{II} and Cu^{I} act together to sort a more complex dynamic library of ligand subcomponents (Scheme 1.13) [37]. When pyridine-2-carbaldehyde, 6-methylpyridine-2-carbaldehyde, ethanolamine, and tris(2-

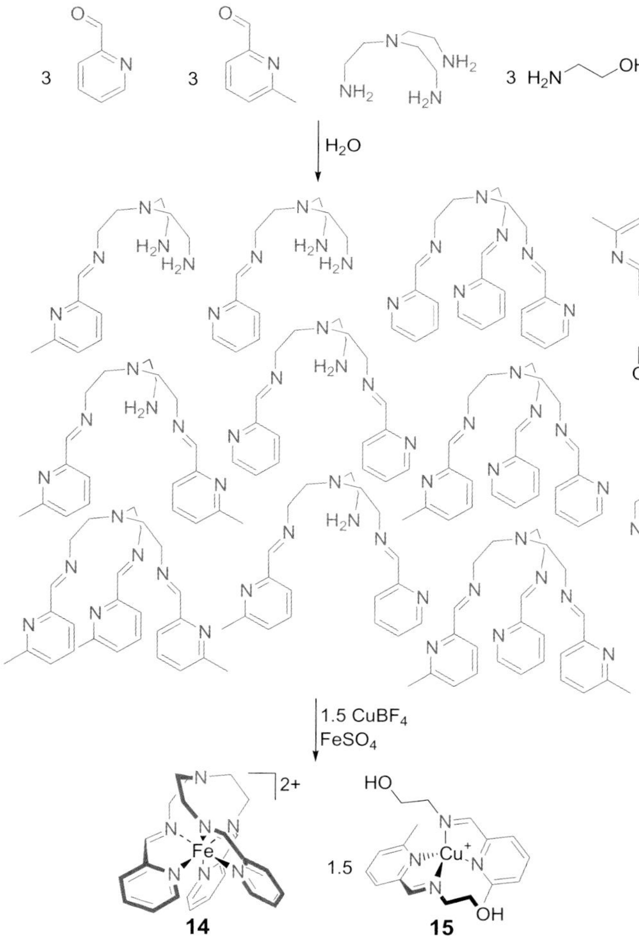

Scheme 1.13 The formation of a dynamic combinatorial library of ligands, and the collapse of this library into complexes **14** and **15** following the addition of Cu^{I} and Fe^{II}.

aminoethyl)amine were mixed together in water, a dynamic library of imines formed in equilibrium with the starting materials. When copper(I) tetrafluoroborate and iron(II) sulfate were added, this dynamic library was observed to collapse, leaving compounds **14** and **15** as the sole remaining products. This thermodynamic sorting process thus directed each building block to its unique destination.

Certain factors play an obvious role in winnowing down the number of observed product structures: The template effect [14] should eliminate all partially formed ligands and ligand subcomponents from the mixture, the chelate effect [40] should favor structures containing ligands that bear the highest number of bound donor atoms possible, and iron(II) and copper(I) should be bound to six and four donor atoms, respectively. Within these bounds, a variety of different product structures might nonetheless be envisaged. We discuss below our investigations of three distinct preferences exhibited by copper(I) and iron(II). These preferences act in concert to select **14** and **15** alone as products of the reaction of Scheme 1.13.

Firstly, copper(I) preferentially formed complexes that incorporated the methylated aldehyde. All three possible products shown in Scheme 1.2 were observed in the reaction of Scheme 1.14. The two aldehydes were incorporated into the product mixture in a molar ratio of 30:70, which indicates a slight thermodynamic preference for the incorporation of the methylated aldehyde, whose more electron-rich character would allow ligands that incorporate this residue to better stabilize the cationic copper(I) center.

Secondly, iron(II) was observed to form octahedral complexes that incorporate the triamine in preference to the monoamine. When complex **16** was mixed with an equimolar amount of triamine in aqueous solution, complex **14** and monoamine were the only products observed (Scheme 1.15). The chelate effect [40] may be understood to drive this substitution: The incorporation of one equivalent of triamine results in the liberation of three equivalents of monoamine, which provided the entropic driving force for this reaction.

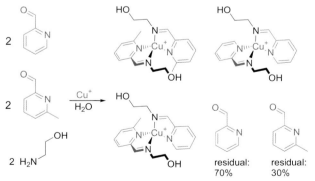

Scheme 1.14 The choice of copper: methylated aldehyde was preferentially incorporated into copper(I) complexes in aqueous solution.

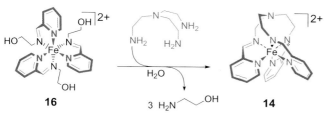

Scheme 1.15 The choice of iron: the entropically driven displacement of monoamine by triamine during the preparation of **14** from **16**.

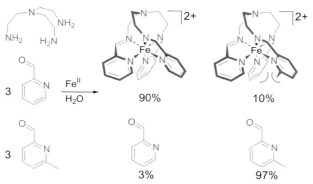

Scheme 1.16 The choice of iron: the preferential incorporation of the less-hindered aldehyde into pseudo-octahedral Fe^{II} complexes.

A third driving force for the observed selectivity, complementing and amplifying copper's choice of the methylated aldehyde, is the preference of iron(II) to incorporate the nonmethylated aldehyde into complexes of type **14**. As shown in Scheme 1.16, the addition of iron(II) to a mixture of triamine and both aldehydes gave a product mixture in which nonmethylated and methylated aldehydes are present in a 3:97 ratio following equilibration. Only the two products shown in Scheme 1.16 were observed in the product mixture. The reaction of Scheme 1.16 thus deviates substantially from a statistical mixture of products; indeed, only two of the expected four products are observed to form. No evidence was found of complexes incorporating two or three equivalents of methylated aldehyde.

When either cobalt(II) or zinc(II) sulfate was used in place of the iron(II) salt in the reaction of Scheme 1.13, mixtures of products were obtained. In the crystal, iron(II) is bound more tightly in **1** (mean r_{Fe-N} = 1.95 Å) [41] than are cobalt(II) (mean r_{Co-N} = 2.15 Å) [41] or zinc(II) (mean r_{Zn-N} = 2.18 Å) [41] in complexes with the same ligand; this observation was correlated with the finding that iron(II) is the only divalent first-row transition metal having a low-spin ground state with this ligand [41]. The groups of Drago [42], Hendrickson [43], and Hauser [44] have investigated the magnetic behavior of the two iron complexes shown in Scheme

1.16, in addition to the other two congeners incorporating 2 and 3 equivalents of methylated aldehyde, respectively. They determined that although **14** remains in the low-spin $^1A_1$ state from 30 to 450 K [43, 44], the complexes containing methylated aldehyde residues undergo spin crossover to the high-spin $^5T_2$ state as the temperature increases. The more such residues a complex contains, the lower the temperature at which it undergoes spin crossover. A steric clash between the methyl groups and facing pyridyl rings, as shown in Scheme 1.16, appears to destabilize the low-spin state with respect to the high-spin state by elongating the Fe–N bonds.

The strong thermodynamic preference of iron(II) to incorporate nonmethylated aldehyde might thus be attributed to the high energetic penalty paid for a steric clash in iron(II) complexes of this type. Since the presence of high-spin iron(II) may be inferred in complexes bearing one or more methylated aldehyde residues, one might also invoke the possibility of a "spin-selection" phenomenon, whereby the formation of short, strong bonds between low-spin iron(II) and sp^2 nitrogen atoms serves as a thermodynamic driving force for the preferential incorporation of sterically unhindered aldehyde.

1.6
Substitution/Reconfiguration

Many of the complexes prepared through subcomponent self-assembly underwent clean substitution chemistry, which may operate both at covalent and coordinative levels. As discussed below, driving forces for such substitutions included the relief of steric encumbrance, the substitution of an electron-poor subcomponent for an electron-rich one, the use of pK_a differentials, and the chelate effect.

1.6.1
New Cascade Reaction

Pseudotetrahedral complexes such as **17** (Scheme 1.17) were observed to possess a particularly rich substitution chemistry [37]. Complex **17** reacted cleanly with *o*-phenylenediammonium to give the covalent substitution product **18** shown in Scheme 1.17. This imine substitution was driven by the same pK_a effect employed in the **1**-to-**2** transformation of Scheme 1.2. In addition, **17** reacted cleanly with copper bis(biquinoline) complex **10** to give the coordinative substitution product **20**. This ligand exchange appears to have been sterically driven: the substitution of one of the encumbering di(imine) ligands for a less bulky biquinoline provided the driving force for this reaction [45].

In contrast with **17**, complex **21** (Scheme 1.18) did not undergo ligand substitution with the copper(I) bis(biquinoline) complex, possibly as a result of the different steric properties of the two complexes. The imine exchange reaction with phenylenediammonium worked well, creating the possibility of a new kind of domino or cascade reaction (Scheme 1.18). The intermediate product **18** (from

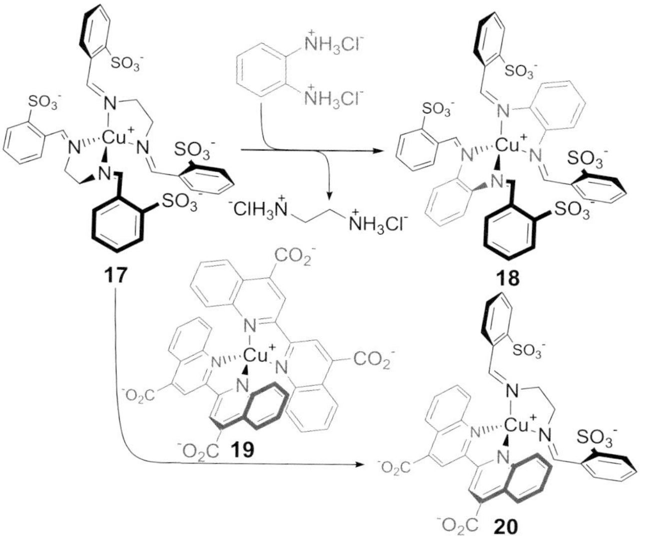

Scheme 1.17 Covalent (above) and coordinative (below) rearrangements of complex **17** to give **18** and **20**, driven by pK_a differences and sterics, respectively.

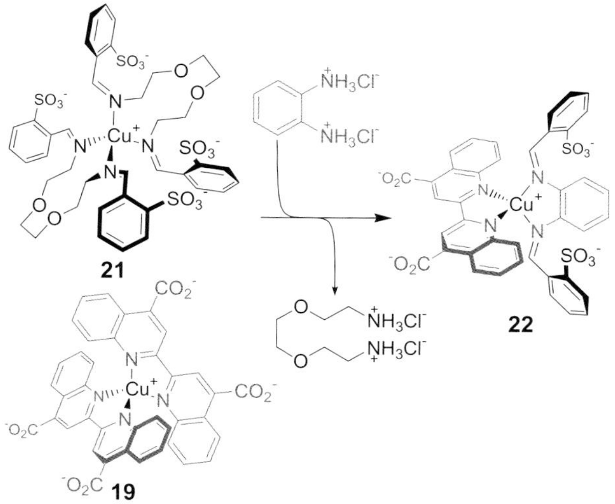

Scheme 1.18 The cascade reaction of **19** and **21**, rearranging on both covalent and coordinative levels to give **22** upon the addition of *o*-phenylenediammonium dichloride.

Scheme 1.17) formed following the reaction between **21** and phenylenediammonium, reacted immediately with **19** to give the final product **22**.

The addition of phenylenediammonium to a mixture of **19** and **21** thus caused two distinct rearrangements to occur: initial (covalent) imine exchange followed immediately by (coordinative) ligand exchange, resulting in the exclusive formation of mixed-ligand complex **22**.

1.6.2
Hammett Effects

The electronic nature of the amine incorporated into these imine complexes should play an important role in determining the stability of Cu^I complexes, and therefore the composition of equilibrium mixtures when several amines compete as subcomponents. To investigate the influence of electronic effects, we ran a series of competition experiments between unsubstituted and substituted anilines [46] (Scheme 1.19).

One equivalent each of unsubstituted aniline, substituted aniline, and pyridine-2-carbaldehyde were mixed in DMSO. Following equilibration, no free aldehyde could be detected: An equilibrium mixture of imines and free anilines was observed in each case. Once the equilibrium had stabilized, half of an equivalent of copper(I) was added, and the equilibrium population of the two free anilines was again measured.

A high-quality correlation was found between the Hammett σ_{para} value [47] of a given aniline and the K_{eq} of the competition between free and substituted aniline, as shown in Figure 1.5. The increased magnitude of ρ following copper coordination indicated that the cationic copper complex was better stabilized by an electron-donating group than was the free ligand, as expected.

The quality of this linear free energy relationship allowed us to predict with confidence the equilibrium constant of a subcomponent substitution reaction

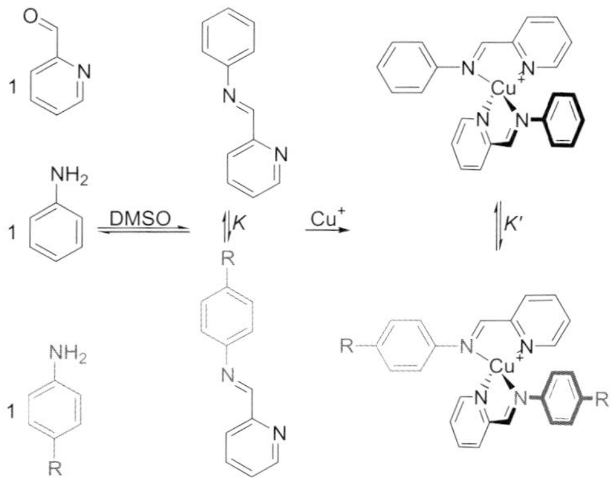

Scheme 1.19 Competition between unsubstituted aniline and 4-substituted anilines (—R = —NMe$_2$, —OH, —OMe, —Me, —SMe, —I, —CO$_2$Et, —Ac) in the absence and presence of Cu^I.

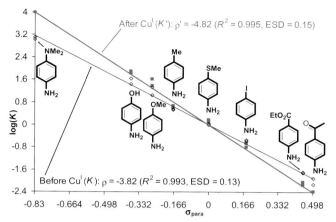

Figure 1.5 Linear free energy relationships that correlate the σ_{para} of 4-substituted anilines with the stability of their 2-pyridylimines and the Cu^I complexes thereof.

between an arbitrary pair of anilines. The large magnitude of ρ also indicated that differently substituted anilines might be used to effect clean transformations between assemblies. For example, in Scheme 1.20 we demonstrate a series of transformations between four distinct structures, ending with metallocycle **23**. The entire sequence could be carried out in the same reaction flask, and the yields of the individual displacement reactions were close to those predicted using the Hammett equation. The driving force for the last displacement, as well as part of the first, is entropic in nature, and may be considered as a special case of the chelate effect [46].

1.6.3
Helicate Reconfigurations

The dicopper double-helicate moiety [31] has exhibited rich and varied substitution chemistry, as discussed below [17, 28, 46]. It is more rigid and structurally better-defined than the mononuclear complexes discussed above, which allows one to use it as a persistent, well-defined tecton [13].

The pK_a-differential-driven chemistry that functions in mononuclear cases (Schemes 1.2, 1.17, and 1.18) also works well in the context of helicates. As shown in Scheme 1.21, helicate **24** was transformed into **3** upon the addition of sulfanilic acid.

Entropy may be harnessed as a driving force in the context of helicates as well as in mononuclear complexes, as evidenced by the **24**-to-**5** and **3**-to-**5** conversions shown in Scheme 1.21. Two distinct hierarchical layers of control over subcomponent substitution may thus be employed in tandem, based upon pK_a differences and the chelate effect.

The entropy-driven conversion of **3** to **5** may be reversed upon lowering the pH (Scheme 1.22). The addition of sulfanilic acid to macrocycle **5** resulted in its con-

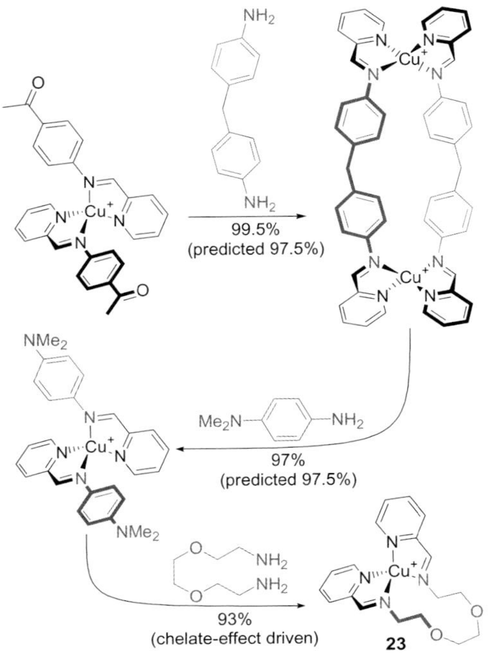

Scheme 1.20 A one-pot series of transformations between four distinct products, bearing alternatively macrocyclic and open topologies, ending with metallomacrocycle **23**.

version to helicate **3**. Basification of this solution through the addition of NaHCO$_3$ resulted in the regeneration of **5**, closing the cycle. By changing the pH, it was thus possible to switch dynamically between the open topology of helicate **3** and the closed topology of macrocycle **5** [28].

Tricopper double-helicate **4** may also be synthesized through subcomponent substitution, starting with the 4-chloroaniline-containing dicopper helicate shown in Scheme 1.23. The electron-poor 4-chloroaniline residue (σ_{para} = 0.23) [47] thus serves as an excellent leaving group in this substitution reaction. Starting with a dicopper helicate containing more electron-rich 4-methoxyaniline residues (σ_{para} = −0.27) [47] we observed only 18% of **4** following equilibration.

1.6.4
Substitution as a Route to Polymeric Helicates

The reaction of dicopper helicate **25** (Scheme 1.24) with o-phenylenediamine produced the dimeric tetracopper helicate **26** in 51% yield. This reaction is the first step of a step polymerization reaction [48]; during the course of the reaction, a substantial amount of insoluble brown material is also produced. The elemental analysis of this insoluble co-product is consistent with that of a higher oligomer

1.6 Substitution/Reconfiguration | 25

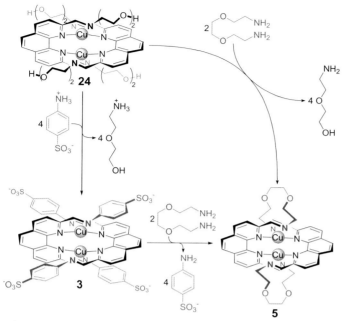

Scheme 1.21 The subcomponent substitution of diamine for both aryl (3→5) and alkyl (24→5) monoamines, complementing the substitution of arylamines for alkylamines (24→3).

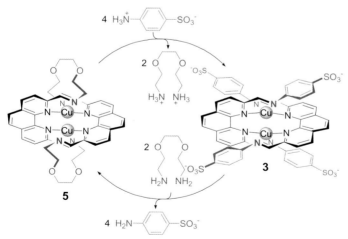

Scheme 1.22 Cycling between **5** and **3** as a function of pH.

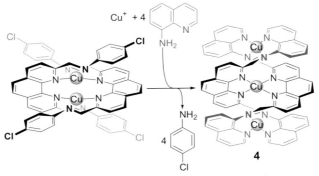

Scheme 1.23 Substitution of aminoquinoline for chloroaniline, generating trimetallic **4** from a chloroaniline-containing dimetallic double-helicate.

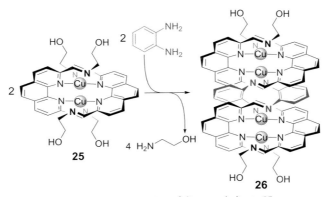

Scheme 1.24 Dimerization reaction of dicopper helicate **25** and o-phenylenediamine to give tetracopper helicate **26**.

or polymer; the mass of this material, formulated as a polymer, corresponds well to the lost mass of subcomponents that are not incorporated into **26**.

Tetracopper helicate **26** thus appears only to be isolable as a result of its greater solubility, with respect to higher oligomers. The use of more-soluble subcomponents might thus allow for the preparation of soluble polymers based upon the stacked-helicate motif of **26**.

Such polymers might be of substantial technological interest as electrically conductive "molecular wires." Density functional (DFT) calculations indicated that the metal-based HOMO of **26** is delocalized across all four of the central copper(I) ions, as shown in Figure 1.6. Calculations also indicate that removal of one electron from this HOMO likewise results in a fully delocalized singly occupied orbital. This delocalization could provide a path for conductivity, allowing an electric current to flow from one end of a helicate polymer strand to the other.

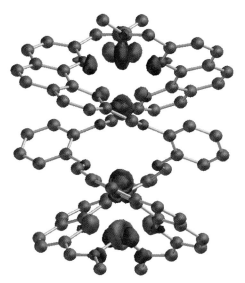

Figure 1.6 The metal-based HOMO of **26**, delocalized across all copper ions, as determined by DFT calculations.

1.7
Conclusion and Outlook

The creation of structural complexity, including topological complexity [19, 22], is feasible using subcomponent self-assembly, and the structures thus made may be induced to reassemble in well-defined ways using a variety of driving forces. The demonstration of directing "nonorthogonal" sets of subcomponent building blocks to come together in well-defined ways also opens up the possibility of linking such subunits together covalently, such that their self-assembly instructions serve as "subroutines" to guide the generation of a more complex superstructure.

We are currently investigating the use of subcomponent self-assembly to prepare new metal-containing polymeric materials. Following the same methodology that allowed the preparation of macrocycle **5** and catenane **6** (Scheme 1.7), further variations in the length, rigidity, and geometry of diamine subcomponents might allow for the generation of double-helical polymers, cyclic catenanes, or perhaps even polymeric catenanes.

1.8
Acknowledgments

This work was carried out by Marie Hutin, David Schultz, David Gérard, and Sonya Torche, in addition to our collaborators Christopher Cramer, Laura Gagliardi,

Gérald Bernardinelli, Christoph Schalley, Jérôme Lacour, Richard Frantz, Damien Jeannerat, André Pinto and Philippe Perrottet. Financial support has been provided by the Walters-Kundert Chairtable trust, the University of Geneva, the Swiss National Science Foundation, the ERA-Chemistry Network, and the Swiss State Secretariat for Education and Research.

References

1 Lehn, J.M. (2002) *Science*, **295**, 2400–3. Whitesides, G.M., Mathias, J.P. and Seto, C.T. (1991) *Science*, **254**, 1312–19.
2 Bahcall, N.A. (1988) *Annual Review of Astronomy and Astrophysics*, **26**, 631–86.
3 Parrish, J.K. and Edelstein-Keshet, L. (1999) *Science*, **284**, 99–101.
4 Lindsey, J.S. (1991) *New Journal of Chemistry*, **15**, 153–80.
5 Rowan, S.J., Cantrill, S.J., Cousins, G.R.L., Sanders, J.K.M. and Stoddart, J.F. (2002) *Angewandte Chemie (International ed. in English)*, **41**, 898–952.
6 Nicolaou, K.C. and Sorensen, E.J. (1996) *Classics in Total Synthesis*, Wiley-VCH Verlag GmbH, Weinheim.
7 Thomas, S. (2002) *Angewandte Chemie (International ed. in English)*, **41**, 48–76.
8 Hof, F., Craig, S.L., Nuckolls, C. and Rebek, J. (2002) *Angewandte Chemie (International ed. in English)*, **41**, 1488–508.
9 Berl, V., Schmutz, M., Krische, M.J., Khoury, R.G. and Lehn, J.M. (2002) *Chemistry – A European Journal*, **8**, 1227–44.
10 Zerkowski, J.A., Seto, C.T. and Whitesides, G.M. (1992) *Journal of the American Chemical Society*, **114**, 5473–5.
11 Lehn, J.M. (1995) *Supramolecular Chemistry: Concepts and Perspectives*, Wiley-VCH Verlag GmbH, Weinheim.
12 Albrecht, M. (2000) *Journal of Inclusion Phenomena and Macrocyclic Chemistry*, **36**, 127–51. Seidel, S.R. and Stang, P.J. (2002) *Accounts of Chemical Research*, **35**, 972–83. Sun, W.-Y., Yoshizawa, M., Kusukawa, T. and Fujita, M. (2002) *Current Opinion in Chemical Biology*, **6**, 757–64. Caulder, D.L. and Raymond, K.N. (1999) *Accounts of Chemical Research*, **32**, 975–82. Holliday, B.J. and Mirkin, C.A. (2001) *Angewandte Chemie (International ed. in English)*, **40**, 2022–43.
13 Hosseini, M.W. (2005) *Accounts of Chemical Research*, **38**, 313–23.
14 Hubin, T.J. and Busch, D.H. (2000) *Coordination Chemistry Reviews*, **200**, 5–52.
15 MacLachlan, M.J. (2006) *Pure and Applied Chemistry*, **78**, 873–88. Schafer, L.L., Nitschke, J.R., Mao, S.S.H., Liu, F.Q., Harder, G., Haufe, M. and Tilley, T.D. (2002) *Chemistry – A European Journal*, **8**, 74–83.
16 Houjou, H., Iwasaki, A., Ogihara, T., Kanesato, M., Akabori, S. and Hiratani, K. (2003) *New Journal of Chemistry*, **27**, 886–9. Childs, L.J., Alcock, N.W. and Hannon, M.J. (2002) *Angewandte Chemie (International ed. in English)*, **41**, 4244–7. Hamblin, J., Childs, L.J., Alcock, N.W. and Hannon, M.J. (2002) *Dalton Transactions*, 164–9.
17 Nitschke, J.R., Schultz, D., Bernardinelli, G. and Gérard, D. (2004) *Journal of the American Chemical Society*, **126**, 16538–43.
18 Hogg, L., Leigh, D.A., Lusby, P.J., Morelli, A., Parsons, S. and Wong, J.K.Y. (2004) *Angewandte Chemie (International ed. in English)*, **43**, 1218–21.
19 Leigh, D.A., Lusby, P.J., Teat, S.J., Wilson, A.J. and Wong, J.K.Y. (2001) *Angewandte Chemie (International ed. in English)*, **40**, 1538–43.
20 Nitschke, J.R. and Lehn, J.M., (2003) *Proceedings of the National Academy of Sciences of the United States of America*, **100**, 11970–4.
Brooker, S., Hay, S.J. and Plieger, P.G. (2000) *Angewandte Chemie (International ed. in English)*, **39**, 1968–70.

21. Nitschke, J.R., Hutin, M. and Bernardinelli, G. (2004) *Angewandte Chemie (International ed. in English)*, **43**, 6724–7.
22. Chichak, K.S., Cantrill, S.J., Pease, A.R., Chiu, S.-H., Cave, G.W.V., Atwood, J.L. and Stoddart, J.F. (2004) *Science*, **304**, 1308–12.
23. Pentecost, C.D., Chichak, S.K., Peters, A.J., Cave, G.W.V., Cantrill, S.J. and Stoddart, J.F. (2007) *Angewandte Chemie (International ed. in English)*, **46**, 218–22.
24. Nitschke, J.R. (2004) *Angewandte Chemie (International ed. in English)*, **43**, 3073–5.
25. Lam, R.T.S., Belenguer, A., Roberts, S.L., Naumann, C., Jarrosson, T., Otto, S. and Sanders, J.K.M. (2005) *Science*, **308**, 667–9. Lehn, J.M. and Eliseev, A.V. (2001) *Science*, **291**, 2331–2.
    Huc, I., Krische, M.J., Funeriu, D.P. and Lehn, J.M. (1999) *European Journal of Inorganic Chemistry*, 1415–20.
    van Gerven, P.C.M., Elemans, J.A.A.W., Gerritsen, J.W., Speller, S., Nolte, R.J.M. and Rowan, A.E. (2005) *Chemical Communications*, 3535–7.
    Star, A., Goldberg, I. and Fuchs, B. (2000) *Angewandte Chemie (International ed. in English)*, **39**, 2685–9.
    Constable, E.C., Housecroft, C.E., Kulke, T., Lazzarini, C., Schofield, E.R. and Zimmermann, Y. (2001) *Dalton Transactions*, 2864–71.
26. Severin, K. (2004) *Chemistry – A European Journal*, **10**, 2565–80.
27. Godoy-Alcantar, C., Yatsimirsky, A.K. and Lehn, J.M. (2005) *Journal of Physical Organic Chemistry*, **18**, 979–85.
28. Hutin, M., Schalley, C.A., Bernardinelli, G. and Nitschke, J.R. (2006) *Chemistry – A European Journal*, **12**, 4069–79.
29. Kamlet, M.J., Abboud, J.L.M., Abraham, M.H. and Taft, R.W. (1983) *The Journal of Organic Chemistry*, **48**, 2877–87.
30. Hosseini, M.W. (2003) *Coordination Chemistry Reviews*, **240**, 157–66.
31. Ziessel, R., Harriman, A., Suffert, J., Youinou, M.-T., De Cian, A. and Fischer, J. (1997) *Angewandte Chemie (International ed. in English)*, **36**, 2509–11.
32. Ameerunisha, S., Schneider, J., Meyer, T., Zacharias, P.S., Bill, E. and Henkel, G. (2000) *Chemical Communications*, 2155–6.
33. Harding, C., McKee, V. and Nelson, J. (1991) *Journal of the American Chemical Society*, **113**, 9684–5.
34. Dietrich-Buchecker, C.O., Sauvage, J.P. and Kintzinger, J.P. (1983) *Tetrahedron Letters*, **24**, 5095–8.
35. Wiberg, K.B. (1986) *Angewandte Chemie (International ed. in English)*, **25**, 312–22.
36. Corbett, P. T., Leclaire, J., Vial, L., West, K.R., Wietor, J.L., Sanders, J.K.M. and Otto, S. (2006) *Chemical Reviews*, **106**, 3652–711.
37. Schultz, D. and Nitschke, J.R. (2005) *Proceedings of the National Academy of Sciences of the United States of America*, **102**, 11191–5.
38. Hutin, M., Franz, R. and Nitschke, J.R. (2006) *Chemistry – A European Journal*, **12**, 4077–82.
39. Hutin, M., Bernardenelli, G. and Nitschke, J.R. (2006) *Proceedings of the National Academy of Sciences of the United States of America*, **103**, 17655–60.
40. Schwarzenbach, G. (1952) *Helvetica Chimica Acta*, **35**, 2344–63.
41. Kirchner, R.M., Mealli, C., Bailey, M., Howe, N., Torre, L.P., Wilson, L.J., Andrews, L.C., Rose, N.J. and Lingafelter, E.C. (1987) *Coordination Chemistry Reviews*, **77**, 89–163.
42. Hoselton, M.A., Wilson, L.J. and Drago, R.S. (1975) *Journal of the American Chemical Society*, **97**, 1722–9.
43. Conti, A.J., Xie, C.L. and Hendrickson, D.N. (1989) *Journal of the American Chemical Society*, **111**, 1171–80.
44. Schenker, S., Hauser, A., Wang, W. and Chan, I.Y. (1998) *The Journal of Chemical Physics*, **109**, 9870–8.
45. Schmittel, M. and Ganz, A. (1997) *Chemical Communications*, **1997**, 999–1000.
46. Schultz, D. and Nitschke, J.R. (2006) *Journal of the American Chemical Society*, **128**, 9887–92.
47. Mcdaniel, D.H. and Brown, H.C. (1958) *The Journal of Organic Chemistry*, **23**, 420–7.
48. Odian, G. (1991) *Principles of Polymerization*, 3rd edn, John Wiley & Sons, Ltd, New York.

2
Molecular Metal Oxides and Clusters as Building Blocks for Functional Nanoscale Architectures and Potential Nanosystems

Leroy Cronin

2.1
Introduction

Lithography has enabled the miniaturization revolution in modern technology, with over 800 million transistors being packed into the latest multicore processor architectures. Despite these great advances in technology, the ability of top-down approaches to access smaller and smaller architectures is limited. Therefore, the use of chemical self-assembly from molecular building blocks holds great promise to enable fabrication on the nanoscale, or even the sub-nanoscale. By controlling molecular organization it will be possible to develop functional nanosystems and nanomachines, ultra-high-capacity information storage materials [1], molecular electronics [2], and sensors [3]. However, there are many fundamental issues that need to be tackled, including the design of systems that can be synthesized or self-assembled in a predetermined manner to form highly complex architectures.

Here we are going to discuss a class of molecules and materials that have nanoscale architectures that can be controlled and begin to suggest how these building blocks may eventually be used for the fabrication of nanoscale devices. These building blocks are based upon a class of inorganic clusters known as polyoxometalates (POMs), and appear to have a great many desirable characteristics applicable for nanoscale assembly. This is because nanoscale polyoxometalate clusters provide an arguably unrivaled structural diversity of molecules displaying a wide range of important physical properties and nuclearities; these cover the range from 6 to 368 metal ions in a single molecule and are assembled under "one-pot" reaction conditions [4]. At the extreme, these cluster molecules are truly macromolecular, rivaling the size of proteins, and are formed by self-assembly processes [5].

The POM clusters tend to be anionic in nature, being based upon metal oxide building blocks with a general formula of MO_x, (where M is Mo, W, V and sometimes Nb and x can be 4, 5, 6 or 7). POM-based materials have a large range of interesting physical properties [6–10] which result from their many structures, the ability to delocalize electrons over the surface of the clusters, and the ability

Tomorrow's Chemistry Today. Concepts in Nanoscience, Organic Materials and Environmental Chemistry.
Edited by Bruno Pignataro
Copyright © 2008 WILEY-VCH Verlag GmbH & Co. KGaA, Weinheim
ISBN: 978-3-527-31918-3

to incorporate heteroanions, electrophiles, and ligands, and to encapsulate guest molecules within a the metal-oxo cage defined by the POM. Further, POM clusters have been shown to exhibit superacidity [6], catalytic activity [6], photochemical activity [7], ionic conductivity [7], reversible redox behavior [8], bistability [7], cooperative electronic phenomena [7], the ability to stabilize highly reactive species [9], and extensive host–guest chemistry [10].

The large number of structural types in polyoxometalate chemistry [11] can be broadly split into three classes. (i) Heteropolyanions: these are metal oxide clusters that include heteroanions such as SO_4^{2-}, PO_4^{3-}. These represent by far the most explored subset of POM clusters, with over 5000 papers being reported on these compounds during the last four years alone. Much of this research has examined the catalytic properties of POMs with great emphasis on the Keggin $\{XM_{12}O_{40}\}$ and the Wells–Dawson $\{X_2M_{18}O_{62}\}$ (where M = W or Mo and X = a tetrahedral template) anions which represent the archetypal systems. In particular W-based POMs are robust and this has been exploited to develop W-based Keggin ions with vacancies that can be systematically linked using electrophiles to larger aggregates [4, 11]. (ii) Isopolyanions: these are composed of a metal-oxide framework, but without the internal heteroatom/heteroanion. As a result, they are often much more unstable than their heteropolyanion counterparts [12]. However they also have interesting physical properties such as high charges and strongly basic oxygen surfaces, which means they are attractive units for use as building blocks [13]. (iii) Mo-blue and Mo-brown reduced POM clusters: these are related to molybdenum blue type species, which were first reported by Scheele in 1783 [14]. Their composition was largely unknown until Müller *et al.* reported, in 1995, the synthesis and structural characterization of a very high-nuclearity cluster $\{Mo_{154}\}$ crystallized from a solution of Mo-blue, which has a ring topology [15]. Changing the pH and increasing the amount of reducing agent along with incorporation of a ligand like acetate facilitates the formation of a $\{Mo_{132}\}$ spherical ball-like cluster [16] and therefore this class of highly reduced POM cluster represents one of the most exciting developments in POM chemistry and with many potential spin-off applications in nanoscience, see Figure 2.1.

Here we start to examine the pivotal role that polyoxometalate clusters can play in the development of nanoscale devices that utilize POM components, and start to conceptualize some example systems in which POM components could have a crucial role [13, 19]. This is because such functional nanosystems can exploit the building block principle already established in this area of chemistry, coupled with the range of physical properties, and the fact that POM systems can really be seen as molecular metal oxides [20]. To demonstrate this point, a number of examples have been selected across the area of POM chemistry, including our contributions, to help highlight new directions and concepts. It should also be noted that metal oxides already play an important role in the electronics and semiconductor industry today and their solid-state properties have been studied extensively [21, 22]. Many of these concepts are not new in isolation, but the possibility of using molecular design in metal oxides to produce

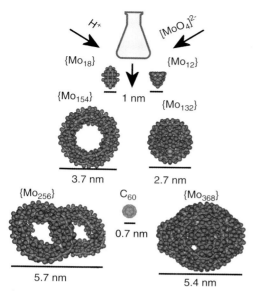

Figure 2.1 Representations of the structures of some Mo-based POM clusters (nuclearity given in subscript), all synthesized under "one-pot–one-step" acidic reaction conditions (space filling) from the well-known and studied $\{M_{12}\}/\{M_{18}\}$ Keggin/Dawson [11, 12] ions to the $\{Mo_{154}\}$ [15]/$\{Mo_{132}\}$ [16] and $\{Mo_{256}\}$ [17]/$\{Mo_{368}\}$ [18] clusters. These clusters are compared (to scale) with C_{60} to demonstrate their macromolecular dimensions [4, 5].

functional systems that exploit size effects, ligand/hetero-ion modification, switchable properties, and cooperative electronic effects will undoubtedly be significant in the quest for functional nanosystems that start to bridge the gap between bottom-up and top-down assembly [23]. This is because POMs can be constructed that bridge large length scales and lithographic techniques (top down) could be used to direct the positioning of clusters that might be built using self-assembly (bottom up).

2.2
From POM Building Blocks to Nanoscale Superclusters

The ability to design new nanoscale systems and architectures requires access to a range of building blocks. In this respect, POM cluster chemistry can offer a large range of building blocks that are conserved between structures and it appears that they have intrinsic properties (high and variable charge and flexible ligand coordination modes) that facilitate the self-assembly of clusters containing many thousands of atoms in solution. In particular, the use of pentagonal-type building groups directed by reduced {Mo$^V_2$} units plays a key role in the synthesis of these systems and the construction of nanoscale architectures [5]. This can be taken further by considering that edge-sharing (condensed) pentagons cannot be used to tile an infinite plane, whereas exactly 12 pentagons are required, in connection with well-defined sets of hexagons, to construct spherical systems such as that observed in the truncated icosahedron – the most spherical Archimedean solid – in polyhedral viruses, or in the geodesic Fuller domes [24]. Indeed, it has been shown [15–18, 20] that the Mo-based pentagonal building blocks allow the generation of very large clusters with nuclearities between 36 and 368 metal atoms in a single cluster molecule, see Figure 2.2. These clusters can be seen to be built using a range of conserved building blocks. For instance the spherical Keplerate cluster [16] can be considered in geometrical terms to be comprised of (Pentagon)$_{12}$(Linker)$_{30}$, where Pentagon = {Mo(Mo)$_5$} and Linker = {Mo$^V_2$O$_4$(OOR)$^+$}, {OMo^V(H$_2$O)}$^{3+}$, {Fe^{III}(H$_2$O)$_2$}$^{3+}$; that is, in the case of the cluster where the linkers are {Mo$^V_2$O$_4$(OAc)$^+$} the overall formula is [Mo$^{VI}_{72}$Mo$^V_{60}$O$_{372}$(MeCO$_2$)$_{30}$(H$_2$O)$_{72}$]$^{42-}$. In the case of the {Mo$_{256}$Eu$_8$} ≡ {Mo$^{VI}_{104}$Mo$^V_{24}$Eu$^{III}_4$O$_{388}$H$_{10}$(H$_2$O)$_{81}$}$_2^{20-}$ system, which comprises two elliptical {Mo$_{128}$Eu$_4$} rings, each ring is composed of 12 pentagonal units [17]. However the elliptical ring has a more complicated set of building blocks than the Keplerate cluster and is formally composed of [{Mo$_1$}$_6${Mo$_2$}$_4${Mo$_8$}$_2${Mo$_7$}$_2${Mo$_9$}$_2$] which is similar to the building blocks found in the archetypal "big-wheel" clusters [15]. As such, the wheel clusters also incorporate a {Mo$_2$} unit where the polyhedra are corner rather than edge sharing (red polyhedra in Figure 2.2). Also, the pentagonal-centered units in the {Mo$_{132}$}/{Mo$_{368}$} spherical clusters are {Mo$_6$} type units (central pentagonal unit with 5 octahedra attached, preserving the fivefold symmetry) whereas the pentagonal center unit in the {Mo$_{256}$} and other wheel clusters has two additional {Mo} units fused to the bottom of the {Mo$_6$} to make a {Mo$_8$} type unit, see Figure 2.2.

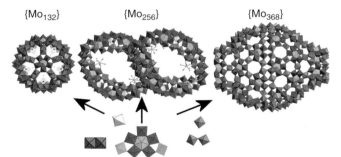

Figure 2.2 Structures of the {Mo$_{132}$}$^{[16]}$ ≡ [Mo$^{VI}_{72}$Mo$^V_{60}$O$_{372}$(MeCO$_2$)$_{30}$(H$_2$O)$_{72}$]$^{42-}$, {Mo$_{256}$}$^{[17]}$ ≡ [Mo$_{256}$Eu$_8$O$_{776}$H$_{20}$(H$_2$O)$_{162}$]$^{20-}$, and {Mo$_{368}$}$^{[18]}$ ≡ [H$_x$Mo$_{368}$O$_{1032}$(H$_2$O)$_{240}$(SO$_4$)$_{48}$]$^{48-}$, clusters shown with polyhedral plots. The structurally conserved building blocks found in these clusters is shown below; the {Mo$_2$} show below on the left (edge sharing) and right hand side (corner sharing), the {Mo$_1$} groups are also shown below and the central pentagonal unit of the {Mo$_6$} unit is shown in light grey.

The $\{Mo_{368}\} \equiv [H_xMo_{368}O_{1032}(H_2O)_{240}(SO_4)_{48}]^{48-}$ is even more complex since it combines both negative and positive curvature [18]. The building blocks can be represented as $\{Mo(Mo_5)\}_8\{Mo(Mo_5)\}'_{32}\{Mo_2\}_{16}\{Mo_2\}'_8\{Mo_2\}''_8\{Mo_1\}_{64}$ with 40 pentagonal units being required to complete the structure and can be considered to be a hybrid between the wheel and ball clusters.

The major problem with this approach lies in establishing routes to produce reactive building blocks present in solution in significant concentrations that can be reliably utilized in the formation of larger architectures without reorganizing to other unknown fragments. Access to such building blocks has been the major limitation in stepwise growth of Mo-based POM clusters compared with the more kinetically inert W-based clusters which have shown a degree of control, as illustrated by the isolation of several W-based POMs such as the large nanoscale $\{W_{148}\} \equiv [Ln_{16}As_{12}W_{148}O_{524}(H_2O)_{36}]^{76-}$ cluster [25]. Such limitations may be circumvented by adopting an approach that kinetically stabilizes the building block in solution, thereby effectively preventing its reorganization to other structure types.

In our work, while developing strategies toward this goal, we have found a new family of polyoxomolybdates [26, 27] based on the $[H_2Mo_{16}O_{52}]^{10-}$ framework which appears to achieve the first part of this goal and allows the isolation of a new structure type by virtue of the cations used to "encapsulate" this unit, thereby limiting its reorganization to a simpler structure, see Figure 2.3. Furthermore, the building block character of this anion is demonstrated when electrophilic transition metal cations M^{2+} (M = Fe, Mn, Co) are added to solutions of this cluster, resulting in $[H_2Mo_{16}M_2O_{52}]^{6-}$ species that can undergo further condensation reactions.

These clusters were trapped during the self-assembly process by bulky organic cations which appear to have restrained the clusters from reorganizing into other well-known structure types. This yields a family of POMs with a range of symmetries and nuclearities and, most importantly, the potential to really "tailor" the physical properties by changing the cluster framework. This approach relies on trapping and stabilizing nonspherical polyanions of low nuclearity and symmetry before their aggregation and rearrangement to more uniform and stable struc-

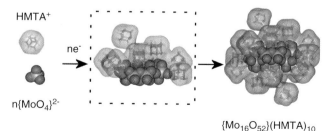

Figure 2.3 A schematic showing the "encapsulation" of the cluster units during the cluster assembly process in the presence of the bulky organo-cation HMTA (hexamethylenetetraamine).

tures. See Figure 2.4 for an example of a cluster trapped using this approach to yield a $\{Mo_{16}\} \equiv [H_2Mo_{16}O_{52}]$ [26, 27].

Furthermore, the organocations can also be used as structure-directing moieties, not only at a molecular level but also allowing the formation of polymers that enable large aggregates to assemble, see Figure 2.5 [28]. The effect of the use of encapsulating cations, here the tetra-n-butylammonium cation (n-Bu$_4$N$^+$), is demonstrated by the reaction of $(n$-Bu$_4$N$)_2[Mo_6O_{19}]$ with silver(I) fluoride in methanol,

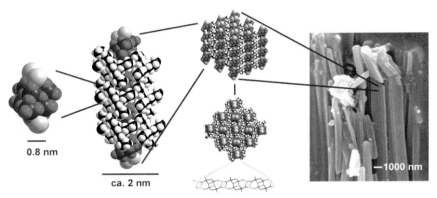

Figure 2.4 Space-filling representations of segments of the linear chain of linked $[Ag^IMo^{VI}_8O_{26}Ag^I]^{2-}$ showing the growth of the structure into linear chains encapsulated by the organic n-Bu$_4$N$^+$ cations and the arrangement of the packed array of these chains, along with a stick representation of the chain framework (Mo: dark grey, O: medium grey, Ag: light grey, C: black, H: white). The organization of the packed linear chains forming microcrystals of the compound are shown on the right SEM image with the crystallographic a axis parallel to the direction of the molecular chains.

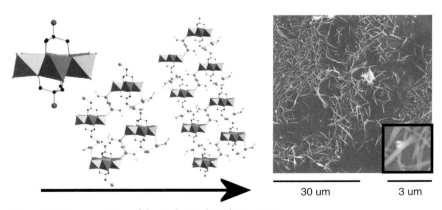

Figure 2.5 Representation of the Anderson-based, $[MnMo_6O_{18}\{(OCH_2)_3CNH_2\}_2]^{3-}$, unit on the left, the network formed by complexation with $\{Ag_2(DMSO)_4\}^{2+}$ in the middle, and the SEM of the material on a silicon substrate.

which ultimately results in the formation of a unique one-dimensional chain structure of the composition: $(n\text{-Bu}_4\text{N})_{2n}[\text{Ag}_2\text{Mo}_8\text{O}_{26}]_n$. Here, the flexible $n\text{-Bu}_4\text{N}^+$ cations wrap almost completely around the linear chain of linked $[\text{Ag}^\text{I}\text{Mo}^\text{VI}{}_8\text{O}_{26}\text{Ag}^\text{I}]^{2-}$ units, see Figure 2.5. In the solid state these strands are packed to a network of collinear, organic "tunnels" that accommodate the polymeric $\{\text{Ag}_2\text{Mo}_8\}_\infty$ anions. The nature of the $\{\text{Ag}_2\}$ linker groups and the Ag coordination environments, however, were found to depend on the reaction conditions, which suggests that the precursors in the reaction solution are not individual $\{\text{Ag}_2\}$ and $\{\text{Mo}_8\}$ groups but, most probably, $\{\text{Ag}(\text{Mo}_8)\text{Ag}\}$-type synthons, see Figure 2.4. Silver linkers are extremely versatile in polyoxometalate chemistry simply due to the poor mismatch between the oxo and silver species.

The mismatch between the ligand preferences of Ag(I) and the Mo-based POMs can be exploited even further by using coordinative groups on the POM. For instance, by derivatizing the Anderson-type cluster via the *tris* ligand [tris(hydroxymethyl)aminomethane] which has three pendant hydroxyl groups that can replace the hydroxide groups on the surface of the Anderson cluster [29], it is possible to produce a range of materials that can be manipulated by the coordination of the Ag(I) species [30]. For instance, the Anderson derivatized with tris and complexed with Ag(I) forms a 1D chain in the solid state where the repeat unit in the chain is built from two *tris*-derived Anderson cluster $[\text{MnMo}_6\text{O}_{18}\{(\text{OCH}_2)_3\text{CNH}_2\}_2^{3-}]$ units connected via a bridging $\{\text{Ag}_2(\text{DMSO})_4\}^{2+}$ unit and the chain is propagated by a single Ag(I) which connects to the nitrogen atom of the *tris* ligands. It is interesting that this compound, when assembled on silicon, forms a 1D wirelike structure, see Figure 2.5. This fibers observed on the surface are >10 μm in length with a diameter of ~0.5 μm. This means that the derivatized Anderson cluster appears to be a robust and useful building block with potential for the use in the self-assembly of functional materials. Given the versatile electronic properties of POMs, this building block approach might become relevant, for example, in the production of spacers of specific dimensionality for use as a skeleton for conducting interconnectors in nanoscale electronic devices, or as electron-beam resists.

It is clear that POM-based clusters have a very interesting range of accessible building blocks that bridge a number of length scales, which allows the construction of complex molecules with a great deal of flexibility and structural variation. The challenge in many POM systems is to understand and control this variation to develop the building block route even further.

2.3
From Building Blocks to Functional POM Clusters

There is little doubt that diverse physical properties common to polyoxometalates places them in an almost unmatched class of materials which could be extremely useful as hybrid materials and nanocomposites [31]. In this section we will focus on POM systems with properties that could be exploited in the development of molecular-scale devices.

2.3.1
Host–Guest Chemistry of POM-based Superclusters

The development of host–guest chemistry based on POM superstructures has been one of the most interesting developments and begins to show possibilities for POMs acting as sensors, storage capsules, and hosts that are able to respond to external stimuli. For instance molecular growth from a $\{Mo_{176}\}$ ring to a $\{Mo_{248}\}$ ring with the inner voids covered with "hub-caps" has been possible [32] as well as the complexation of a metalloporphyrin within the cavity of the $\{Mo_{176}\}$ wheel [33]. A similar assembly with a $\{W_{48}\}$ cluster has also been reported but this time including a 1st row transition metal fragment inside [34].

The Keplerate $\{Mo^{VI}_{72}Mo^{V}_{60}L_{30}\}^{n-}$ ($n = 42$ when L = acetate, $n = 72$ when L = sulfate) ball cluster provides an ideal framework to extend these ideas as it is a spherical cluster with a high charge and an accessible inner chamber with a large volume, and the nature of the surface of the inner chamber can be tuned as well as the pores. Investigations of this cluster taking up various cations such as lithium have given insight into basic principles of cation transport through "molecular pores." This was investigated using porous Keplerates with sulfate ligands on the inner surface; the cluster behaves as a semipermeable inorganic membrane open for H_2O and small cations [35]. Similar studies of the uptake/release of cations by a capsule in solution may be extended to investigate nanoscale reactions in solutions as well as a large variety of cation-transport phenomena, see Figure 2.6. The pores shown in the Keplerate have the form $\{Mo_9O_9\}$ and provide a structural motif rather similar to that of the classical crown ethers.

This comparison is even more striking for the $\{W_{36}\}$-based cluster with the formula $\{(H_2O)_4K\subset[H_{12}W_{36}O_{120}]\}^{11-}$ and includes the threefold symmetric cluster anion $[H_{12}W_{36}O_{120}]^{12-}$, see Figure 2.7. Interestingly, the cluster anion complexes a potassium ion at the center of the $\{W_{36}\}$ cluster in a $\{O_6\}$ coordination environment [36]. The $\{W_{36}\}$ structure consists of a ring of six basal W positions, an additional W position in the center of this ring, and four apical W positions in a butterfly configuration. Every W center has a distorted WO_6 octahedral coordina-

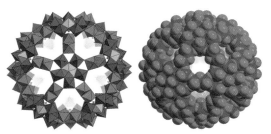

Figure 2.6 Polyhedral structure (left) and space-filling structure (right) of the $\{Mo^{VI}_{72}Mo^{V}_{60}L_{30}\}^{n-}$ ball cluster. The space-filling structure is looking directly down the one of the $\{Mo_9O_9\}$ pores.

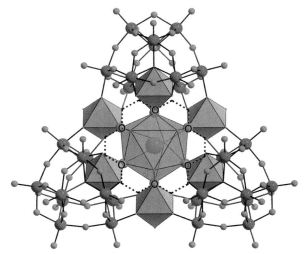

Figure 2.7 Representation on the $\{(H_2O)_4K\subset[H_{12}W_{36}O_{120}]\}^{11-}$ cluster with the central potassium ion shown behind the central polyhedra. The framework of 18-C-6 is superimposed onto the central $\{O_6\}$ moiety to scale and the six W-groups which each donate oxygen ligands to coordinate to the potassium ion are shown in polyhedral representation.

tion geometry with one terminal W=O (W—O ~1.70 Å) extending away from the cluster; this arrangement maps extremely well onto the structure of the crown ether 18-C-6. The implications for the development of this system in a similar fashion to the crown ethers are interesting, especially the possibilities for discrimination and sensing of metal ions using this cluster framework [36].

2.3.2
Magnetic and Conducting POMs

The development of POM-based clusters incorporating paramagnetic centers is an interesting goal since it is possible to utilize existing building blocks/clusters to generate very large magnetic molecules. In fact it has been shown that it is possible to substitute the $\{Mo_2\}$ "linker" groups present in the Keplerate $\{Mo_{132}\}$, (Pentagon)$_{12}$(Linker)$_{30}$ species with Fe^{III} to yield a $\{Mo_{72}Fe_{30}\}$ cluster with the formula $[Mo_{72}Fe_{30}O_{252}(CH_3COO)_{10}\{Mo_2O_7(H_2O)\}\{H_2Mo_2O_8(H_2O)\}_3(H_2O)_{91}]$ [37]. This cluster is smaller than the parent $\{Mo_{132}\}$ structure with an outer diameter of ~25 Å and an inner diameter of ~18 Å, see Figure 2.8. Further, the $\{Mo_{72}Fe_{30}\}$ cluster comprises only Mo^{VI} atoms whereas the $\{Mo_{132}\}$ cluster contains 60 reduced Mo^V centers (the 30 linking $\{Mo_2\}$ units are reduced). The presence of the Fe(III) centers, combined with the weak antiferromagnetic exchange between these centers, means that there are 30 mainly uncorrelated spins 5/2 at room

Figure 2.8 Polyhedral representations of the structures of the {Mo$_{132}$} (left) and {Mo$_{72}$Fe$_{30}$} (right) clusters to scale. The icosidodecahedron which is formed by connecting the 30 Fe centers is shown.

temperature and the cluster therefore behaves like a paramagnet with 150 unpaired electrons. The Fe(III) centers of the cluster span an icosidodecahedron and the extremely rich and interesting magnetic properties have been investigated using a simple Heisenberg model [38]. In this respect the {Mo$_{72}$Fe$_{30}$} has been termed a mesoscopic paramagnet for which classical behavior extends down to extraordinarily low temperatures.

The formation of hybrid materials based on POMs with stacks of partially oxidized p-electron donor molecules of tetrathiafulvalene (TTF) has been accomplished to yield conducting POM-based materials. This is interesting because the inorganic POM anion can act as a structural spacer unit, incorporate additional functionality such as a scaffold for paramagnetic ions or to act as an electron acceptor [27]. This area is progressing rapidly with the compounds based on [BEDT-TTF]$_5$[H$_3$V$_{10}$O$_{28}$] [39] and [BEDT-TTF]$_6$[Mo$_8$O$_{26}$] [40] (BEDT-TTF = bis(ethylenedithio)tetrathiafulvalene) which behave as metals down to 50 and 60 K with room temperature conductivities of 360 and 3 S cm^{-1}, respectively. In addition, a POM radical salt with metallic behavior down to 2 K has been synthesized [41]. The compound is based on [BEDOTTF]$_6$K$_2$[BW$_{12}$O$_{40}$] and is formed from [BW$_{12}$O$_{40}$]$^{5-}$ and the organic radical (BEDO-TTF) (= bis(ethylenedioxo) tetrathiafulvalene). The realization of POM-organic conducting hybrids means that devices incorporating both POM clusters and organic conductors and polymers are also accessible.

2.3.3
Thermochromic and Thermally Switchable POM Clusters

In our attempts to design functional clusters we have focused on substitution of the heteroanions within the Wells-Dawson structure to create nonconventional Dawson clusters incorporating *two* pyramidal anions. Our design rationale was based on the idea that such clusters may exhibit unprecedented properties arising from the intramolecular electronic interactions between the encapsulated anions (in this case we aimed to engineer between S···S atoms of two encapsulated sulfite ions), thus providing a novel route to manipulate the physical properties of the {Mo$_{18}$} Dawson-type clusters. The synthesis of these clusters was accomplished by extending our previous work utilizing organocations and allowed the isolation of the α-[Mo$^{VI}_{18}$O$_{54}$(SO$_3$)$_2$]$^{4-}$ (type 1) which incorporates the targeted two pyramidal sulfite SO$_3^{2-}$ anions as the central cluster templates. This compound showed thermochromic behavior between 77 K (pale yellow) and 500 K (deep red), see Figure 2.9 [42].

In an extension of this work to W-based Dawson-like clusters we succeeded in synthesizing the analogous polyoxotungstate clusters incorporating the sulfite anion [43], [W$^{VI}_{18}$O$_{54}$(SO$_3$)$_2$]$^{4-}$ (type 1) the isostructural tungstate analog to the {Mo$_{18}$} example, and [W$^{VI}_{18}$O$_{56}$(SO$_3$)$_2$(H$_2$O)$_2$]$^{8-}$ (type 2), see Figure 2.10.

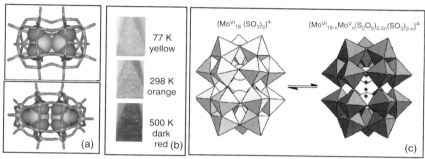

Figure 2.9 (a) The crystal structure of the new type of Sulfite-Dawson above, compared to the known sulfate-Dawson below, showing the S···S interaction. (b) are photographs of the thermochromic material at 77, 298 and 500 K. (c) Shows two polyhedral views of the low temperature sulfite-Dawson (on the left) at 77 K and the resulting high temperature sulfite-Dawson (on the right) at 500 K along with a possible mechanism for the process resulting from the formation of a partial bond between the SO$_3^{2-}$ which would be accompanied by electron release to the cluster shell (S···S in space filling to show the interaction between the two sulfur atoms).[40]

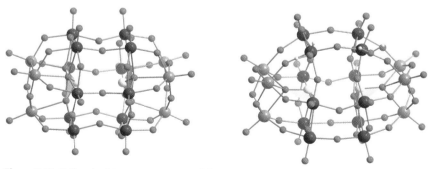

Figure 2.10 Ball-and-stick representations of the structures of the D_{3h}-symmetric α-[W$_{18}$O$_{54}$(SO$_3$)$_2$]$^{4-}$ (type 1-left) and the C_{2v}-symmetric [W$_{18}$O$_{56}$(SO$_3$)$_2$(H$_2$O)$_2$]$^{8-}$ (type 2-right) cluster anions (W: equatorial {W$_6$} – dark grey, capping {W$_3$} – light grey; O: small grey spheres; S: light grey sphere at the apex of the polyhedra which represent the pyramidal SO$_3^{2-}$ units).

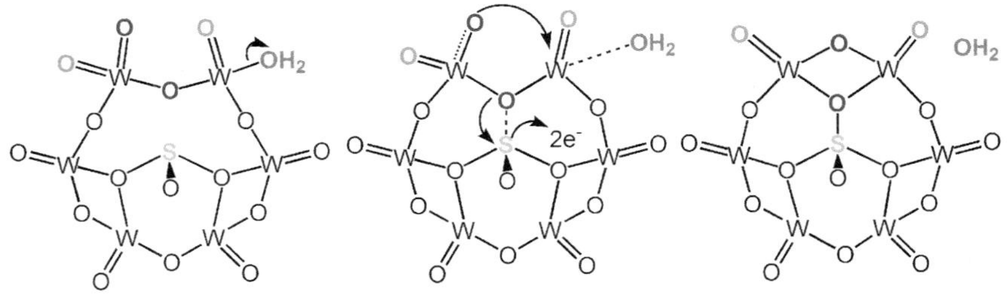

Figure 2.11 Mechanism of the internal sulfite to sulfate oxidation showing the movement of the oxygen atoms (shown faded). The equatorial cluster "belts" and the central sulfur position of a $\{W_9(SO_3)\}$ fragment is illustrated (S atoms also faded).

Comparison of the type 1 and type 2 $\{W_{18}\}$-Dawson structures shows that $[W^{VI}{}_{18}O_{54}(SO_3)_2]^{4-}$ also engineers the short $S\cdots S$ interaction, whereas the $[W^{VI}{}_{18}O_{56}(SO_3)_2(H_2O)_2]^{8-}$ cluster contains two sulfite anions that are locked into a different binding mode and appear to expand the upper part of the cage. This cluster also undergoes an interesting reaction when heated whereby a structural rearrangement allows the two embedded pyramidal sulfite ($S^{IV}O_3{}^{2-}$) anions to release up to four electrons to the surface of the cluster and results in the sulfate-based, deep blue, mixed-valence $[W_{18}(SO_4)_2]^{8-}$ cluster. Thus the cluster type 2 appears already surprisingly well prearranged for an internal reorganization and a concurrent internal redox reaction, in which the encapsulated sulfite anions act as embedded reducing agents and are oxidized to sulfate when heated to over 400 °C, see Figure 2.11. In the course of this reaction, a maximum of four electrons could be transferred to the metal-oxide framework, causing a color change from colorless ($\{W^{VI}{}_{18}\}$) to blue ($\{W^{VI}{}_{14}W^{V}{}_4\}$). The overall reaction is accompanied by the release of the two coordinated water ligands of the W centers, so the following reaction occurs:

$$[W^{VI}{}_{18}O_{56}(S^{IV}O_3)_2(H_2O)_2]^{8-} \rightarrow \alpha\text{-}[W^{VI}{}_{14}W^{V}{}_4O_{54}(S^{VI}O_4)_2]^{8-} + 2H_2O \qquad (2.1)$$

In this cluster system the enclosure of sulfite anions with a "correct" orientation transforms the anions from "innocent" structural templates to electronically reactive, functional units. These can now release electrons to the cluster shell upon activation by heat – the sulfite groups in type 2 clusters are "activated" whereas those in type 1 are not.

2.4
Bringing the Components Together – Towards Prototype Polyoxometalate-based Functional Nanosystems

The gap between concepts in molecular design to produce polyoxometalate integrated nanosystems or molecular-scale devices is vast due to the problem of fabrication and control of molecular orientation. Molecule-by-molecule assembly is clearly a great challenge; therefore, the design of self-organizing and self-assembling systems utilizing ideas and inspiration from supramolecular chemistry to form functional molecular systems that can be connected from the molecule to the macroscale world is highly desirable [44]. However, the route to achieve such a grand aim is still unclear and many scientific fields are converging on the development of nanoscale and molecular electronics and interdisciplinary approaches are being developed to address the significant scientific and technical barriers. Clearly the design and synthetic approaches to polyoxometalates, and the fact these clusters can be constructed over multiple length scales, along with their almost unmatched range of physical properties means that they are great candidates to be used as both the scaffold and the functional part of any nanodevice. There is therefore great scope to investigate the self-assembly of functional POM systems on surfaces and in the crystalline state to produce architectures that can be fabricated to form a polyoxometalate-based device. Indeed, recent work in the production of thin films of polyoxometalate clusters [45], and the use of POM clusters in "nanocasting" [46] are examples where the cross-disciplinary approach is beginning to utilize the potential of this class of clusters. One possible approach to the fabrication of POM may, for instance, utilize lithographic techniques to prepare patterned substrates for the formation of "functional" polyoxometalate clusters, or even utilize the POM cluster in the growth of nanoscale connects that can be directed to individual electrodes patterned using lithography [47].

In one such approach we seek to exploit the potential signal transduction properties of the thermochromic Dawson [40] polyoxometalates and combine this with a fluorescent POM-hybrid to produce a device that could respond optically as a function of the local environment, see Figure 2.12. In this example, the clusters would be positioned on a gold surface using SAMs (Self-assembled monolayer) with a cationic tail and local positional control could be aimed using self-assembly, or even by means of an atomic force microscope tip.

The challenge now is to design individual POM cluster molecules that can interact both with each other and with the macroscale (as shown in Figure 2.12), in a desired fashion in response to inputs and environmental effects, so that a functioning molecular system is really constructed.

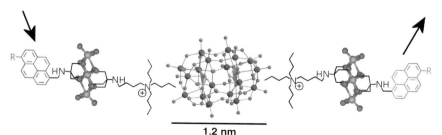

1.2 nm

Figure 2.12 Schematic of a surface assembly of a pyrene-derivatized Anderson (left) that absorbs in the blue arranged next to a $[Mo^{VI}_{18}O_{54}(SO_3)_2]^{4-}$ cluster species which has thermochromic properties. On the right there is a pyrene-derivatized Anderson cluster hybrid that fluoresces. The aim is to use the thermochromic $\{Mo_{18}\}$, which is positioned between the two pyrene-Anderson units, to control the response of the system to light.

2.5
Acknowledgments

We are grateful to the University of Glasgow, WestCHEM, the Leverhulme Trust, the Royal Society and the EPSRC for supporting this work.

References

1. Badjic, J.D., Balzani, V., Credi, A., Silvi, S. and Stoddart, J.F. (2004) *Science*, **303**, 1845.
2. Collier, C.P., Wong, E.W., Belohradský, M., Raymo, F.M., Stoddart, J.F., Kuekes, P.J., Williams, R.S. and Heath, J.R. (1999) *Science*, **285**, 391–4.
3. Yoshimura, I., Miyahara, Y., Kasagi, N., Yamane, H., Ojida, A. and Hamachi, I. (2004) *Journal of the American Chemical Society*, **126**, 12204–5. Maue, M. and Schrader, T. (2005) *Angewandte Chemie (International Ed. in English)*, **44**, 2265–70.
4. Long, D.-L., Burkholder, E. and Cronin, L. (2007) *Chemical Society Reviews*, **36**, 105–21.
5. Cronin, L. (2002) The potential of pentagonal building blocks, in *Inorganic Chemistry Highlights* (eds G. Meyer, D. Naumann and L. Wesemann), Wiley-VCH Verlag GmbH, Weinheim, pp. 113–21.
6. Neumann, R. and Dahan, M. (1997) *Nature*, **388**, 353–5. Mizuno, N. and Misono, M. (1998) *Chemical Reviews*, **98**, 199–218.
7. Katsoulis, D.E. (1998) *Chemical Reviews*, **98**, 359–87. Yamase, T. (1998) *Chemical Reviews*, **98**, 307–25.
8. Rüther, T., Hultgren, V.M., Timko, B.P., Bond, A.M., Jackson, W.R. and Wedd, A.G. (2003) *Journal of the American Chemical Society*, **125**, 10133–43.
9. Anderson, T.M., Neiwert, W.A., Kirk, M.L., Piccoli, P.M.B., Schultz, A.J., Koetzle, T.F., Musaev, D.G., Morokuma, K., Cao, R. and Hill, C.L. (2004) *Science*, **306**, 2074–7.
10. Müller, A., Das, S.K., Talismanov, S., Roy, S., Beckmann, E., Bögge, H., Schmidtmann, M., Merca, A., Berkle, A., Allouche, L., Zhou, Y. and Zhang, L. (2003) *Angewandte Chemie (International Ed. in English)*, **42**, 5039–44.
11. Pope, M.T. (1987) Isopolyanions and Heteropolyanions, in *Comprehensive Coordination Chemistry*, (eds G. Wilkinson, R.D. Gillard and J.A. McCleverty), Pergamon Press, Vol. 3, pp. 1023–58.
12. Cronin, L. (2004) High nuclearity polyoxometalate clusters, in *Comprehensive Coordination* (eds I.J.A. McCleverty and T.J. Meyer), Elsevier, Amsterdam, Vol. 7, pp. 1–57.
13. Long, D.L. and Cronin, L. (2006) *Chemistry–A European Journal*, **12**, 3698.
14. Scheele, W. (1971) *Sämtliche Physische und Chemische Werke*. (ed. D.S.F. Hermbstädt) M. Sändig oHG, Niederwalluf/Wiesbaden, Vol. 1, pp. 185–200 (reprint, original 1783).
15. Müller, A., Krickemeyer, E., Meyer, J., Bögge, H., Peters, F., Plass, W., Diemann, E., Dillinger, S., Nonnenbruch, F., Randerath, M. and Menke, C. (1995) *Angewandte Chemie (International Ed. in English)*, **34**, 2122–4.
16. Müller, A., Krickemeyer, E., Bögge, H., Schmidtmann, M. and Peters, F. (1998) *Angewandte Chemie (International Ed. in English)*, **37**, 3360. Müller, A., Krickemeyer, E., Bögge, H., Schmidtmann, M. and Peters, F. (1998) *Angewandte Chemie (International Ed. in English)*, **37**, 3359–63.
17. Cronin, L., Beugholt, C., Krickemeyer, E., Schmidtmann, M., Bögge, H., Koegerler, P., Luong, T.K.K. and Müller, A. (2002) *Angewandte Chemie (International Ed. in English)*, **41**, 2805–8.
18. Müller, A., Beckmann, E., Bögge, H., Schmidtmann, M. and Dress, A. (2002) *Angewandte Chemie (International Ed. in English)*, **41**, 1162–7.

19 Kögerler, P. and Cronin, L. (2005) *Angewandte Chemie (International Ed. in English)*, **44**, 844–5.
20 Pope, M.T. and Müller, A. (eds) (1994) *Polyoxometalates. From Platonic Solids to Anti-retroviral Activity*, Kluwer Academic Publishers, Dordrecht, The Netherlands.
21 Maekawa, S., Tohyama, T., Barnes, S.E., Ishihara, S., Koshibae, W. and Khaliullin, G. (2004) *Physics of Transition Metal Oxides*, Springer, Berlin, pp. 1–137.
22 Cox, P.A. (1992) *Transition Metal Oxides: An Introduction to Their Electronic Structure and Properties*, Clarendon press, Oxford.
23 Lehn, J.M. (2002) *Proceedings of the National Academy of Sciences of the United States of America*, **99**, 4763–8.
24 Kroto, H.W., Heath, J.R., O'Brien, S.C., Curl, R.F. and Smalley, R.E. (1985) *Nature*, **318**, 162–3.
25 Wassermann, K., Dickman, M.H. and Pope, M.T. (1997) *Angewandte Chemie (International Ed. in English)*, **36**, 1445–8.
26 Long, D., Kögerler, P., Farrugia, L.J. and Cronin, L. (2003) *Angewandte Chemie (International Ed. in English)*, **42**, pp. 4180–3.
27 Long, D.L., Kögerler, P., Farrugia, L.J. and Cronin, L. (2005) *Dalton Transactions (Cambridge, England: 2003)*, 1372–80.
28 Abbas, H., Pickering, A.L., Long, D.-L., Kögerler, P. and Cronin, L. (2005) *Chemistry – A European Journal*, **11**, 1071–8.
29 Favette, S., Hasenknopf, B., Vaissermann, J., Gouzerh, P. and Roux, C. (2003) *Chemical Communications*, 2664–5.
30 Song, Y.-F., Abbas, H., Ritchie, C., McMillian, N., Long, D.-L., Gadegaard, N. and Cronin, L. (2007) *Journal of Materials Chemistry*, DOI: 10.1039/b617830h.
31 Yamase, T. and Pope, M.T. (eds) (2002) *See in Polyoxometalate Chemistry for Nano-Composite Design*, Kluwer, New York.
32 Müller, A., Shah, S.Q.N., Bögge, H. and Schmidtmann, M. (1999) *Nature*, **397**, 48–50.
33 Tsuda, A., Hirahara, E., Kim, Y.-S., Tanaka, H., Kawai, T. and Aida, T. (2004) *Angewandte Chemie (International Ed. in English)*, **43**, 6327–31.
34 Mal, S.S. and Kortz, U. (2005) *Angewandte Chemie (International Ed. in English)*, **44**, 3777–80.
35 Müller, A., Rehder, D., Haupt, E.T.K., Merca, A., Bögge, H., Schmidtmann, M. and Heinze-Brückner, G. (2004) *Angewandte Chemie (International Ed. in English)*, **43**, 4466–70.
36 Long, D.-L., Abbas, H., Kögerler, P. and Cronin, L. (2004) *Journal of the American Chemical Society*, **126**, 13880–1.
37 Müller, A., Sarkar, S., Shah, S.Q.N., Bögge, H., Schmidtmann, M., Sarkar, S., Kögerler, P., Hauptfleisch, B., Trautwein, A.X. and Schünemann, V. (1999) *Angewandte Chemie (International Ed. in English)*, **38**, 3238–41.
38 Müller, A., Luban, M., Schröder, C., Modler, R., Kögerler, P., Axenovich, M., Schnack, J., Canfield, P., Bud'ko, S. and Harrison, N. (2001) *Chemphyschem: A European Journal of Chemical Physics and Physical Chemistry*, **2**, 517–21.
39 Coronado, E., Galán-Mascarós, J.R., Giménez-Saiz, C., Gómez-García, C.J., Martínez-Ferrero, E., Almeida, M. and Lopes, E.B. (2004) *Advanced Materials (Deerfield Beach, Fla.)*, **16**, 324–7.
40 Lapinski, A., Starodub, V., Golub, M., Kravchenko, A., Baumer, V., Faulques, E. and Graja, A. (2003) *Synthetic Metals*, **138**, 483–9.
41 Coronado, E., Giménez-Saiz, C., Gómez-García, C.J. and Capelli, S.C. (2004) *Angewandte Chemie (International Ed. in English)*, **43**, 3022–5.
42 Long, D.L., Kögerler, P. and Cronin, L. (2004) *Angewandte Chemie (International Ed. in English)*, **43**, 1817–20. Baffert, C., Boas, J.F., Bond, A.M., -L. Long, P., Kögerler D., Pilbrow, J.R. and Cronin, L. (2006) *Chemistry – A European Journal*, **12**, 8472–83.
43 Long, D.-L., Abbas, H., Kögerler, P. and Cronin, L. (2005) *Angewandte Chemie (International Ed. in English)*, **44**, 3415–9.
44 Nakamura, T., Matsumoto, T., Tada, H. and Sugiura, K.-I. (eds) (2003) *Chemistry of Nanomolecular Systems: Towards the*

Realization of Nanomolecular Devices, Springer Series in Chemical Physics 70, Springer-Verlag, New York.

45 Liu, S., Kurth, D.G., Möhwald, H. and Volkmer, D. (2002) *Advanced Materials (Deerfield Beach, Fla.),* **14**, 225–8.

46 Polarz, S., Smarsly, B., Göltner, C. and Antonietti, M. (2000) *Advanced Materials (Deerfield Beach, Fla.),* **12**, 1503–7.

47 Song, Y., Long, D.-L. and Cronin, L. (2007) *Angewandte Chemie (International Ed. in English),* **46**, 1340–4.

3
Nanostructured Porous Materials: Building Matter from the Bottom Up

Javier García-Martínez

3.1
Introduction

New synthetic techniques based on weak interactions are continually being developed to gain more precise control over the organization of solids from the nanometer to the final size of the piece [1–7]. Template-assisted synthesis, self-assembly, micromolding, and biomimetic methods are becoming widely used in the fabrication of materials with controlled porosity. All these new synthetic tools greatly improve the performance of materials built from the bottom up in fields such as catalysis, separation, adsorption, sensors, and biomedicine. For example, bimodal micro-mesoporous materials exhibit enhanced intraparticle mass transport allowing for processing of bulkier molecules. Additional control over the solid structure, such as pore shape, connectivity, and dimensionality, provides for further command over reaction selectivity [5].

The bottom-up techniques described herein are based on the use of nanosize building blocks to fabricate precisely organized solids at various scales. The final architecture of the solid, and the way these blocks combine with each other, can be conveniently adjusted by the synthesis conditions, the selection and modification of these nanoblocks, and their chemical functionality. The spontaneous arrangement of individual nanoblocks is generally obtained via self-assembly through weak interactions. The control over the organization of these components allows for the incorporation of nanoparticles, biomolecules, or chemical functionalities inside the solid structure in highly precise locations.

The aim of this chapter is to describe the synthetic strategies used for the judicious organization of building blocks to produce materials with controlled structures ranging from the atomic scale (crystalline materials) through the nanometer range (nanostructured solids) all the way to the final shape of the material.

3.2
Synthesis by Organic Molecule Templates

The use of organic molecules as templates to induce and direct the formation of solids around them is a well-established technique, widely used in the synthesis of zeolites and related materials. These templates, typically quaternary amines, maintain their original shape and size during the synthesis of nanostructured material (Figure 3.1a). In a subsequent step, the template is removed (either by calcination or by chemical extraction), leaving a microporous crystalline material [2, 3, 5]. The requirements that an organic molecule must fulfill to be used as a template are: (i) chemical stability under the reaction conditions (typically high pH and hydrothermal treatment); (ii) specific, although weak, interaction with the solid precursor; and (iii) easy removal after the synthesis (the collapse of the structure during the removal of the template is not infrequent).

In some cases, the spatial repetition of basic construction blocks is stimulated by interactions between template molecules and between the template molecule and the inorganic precursor. This repetition yields ordered porous solids such as the zeolite shown in Figure 3.2 [8]. As mentioned, zeolites are a prominent example of materials prepared using molecular templates. These are crystalline aluminosilicates whose framework is organized in cavities or channels of sizes ranging from 0.3 to 1.5 nm. They are the archetypical example of microporous crystalline molecular sieves. Because of their large surface area (hundreds of square meters per gram) and extremely narrow pore sieve distribution, zeolites are able to adsorb a wide variety of small and medium-sized molecules with high selectivity. Zeolites are made of TO_4 tetrahedra (T being Si or Al) with the T atom located at the center and each of the four apical oxygens shared with an adjacent tetrahedron. The isomorphic substitution of Si(IV) by Al(III) produces a negative charge in the framework that is compensated by a cation. These cations are very labile and, in general, they can easily be exchanged. When this cation is a proton, the zeolite exhibits Brönsted acidity able to catalyze typical organic acid-catalyzed

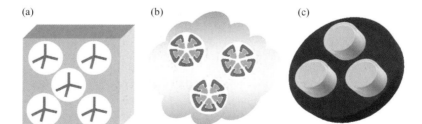

Figure 3.1 Schematic representation of three nanomaterial construction techniques: (a) Use of molecular templates (red) that guide the formation of the crystalline solid (yellow) while maintaining their original shape during the synthesis. (b) Self-assembly of individual components (red and green) into supramolecular structures around which the amorphous solid (yellow) is formed. (c) Use of molds (blue) to generate solids (yellow) that replicate the cavities by restricting their growth.

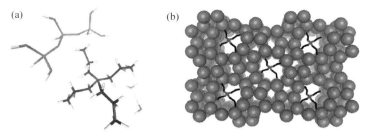

Figure 3.2 Two steps in the synthesis of the ZSM-5 zeolite: (a) template-inorganic precursor interaction and (b) periodic repetition of the inorganic template–solid hybrid, source of the crystalline framework.

reactions such as cracking, alkylation, and isomerization. Because of all these properties – molecular sieving, ion-exchange capability, and acidity – zeolites are one of the most widely used and versatile materials in the chemical industry [5, 9–12].

Zeolites are typically prepared under autogenous pressure (hydrothermal treatment) at a alkaline pH [13]. Under these conditions, the interaction of the silicate anions and the cationic amines used as templates is favored. The substitution of hydroxy by fluoride ions as a mineralizing agent greatly expands the pH range in which a zeolite can be prepared. The fluoride route allows for using organic templates which are not stable under alkaline conditions and the introduction of heteroatoms in the zeolitic framework (zeotypes) which are not soluble at high pH [10–11].

Quaternary amines, such as tetraalkylammonium bromides and hydroxides (the alkyl group being C_1 to C_4) are the typical zeolite templates. Quaternary amines fulfill the above-mentioned requirements of stability, specific interaction with the precursor (electrostatic interaction between quaternary amines and silicate), and easy removal (by calcination).

The role of the template in the synthesis is not merely as a porogen: on the contrary, it is also responsible for many key functions [5, 9, 10]. The template (typically cationic) balances the negative charge that characterizes zeolitic framework, due to the isomorphic substitution of Si(IV) by Al(III), prearranges the secondary building units (SBUs) toward the zeolitic framework, improves the gel synthesis conditions, especially the solubility of the silica precursors, and favors the thermodynamics of the reaction by stabilizing the porous zeolite framework.

Both naturally occurring zeolites and the first synthetic zeolites (prepared without the aid of organic templates) contain a relatively high amount of aluminum in their framework (low Si/Al ratio zeolites). Although this is advantageous for some applications, the higher stability of the silicon-rich zeolites made desirable the synthesis of zeolites with high a Si/Al ratio. This was first achieved by Wadlinger *et al.* who prepared zeolite beta in 1967 using a quaternary amine, tetraethylammonium hydroxide (TEAOH) [14]. In year 1978, Flanigen *et al.*

prepared silicalite, a purely siliceous version of the ZSM-5 zeolite, by using tetrapropylammonium hydroxide (TPAOH) as a template [15]. This synthesis strategy has proved extremely versatile.

The use of templates to control the porosity of solids is not limited to small organic molecules. Alternative templates include: dendrimers [16, 17], polymers [18], hard templates such as nanoparticle colloidal suspensions [19] and latex spheres [20] or even biological materials like butterfly wings [21], DNA [22] or viruses [23].

3.3
Synthesis by Molecular Self-Assembly: Liquid Crystals and Cooperative Assembly

Self-assembly is a spontaneous organization process of individual components into complex structures, usually highly symmetric, driven by weak interactions. Molecular self-assembly differs from the first technique described (molecular templates) in that the organic molecules used to organize the final nanostructured solid do not maintain their size and shape during synthesis, as shown in Figure 3.1b. In this case, the molecules are organized in supramolecular entities, either before the formation of the nanomaterial (liquid crystal mechanism, in which the solid grows around a previously ordered structure [24]) or during that formation (cooperative mechanism, where the interaction between the organic template and the inorganic precursors induces the solid to arrange and precipitate [25]). In both cases, these organic molecules, which are responsible for the assembly, are amphiphilic, that is, formed by at least two domains with differentiated properties, in general one hydrophilic and one hydrophobic [16]. This feature is typical of surfactants, and induces the supramolecular arrangement or self-assembly of the molecules to form micelles of various geometries (Figure 3.3).

Structures originated by molecular self-assembly are usually larger (on the order of several nanometers, yielding mesoporous materials, Figure 3.4) than those obtained from organic templates (typically microporous, pore size <2 nm) [5]. The large size of the mesopore (2–50 nm) facilitates the access of reactants to the interior of the solid. This allows for processing of bulky molecules that cannot access the narrower porosity of microporous materials, like zeolites. Control of the synthesis

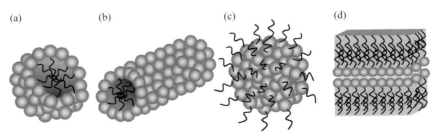

Figure 3.3 Some micelle structures: (a) sphere, (b) cylinder, (c) inverse micelle, and (d) laminar.

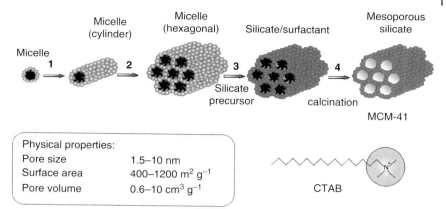

Figure 3.4 Plausible mechanism of the formation of MCM-41 by a liquid-crystal model: (1) formation of cylindrical surfactant micelles, (2) hexagonal assembly, (3) formation of the silica around the micelles, and (4) surfactant elimination.

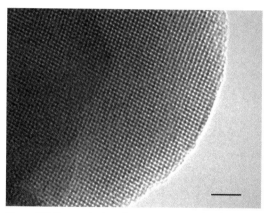

Figure 3.5 Transmission electron micrograph of a mesoporous silica MCM-48 with cubic porous geometry. The scale bar represents 25 nm.

conditions (surfactant concentration, for instance) directs the structure in which the mesopores are organized. Using such strategies, nanostructured materials with hexagonal (MCM-41), cubic (MCM-48) and laminar (MCM-50) structures have been prepared [9]. (MCM corresponds to *Mobile Crystalline Material* [2, 12].

Figure 3.5 shows the organized mesoporous structure of one of these materials. The wide variety of organic molecules with self-assembling properties allows for the synthesis of a myriad of solids with controlled porosity. Cationic surfactants, like the hexadecyltrimethylammonium used to synthesize MCM-41 [26], or three-block copolymers (hydrophilic–hydrophobic–hydrophilic) are two good examples.

The use of such self-assembling polymeric molecules allows for generating superstructures with very large pores (several nanometers in diameter). A good example of this family of porous materials is SBA-15 (SBA = *Santa Barbara amorphous*), a mesoporous silica with hexagonal arrangement whose pore size may reach 30 nm due to the use of large block copolymers as surfactants [27]. The use of swelling agents, which dissolve inside the micelle (hydrophobic zone), allows for further increasing the pore diameter. As a counterpart to the increased accessibility, surfactant-templated materials are typically amorphous, which greatly limits their stability, acidity, and ion-exchange capacity [26].

Different inorganic precursors have been used for the synthesis of mesostructured solids [5, 9–12]. Among them, metallic alkoxides are specially suitable because they can be obtained with high purity and they hydrolyze slowly, allowing for an adequate interaction between the template and the inorganic precursors, and because of the large assortment of commercially available metal alkoxides with many different functional groups. These can be used to decorate the surface of solids by adding an organic-containing metal alkoxide ((R'O)$_3$Si-R, where R contains the desired functionality) to the synthesis mixture (Figure 3.6e) [28]. In this case, the surfactant should be removed by extraction instead of calcination, so the organic groups are left intact on the silica surface. Alternatively, the surface can

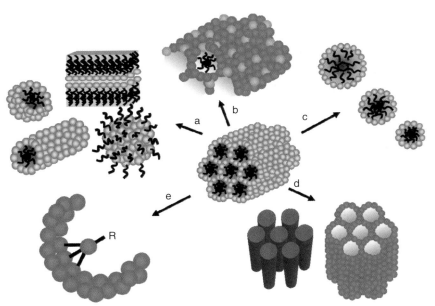

Figure 3.6 Some synthetic possibilities of self-assembly with surfactants: (a) different micelle geometries can be obtained and used to produce many pore architectures, (b) isomorphic substitution of Si by other elements, (c) pore size control using surfactants of different length or swelling agents, (d) use of mesoporous silica as a mold for casting nanostructures such as metal nanowires and (e) functionalization of the surface with trialkoxysilanes.

be functionalized after the calcination by hydrolyzing the organic-containing metal alkoxide ($(R'O)_3Si$-R) on the surface of the mesoporous material, leaving the organic groups covalently bonded to the surface of the solid. These surface organic groups can be used to heterogenize homogeneous catalysts or immobilize biomolecules, biosensors, or species with optical, electrical, or magnetic properties. As an example, this strategy has been used to introduce strong acidity into MCM-41 without the need to add aluminum to its framework—which reduces its stability—by using mercaptotrialkoxysilanes, and the subsequent oxidation of the thiol group to sulfonic acid [29, 30]. The acidity of such materials is sufficient to catalyze esterification [29] or condensation [30] reactions.

Although most common ordered mesoporous solids are based on silica, in recent years mesoporous materials of many different compositions have been prepared containing, for example, titanium, zirconium, niobium, iron, and magnesium, among many others [31]. One reason for the prevalence of silica in ordered mesoporous materials is that silicon alkoxides hydrolyze slowly, allowing an adequate interaction with surfactant molecules. On the other hand, some metallic oxides are unstable while removing the surfactant, which causes the collapse of the structure. The isomorphic substitution of some of its silicon atoms by other elements in already-formed silica nanostructures (Figure 3.6b) is an alternative strategy that allows for introducing new properties into mesoporous materials (for example, redox activity).

Self-assembly with metal alkoxides has also been used to incorporate many organic groups inside the pore walls of ordered mesoporous materials [32, 33]. These hybrid organic–inorganic nanostructures, such as periodic mesoporous organosilicas (PMOs), are prepared by replacing controlled amounts of the tetraalkoxysilane used as metal oxide source, by bridged organosilane precursors [$(R'O)_3Si$—R—$Si(OR')_3$], where R is an organic group that directly links two silicon atoms through silicon–carbon covalent bonds. The condensation of the alkoxides around the self-assembled surfactant produces an ordered mesoporous metal oxide with the R group homogeneously inserted in the pore walls, as shown in Figure 3.7. Hybrid organic–inorganic nanostructured materials have narrow pore distribution, high stability, and tunable composition and porous structure [34]. To date, PMOs with hexagonal, cubic, and wormlike porosity have been reported. Many organic groups have been incorporated in PMOs at basic, neutral, and acidic pH values. Applications in catalysis, separation, sensors, and optical and electro-optical devices for communication and data storage have been suggested for this rapidly growing family of materials [32].

It is worth mentioning the particular case in which R is a benzene group. Inagaki *et al.* reported the synthesis of the first ordered phenylene-bridged hybrid mesoporous organosilica from the assembly of [$(EtO)_3Si$—Ph—$Si(OEt)_3$] and trimethylalkylammonium chloride surfactants [35]. In addition, to the periodic array of mesopores, which repeat each 5.25 nm, these materials exhibit atomic scale periodicity in their pore walls with a spacing of 0.76 nm along the direction of the mesopores. This double periodicity causes the presence of x-ray diffraction (XRD) peaks both at low angles due to the hexagonally ordered mesoporosity and at high

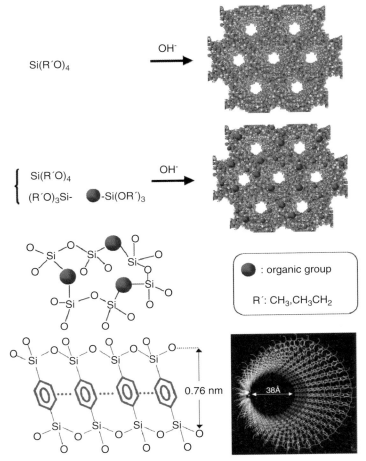

Figure 3.7 Schematic of the incorporation of organic groups in the pore walls of mesoporous silica using bridged organosilane precursors (top). Schematic of the structure of a phenylene-bridged hybrid mesoporous organosilica with both atomic and mesoporous periodicity (bottom) [36].

angles, corresponding to the periodicity at atomic scale (crystallinity). The crystal lattice of this material was observed by transmission electron microscopy (TEM). The benzene rings are stacked via π–π interaction and aligned around the mesopore, forming alternate hydrophobic (benzene) and hydrophilic (silica) layers (see Figure 3.7 bottom). This crystalline hydrophobic–hydrophilic mesoporous material has been suggested to enable structural orientation of guest molecules or cluster as a tool to obtain more selective catalysts [35].

Inspired by the use of bridged organosilane precursors to homogenously introduce chemical functionalities in the walls of mesoporous materials, our group has

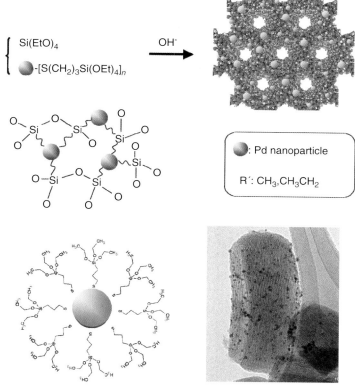

Figure 3.8 Schematic of the incorporation of metal nanoparticles in the pore walls of mesoporous silica using trialkoxysilane-functionalized nanoparticles (top). Transmission electron micrograph of a Pd nanoparticle incorporated in periodic mesoporous silica (bottom) [36].

reported the synthesis of metal-incorporated periodic mesoporous silica by the use of trialkoxysilane-functionalized metal nanoparticles [36]. In a similar way to the preparation of PMOs, and as shown in Figure 3.8, palladium nanoparticles covalently bonded to propyltriethoxysilane were added to a typical MCM-41 synthesis solution. Tetraalkoxysilane (TEOS) was used as the silica source. The condensation of the silane-functionallized nanoparticles and the TEOS produced a MCM-41 type material with Pd nanoparticles homogeneously incorporated in its pore walls (see Figure 3.8). The calcined material exhibit XRD at low angles and a type IV isotherm characteristic of MCM-41. The use of metal-incorporated periodic mesoporous materials in catalysis is expected to reduce agglomeration and leaching during reaction.

The preparation of mesoporous zeolites has long been a desired goal, since it can provide for more accessible catalysts. Recently, Pinnavaia et al. made a

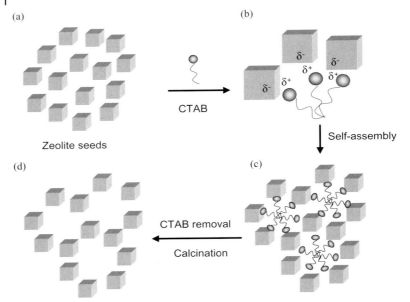

Figure 3.9 Schematic of the process used by Pinnavaia *et al.* for the preparation of steam-stable aluminosilicate nanostructures assembled from zeolite seeds. (a) Nanosize zeolites (zeolite seeds); (b) zeolite seeds assembling around the surfactant (CTAB); (c) assembled structure and, (d) calcined micro-mesoporous steam-stable aluminosilicate.

significant contribution in this field by self-assembling nanosize zeolites with surfactants typically used in the synthesis of mesoporous silica [37] (Figure 3.9). These hybrid micro-mesoporous materials exhibit enhanced catalytic activity for bulky molecules over zeolites, and better hydrothermal stability than MCM-41. However, because of the very small size of the zeolite seeds, the final material is not crystalline, lacking the XRD peaks characteristic of the zeolites and, therefore, most of their catalytic activity and hydrothermal stability [37].

The synthesis of nanostructured carbon using aliphatic alcohols as self-assembling molecules has demonstrated that this strategy can be extended beyond metal oxide-based materials [38]. Recently, we have reported the synthesis of a novel carbon material with tunable porosity by using a liquid-crystalline precursor containing a surfactant and a carbon-yielding chemical, furfuryl alcohol. The carbonization of the cured self-assembled carbon precursor produces a new carbon material with both controlled porosity and electrical conductivity. The unique combination of both features is advantageous for many relevant applications. For example, when tested as a supercapacitor electrode, specific capacitances over 120 F/g were obtained without the need to use binders, additives, or activation to increase surface area [38]. The proposed synthesis method is versatile and economically attractive, and allows for the precise control of the structure.

3.4
Spatially Constrained Synthesis: Foams, Microemulsions, and Molds

Both the molecular template and the self-assembly techniques presented above have limited control over the final shape of the solid, since this is generally obtained in the form of a powder, fibers, or thin films. It is possible, however, to control the shape and size of solids by combining the former techniques with techniques that restrict the volume in which the synthesis takes place. The final goal is to have control over the solids at the molecular as well as macroscopic level, in order to have in a single material properties emerging from several levels of scale. Such structures are referred to as hierarchical [2, 6].

Some of the various techniques used to restrict particle growth to control materials' final size and, to some extent, shape are described below.

3.4.1
Microemulsions

Microemulsions are stable, clear suspensions of two immiscible liquids and a surfactant. The surfactant forms a monolayer with its hydrophilic head dissolved in the water and its hydrophobic tail in the oil. The ratio between the three and the addition of salts, other liquids, or co-surfactants allow for fine tuning of the size of the droplets, which typically range from 5 to 100 nm.

The synthesis of many nanomaterials can be conveniently restricted to the interior of small aqueous droplets suspended in a hydrophobic organic solvent using suitable surfactants. This two-phase system, known as an inverse microemulsion, allows for controlling the size of the aqueous nanospheres, and therefore the final size of the solid. Using this technique, Dutta *et al.* managed to prepare nanosize microporous zincophosphates with controlled morphology and tunable crystal size [39]. The two inorganic components (zinc and phosphate) are dissolved in separate aqueous droplets and react via mass transfer through the surfactant (scheme shown in Figure 3.10a). This causes a preferential growth of some specific crystal faces. The authors claim that this technique provides a means of controlling both the size and the morphology of the microporous zincophosphate crystals.

3.4.2
Capping Agents

Two of the disadvantages of the microemulsion technique is its low yield and the use of high amounts of organic solvents. Capping agents allow for controlling the size of the particles not by forming a layer between two immiscible phases but by strongly bonding to the surface of the solid as it is being formed, effectively limiting its growth, as schematically presented in Figure 3.10b. Capping agents are formed by a reactive head (for example, a thiol group), and a hydrophobic tail (like a long alkane chain). The thiol group covalently bonds to the surface of the particle

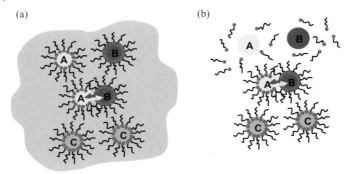

Figure 3.10 (a) Schematic of an inverse microemulsion in which two chemicals (A and B) dissolved in aqueous droplets react to produce a solid (C) whose size is controlled by the size of the droplet in which it is formed. (b) Schematic of two chemicals dissolved in water (A and B) reacting to form a solid (C) whose size is controlled by the presence of a capping agent, which bonds strongly to its surface, limiting its growth.

(typically a metal or metal oxide), whereas the aliphatic chain separates the various nanoparticles, keeping them suspended.

This technique is widely used for the fabrication of quantum dots. This can be illustrated by the preparation of near-monodisperse semiconductor CdSe nanocrystallites with diameters ranging from 1.2 to 11.5 nm by rapid injection of organometallic reagents (dimethylcadmium and tributylphosphine selenide) into a hot coordinating solvent (trioctylphosphine oxide) that acts as a capping agent [40]. This reaction can be carried out in macroscopic quantities in a single step.

3.4.3
Foams

Trapped bubbles of gas in liquids or solids can be also used to control the final shape of solids. Cellular materials are especially attractive for various applications because of their controllable combination of strength, low density, thermal and sound insulation, large surface area, and gas permeability. Many natural materials, such as wood and bone, exhibit a cellular structure and various artificial cellular solids among the most widely used materials. They can be prepared using foams as porogens. For example, ceramic cellular architectures can be obtained by inserting fluorotrichloromethane (Freon) foams into a silica sol containing appropriate surfactants at basic pH [41]. The cellular silica materials prepared using this technique exhibit remarkable low density and good mechanical strength.

All these techniques can be combined to obtain more precise control over several scales, as done by Stucky *et al.* to create hollow microcavities in mesoporous silica. In order to do that, the inorganic precursor (silicon alkoxide) was dissolved in the oil droplets of a regular (not inverse) microemulsion, while the surfactants were in the aqueous phase. Both the mesopore (produced by the surfactant) and microcavity (formed by the oil droplets) sizes could be controlled by varying the dimensions of the surfactant, its concentration, the nature of the organic phase, the temperature, and the pH [42].

3.4.4
Molds

A third possibility for the synthesis of nanomaterials in constrained volumes is the use of molds (Figure 3.1c). Advantages of this method include its simplicity, versatility, and precise control over the shape of the solid, even with intricate forms. An elegant example of this strategy is the preparation of zeolites which precisely replicate the complex microstructure of wood. To do this, Dong et al. [43] infiltrated a zeolite synthesis solution into a wood sample. After the necessary hydrothermal treatment, and subsequent calcination to remove the template as well as the wood, a zeolitic structure was obtained that reproduced with full detail and fidelity the wooden sample used as a mold.

Despite the versatility of the use of molecular and supramolecular templates for preparing materials with controlled porosity, this strategy requires compatibility between the organic template and the inorganic precursor in terms of interaction and stability with respect to pH, temperature, solvent, and, in some cases, autogenous pressure [5, 9–12]. Furthermore, the removal of the template is not a trivial task, since porosity can collapse, especially in the case of some metallic oxides [31]. To avoid these drawbacks, inorganic templates (often referred to as rigid or hard templates) have been used to generate negative replicas of their structures. Porous silica and carbon are the most common hard templates used to prepare metal nanoparticles and nanowires of controlled size (see Figure 3.6d) or negative replicas of their structure (nanocasting) [44–47]. This technique produces especially narrow distributions of pore and particle size [48]. Nanocasting has been used for the preparation of ordered mesoporous carbon (CMK) [49] using MCM-48 (cubic) [50] and SBA-15(hexagonal) [51] as templates through the polymerization and posterior carbonization of furfurylic alcohol or other carbon precursors on the surface of the silica template. Nevertheless, the use of molds and hard templates presents several drawbacks, such as the need for sufficiently rigid and continuous material to allow for mold removal without structural collapse, the preparation of the mold itself, and its low efficiency [48].

3.5
Multiscale Self-Assembly

The simultaneous combination of the all three aforementioned techniques allows for a precise control over the structure of materials at several scales [52]. One of the most useful characteristics of molecular self-assembly is that it can take place simultaneously at multiple scales, producing highly hierarchical structures. This allows for the programmed organization of molecules, biological structures, and nanoparticles in the final architecture of the material in a bottom-up fashion [5].

The key to controlling multiscale self-assembly is based on: (i) the existence of previous individual components; (ii) the weak–noncovalent–interactions between them; and (iii) the dynamic formation of multiple suprastructures of which the most favored is that which minimizes its energy by a maximum number of interactions between individual components [48]. For this reason, the ultimate structure is predefined by various parameters of the initial components such as functionality, surface chemistry, shape, and size.

Hierarchical self-assembly starts with the integration of individual components into complex structures, which in turn organize themselves to form higher-level architectures. This spontaneous assembly continues in a hierarchical way until the solid is completely built. This phenomenon – common to supramolecular chemistry and many biological systems – [53] produces solids with new properties that are not present in its original components.

A good example of the integration of the techniques described is the self-assembly of supramolecular entities to be used as templates [54]. For example, molecular templates used to produce zeolites have several restrictive requirements: (i) adequate hydrophobicity, (ii) no tendency to form complexes with the solvent, (iii) as many van der Waals interactions with the internal zeolite surface as possible, and i(v) the ability to efficiently occupy the cavity of the zeolite [5, 9–12]. Recently, Professor A. Corma's group [55] showed that constructing large templates starting from smaller ones effectively overcomes these limitations. The self-assembly of two polyaromatic templates by means of the formation of a cationic dimer through π–π interactions yields a larger template that can directly be used to synthesize zeolites (see Figure 3.11). This allowed the preparation for the first time of an LTA zeolite without addition of aluminum (ITQ-29). This is a stable

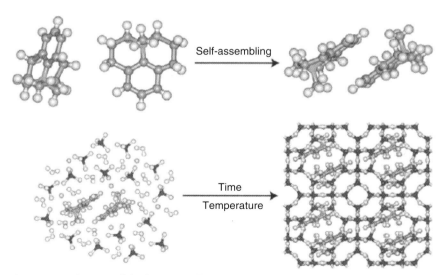

Figure 3.11 Schematic of the formation of ITQ-29, a purely siliceous version of the widely used zeolite A, using a large cationic dimer as a template [55].

material with good adsorption and gas separation properties in the presence of water and other polar substances. The deliberate addition of aluminum to ITQ-29 has been suggested to produce catalysts with good selectivity toward linear products in olefin cracking [55].

An excellent example that shows the potential of combining various bottom-up techniques is the joint work by Whitesides and Stucky [4]. Hierarchical metallic oxides were produced by combining (i) sol-gel self-assembly of neutral surfactants, (ii) spherical polystyrene templates, and (iii) molds with micrometric cavities (micromolding). Figure 3.12 shows how the described materials are hierarchically organized at several scales ranging from a few nanometers to hundreds of micrometers.

Micromolds can be used not only to control the size of nanomaterials but also as an additional tool to control the nanostructure of solids by spatial confinement [56]. An illustrative example of this strategy is the growth of co-polymer templated silica inside the nanochannels of porous anodic alumina with diameters ranging from 18 nm to 80 nm. These cylinders were used as combinatorial nanoscale test tubes to study the effect of confinement on the nanostructure of ordered mesoporous silica. A wide variety of circular nanostructures was obtained after the dissolution of the alumina mold with phosphoric acid (see Figure 3.13). The formation of helical structures is especially interesting because they were produced by spatial confinement and not by using chiral molecules. The large mesopores the silica formed were backfilled with silver by electrochemical deposition, yielding tightly coiled silver nanowires that replicate the internal structure of the helical silica.

3.6
Biomimetic Synthesis: Toward a Multidisciplinary Approach

Nature produces biological structures which are not only organized at multiple scales but, in many ways, adapted to the environment [53]. They are formed in mild conditions – usually neutral pH and room temperature – and at low reactant concentrations. Furthermore, these processes are characterized by a high energetic efficiency and the continuous reuse of materials. This makes us realize that, despite our great advances in the preparation of synthetic materials, nature still has plenty to teach us. As an example, in Figure 3.14 some natural opals and man-made inverse opals are presented. The ordered packing of silica microspheres or cavities, respectively, produces interesting optical properties.

As in many other cellular processes, the formation of rigid biological structures is controlled by specifically designed proteins which catalyze the formation of biomaterials with high efficiency and selectivity. The exoskeletons of many marine microorganisms, such as diatoms or radiolaria, are good examples of naturally occurring inorganic biostructures [58]. These exoskeletons provide mechanical stability and protection and can be extremely complex and varied; they are generally symmetric and specific to each species [58].

Three families of proteins have been identified in the organic matrix of the cellular wall of marine microorganisms [59], among which silaffins are responsible

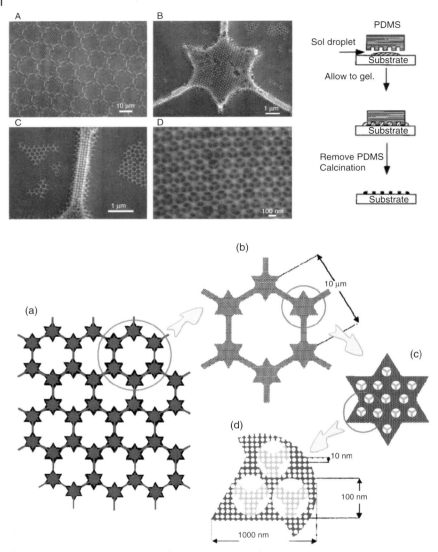

Figure 3.12 (Upper) Transmission electron micrographs of hierarchical materials prepared using micromolding and molecular self-assembly (top) [4]. (Lower) Schematic of a hierarchically organized silica prepared by combining various synthetic techniques. (a) Ordered template prepared by micromolding (stamps) at increasing magnification from A to D. These structures range over a hundred micrometers. (b) Repetition element of the micrometric template. (c) Zoom of the former element with the 100 nm pores prepared by spherical latex templates. (d) Detail showing the meso- and macropores over a cubic superstructure generated by self-assembly of neutral surfactants.

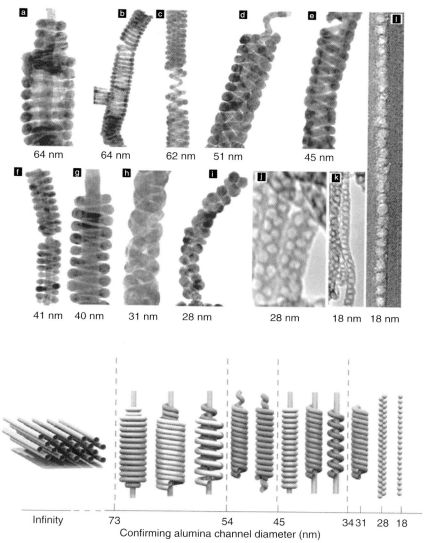

Figure 3.13 (Top) TEM micrographs of (a–i) silver nanocoils prepared by backfilling, (j–k) helical mesoporous silica grown inside nanochannels of different sizes (indicated below each micrograph). (Bottom) Schematic of differing confined nanostructural evolution with decreasing diameter of the nanochannels [56].

Figure 3.14 Detail of some natural opals from the Natural History Museum of Vienna. The iridescence of this semiprecious stone of hydrated silica is due to the ordered packaging of the silica microspheres as shown on the left bottom corner. This structure can be artificially reproduced and used as molds to construct a 3D network of cavities (top right) [57], which is known as inverse opal. This periodic cavity structure is the basis for photonic crystals with the property of affecting the propagation of electromagnetic waves according to the size and organization of the cavities.

for inducing and regulating the formation of silica at room temperature and neutral pH. The simultaneous presence of positively and negatively charged species in silaffins causes their self-assembly via intermolecular electrostatic interactions, yielding a supramolecular template that induces the formation of silica [60]. In recent years, a growing interest has emerged in the preparation of synthetic analogues to silaffins [45, 59]. Combinatorial techniques allow for the fast screening of polypeptides that efficiently induce the formation of silica in physiological conditions [61], which were recently shown useful for generating 2D and 3D structures based on silica [62].

Using an elegant approach, Che et al. prepared chiral mesoporous silica using bio-inspired surfactants [63]. The trimethylammonium group of the quaternary amine used as a surfactant in the synthesis of MCM-41 (CTAB) was replaced by L-alanine. The chirality of the amino acid in the polar head of the surfactant induces chirality in the micelle used as template (see Figure 3.15). This simple modification in the surfactant allowed the preparation of the first chiral mesoporous silica with tunable pore size and ordered porosity. A key step in this synthesis is the transfer of the chirality from the surfactant to the solid, which was accomplished by electrostatic interaction between the terminal amino acid and the

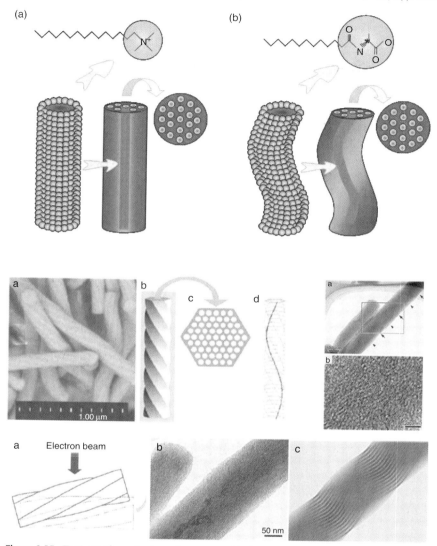

Figure 3.15 (Top) (a) The cationic surfactant hexadecyltrimethylammonium forms tubular micelles that self-assemble in hexagonal superstructures. These are then used as templates to induce controlled mesoporosity in MCM-41 silica. (b) A similar surfactant, with the amino acid alanine as polar group, assembles into tubular chiral structures to produce chiral replicas of MCM-41. (Bottom) Scanning and transmission electron micrographs of chiral mesoporous silica and some schematics of their structures [63].

inorganic precursor. The authors suggested using these materials for asymmetric catalysis and separation of enantiomers.

Biomimetic synthesis may also take advantage of the combination of templates and spatially constrained synthesis techniques, for instance to create inorganic replicas of living organisms. The tobacco mosaic virus is a cylinder of 300 nm length and 18 nm diameter consisting of 2130 identical proteins along a dextrorotatory helix associated with an RNA (ribonucleic acid) thread [64]. These structures are chemically and thermally stable, and their surface is made of lateral chains of particular amino acids. Such favorable properties makes tobacco mosaic virus an ideal biomold for preparation of nanostructured solids based on silica and other metallic oxides and sulfides [65, 66], or metallic nanoparticles [67, 68] The biological nature of the mold or template opens the possibility of using recombinant technology to design biological surfaces with the desired properties.

The limit of the nanometer scale can be surpassed by combining biomolds with other techniques to generate larger hierarchical structures. An attractive example of a multidisciplinary effort to prepare new bio-inspired materials was described by Davis et al [69]. Multicellular filaments from bacteria (biomold) were combined with cationic surfactants to produce ordered meso-macroporous materials by self-assembly. Figure 3.16 shows how the bacterium *Bacillus subtilis* spontaneously organizes itself into multicellular filamentary superstructures several millimeters in length and around 200 μm. Infiltration of a solution containing both the silica precursor (tetraorthosilicate) and the surfactant (CTAB)

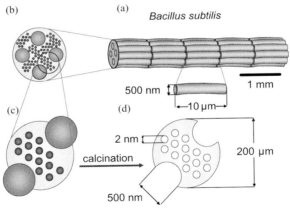

Figure 3.16 (a) A multicellular filament of the bacterium *Bacillus subtilis* several millimeters in length and about 200 μm thick. (b) Transverse section of the filament with the bacteria used as a biomold (green), the intercellular space (yellow) and the surfactant micelles (red). (c) Detail of the hexagonally self-assembled surfactant micelles. (d) After calcination, silica monoliths several millimeters long with macropores of 500 nm in diameter surrounded by pore walls made of hexagonally arranged 2 nm mesopores (MCM-41).

was used to produce the hexagonal mesoporous silica MCM-41 inside the intercellular spaces. The posterior calcination of the material yields monoliths several millimeters in length with macropores of hundreds of nanometers with pore walls made of MCM-41-type mesoporous silica, as schematically shown in Figure 3.16.

Besides biological structures, porous solids can be used to encapsulate biological material. This synthetic alternative enables immobilization of enzymes which maintain their catalytic activity [70, 71]. These heterogenized biocatalysts can easily be separated from the reaction medium and reused without significant loss in activity. Silica precipitation occurs only under aggressive conditions, such as alkaline pH, which are not compatible with the stability of the biomolecules. To overcome this limitation, silaffins and other polypeptides or polyamines can be used to induce the formation of silica under milder conditions. For example, the enzyme butyrylcholinesterase has been immobilized on silica with 90% efficiency by the use of synthetic P5 peptide, which is one of the repetitive units of the natural polypeptide silaffin. The bioinorganic compound material presents higher thermal stability than the enzyme in solution. Several tests in fixed-bed and fluidized-bed reactors displayed the sustained catalytic activity of the immobilized material, even after successive uses. This is a general method that can be extended to other enzymes.

Biomolecules can also be used to obtain further control over the nanostructure of solids. Our group recently showed that the addition of α-L-phosphatidylcholine (lecithin) to a typical MCM-41 synthesis solution produces circular ordered mesoporous silica with narrow pore size distribution and large surface area (Figure 3.17) [72]. The addition of a small concentration of lecithin (lecithin/CTAB molar ratio = 0.02) causes the pores of MCM-41 to curve to form hexagonally ordered, circularly shaped mesopores that concentrically organize into beautiful structures that resemble certain natural biostructures, such as the exoskeletons of some radiolaria. Increasing concentrations of lecithin induce the transformation from a hexagonal to a lamellar nanostructure (Figure 3.17 bottom). Biomolecules which act as natural surfactants, for example lecithin in cell membranes, are useful chemicals for gaining further control over the porous architecture of solids as well as a good example of the contributions that biology is making to nanotechnology.

The versatility of the synthetic techniques described, combined with the enormous creativity shown by many research groups, makes us believe that exciting new materials will be described in the coming years. Some of the materials described are currently being applied in such diverse fields as catalysis, separation, drug delivery, gas adsorption, and energy production and storage [3].

Nature will always be an excellent source of inspiration, not only because of the complexity and perfection of living biostructures but also because they are fabricated in mild conditions, minimize energy consumption and recycle the materials repeatedly. This gives hope of move forward to the construction of new and better materials making more efficient use of natural resources.

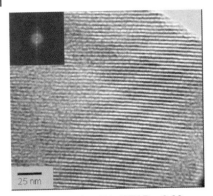
Straight: Lecithin/CTAB = 0.00

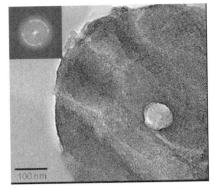
Hexagonal: Lecithin/CTAB = 0.02

Lamellar: Lecithin/CTAB = 0.10

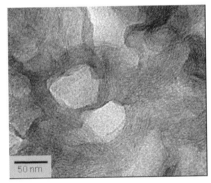
Tortuous: Lecithin/CTAB = 0.30

Figure 3.17 (Top) Transmission electron micrographs of mesoporous silica materials prepared by adding increasing amounts of lecithin to a typical MCM-41 synthesis solution. This biomolecule induces the formation of circular-shaped mesopores. (Bottom) A plausible scheme of the structures of materials prepared at two different lecithin/CTAB ratios: (a) hexagonal concentrically assembled mesopores (lecithin/CTAB = 0.02) and (b) lamellar concentrically assembled mesopores (lecithin/CTAB = 0.10).

3.7
Acknowledgments

J.G.M. thanks Guillermo Rus for his valuable help during the writing of this chapter and the Ministry of Education and Science (Project CTQ2005-09385-CO3-02 and Programa Ramón y Cajal) for financial support.

References

1. García-Martínez, J. (2006) *Anales de Quimica*, **102**, 5–12.
2. Soler-Illia, G.J.A.A., Sanchez, C. Lebeau, B. and Patarin, J. (2002) *Chemical Reviews*, **102**, 4093–138.
3. Mann, S., Burkett, S.L., Davis, S.A., Fowler, C.E., Mendelson, N.H., Sims, S.D., Walsh, D. and Whilton, N.T. (1997) *Chemistry of Materials: A Publication of the American Chemical Society*, **9**, 2300–10.
4. Yang, P., Deng, T., Zhao, D., Feng, P., Pine, D., Chmelka, B.F., Whitesides, G.M. and Stucky, G.D. (1998) *Science*, **282**, 2244–6.
5. Corma, A. (1997) *Chemical Reviews*, **97**, 2373–419.
6. Ozin, G. and Arsenault, A. (2005) *Nanochemistry: A Chemical Approach to Nanomaterials*, Royal Society of Chemistry, London.
7. WTEC Panel Report on Nanostructure Science and Technology: R&D Status and Trends in Nanoparticles, Nanostructured Materials and Nanodevices. www.nano.gov (accessed 26/08/07).
8. Burkett, S.L. and Davis, M.E. (1994) *The Journal of Chemical Physics*, **98**, 4647–53.
9. García Martínez, J. and Pérez Pariente, J. (eds) (2003) *Materiales Zeolíticos: Síntesis, Propiedades y Aplicaciones*, Publicaciones Universidad de Alicante, Alicante.
10. Auerbach, S.M., Carrado, K.A. and Dutta, P.K. (eds) (2003) *Handbook of Zeolite Science and Technology*, Marcel Dekker, New York.
11. van Bekkum, H., Flanigen, E.M. and Jansen, P.A. (eds) (2001) *Introduction to Zeolite Science and Practice*, Studies in Surface Science and Catalysis 137, Elsevier, Amsterdam.
12. International Zeolite Association. www.iza-online.org (accessed 26/08/07).
13. Cundy, C.S. and Cox, P.A. (2005) *Microporous and Mesoporous Materials*, **82**, 1–78.
14. Wadlinger, R.L., Kerr, G.T. and Rosinski, E.J. (1967) US Patent 3.308069.
15. Flanigen, E.M. *et al.* (1978) *Nature*, **271**, 512–6.
16. Knecht, M.R., Sewell, S.L. and Wright, D.W. (2005) *Langmuir: The ACS Journal of Surfaces and Colloids*, **21** (5), 2058–61.
17. Galliot, C., Larré, C., Caminade, A.-M. and Majoral, J.-P. (1997) *Science*, **277**, 1981–4.
18. Soler-Illia, G.J.A.A., Crépaldi, E.L., Grosso, D. and Sanchez, C. (2003) *Current Opinion in Colloid and Interface Science*, **8**, 109–126.
19. Stein, A. (2001) *Microporous and Mesoporous Materials*, **4445**, 227–39.
20. Imhof, A. and Pine, D.J. (1997) *Nature*, **389**, 948–51.
21. Beckwith, D., Christou, R., Cook, G., Goltner, C.G. and Timms, P.L. (2001) Proceedings of Silica 2001, Second International Conference on Silica.
22. Mertig, M., Ciacchi, L.C., Seidel, R., Pompe, W. and De Vita, A. (2002) *Nano Letters*, **2**, 841–4.
23. Douglas, T., Strable, E., Willits, D., Aitouchen, A. and Libera, M. (2002) *Advanced Materials (Deerfield Beach, Fla.)*, **14**, 415–8.
24. Seddon, J.M. and Raimondi, M.E. (2000) *Molecular Crystals and Liquid Crystals*, **347**, 221–9.
25. Sanchez, C., Soler-Illia, G.J.A.A., Ribot, F., Lalot, T., Mayer, C.R. and Cabuil, V. (2001) *Chemistry of Materials: A Publication of the American Chemical Society*, **13**, 3061–83.
26. Kresge, C.T., Leonowicz, M.E., Roth, W.J., Vartuli, J.C. and Beck, J.S. (1992) *Nature*, **359**, 710–2.

27 Zhao, D., Feng, J., Huo, Q., Melosh, N., Fredrickson, G.H., Chmelka, B.F. and Stucky, G.D. (1998) *Science*, **279**, 548–52.

28 Maschmeyer, T., Rey, F., Sankar, G. and Thomas, J.M. (1995) *Nature*, **378**, 159–62.

29 a. Díaz, I., Márquez Álvarez, C., Mohino, F., Pérez Pariente, J. and Sastre, E. (2000) *Journal of Catal*, **193**, 283–94.
b. Díaz, I., Márquez Álvarez, C., Mohino, F., Pérez Pariente, J. and Sastre, E. (2000) *Journal of Catal*, **193**, 295–302.
c. Díaz, I., Mohino, F., Pérez Pariente, J. and Sastre, E. (2001) *Applied Catalysis*, **205**, 1930. d. Boveri, M., Aguilar Pliego, J., Pérez Pariente, J. and Sastre, E.C. (2005) *Catal Today*, **107**, 868–73.

30 Yang, Q., Liu, J., Yang, J., Kappor, M.P., Inagaki, S. and Li, C. (2004) *Journal of Catal*, **228**, 265–72.

31 Ying, J.Y., Mehnert, C.P. and Wong, M.S. (1999) *Angewandte Chemie (International Ed. in English)*, **38**, 56–77.

32 Kapoor, M.P. and Inagaki, S. (2006) *Bulletin of the Chemical Society of Japan*, **79** (10), 1463–75.

33 Asefa, T., MacLachlan, M.J., Coombs, N. and Ozin, G.A. (1999) *Nature*, **402**, 867–71.

34 Bion, N., Ferreira, P., Valente, A., Goncalves, I.S. and Rocha, J. (2003) *Journal of Materials Chemistry*, **13**, 1910–3.

35 Inagaki, S., Guan, S., Ohsuna, T. and Terasaki, O. (2002) *Nature*, **416**, 304–7.

36 García-Martínez, J., Linares, N., Sinilbaldi, S., Coronado, E. and Ribera, A. *Microporous and Mesoporous Materials*, (submitted).

37 a. Liu, Y., Zhang, W. and Pinnavaia, T.J. (2001) *Angewandte Chemie (International Ed. in English)*, **40** (7), 1255–8. b. Liu, H., Wang, Y. and Pinnavaia, T.J. (2006) *The Journal of Physical Chemistry B*, **110** (10), 4524–6. c. Liu, Y. and Pinnavaia, T.J. (2004) *Journal of Materials Chemistry*, **14**, 1099–103.

38 Ying, J., Martinez, Garcia, J., and Lancaster, T. (2005) (Massachusetts Institute of Technology), US Patent 60/556976.

39 Dutta, P.K., Jakupca, M., Reddy, K.S.N. and Salvati, L. (1995) *Nature*, **374**, 44–6.

40 Murray, C.B., Norris, D.J. and Bawedi, M.G. (1993) *Journal of the American Chemical Society*, **115**, 8706–15.

41 Wu, M., Fujiu, T. and Messing, G.L. (1990) *Journal of Non-Crystalline Solids*, **121**, 407–12.

42 Schacht, S., Huo, Q., VoightMartin, I.G., Stucky, G.D. and Schuth, F. (1996) *Science*, **273**, 768–71.

43 Dong, A., Wang, Y., Tang, Y., Ren, N., Zhang, Y., Yue, Y. and Gao, Z. (2002) *Advanced Materials (Deerfield Beach, Fla.)*, **14**, 926–9.

44 Tian, B., Che, S., Liu, Z., Liu, X., Fan, W., Tatsumi, T., Terasaki, O. and Zhao, D. (2003) *Chemical Communications*, 2726–7.

45 Eliseev, A.A., Napolskii, K.S., Lukashin, A.V. and Tretyakov, Y.D. (2004) *Journal of Magnetism and Magnetic Materials*, 1609–11.

46 Sakamoto, Y., Fukuoka, A., Higuchi, T., Shimomura, N., Inagaki, S. and Ichikawa, M. (2004) *The Journal of Physical Chemistry. B*, **108**, 853–858.

47 Adhyapak, P.V., Karandikar, P., Vijayamohanan, K., Athawale, A.A. and Chandwadkar, A.J. (2004) *Materials Letters*, **58**, 1168–71.

48 Lu, A.H. and Schüth, F. (2005) *Comptes Rendus Chimie*, **8**, 609–20.

49 Ryoo, R., Joo, S.H., Kruk, M. and Jaroniec, M. (2001) *Advanced Materials (Deerfield Beach, Fla.)*, **13**, 677–81.

50 Ryoo, R., Joo, S.H. and Jun, S. (1999) *The Journal of Physical Chemistry. B*, **103**, 7743–6.

51 Jun, S., Joo, S.H., Ryoo, R., Kruk, M., Jaroniec, M., Liu, Z., Oksuna, T. and Terasaki, O. (2000) *Journal of the American Chemical Society*, **122**, 10712–3.

52 Whitesides, G.M. and Grzybowski, B. (2002) *Science*, **295**, 2418–21.

53 Ball, P. (1999) *The Self Made Tapestry: Pattern Formation in Nature*, Oxford University Press, Oxford.

54 Sanchez, C., Soler-Illia, G.J.A.A., Ribot, F. and Grosso, D. (2003) *Comptes Rendus Chimie*, **6**, 1131–51.

55 a. Corma, A., Rey, F., Rius, J., Sabater, M.J. and Valencia, S. (2004) *Nature*, **431**, 287–90. b. Corma, A., Rey, F. and Valencia, S. (2005) ES Patent 2/245588.

56. Chu, Y., Cheng, G., Katsov, K., Sides, S.W., Wang, J., Tang, J., Fredrickson, G.H., Moskovits, M. and Stucky, G.D. (2004) *Nature Materials*, **3** (*11*), 816–22.
57. Vlasov, Y.A., Bo, X.-Z., Sturm, J.C. and Norris, D.J. (2001) *Nature*, **414**, 289–93.
58. a. Heaps Pickett, J. (1990) Progress in Phycological Research, in *The Cell Biology of Diatom Valve Formation* (eds F.E. Round and D.J. Chapman), Biopress, Bristol. b. Round, F.E., Crawford, R.M. and Mann, D.G. (1990) *Diatoms: Biology and Morphology of the Genera*, Cambridge University Press, Cambridge.
59. Foo, C.W.P., Huang, J. and Kaplan, D.L. (2004) *Trends in Biotechnology*, **22**, 577–85.
60. Kroger, N., Lorenz, S., Brunner, E. and Sumper, M. (2002) *Science*, **298**, 584–6.
61. Naik, R.R., Brott, L., Clarson, S.J. and Stone, M.O. (2002) *Journal of Nanoscience and Nanotechnology*, **2**, 95–100.
62. Naik, R.R., Whitlock, P.W., Rodriguez, F., Brott, L.L., Glawe, D.D., Clarson, S.J. and Stone, M.O. (2003) *Chemical Communications*, 238–9.
63. Che, S., Liu, Z., Ohsuna, T., Sakamoto, K., Terasaki, O. and Tatsumi, T. (2004) *Nature*, **429**, 281–4.
64. Stubbs, G. (1990) *Seminars in Virology*, **1**, 405–12.
65. Shenton, W., Douglas, T., Young, M., Stubbs, G. and Mann, S. (1999) *Advanced Materials (Deerfield Beach, Fla.)*, **11**, 253–6.
66. Fowler, C.E., Shenton, W., Douglas, T., Stubbs, G. and Mann, S. (2001) *Advanced Materials (Deerfield Beach, Fla.)*, **13**, 1266–9.
67. Knez, M., Sumser, M., Bittner, A.M., Wege, C., Jeske, H., Kooi, S., Burghard, M. and Kern, K. (2002) *Journal of Electroanalytical Chemistry*, **522**, 70–4.
68. Dujardin, E., Peet, C., Stubbs, G., Culver, J.N. and Mann, S. (2003) *Nano Letters*, **3**, 413–7.
69. Davis, S.A., Burkett, S.L., Mendelson, N.H. and Mann, S. (1997) *Nature*, **385**, 420–3.
70. Luckarift, H.R., Spain, J.C., Naik, R.R. and Stone, M.O. (2004) *Nature Biotechnology*, **22**, 211–3.
71. Kim, J., Park, J.K. and Kim, H.K. (2004) *Colloids and Surfaces. A*, **241**, 113–7.
72. García Martínez, J., Domínguez, S. and Brugarolas, P. (2007) *Microporous and Mesoporous Materials*, **100** (1–3), 63–9.

4
Strategies Toward Hierarchically Structured Optoelectronically Active Polymers

Eike Jahnke and Holger Frauenrath

Many biomaterials derive their remarkable and often extraordinary properties from hierarchical structure formation, that is, the presence and control of structural order on different levels, such as the molecular, supramolecular, nanoscopic, and microscopic length scales. Accordingly, the preparation of hierarchically structured synthetic polymer materials has been recognized as an important field of research [1, 2]. Nevertheless, the "state of the art" in controlling structure and order over many levels in the structural hierarchy is still found in biological systems themselves, which often combine "bottom-up" and "top-down" approaches [3]. Thus, the information to adopt certain higher structures is programmed on the molecular level while, at the same time, the system is carefully guided to find the desired structure among the manifold of energetically similar possibilities from the macroscopic level, that is, by means of sophisticated processing procedures. Recent advances in the field of optoelectronic devices aimed at applications in the life sciences, for example, the development of new generations of biomedical devices, make it appear advantageous to develop a set of optoelectronically active synthetic materials that offer an accessible interface to the biomaterials they are supposed to interact with. Hence, they should not only offer the diverse chemical functionality for such an interaction on the molecular level, but also display a similar degree of hierarchical structural order. As will be shown in this review, recent successful attempts to prepare such materials were based on strategies that aimed to combine modern organic and polymer chemistry [4], as well as materials science and supramolecular chemistry [5].

4.1
Hierarchically Structured Organic Optoelectronic Materials via Self-Assembly

Supramolecular self-assembly has recently been extensively investigated as a tool to create nanostructured or hierarchically structured optoelectronically active materials from monodisperse, conjugated oligomers because such π-conjugated organic materials are interesting candidates for the fabrication of organic electronic devices [6]. Usually, π-conjugated oligomers processed by chemical vapor deposition (CVD) [7] give rise to highly ordered domains and, consequently, superior optoelectronic properties due to their well-defined chemical structure. However, CVD methods are inferior to solution processing when it comes to larger devices in terms of costs [8]. The "supramolecular approach" attempts to alleviate this disadvantage by providing a pathway toward highly ordered, nanostructured arrays of π-conjugated oligomers from solution [5]. A particularly interesting recent example was reported by Meijer and co-workers, who self-assembled oligo(phenylene vinylene)s (OPVs) carrying nucleobases using complementary oligonucleotides and obtained double-helical fibrillar aggregates (Figure 4.1a) [9]. Another approach toward a helical assembly of OPVs utilized a foldamer based on the ureidophthalimide motif to which chirally substituted OPVs were attached as the side-chains and organized into six blades helically twisted around the foldamer backbone (Figure 4.2c) [10]. Likewise, chiral OPVs with terminal ureido-s-triazine units were observed to form dimers via quadruple hydrogen bonding which further organized into helical columnar structures by π–π stacking (Figure 4.1b) [11, 12]. Similarly interesting materials were obtained from molecules comprising a perylene bisimide dye hydrogen-bonded to two OPV units, which can be regarded as donor–acceptor–donor dye arrays, were shown to exhibit photoinduced electron transfer capabilities, and gave rise to fibrillar features suitable for the manufacturing of semiconducting devices [13, 14]. The self-assembly of oligo(thiophene)s (OTs) has also been extensively studied in recent years because well-ordered OTs are interesting organic semiconductors [15]. Thus, OTs with chiral oligo(ethylene oxide) side-chains formed chiral superstructures in organic solvents [16–20]. While their self-assembly was mainly driven by π–π stacking interactions, Shinkai *et al.* attempted to enhance the intermolecular forces and obtained efficient organogelators from a cholesteryl amide-substituted OT [21]. The gels exhibited thermochromism attributed to a change in conjugation length upon aggregation. OTs have also been incorporated into block copolymers with poly(ethylene oxide) [22–26] or poly(styrene) [27–29], giving rise to ordered nanoscopic domains when processed from selective solvents.

4.2
Toward Hierarchically Structured Conjugated Polymers via the Foldamer Approach

The above examples highlight the power of supramolecular self-assembly as a synthetic method for organizing short π-conjugated oligomers into hierarchically structured optoelectronically active materials. Conjugated polymers would be an interesting alternative because they are better processable from solution, for example, by ink-jet printing [33]. However, problems in purification and a typically lower degree of internal order often lead to inferior electronic properties compared to vapor-deposited thin layers of monodisperse oligomers [34]. For this reason,

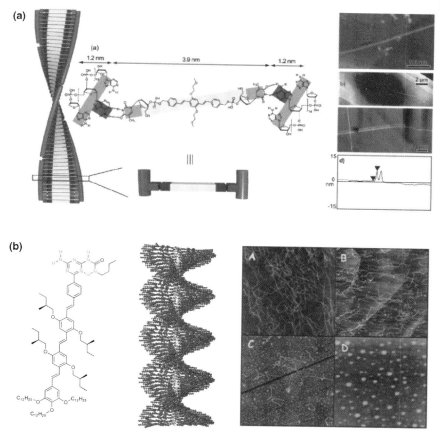

Figure 4.1 (a) Structural model and SFM images of nucleobase-equipped OPV self-assembled by means of an interaction with oligonucleotides. (b) Model and SFM images of supramolecular aggregates of an OPV carrying a head group capable of hydrogen bonding. Reprinted with permission from [9] and [12].

investigations toward hierarchically structured conjugated polymers are an attractive field of research.

In the realm of conventional (nonconjugated) synthetic polymers, the most successful examples of hierarchically structured polymers resulted from recent attempts to transfer the concept of "foldamers" [35, 36] to the world of high-molecular-weight polymer materials. Contrary to other examples of synthetic polymers with controlled helicity [37] in which the higher structure formation originated from controlled polymer tacticity combined with sterically demanding side-chains, these high-molecular-weight foldamers contain "sticky sites" in nonadjacent

76 | *4 Strategies Toward Hierarchically Structured Optoelectronically Active Polymers*

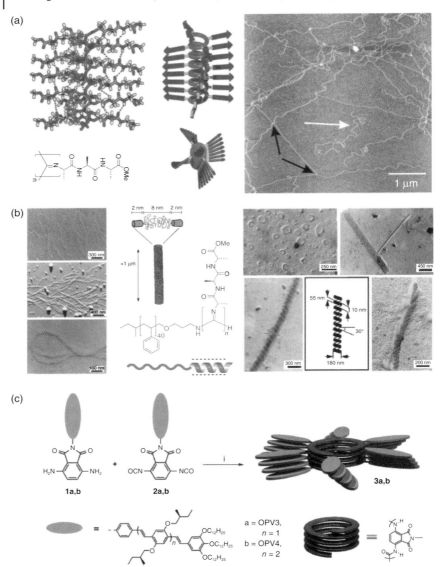

Figure 4.2 (a) Structural model of a poly(isocyano dipeptide) foldamer, and an SFM image of the polymers obtained with acid initiators. (b) "Superamphiphile" aggregates from poly(styrene)–poly(isocyanide) block copolymers PS_{40}-b-PIAA20 (left) and PS40-b-PIAA10 (right). (c) Phthalimido-based OPV foldamer. Reprinted with permission from [30], [31], [32] and [10].

repeating units. The latter induce the formation of stable folded conformations due to cooperative intrachain supramolecular interactions and, thus, mimic the folding mechanism observed in biopolymers such as proteins. The preparation of oligopeptide-substituted poly(isocyanide)s reported by Nolte, Cornelissen, and colleagues represents a particularly beautiful example [30, 38]. The polymers were shown to attain a stable tertiary structure, that is, a 4_1-helical backbone stabilized by the oligopeptide side-chains aligned in four "blades" of helically twisted β-sheet arrays. In some cases, extremely high-molecular-weight polymers were obtained by simple acid addition, and the authors interpreted their results in terms of an efficient pre-organization of the monomers prior to polymerization [39, 40]. Block copolymers from charged poly(isocyanide) segments and poly(styrene) formed "superamphiphiles" that spontaneously self-assembled into a large variety of superstructures such as superhelices, vesicles, and micellar rods in aqueous solution depending on the pH [32]. Block copolymers with poly(γ-benzyl-L-glutamate) segments, on the other hand, were reported to feature a secondary structure diversity unprecedented in the realm of synthetic polymers [41].

However, the particular synthetic requirements in the preparation of conjugated polymers have thus far severely limited the number of similarly hierarchically structured examples. Pu et al. reported different types of conjugated polymers with fixed main-chain chirality containing binaphthyl units in their backbone which exhibited, for example, nonlinear optical activity or were used as enantioselective fluorescent sensors [42–46]. Some chirally substituted poly(thiophene)s were observed to form helical superstructures in solution [47–51]. Okamoto and coworkers reported excess helicity in nonchiral, functional poly(phenyl acetylene)s upon supramolecular interactions with chiral additives, and they were able to induce a switch between "unordered" forms as well as helical forms with opposite helical senses [37, 52, 53].

The largest body of investigations in the field deals with different types of poly(acetylene) foldamers. Thus, Masuda et al. investigated the polymerization of N-propargyl amides which they found to proceed with high cis stereoselectivity [54–58]. The rigidity of the obtained polymers increased upon the incorporation of homochiral side-chains, and the chiral polymers exhibited a large optical rotation with opposite sign as compared to the monomers, as well as a strong CD effect for the main-chain chromophore. The authors attributed these observations to the presence of stable folded conformations due to cooperative intramolecular hydrogen bonding. They also investigated related functionalized polymers and typically observed a complex and stimulus-responsive temperature-, solvent-, and pH-dependent helicity [59, 60]. Masuda et al. also reported on the related polymerization of chiral and achiral alkyl poly(propargyl carbamate)s [61], G2 lysine dendron functionalized phenylacetylenes [62], and P-chiral N-propargyl phosphonamides [63, 64]. Furthermore, an attempt toward a photoswitchable poly(acetylene) was undertaken using glutamic propargyl amide with azobenzene side-chains, but UV irradiation led only to a decrease in helicity which was not recovered upon re-isomerization [65, 66]. More recently, Tang and co-workers performed similar investigations on amino acid substituted phenylacetylenes [67–69], Percec, Meijer, and colleagues prepared dendronized polyacetylenes [70], and Kakuchi *et al.* reported on poly(phenyl acetylene)s with crown ether side groups which were used to achieve a chemo- and thermoresponsive helix-coil transition [71].

4.3
"Self-Assemble, then Polymerize" – A Complementary Approach and Its Requirements

Most examples of hierarchically structured synthetic polymers have been prepared using the same general approach, which can be described as *"polymerize, then fold into a hierarchical structure."* Alternatively, one may envision reversing the order of the required steps and develop a novel complementary strategy, which may be paraphrased as *"self-assemble into a hierarchical structure, then polymerize"* (Figure 4.3). This approach would combine the supramolecular pre-organization of monomers prior to polymerization [40, 72] and covalent capture as a useful concept for the fixation of supramolecular materials [73]. Thus, appropriately functionalized macromonomers are envisaged to self-assemble into well-defined supramolecular polymers with a finite number of strands and the propensity to hierarchical structure formation in solution. They are then to be converted into functional, multistranded conjugated polymers with retention of their previously assembled hierarchical structure. This strategy may thus offer "structural self-healing" as one of the crucial advantages of the supramolecular self-assembly of oligomers in the preparation of high-molecular-weight conjugated polymers. This would be beneficial because one severe limitation of the "foldamer approach" is the poor control over the polymer primary structure even with modern polymerization methods.

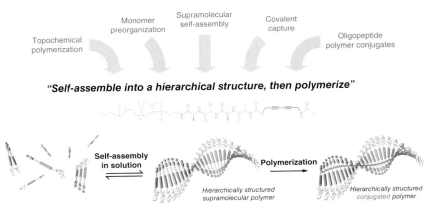

Figure 4.3 Schematic representation of the *"self-assemble, then polymerize"* approach for the preparation of hierarchically structured conjugated polymers.

Efficient and defect-free folding of the polymer will, at least, require a control of the regioselectivity of monomer incorporation and polymer tacticity. If, for example, the degree of tacticity control is poor, the stereochemical defects are irreversibly incorporated into the polymer, and the subsequent folding may fail or produce defective structures. By contrast, dynamic self-assembly may allow defects to be corrected and the hierarchical structure to be controlled or fine-tuned using external parameters (solvent, additives, temperature) *prior* to covalent fixation by polymerization.

For this strategy to be applied successfully, two prerequisites will have to be fulfilled. First of all, a reliable supramolecular synthon is needed that will give rise to well-defined supramolecular polymers with predictable higher-order structure formation. Secondly, the applied polymerization methodology must produce conjugated polymers under conditions compatible with the self-assembly process. In our own investigations, we have chosen macromonomers based on β-sheet-forming oligopeptide–polymer conjugates as the supramolecular synthon and the topochemical diacetylene polymerization as the polymerization methodology because they are mutually compatible, and this combination perfectly fulfills the above requirements.

4.3.1
Topochemical Polymerization Using Self-Assembled Scaffolds

The UV-induced topochemical diacetylene polymerization appears to be the ideal polymerization methodology because it is atom-efficient and initiator- as well as catalyst-free, and the obtained poly(diacetylene)s are conjugated, photoconductive polymers. It was first performed in single-crystalline samples of diacetylene derivatives, as reported by Wegner et al. [74–78]. The reaction proceeds in the sense of a *trans*-stereospecific 1,4-polyaddition along a unique crystal direction and is strictly controlled by the crystal packing parameters [79]; only if an identity period $d \approx 4.9$ Å and an inclination angle $\phi \approx 45°$ between the diacetylene axis and the packing axis are established in the monomer crystals are adjacent molecules placed at a distance compatible with the length of the polymer repeat unit and the polymerization may take place with only minimal packing rearrangements. As the topochemical diacetylene polymerization is possible whenever these criteria are fulfilled, it is not restricted to 3D single crystals and has been performed in other types of ordered phases as well, for example, in self-assembled mono-, bi-, and multilayers and Langmuir–Blodgett films of diacetylene containing amphiphiles [80–84]. Polymerized bilayers of this type have been used in the preparation of materials aimed at applications in the field of biosensing, based on the well-known solvato-, thermo- and mechanochromism of the poly(diacetylene) backbones [85]. Other types of self-assembled scaffolds have also been employed, such as 1D lamellar structures in self-assembled monolayers on surfaces [86–88], columnar liquid crystal (LC) phases of discotic monomers [89–91], vesicles and related self-assembled structures in the submicrometer range [92–100], as well as organogels from hydrogen-bonded 1D aggregates in solution [101–105].

These examples serve to highlight that supramolecular self-assembly and topochemical diacetylene polymerizations are a perfect match. Topochemical diacetylene polymerizations are an advantageous means of covalent capture for the reasons outlined above. The required order may, on the other hand, be provided by supramolecular self-assembly, which extends the scope beyond single-crystalline monomers. This aspect becomes particularly important in the case of functional monomers in order to address specific applications. However, in contrast to previous investigations, the targeted preparation of hierarchically structured poly(diacetylene)s with a defined, finite number of strands required the presence of equally well-defined, uniform supramolecular polymers [106] with the propensity to form predictable superstructures, instead of micellar or vesicular 1D aggregates.

4.3.2
Self-Assembly of β-Sheet Forming Oligopeptides and Their Polymer Conjugates

The choice of β-sheet-forming oligopeptide–polymer conjugates as the scaffold for the topochemical diacetylene polymerization was based on the fact that a variety of previous reports had shown that they were reliable supramolecular synthons, giving rise to uniform 1D aggregates with a high aspect ratio and the propensity to higher structure formation in solution. Furthermore, the packing distance between adjacent β-strands of $d = 4.8$ Å appeared to be perfectly appropriate for a successful topochemical diacetylene polymerization.

The self-assembly of β-sheet-forming amyloid peptides [107, 108] and synthetic peptidomimetics [109] as well as corresponding oligopeptide–polymer conjugates has recently received increasing attention, fueled by their relation to neurodegenerative diseases [110, 111]. For example, oligopeptide-based nanostructures were used in the manufacture of gold nanowires [112, 113]. Nanoscopic, helically twisted ribbons were obtained from oligo(ethylene oxide) segments attached to oligopeptides (Figure 4.4b) [114]. Börner *et al.* recently prepared well-defined nanoscopic tubular fibrils from cyclic peptides equipped with poly(butyl acrylate) segments [115]. Block copolymers containing oligopeptide segments inspired by fibrous proteins such as spider dragline silk showed nanophase segregation and exhibited remarkable mechanical properties [116–119].

Boden and co-workers investigated the hierarchical self-organization of short designed oligopeptides in protic environments into uniform 1D nanoscopic objects with defined superstructures. For example, the 24-residue peptide K24 derived from the β-sheet-forming transmembrane domain of a protein was designed to form gels in protic solvents via antiparallel β-sheet formation [120, 121]. Transmission electron microscopy measurements proved the existence of nanoscopic tapes with a width on the order of the molecules' extended length and a monomolecular thickness. The authors also investigated a set of *de novo* designed 11-residue oligopeptides which were efficient gelators in protic solvents and gave rise to left-handed helically bent single tapes, helically twisted ribbons (double tapes) or even

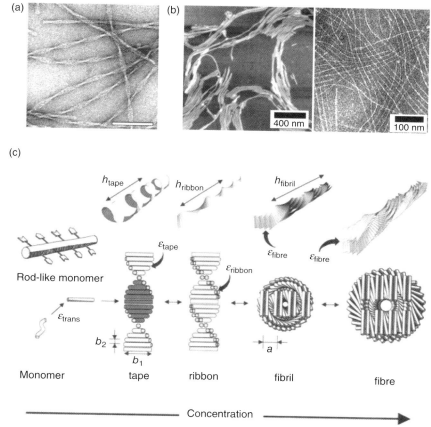

Figure 4.4 (a) TEM image of fibrils from a designed 11-residue oligopeptide (scale bar = 100 nm). (b) SFM and TEM images of nanoscopic fibrillar features from a PEO–oligopeptide conjugate. (c) Model for the hierarchical self-organization of oligopeptides in protic media into tapes, ribbons, fibrils, and fibers. Reprinted with permission from [123], [125], and [122].

higher aggregates depending on the presence of favorable electrostatic or hydrophobic interstrand interactions, controlled lateral recognition of adjacent β-strands, and controlled amphiphilicity of the obtained β-sheet surfaces (Figure 4.4a) [120–124].

Based on their experimental findings, the authors developed a remarkably intuitive generalized model for the observed stepwise hierarchical self-organization of oligopeptides into defined supramolecular aggregates (Figure 4.4c) [122]. Thus, the oligopeptides attain a chiral rodlike configuration in solution and then self-assemble into long single β-sheet *tapes* exhibiting a left-handed twist. The latter originates from the right-handed twist of the monomers, which in turn finds its origin in the chirality of the natural amino acids. If the tapes' surfaces are amphiphilic, the tapes may either curl (bend) into tubular helices or form *ribbons* (double tapes) via β-sheet stacking in order to hide the hydrophobic surface from the protic environment. As the surfaces of the ribbons are identical, the latter will no longer bend but will maintain their twist. The ribbons may further aggregate into *fibrils* via β-sheet stacking, but the associated aggregation enthalpy is smaller than for the ribbon formation. The latter gains an additional driving force due to the hydrophobic effect originating from the amphiphilicity of the β-sheet surfaces. Furthermore, the fibril formation is accompanied by an additional energetic penalty because of the required geometric adaptation: that is, the reduction of the twist angle. Finally, fibrils may form *fibers* via edge-to-edge attraction. An important conclusion from this model is that the inherent helical twisting of β-sheet-based fibrillar aggregates in solution will prohibit "unlimited" stacking. Thus, it is the interplay of aggregation enthalpy and the energetic penalty associated with bending and readjusting the helix geometry upon aggregation which is the main factor responsible for the formation of well-defined aggregates with a finite number of supramolecular strands and a finite width at every level of self-organization.

4.4
Macromonomer Design and Preparation

Based on the above findings concerning the supramolecular self-organization of oligopeptides and their polymer conjugates as well as the topochemical diacetylene polymerization using self-assembled scaffolds, we have chosen the following macromonomer design for our own investigations (Figure 4.5) [126, 127]:

1. A short oligo(L-alanine) segment was envisioned to induce self-assembly into high-aspect-ratio 1D aggregates in solution via the formation of β-sheets.
2. A hydrophobic, flexible polymer segment was designed to provide good solubility in organic solvents and prevent the formation of higher aggregates or even insoluble material resulting from stacking of the β-sheets. For this purpose, hydrogenated poly(isoprene) was chosen because of its inherent nonuniformity in chain length and constitution.
3. A diacetylene moiety served as the polymerizable unit and was incorporated into the hydrogen-bonding array via short flexible linkers such that minor changes of the geometry during the polymerization might be compensated.
4. Finally, different types of end groups were included in order to help control the hierarchical structure formation and introduce additional functionalities for specific applications at a later stage.

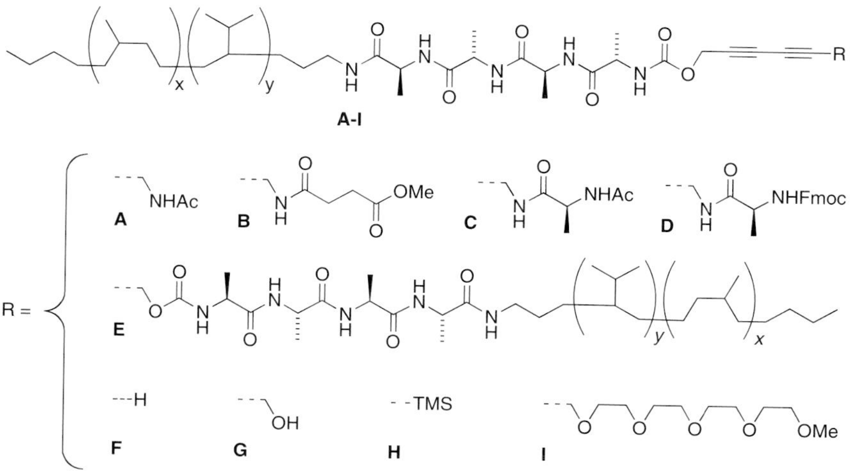

Figure 4.5 Macromonomers **A–I**; derivatives **A–E** have additional N—H⋯O=C hydrogen-bonding sites in their end groups, whereas **F–I** do not.

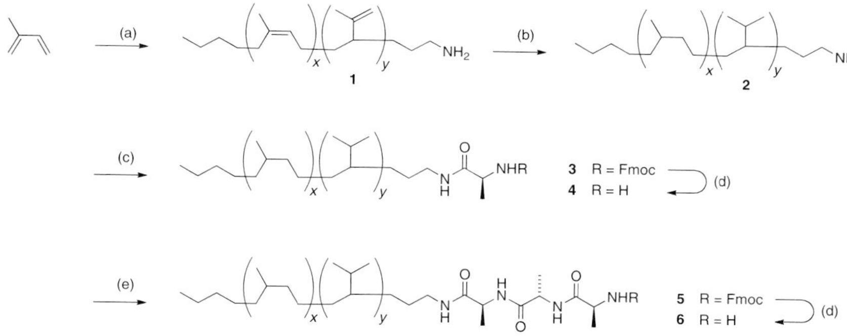

Scheme 4.1 Synthesis of oligopeptide conjugates. Reaction conditions and reagents: (a) n-BuLi, THF, −78 °C→0 °C; 1-(3-bromopropyl)-2,2,5,5-tetramethyl-1,1-aza-2,5-disilacyclopentane; THF/HCl; (b) H$_2$ (100 bar), Pd/C, toluene, 80 °C, 3 days; (c) Fmoc-Ala-OH, EDCI/HOBt, TEA, DCM, −40 °C→r.t.; (d) piperidine, chloroform; (e) Fmoc-Ala-Ala-OH, PyBOP, DIEA, DCM, r.t.

Amine-terminated poly(isoprene) **1** (Scheme 4.1) with a controlled molecular weight as well as narrow molecular weight distribution was prepared by living anionic polymerization of isoprene initiated with n-BuLi in THF at −78 °C to 0 °C and quenching with a protected 3-bromopropylamine derivative. The degree of polymerization was controlled by the monomer/initiator ratio, which was typically chosen to be 9:1 and led to an average degree of polymerization P_n = 9–10. A ratio of 1,4- to 1,2-addition of approximately 1:4 was usually observed. The olefin

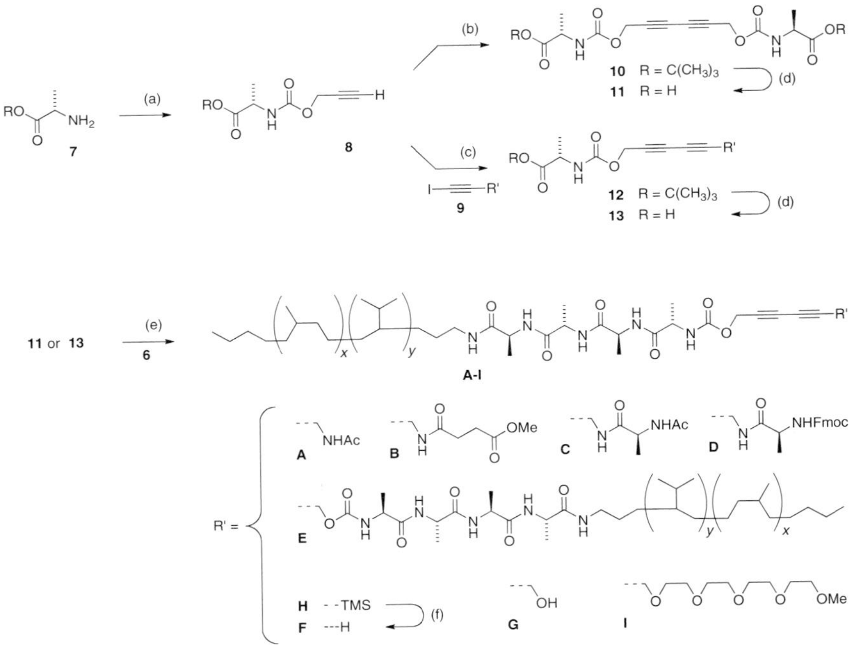

Scheme 4.2 Synthesis of macromonomers **A–I**. Reaction conditions and reagents: (a) POC-Cl, TEA, DCM; (b) O_2, CuCl, TMEDA, acetone; (c) Pd(Cl_2(PPh_3)$_2$ (2 mol%), CuI (10 mol%), DIPA, 0 °C, THF, N_2/H_2 atmosphere; (d) TFA, DCM; (e) EDCI/HOBt, TEA, DCM, −40 °C→r.t.; (f) TBAF, THF.

functions were then removed by high-pressure hydrogenation in order to avoid side reactions in subsequent steps, so that the hydrogenated poly(isoprene) **2** was obtained. The oligopeptide segment was subsequently introduced by repetitive solution-phase peptide coupling and deprotection protocols using Fmoc-Ala-OH and Fmoc-Ala-Ala-OH building blocks, yielding the intermediate **6** from which a variety of functionalized compounds was prepared (Scheme 4.1).

Thus, a series of differently functionalized iodoacetylene building blocks **9** was synthesized which were then coupled to the protected L-alanine N-propargyl carbamate **8** under Sonogashira conditions with acceptable to good yields (Scheme 4.2) [128]. Analogously, a homocoupling of **8** under Hay conditions afforded the symmetric diacetylene **10**. The diacetylene compounds **10** and **12** were then deprotected using TFA and afterwards coupled to intermediate **6** in a last peptide coupling reaction yielding the target macromonomers **A–E** and **G–I**. Finally, the hydrogen-terminated compound **F** was obtained from the TMS-protected derivative **H** by desilylation with TBAF.

In summary, the advantages of the chosen synthetic strategy are its modularity and the utilization of well-established and convenient peptide chemistry procedures. Thus, a variety of macromonomers could be prepared on the multi-gram

4.5
Hierarchical Self-Organization in Organic Solvents

scale beginning from a small number of building blocks and chemical transformations, and a further diversification of the target molecules is easily possible.

The aggregation of macromonomers **A–H** into secondary structures in organic solution was investigated by solution phase IR spectroscopy. It should be noted that detailed conclusions from IR spectra have to be drawn with some caution because of the conflicting and contradictory assignments of IR bands to protein secondary structures in the literature [129] and because of the aggregation in organic media in the present work. However, the following interpretations were independently confirmed by solid-state REDOR and DOQSY NMR experiments on ^{13}C- and ^{15}N-labeled compounds synthesized specifically for this purpose [130] and also unambiguously supported by the topochemical diacetylene polymerization results (see below). Several distinct regions in the IR spectra of oligopeptides contain information about the adopted secondary structures, that is, the amide A (v_{N-H}) absorptions at 3400–3200 cm^{-1}, the amide I ($v_{C=O}$) region located at 1700–1600 cm^{-1}, the amide II bands at 1550–1500 cm^{-1}, and the amide III range at 1300–1200 cm^{-1}. A thorough analysis of the observed absorptions in these regions revealed that all macromonomers showed the expected predominant formation of β-sheet-type secondary structures. Only the TMS functionalized macromonomer **H** and the oligo(ethylene glycol) equipped macromonomer **I** showed mixtures of β-sheets and other types of secondary structures. More importantly, a detailed comparison revealed systematic and distinct differences between macromonomers **A–E** that contained additional N—H $\cdots$ O=C type hydrogen bonding sites in their end groups, and the other compounds (Figure 4.6). Thus, an unusual combination of amide I and amide II bands in the IR spectra of macromonomers **A–E** led to the conclusion that the latter self-assembled into β-sheets exhibiting a *parallel* orientation of the β-strands [131]. The fact that the IR spectra of the symmetric dimer **E** exhibited the same features as those of **A–D** may be taken as further circumstantial evidence for this interpretation. By contrast, compounds **F** and **G** gave rise to IR spectra in excellent agreement with literature examples of *antiparallel* β-sheets.

This control of secondary structure formation exerted by the nature of the end groups is remarkable given the otherwise close structural relation of the different macromonomers. Generally speaking, the formation of *antiparallel* β-sheets should be preferred due to the more optimal hydrogen bond geometries, the cancellation of dipole moments, and the sterically more favorable alternating placement of the attached polymer segments. However, the role of the additional x N—H $\cdots$ O=C type hydrogen bonds in the end groups of **A–E** can be rationalized straightforwardly, since it leads to an overall *nonequidistant* placement of such hydrogen-bonding sites in the macromonomers. As a consequence, the maximum number

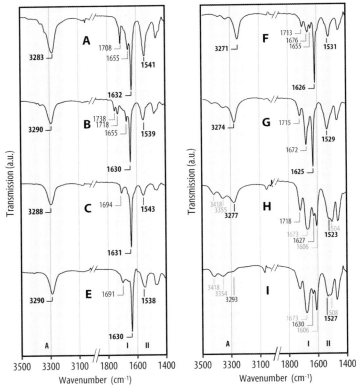

Figure 4.6 IR spectra of macromonomers **A–I**. (Left) Macromonomers **A–E** exhibit only one band for the amide A absorptions between 3283 and 3290 cm^{-1}, a very sharp and greatly predominating main amide I absorption at 1630–1632 cm^{-1} with a half-height width of about 11 cm^{-1}, and an amide II absorption at 1538–1543 cm^{-1}, attributed to predominantly *parallel* β-sheets. (Right) **F** and **G** exhibit amide I bands at 1625–1627 cm^{-1} and amide II absorptions at around 1530 cm^{-1} in agreement with *antiparallel* β-sheet aggregates; **H** and **I** gave rise to mixtures of β-sheet and other secondary structures. Reprinted with permission from [131].

of (5 + x) hydrogen bonds can only be achieved with a *parallel* β-strand orientation which appears to overcompensate the above factors [131].

The higher structure formation based on the β-sheet formation was investigated using transmission electron microscopy (TEM) and scanning force microscopy (SFM) [126, 127]. Both TEM and SFM images of samples of **A–E** revealed fibrillar features which extended over a few or even several dozens of micrometers, were remarkably straight, and appeared to have uniform diameters on the order of a few nanometers upon qualitative inspection (Figure 4.7). By contrast, no such features were observed in TEM images of **F–I**, that is, those molecules without hydrogen bonding sites in the end groups which aggregated via *antiparallel* β-sheet formation.

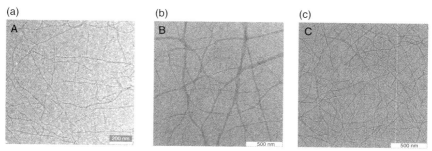

Figure 4.7 Representative examples of fibrillar features observed in TEM images of macromonomers **A-C**; no such features were observed in TEM images of compounds **F–I** [126].

Scanning force microscopy images of the NHAc terminated macromonomer **A** with $x = 1$ additional hydrogen bond in its end group revealed the presence of fibrils with an apparent height of about 5 nm and an estimated width on the order of 5–6 nm after correction for the SFM tip radius. These fibrillar features were found to be right-handed double-helices with a periodicity of about 18 nm that were formed from two smooth ribbonlike substructures (Figure 4.8a). The width of the latter was determined to be on the order of 10–14 nm, which would be approximately twice the extended length of **A**. SFM images of the NHSucOMe terminated macromonomer **B** showed very similar fibrillar structures which were even longer and more rigid (Figure 4.8b) [131]. Their apparent height and estimated width were both on the order of 6 nm, indicating a close to circular cross-section, and they were also right-handed helices with a periodicity of about 18 nm. The AlaNHAc-terminated macromonomer **C**, which comprises $x = 2$ additional hydrogen-bonding sites in its end group, gave rise to even more rigid fibrillar features with an apparent height and estimated width comparable to those from **A** and **B** [131]. However, the aggregates were uniform *left-handed* single-helices, and only occasionally formed (left-handed) double-helical bundles. These helices exhibited a complex fine structure with a periodicity of approximately 120 nm in which elevated "turns," reminiscent of the helical structure of **A** and **B** alternated with two longer flat segments separated by a second type of "twist" (Figure 4.8c). With the help of height profiles and phase images, the fibrillar features formed from **C** were tentatively interpreted as *bent and twisted* ribbons. Almost the same type of complex left-handed single-helical fibrils was observed in the case of **D** (not shown). By contrast, no helical aggregates but smooth, flat tapes with an apparent height of about 2.5 nm and an approximate width of 7 nm were observed in SFM images of the symmetric dimer **E** (Figure 4.8d). Their width was a little less than the extended length of one molecule, and phase images suggested that they had a dumbbell-shaped cross-section [131]. Finally, SFM images of derivatives **F–I** showed the presence of much shorter, thinner and comparably flexible fibrillar features, and also substantial amounts of

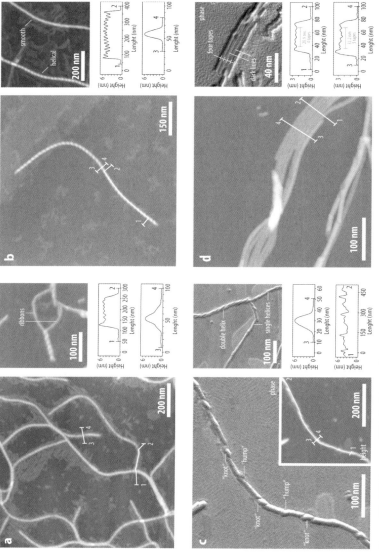

Figure 4.8 SFM images of (a) **A**, (b) **B**, and (c) **C** on a monolayer of tricontanoic acid on HOPG, as well as an SFM image of (d) **E** on HOPG. **A** formed right-handed double-helical fibrils with an apparent height of 5.8 (±0.4) nm and a periodicity of 18.1 (±1.0) nm. Very similar fibrils were obtained from **B**, with an apparent height of 5.7 (±0.4) nm and a periodicity of 18.6 (±0.9) nm. **C** gave rise to left-handed single-helices which only occasionally formed bundles and had a complex periodic fine structure with an identity period of 120 (±18) nm; by contrast, **E** formed flat tapes with an apparent height of 2.5 (±0.3) nm and an approximate width of about 7 nm. Reprinted with permission from [131].

nonfibrillar material (not shown). These results imply that derivatives **F–I** display a significantly smaller tendency to form fibrillar aggregates and that the latter are less stable than the fibrillar features obtained from macromonomers **A–E** which formed *parallel* β-sheet secondary structures.

In summary, stable fibrillar aggregates were obtained from all macromonomers **A–E**, which had end groups capable of N—H ··· O═C type hydrogen bonding. In all cases, these aggregates had a uniform diameter of a few nanometers, that is, on the same length scale as the molecular dimensions. Hence, they appeared to be well-defined supramolecular polymers (rather than micellar or vesicular structures) with the propensity to higher-order structure formation. The exact nature of the latter was found to depend on the number of additional N—H ··· O═C type hydrogen-bonding sites in the molecules' end groups.

4.6
A General Model for the Hierarchical Self-Organization of Oligopeptide–Polymer Conjugates

On the basis of the experimental findings, the formation of well-defined supramolecular polymers with a finite number of strands can be described in detail by extending the model for the hierarchical self-organization of oligopeptides in protic media proposed by Boden and Fishwick [122] (see above) toward the self-assembly of amphiphilic oligopeptide–polymer conjugates in organic solution [131]. In the latter case, hydrogen bonding responsible for β-sheet formation should be stronger due to the lack of competition with the solvent. At the same time, the mutual attraction of the tapes or ribbons ought to be noticeably reduced in the absence of the hydrophobic effect. Finally, the diacetylene moiety may be regarded as a nonpeptidic spacer which interrupts the hydrogen-bonding array, and the attachment of an amorphous, hydrophobic polymer segment introduces a novel element of phase separation and molecular disorder unknown in the realm of pure oligopeptides.

Hence, the formation of comparably stable single β-sheet *tapes* may be safely assumed as the first step in the self-organization of all macromonomers. In the case of compounds **F–I** (Figure 4.9b), which were found to favor an *antiparallel* aggregation, the resulting tapes are equally lined with the hydrophobic polymer segments grafted to both tape edges in an alternating fashion, and the oligopeptides' inherent dipole moment components in β-strand direction cancel each other out. Consequently, this mode of assembly may be referred to as *apolar*. The attached polymer segments are able to completely wrap the whole tape into a "hydrophobic cushion" and, thus, shield the polar peptide core from the hydrophobic medium. This renders any sort of β-sheet stacking into ribbons or other higher aggregates unfavorable because it would restrict the polymer segments grafted to the tape edges from exploring the space above and below the tapes. The required chain extension would only be favorable if the associated entropic penalty were overcompensated by an enthalpic contribution, for example, from a partial crystallization

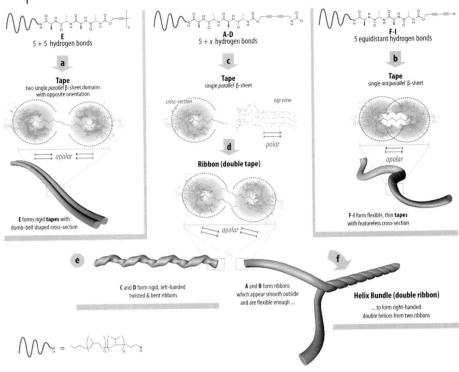

Figure 4.9 Model for the hierarchical self-organization of **A–I**. (a) **E** forms tapes comprising two covalently linked single *parallel* β-sheet domains with opposite orientation. (b) **F–I** aggregate into more flexible and less stable single *antiparallel* β-sheet tapes. (c) Tapes obtained by *parallel* β-sheet formation from **A–D** are *polar* structures which (d) further self-organize into ribbons. (e) Left-handed twisted and bent ribbons from **C** and **D** are more rigid in comparison to (f) ribbons from **A** and **B** which are flexible enough to form double-helical bundles. Reprinted with permission from [131].

of the polymer segments. However, the latter can be excluded for the chosen hydrogenated poly(isoprene) segments, which are unable to pack due to their non-uniformity in molecular weight, tacticity and branching. More illustratively speaking, the β-sheet surfaces and edges are "covered with grease" from the polymer segments, which prohibits further aggregation via stacking or edge-to-edge attraction. Thus, the decoration of the tapes with hydrophobic polymer segments plays a role which is, in a sense, similar to the electrostatic repulsion between tapes containing charged amino acid residues in protic environments, although the thermodynamic origin of the effect is different, that is, entropic in nature.

The symmetric dimer **E** had been shown to form *parallel* β-sheet secondary structures. Due to its symmetry, however, macromonomer **E** contains two oligopeptide segments with opposite directionality covalently attached to the

4.6 A General Model for the Hierarchical Self-Organization of Oligopeptide–Polymer Conjugates

central diacetylene unit, so that single β-sheet tapes would incorporate two separate *parallel* β-sheet domains with an *opposite* β-strand orientation (Figure 4.9a). As a result, the mode of aggregation is *apolar* in this case, as well, because the different domains' dipole moment components perpendicular to the tape axis cancel out and both tape edges are equally decorated with hydrophobic polymer segments. Accordingly, a further organization into higher structures is prohibited. Of course, the resulting tapes are wider than the ones formed by **F–I** and also substantially more rigid as a consequence of the increased (doubled) number of hydrogen bonds [131].

In the cases of macromonomers **A–D**, a more complex situation is encountered. The formation of single β-sheet tapes with a *parallel* β-strand orientation must result in a net dipole component perpendicular to the tape axis so that the mode of self-assembly must be regarded as *polar* (Figure 4.9c) [131]. Furthermore, only one tape edge is decorated with polymer chains whereas the other one is bare, leading to a steric mismatch. Finally, a simple molecular model helps to illustrate that the chosen chain length of $P_n \approx 10$ of the attached hydrogenated poly(isoprene) segments is too short to wrap the entire tapes and, thus, shield the polar cores from the hydrophobic medium. The combination of these factors is believed to be responsible for the formation of *ribbons* composed of two (partially) stacked tapes with opposite chain orientation (Figure 4.9d). This would also be in agreement with the experimentally observed width of the ribbon substructures in the double-helical aggregates on the order of twice the molecules' extended length. By way of this ribbon formation, the system is able to compensate the dipole moment components perpendicular to the tapes' axes and, additionally, wrap the whole ribbon structure with the hydrophobic polymer segments [131].

Obviously, these arguments equally hold for all macromonomers **A–D** but the observed fibrillar aggregates differed significantly. Macromonomers **C** and **D** gave rise to *left-handed* single-helical features with a complex fine structure which, based on the model proposed by Boden and Fishwick, may directly be interpreted as left-handed helically twisted and bent ribbons with an approximately ellipsoidal cross-section. These ribbons are held together by (5 + 2) hydrogen bonds per tape and are apparently too stiff to easily accommodate a further aggregation into fibrils (Figure 4.9e). By contrast, **A** and **B** formed *right-handed* double-helices constituted from two smooth ribbon substructures. Apparently, the smaller number of 5 + 1 hydrogen bonds provides enough flexibility to accommodate double helix formation (Figure 4.9f). The absence or presence of residual dipole moments in the tape direction as a consequence of the even and uneven numbers of hydrogen bonds may be an additional factor contributing to the observed differences between **A** and **B** as well as **C** and **D**, respectively.

While the double-helices formed from **A** and **B** may formally be regarded as *fibrils* because they are constructed from two ribbons, it is important to note that the formation mechanism would be entirely different from the one described for pure oligopeptides [122]. In the latter case, the stacking of β-sheets is the driving force for fibril formation, which implies that the helix sense of ribbons and fibrils must be identical. Whereas such β-sheet stacking was assumed to be suppressed by the soft polymer shells around the tapes, interdigitation of the latter still allows for the formation of *helix bundles*. The double-helices may, hence, be regarded as soft cylindrical structures with four helically wound β-sheet tapes inside. Such a loose helix bundle formation would not impose any predictable geometric constraints and the resulting superstructure may, therefore, be left- or right-handed. The presumed helix sense inversion on different levels of a hierarchical structure (i.e. right-handed β-strands, left-handed ribbons, right-handed fibrils) is well in line with literature examples [32, 132].

4.7
Conversion to Conjugated Polymers by UV Irradiation

The topochemical diacetylene polymerization does not serve only to convert the supramolecular polymers into covalent, conjugated polymers. It is also a sensitive probe for the internal structure of the β-sheet secondary structures present in solution and, thus, helps to confirm the interpretation of the IR spectra presented above. Only in the case of a *parallel* β-strand orientation are the diacetylenes aligned with a distance and geometry appropriate for the formation of a poly(diacetylene), which ought to be accompanied with a dramatic color change from colorless to purple or red and give rise to characteristic UV spectra with absorption maxima at 500–600 nm. Accordingly, dilute solutions of **A–E** in DCM (i.e. all macromonomers with additional N—H $\cdots$ O=C type hydrogen bonding sites in their end groups) showed the expected color change within a few seconds of UV irradiation (Figure 4.10b, c). UV and Raman spectra, as well as solid-state CP-MAS ^{13}C NMR spectra provided the evidence for successful conversion into poly(diacetylene)s. Interestingly, the obtained polymeric material did not precipitate from solution or form gels, and the polymers could even be dried and easily redissolved in chlorinated solvents. By contrast, solutions or gels of the compounds **F–I** in DCM showed only a color change to yellow upon prolonged UV irradiation. Not surprisingly, the corresponding UV spectra of these derivatives, which had been found to favor an antiparallel β-sheet type aggregation, showed no sign of poly(diacetylene) formation (Figure 4.10a) [126, 127]. Furthermore, it turned out that not only the polymerizability was controlled by the absence or presence of additional N—H $\cdots$ O=C type hydrogen bonding sites; their total number also appeared to determine the reactivity in the topochemical diacetylene polymerization (Figure 4.10d). Thus, plots of the maximum extinction coefficients ε_{max} as a function of irradiation time revealed a clear structure–reactivity relationship, with **E** (5 + 5 hydrogen bonds) being most reactive, followed by **D** and **C** (5 + 2 hydrogen bonds), as well as, finally, **B** and **A** (5 + 1 hydrogen bonds) [131].

Scanning force microscopy imaging provided further evidence for the successful conversion of the supramolecular polymers into covalent, conjugated polymers with retention of their hierarchical structure. First of all, SFM images obtained from any of the polymerizable macromonomers **A–E** looked virtually identical before and after polymerization. However, while the addition of a small amount of a deaggregating cosolvent such as hexafluoroisopropanol (HFIP) to the sample

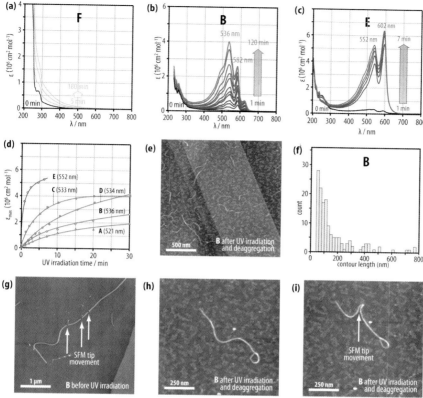

Figure 4.10 UV-induced topochemical polymerization. (a) UV spectra of compounds **F–I** did not show any indication of poly(diacetylene) formation. (b, c) Macromonomers **A–D** were successfully converted into poly(diacetylene)s. (d) The reactivity toward topochemical polymerization was related to the total number of hydrogen bonds. (e) Upon addition of HFIP to a sample solution in DCM after UV irradiation, helical fibrillar features of up to several hundred nanometers in length remained intact. (f) In the case of **B**, their average contour length after 2 hours of UV irradiation was about 106 nm. (g) Manipulation with the SFM tip destroyed the non-irradiated self-assembled fibrils from **B**; but (h,i) SFM tip manipulation was possible after UV-induced polymerization. Reprinted with permission from [131].

solutions in DCM led to a complete disappearance of fibrillar structures in all cases before UV irradiation, intact fibril sections with helical fine structure and an average contour length above 100 nm remained after UV irradiation (Figure 4.10e) [131]. Furthermore, these fibrils could be manipulated with the SFM tip, which had only destroyed the self-assembled fibrils before UV induced polymerization (Figures 4.10g–i). Consequently, it must be concluded that the tapes, ribbons, and helix bundles formed from macromonomers **A–E** had been converted into uniform conjugated polymers with a single-stranded, double-helical, or quadruple-helical quaternary structure, respectively (Figure 4.11).

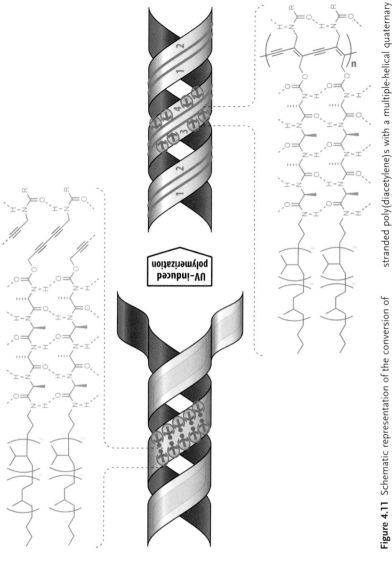

Figure 4.11 Schematic representation of the conversion of the supramolecular polymers into conjugated polymers under retention of the hierarchical structure, leading to four-stranded poly(diacetylene)s with a multiple-helical quaternary structure in the case of **A**.

4.8
Conclusions and Perspectives

In summary, the reliable formation of well-defined, high-molecular-weight supramolecular polymers from the chosen amphiphilic oligopetide-polymer conjugates can be attributed to the interplay of the crystallization enthalpy of the oligopeptides, the elastic energy of helix pitch adjustment [122], and the chain extension entropy of the grafted polymer segments. The exact nature of the observed hierarchical structures, on the other hand, was found to be controlled by a simple set of parameters, that is, the nonequidistant placement of N—H $\cdots$ O=C type hydrogen bonds, their distribution, and their total number. Thus, the subsequent conversion of the supramolecular polymers into poly(diacetylene)s under retention of their higher-order structures offers a rational pathway toward conjugated polymers with a single-stranded, double-helical, or quadruple-helical quaternary structure. Such "twisted pair"-like molecular wires offer a platform for investigation of intermolecular interactions between individual conjugated polymer strands and a means to control their optoelectronic properties. Furthermore, with detailed understanding of the underlying self-assembly process, it will be possible to prepare multifunctional conjugated polymers for sensing applications via co-assembly of different macromonomers. Substitution of the aliphatic polymer segments with, for example, poly(ethylene oxide) may provide a pathway to similar biocompatible polymers, which would represent the next step toward organic optoelectronic materials aimed at applications in the life sciences.

4.9
Acknowledgments

The authors would like to thank Paul Kreutzkamp, Dr. Nikolai Severin, and Prof. Jürgen P. Rabe (Humboldt University Berlin, Germany) for fruitful collaboration, Dr. Jürgen Hartmann (Max Planck Institute of Colloids and Interfaces, Golm, Germany) for TEM measurements, and Prof. A. Dieter Schlüter (ETH Zurich, Switzerland) for his generous support. Funding from Fonds der Chemischen Industrie (Fonds-Stipendium, E. Jahnke), Schweizerischer Nationalfonds (SNF-Projekt 200021-113509) is gratefully acknowledged.

List of Abbreviations

CVD	chemical vapor deposition
OPV	oligo(phenylene vinylene)
OT	oligo(thiophene)
SFM	scanning force microscopy
CD	circular dichroism
LC	liquid crystal, liquid-crystalline

PEO	poly(ethylene oxide)
TEM	transmission electron microscopy
THF	tetrahydrofuran
DCM	dichloromethane
DCE	dichloroethane
Ala	alanine
Fmoc	fluorenylmethyloxycarbonyl
TEA	triethylamine
DIEA	diethylisopropylamine
DIPA	diisopropylamine
EDCI	1-ethyl-3-(3-dimethylaminopropyl)carbodiimide
HOBt	1-hydroxy-1H-benzotriazol
PyBOP	1H-benzotriazol-1-yloxytris(pyrrolidino)phosphonium hexafluorophosphate
POC	propargyloxycarbonyl
TMEDA	tetramethyl ethylene diamine
TFA	trifluoroacetic acid
TBAF	tetrabutylammonium fluoride
TMS	trimethylsilyl
IR	infrared
UV	ultraviolet
REDOR	rotational echo double resonance (NMR spectroscopy)
DOQSY	double-quantum (NMR) spectroscopy
Suc	succinyl
HOPG	highly oriented pyrolytic graphite
CP-MAS	cross-polarization magic-angle spinning (NMR spectroscopy)

References

1 Tirrell, D. (1994) Hierarchical Structures in Biology as a Guide for New Materials Technology, Tech Report, Washington.
2 Muthukumar, M., Ober, C.K. and Thomas, E.L. (1997) *Science*, **277**, 1225.
3 Lakes, R. (1993) *Nature*, **361**, 511.
4 Hawker, C.J. and Wooley, K.L. (2005) *Science*, **309**, 1200.
5 Meijer, E.W. and Schenning, A.P.H.J. (2002) *Nature*, **419**, 353.
6 Hoeben, F.J.M., Jonkheijm, P., Meijer, E.W. and Schenning, A.P.H.J. (2005) *Chemical Reviews*, **105**, 1491.
7 Martin, R.E. and Diederich, F. (1999) *Angewandte Chemie (International Ed. in English)*, **38**, 1351.
8 Gelinck, G.H., Huitema, H.E.A., van Veenendaal, E., Cantatore, E., Schrijnemakers, L., Van der Putten, J.B.P.H., Geuns, T.C.T., Beenhakkers, M., Giesbers, J.B., Huisman, B.-H., Meijer, E.J., Benito, E.M., Touwslager, F.J., Marsman, A.W., Van Rens, B.J.E. and de Leeuw, D.M. (2004) *Nature Materials*, **3**, 106.
9 Iwaura, R., Hoeben, F.J.M., Masuda, M., Schenning, A.P.H.J., Meijer, E.W. and Shimizu, T. (2006) *Journal of the American Chemical Society*, **128**, 13298.
10 Sinkeldam, R.W., Hoeben, F.J.M., Pouderoijen, M.J., De Cat, I., Zhang, J., Furukawa, S., De Feyter, S., Vekemans, J.A.J.M. and Meijer, E.W. (2006) *Journal of*

the American Chemical Society, **128**, 16113.

11 Schenning, A.P.H.J., Jonkheijm, P., Peeters, E. and Meijer, E.W. (2001) *Journal of the American Chemical Society*, **123**, 409.

12 Jonkheijm, P., Hoeben, F.J.M., Kleppinger, R., Van Herrikhuyzen, J., Schenning, A.P.H.J. and Meijer, E.W. (2003) *Journal of the American Chemical Society*, **125**, 15941.

13 Würthner, F., Chen, Z., Hoeben, F.J.M., Osswald, P., You, C.-C., Jonkheijm, P., von Herrikhuyzen, J., Schenning, A. P.H.J., van der Schoot, P.P.A.M., Meijer, E.W., Beckers, E.H.A., Meskers, S.C.J. and Janssen, R.A.J. (2004) *Journal of the American Chemical Society*, **126**, 10611.

14 Schenning, A.P.H.J., van Herrikhuyzen, J., Jonkheijm, P., Chen, Z., Würthner, F. and Meijer, E.W. (2002) *Journal of the American Chemical Society*, **124**, 10252.

15 Sirringhaus, H., Tessler, N. and Friend, R.H. (1998) *Science*, **280**, 1741.

16 Kilbinger, A.F.M., Schenning, A.P.H.J., Goldoni, F., Feast, W.J. and Meijer, E. W. (2000) *Journal of the American Chemical Society*, **122**, 1820.

17 Schenning, A.P.H.J., Kilbinger, A.F.M., Biscarini, F., Cavallini, M., Cooper, H. J., Derrick, P.J., Feast, W.J., Lazzaroni, R., Leclere, P., McDonell, L.A., Meijer, E.W. and Meskers, S.C.J. (2002) *Journal of the American Chemical Society*, **124**, 1269.

18 Leclere, P., Surin, M., Viville, P., Lazzaroni, R., Kilbinger, A.F.M., Henze, O., Feast, W.J., Cavallini, M., Biscarini, F., Schenning, A.P.H.J. and Meijer, E.W. (2004) *Chemistry of Materials*, **16**, 4452.

19 Surin, M., Lazzaroni, R., Feast, W.J., Schenning, A.P.H.J., Meijer, E.W. and Leclere, P. (2004) *Synthetic Metals*, **147**, 67.

20 Westenhoff, S., Abrusci, A., Feast, W.J., Henze, O., Kilbinger, A.F.M., Schenning, A.P.H.J. and Silva, C. (2006) *Advanced Materials*, **18**, 1281.

21 Fujita, S.-I. Kawano, N. and Shinkai, S. (2005) *Chemistry – A European Journal*, **11**, 4735.

22 Kilbinger, A.F.M. and Feast, W.J. (2000) *Journal of Materials Chemistry*, **10**, 1777.

23 Henze, O., Fransen, M., Jonkheijm, P., Meijer, E.W., Feast, W.J. and Schenning, A.P.H.J. (2003) *Journal of Polymer Science. Part A: Polymer Chemistry*, **41**, 1737.

24 Henze, O., Parker, D. and Feast, W.J. (2003) *Journal of Materials Chemistry*, **13**, 1269.

25 Henze, O. and Feast, W.J. (2003) *Journal of Materials Chemistry*, **13**, 1274.

26 Cik, G., Vegh, Z., Sersen, F., Kristin, J., Lakatos, B. and Fejdi, P. (2005) *Synthetic Metals*, **149**, 31.

27 Li, W., Maddux, T. and Lu, Y. (1996) *Macromolecules*, **29**, 7329.

28 Hempenius, M.A., Langeveld-Voss, B.M.W., van Haare, J.A.E.H., Janssen, R.A.J., Sheiko, S.S., Spatz, J.P., Möller, M. and Meijer, E.W. (1998) *Journal of the American Chemical Society*, **120**, 2798.

29 Liu, J., Sheina, E., Kowalewski, T. and McCullough, R.D. (2002) *Angewandte Chemie (International Ed. in English)*, **41**, 329.

30 Cornelissen, J.J.L.M., Donners, J.J.J.M., de Gelder, R., Graswinckel, W.S., Metselaar, G.A., Rowan, A.E., Sommerdijk, N.A.J.M. and Nolte, R.J.M. (2001) *Science*, **293**, 676.

31 Samori, P., Ecker, C., Goessl, I., de Witte, P.A.J., Cornelissen, J.J.L.M., Metselaar, G. A., Otten, M.B.J., Rowan, A.E., Nolte, R. J.M. and Rabe, J.P. (2002) *Macromolecules*, **35**, 5290.

32 Cornelissen, J.J.L.M., Fischer, M., Sommerdijk, N.A.J.M. and Nolte, R.J.M. (1998) *Science*, **280**, 1427.

33 Holdcroft, S. (2001) *Advanced Materials*, **13**, 1753.

34 Dimitrakopoulos, C.D. and Malenfant, P.R.L. (2002) *Advanced Materials*, **14**, 99.

35 Gellman, S.H. (1998) *Accounts of Chemical Research*, **31**, 173.

36 Hill, D.J., Mio, M.J., Prince, R.B., Hughes, T.S. and Moore, J.S. (2001) *Chemical Reviews*, **101**, 3893.

37 Nakano, T. and Okamoto, Y. (2001) *Chemical Reviews*, **101**, 4013.

38 Cornelissen, J.J.L.M., Graswinckel, W.S., Adams, P.J.H.M., Nachtegaal, G.H.,

Kentgens, A.P.M., Sommerdijk, N.A.J.M. and Nolte, R.J.M. (2001) *Journal of Polymer Science. Part A: Polymer Chemistry*, **39**, 4255.

39 Cornelissen, J.J.L.M., Graswinckel, W.S., Rowan, A.E., Sommerdijk, N.A.J.M. and Nolte, R.J.M. (2003) *Journal of Polymer Science. Part A: Polymer Chemistry*, **41**, 1725.

40 Metselaar, G.A., Cornelissen, J.J.L.M., Rowan, A.E. and Nolte, R.J.M. (2005) *Angewandte Chemie (International Ed. in English)*, **44**, 1990.

41 Kros, A., Jesse, W., Metselaar, G.A. and Cornelissen, J.J.L.M. (2005) *Angewandte Chemie (International Ed. in English)*, **44**, 4349.

42 Zhang, H.-C. and Pu, L. (2003) *Tetrahedron*, **59**, 1703.

43 Pu, L. (2000) *Macromolecular Rapid Communications*, **21**, 795.

44 Zheng, L., Urian, R.C., Liu, Y., Jen, A.K.Y. and Pu, L. (2000) *Chemistry of Materials*, **12**, 13.

45 Pu, L. (1999) *Chemistry – A European Journal*, **5**, 2227.

46 Pu, L. (1998) *Chemical Reviews*, **98**, 2405.

47 Langeveld-Voss, B.M.W., Janssen, R.A.J., Christiaans, M.P.T., Meskers, S.C.J., Dekkers, H.P.J.M. and Meijer, E.W. (1996) *Journal of the American Chemical Society*, **118**, 4908.

48 Langeveld-Voss, B.M.W., Christiaans, M.P.T., Janssen, R.A.J. and Meijer, E.W. (1998) *Macromolecules*, **31**, 6702.

49 Langeveld-Voss, B.M.W., Waterval, R.J.M., Janssen, R.A.J. and Meijer, E.W. (1999) *Macromolecules*, **32**, 227.

50 Lermo, E.R., Langeveld-Voss, B.M.W., Janssen, R.A.J. and Meijer, E.W. (1999) *Chemical Communications*, 791.

51 Matthews, J.R., Goldoni, F., Schenning, A.P.H.J. and Meijer, E.W. (2005) *Chemical Communications*, 5503.

52 Yashima, E., Matsushima, T. and Okamoto, Y. (1997) *Journal of the American Chemical Society*, **119**, 6345.

53 Yashima, E., Maeda, K. and Okamoto, Y. (1999) *Nature*, **399**, 449.

54 Nomura, R., Tabei, J. and Masuda, T. (2001) *Journal of the American Chemical Society*, **123**, 8430.

55 Gao, G., Sanda, F. and Masuda, T. (2003) *Macromolecules*, **36**, 3932.

56 Gao, G., Sanda, F. and Masuda, T. (2003) *Macromolecules*, **36**, 3938.

57 Zhao, H., Sanda, F. and Masuda, T. (2004) *Macromolecules*, **37**, 8888.

58 Tabei, J., Shiotsuki, M., Sanda, F. and Masuda, T. (2005) *Macromolecules*, **38**, 9448.

59 Sanda, F., Araki, H. and Masuda, T. (2004) *Macromolecules*, **37**, 8510.

60 Sanda, F., Araki, H. and Masuda, T. (2005) *Macromolecules*, **38**, 10605.

61 Sanda, F., Nishiura, S., Shiotsuki, M. and Masuda, T. (2005) *Macromolecules*, **38**, 3075.

62 Zhao, H., Sanda, F. and Masuda, T. (2006) *Macromolecular Chemistry and Physics*, **207**, 1921.

63 Yue, D., Fujii, T., Terada, K., Tabei, J., Shiotsuki, M., Sanda, F. and Masuda, T. (2006) *Macromolecular Rapid Communications*, **27**, 1460.

64 Yue, D., Shiotsuki, M., Sanda, F. and Masuda, T. (2007) *Polymer*, **48**, 68.

65 Sanda, F., Teraura, T. and Masuda, T. (2004) *Journal of Polymer Science: Part A, General Papers*, **42**, 4641.

66 Zhao, H., Sanda, F. and Masuda, T. (2006) *Polymer*, **47**, 2596.

67 Li, B.S., Cheuk, K.K.L., Ling, L., Chen, J., Xiao, X., Bai, C. and Tang, B.Z. (2003) *Macromolecules*, **36**, 77.

68 Cheuk, K.K.L., Lam, J.W.Y., Lai, L.M., Dong, Y. and Tang, B.Z. (2003) *Macromolecules*, **36**, 9752.

69 Li Bing, S., Kang Shi, Z., Cheuk Kevin, K.L., Wan, L., Ling, L., Bai, C. and Tang Ben, Z. (2004) *Langmuir*, **20**, 7598.

70 Percec, V., Aqad, E., Peterca, M., Rudick, J.G., Lemon, L., Ronda, J.C., De, B.B., Heiney, P.A. and Meijer, E.W. (2006) *Journal of the American Chemical Society*, **128**, 16365.

71 Sakai, R., Otsuka, I., Satoh, T., Kakuchi, R., Kaga, H. and Kakuchi, T. (2006) *Macromolecules*, **39**, 4032.

72 Frauenrath, H. (2005) *Progress in Polymer Science*, **30**, 325.

73 Mueller, A. and O'Brien, D.F. (2002) *Chemical Reviews*, **102**, 727.

References

74 Wegner, G. (1969) *Zeitschrift Für Naturforschung Teil B: Chemie, Biochemie, Biophysik, Biologie* **24**, 824.
75 Wegner, G. (1971) *Die Makromolekulare Chemie*, **145**, 85.
76 Wegner, G. (1972) *Die Makromolekulare Chemie*, **154**, 35.
77 Enkelmann, V. and Wegner, G. (1977) *Angewandte Chemie (International Ed. in English)*, **89**, 416.
78 Enkelmann, V., Leyrer, R.J. and Wegner, G. (1979) *Die Makromolekulare Chemie*, **180**, 1787.
79 Enkelmann, V. (1984) *Advances in Polymer Science*, **63**, 91.
80 Tieke, B., Wegner, G., Naegele, D. and Ringsdorf, H. (1976) *Angewandte Chemie (International Ed. in English)*, **15**, 764.
81 Day, D. and Ringsdorf, H. (1978) *Journal of Polymer Science. Part B: Polymer Letters*, **16**, 205.
82 Nezu, S. and Lando, J.B. (1995) *Journal of Polymer Science. Part A-1, Polymer Chemistry*, **33**, 2455.
83 Britt, D.W., Hofmann, U.G., Moebius, D. and Hell, S.W. (2001) *Langmuir*, **17**, 3757.
84 Batchelder, D.N., Evans, S.D., Freeman, T.L., Haeussling, L., Ringsdorf, H. and Wolf, H. (1994) *Journal of the American Chemical Society*, **116**, 1050.
85 Charych, D.H., Nagy, J.O., Spevak, W. and Bednarski, M.D. (1993) *Science*, **261**, 585.
86 Okawa, Y. and Aono, M. (2001) *Nature*, **409**, 683.
87 Miura, A., De Feyter, S., Abdel-Mottaleb, M.M.S., Gesquiere, A., Grim, P.C.M., Moessner, G., Sieffert, M., Klapper, M., Müllen, K. and De Schryver, F.C. (2003) *Langmuir*, **19**, 6474.
88 Sullivan, S.P., Schnieders, A., Mbugua, S.K. and Beebe, T.P. (2005) *Langmuir*, **21**, 1322.
89 Chang, J.Y., Baik, J.H., Lee, C.B., Han, M.J. and Hong, S.-K. (1997) *Journal of the American Chemical Society*, **119**, 3197.
90 Chang, J.Y., Yeon, J.R., Shin, Y.S., Han, M.J. and Hong, S.-K. (2000) *Chemistry of Materials*, **12**, 1076.
91 Lee, S.J., Park, C.R. and Chang, J.Y. (2004) *Langmuir*, **20**, 9513.
92 Georger, J.H., Singh, A., Price, R.R., Schnur, J.M., Yager, P. and Schoen, P.E. (1987) *Journal of the American Chemical Society*, **109**, 6169.
93 Singh, A., Thompson, R.B. and Schnur, J.M. (1986) *Journal of the American Chemical Society*, **108**, 2785.
94 Okada, S., Peng, S., Spevak, W. and Charych, D. (1998) *Accounts of Chemical Research*, **31**, 229.
95 Jonas, U., Shah, K., Norvez, S. and Charych, D.H. (1999) *Journal of the American Chemical Society*, **121**, 4580.
96 Lee, S.B., Koepsel, R., Stolz, D.B., Warriner, H.E. and Russell, A.J. (2004) *Journal of the American Chemical Society*, **126**, 13400.
97 Svenson, S. and Messersmith, P.B. (1999) *Langmuir*, **15**, 4464.
98 Cheng, Q., Yamamoto, M. and Stevens, R.C. (2000) *Langmuir*, **16**, 5333.
99 Frankel, D.A. and O'Brien, D.F. (1991) *Journal of the American Chemical Society*, **113**, 7436.
100 Fuhrhop, J.H., Blumtritt, P., Lehmann, C. and Luger, P. (1991) *Journal of the American Chemical Society*, **113**, 7437.
101 George, M. and Weiss, R.G. (2003) *Chemistry of Materials*, **15**, 2879.
102 Bhattacharya, S. and Acharya, S.N.G. (1999) *Chemistry of Materials*, **11**, 3121.
103 Masuda, M., Hanada, T., Okada, Y., Yase, K. and Shimizu, T. (2000) *Macromolecules*, **33**, 9233.
104 Dautel, O.J., Robitzer, M., Lere-Porte, J.P., Serein-Spirau, F. and Moreau, J.J.E. (2006) *Journal of the American Chemical Society*, **128**, 16213.
105 Yuan, Z., Lee, C.-W. and Lee, S.-H. (2004) *Angewandte Chemie (International Ed. in English)*, **43**, 4197.
106 Brunsveld, L., Folmer, B.J.B., Meijer, E.W. and Sijbesma, R.P. (2001) *Chemical Reviews*, **101**, 4071.
107 Binder, W.H. and Smrzka, O.W. (2006) *Angewandte Chemie (International Ed. in English)*, **45**, 7324.
108 Gilead, S. and Gazit, E. (2005) *Supramolecular Chemistry*, **17**, 87.
109 Lashuel, H.A., LaBrenz, S.R., Woo, L., Serpell, L.C. and Kelly, J.W. (2000)

Journal of the American Chemical Society, **122**, 5262.

110 Selkoe, D.J. (2003) *Nature*, **426**, 900.

111 Makin, O.S., Atkins, E., Sikorski, P., Johansson, J. and Serpell, L.C. (2005) *Proceedings of the National Academy of Sciences of the United States of America*, **102**, 315.

112 Scheibel, T., Parthasarathy, R., Sawicki, G., Lin, X.-M., Jaeger, H. and Lindquist, S.L. (2003) *Proceedings of the National Academy of Sciences of the United States of America*, **100**, 4527.

113 Djalali, R., Chen, Y.-f. and Matsui, H. (2003) *Journal of the American Chemical Society*, **125**, 5873.

114 Collier, J.H. and Messersmith, P.B. (2004) *Advanced Materials*, **16**, 907.

115 ten Cate, M.G.J., Severin, N. and Boerner, H.G. (2006) *Macromolecules*, **39**, 7831.

116 Winningham, M.J. and Sogah, D.Y. (1997) *Macromolecules*, **30**, 862.

117 Rathore, O. and Sogah, D.Y. (2001) *Journal of the American Chemical Society*, **123**, 5231.

118 Rathore, O. and Sogah, D.Y. (2001) *Macromolecules*, **34**, 1477.

119 Yao, J., Xiao, D., Chen, X., Zhou, P. Yu, T. and Shao, Z. (2003) *Macromolecules*, **36**, 7508.

120 Aggeli, A., Bell, M., Boden, N., Keen, J.N., McLeish, T.C.B., Nyrkova, I., Radford, S.E. and Semenov, A. (1997) *Journal of Materials Chemistry*, **7**, 1135.

121 Aggeli, A., Bell, M., Boden, N., Keen, J.N., Knowles, P.F., McLeish, T.C., Pitkeathly, M. and Radford, S.E. (1997) *Nature*, **386**, 259.

122 Aggeli, A., Nyrkova, I.A., Bell, M., Harding, R., Carrick, L., McLeish, T.C.B., Semenov, A.N. and Boden, N. (2001) *Proceedings of the National Academy of Sciences of the United States of America*, **98**, 11857.

123 Fishwick, C.W.G., Beevers, A.J., Carrick, L.M., Whitehouse, C.D., Aggeli, A. and Boden, N. (2003) *Nano Letters*, **3**, 1475.

124 Whitehouse, C., Fang, J., Aggeli, A., Bell, M., Brydson, R., Fishwick Colin, W.G., Henderson Jim, R., Knobler Charles, M., Owens Robert, W., Thomson Neil, H., Smith, D.A. and Boden, N. (2005) *Angewandte Chemie (International Ed. in English)*, **44**, 1965.

125 Eckhardt, D., Groenewolt, M., Krause, E. and Börner, H.G. (2005) *Chemical Communications*, 2814.

126 Jahnke, E., Lieberwirth, I., Severin, N., Rabe, J.P. and Frauenrath, H. (2006) *Angewandte Chemie (International Ed. in English)*, **45**, 5383.

127 Jahnke, E., Millerioux, A.-S., Severin, N., Rabe, J.P. and Frauenrath, H. (2007) *Macromolecular Bioscience*, **7**, 136.

128 Jahnke, E., Fesser, P., Hoheisel, T., Weiss, J. and Frauenrath, H. Manuscript in preparation.

129 For good reviews of this issue, see Singh, B.R. (2000) in *Infrared Analysis of Peptides and Proteins* (ed. B.R. Singh), ACS Symposium Series, **750**, pp. 2-37; S. Krimm, *ibid.* 38-53; P.I. Haris, *ibid.* 54-95.

130 Jahnke, E., van Beek, J., Verel, R., Arnold, A., Meier, B. and Frauenrath, H. Manuscript in preparation.

131 Jahnke, E., Kreutzkamp, P., Severin, N., Rabe, J.P., Frauenrath, H. *Advanced Materials* in press.

132 Malashkevich, V.N., Kammerer, R.A., Efimov, V.P., Schulthess, T. and Engel, J. (1996) *Science*, **274**, 761.

5
Mimicking Nature: Bio-inspired Models of Copper Proteins
Iryna A. Koval, Patrick Gamez and Jan Reedijk

5.1
Environmental Pollution: How Can "Green" Chemistry Help?

The rapid technological development in the twentieth century has drastically changed our lifestyle in recent years. We can travel between continents in less than 24 hours, share and receive information within seconds, and use machines to assist us in our work, and our daily lives cannot be imagined without synthetic chemical materials – plastics, polymers, and ceramics. However, these inventions have brought a major drawback with them: environmental pollution and loss of sustainability is becoming an increasingly important issue, which, unless addressed urgently, may well have disastrous consequences in the future. Chemical plants, motor vehicles, oil refineries, power plants, and heavy industry all contribute to the contamination of air, soil and water with various chemicals, many of which are very harmful to living organisms. These pollutants can also be the cause of many diseases, such as cancers, immune diseases, allergies and asthma.

For these reasons, much effort in modern chemical research is put into the search for new environmentally friendly processes for industrial applications, which are necessary for sustainable development of industrial chemistry. One way to do this is to look at how Nature performs various biotransformations and use this knowledge as inspiration to develop new bio-inspired and clean procedures. The principle of learning from Nature, often referred to as "biomimetism," dates back many years: for example, Leonardo da Vinci used detailed studies of the flight of birds for his design of flying machines. Nowadays, a discipline called "biomimetic chemistry" deals with the development of synthetic models of various biomolecules, for example, proteins and enzymes. In particular, the modeling of the latter compounds has received significant attention in the past years, as enzymes are used by Nature to catalyze a vast variety of chemical reactions and perform these conversions highly selectively and with high yields. Therefore, the understanding of their functionality and structure–activity relationships is believed to be of great importance for the development of new "green" catalysts for use in the chemical industry.

Tomorrow's Chemistry Today. Concepts in Nanoscience, Organic Materials and Environmental Chemistry.
Edited by Bruno Pignataro
Copyright © 2008 WILEY-VCH Verlag GmbH & Co. KGaA, Weinheim
ISBN: 978-3-527-31918-3

This chapter will outline our research on synthetic model compounds of a copper-containing enzyme called catechol oxidase, and on studies aimed at understanding its mechanism of action. The ability of catechol oxidase to process molecular dioxygen at ambient conditions and to utilize it to perform the selective oxidation of catechols (o-diphenols) makes this enzyme a very interesting candidate for biomimetic studies. In addition, a general overview of different types of active sites found in copper proteins and their functions in living organisms, as well as highlights of other studies devoted to the modeling of this enzyme in the past will be presented in this chapter.

5.2
Copper in Living Organisms

Copper is present in many proteins, occurring in almost all known living organisms. In biological systems, copper is the third most abundant transition-metal element after iron and zinc. Copper-containing proteins are usually involved as redox catalysts in a range of biological processes, such as electron transfer or oxidation of various organic substrates. In general, four major functions of such proteins can be distinguished: (i) metal ion uptake, storage, and transport; (ii) electron transfer; (iii) dioxygen uptake, storage, and transport; and (iv) catalysis.

For these types of reactivity, the copper ions in the proteins usually change their oxidation state from I to II and vice versa. The coordination geometries adopted by copper ions vary with the oxidation state. For instance, Cu^I ions prefer linear, trigonal, and tetragonal geometries, whereas Cu^{II} ions prefer square planar, trigonal bipyramidal, and tetragonal/octahedral geometries. The coordination geometries of the copper ions in proteins are usually in between these preferences, dictated by the rigid protein backbone and side-chains.

Before they were characterized by x-ray crystallography, the classification of the structures of copper proteins was initially based on the spectroscopic features of their active site in the oxidized state. The tremendous development of crystallographic and spectroscopic techniques in recent years has enabled the identification of as many as seven different types of active sites in these proteins: type 1, type 2, type 3, type 4, Cu_A, Cu_B and Cu_Z. The characteristics of these metal sites are briefly described below.

5.2.1
Type 1 Active Site

The copper proteins with a type 1 active site are commonly known as "blue copper proteins" due to their intense blue color in the Cu^{II} state. They are usually participants in electron transfer processes, and the best-known representatives of this class include plastocyanin, azurin and amicyanin [1]. The copper center in the type 1 active site is surrounded by two nitrogen donor atoms from two

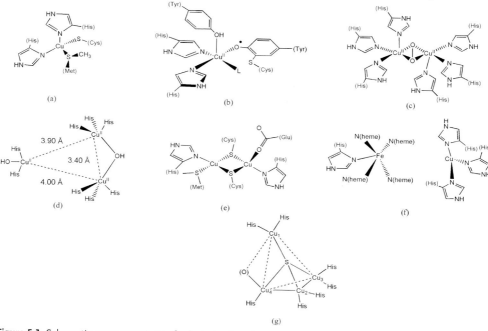

Figure 5.1 Schematic representations of selected active sites of the copper proteins: plastocyanin [56] (type 1, a); galactose oxidase [57] (type 2, b); *oxy* hemocyanin [58] (type 3, c); ascorbate oxidase [10] (type 4, or multicopper site, d); nitrous oxide reductase [59] (Cu$_A$ site, e); cytochrome *c* oxidase [15] (Cu$_B$ site, f); and nitrous oxide reductase (Cu$_Z$ site, g) [16].

histidine residues of a protein backbone, a sulfur atom from a cysteine residue, and a weakly coordinated donor atom from, in most cases, a methionine residue (Figure 5.1a). The blue color is caused by a strong absorption at ~600 nm, corresponding to an LMCT (ligand to metal charge transfer) transition from a cysteine sulfur to the copper(II) ion [2].

5.2.2
Type 2 Active Site

The copper proteins containing the type 2 active site are also known as "normal" copper proteins, because their spectroscopic features are similar to those of common Cu^{II} coordination compounds. The copper ion in these proteins is surrounded by four N and/or O donor atoms in either square-planar or distorted tetrahedral geometry [3, 4]. Examples of proteins with this active site include copper-zinc superoxide dismutase, dopamine-β-hydroxylase, phenylalanine hydroxylase, and galactose oxidase (Figure 5.1b) [5]. The proteins of this class are often involved in catalysis, such as disproportionation of O_2^- superoxide anion, selective hydroxylation of aromatic substrates, C—H activation of benzylic substrates and oxidation of primary alcohols.

5.2.3
Type 3 Active Site

The type 3 active sites in copper proteins feature a dinuclear copper core, each ion being coordinated by three nitrogen atoms of histidine residues [3, 6]. In the oxidized state, the two copper ions are strongly antiferromagnetically coupled, leading to the so-called EPR-silent behavior. This class is represented by three proteins: hemocyanin, tyrosinase, and catechol oxidase. A characteristic feature of the proteins with this active site is their ability to reversibly bind dioxygen under ambient conditions. Hemocyanin (Figure 5.1c) is responsible for dioxygen transport in certain mollusks and arthropods, whereas tyrosinase and catechol oxidase use dioxygen to perform the oxidation of phenolic substrates to catechols (tyrosinase) and subsequently to *o*-quinones (tyrosinase and catechol oxidase), which later undergo polymerization to the dark pigment melanin.

5.2.4
Type 4 Active Site

Some proteins contain more than one copper site, and are therefore among the most complicated and least understood of all. The active site known as type 4 is usually composed of a type 2 and a type 3 active site, together forming a trinuclear cluster. In some cases, such proteins also contain at least one type 1 site and are in this case termed multicopper oxidases, or blue oxidases [3]. Representatives of this class are laccase (polyphenol oxidase) [7–9], ascorbate oxidase (Figure 5.1d) [10], and ceruloplasmin [11], which catalyze a range of organic oxidation reactions.

5.2.5
The Cu$_A$ Active Site

This type of active site is also known as a mixed-valence copper site. Similarly to the type 3 site, it contains a dinuclear copper core, but both copper ions have a formal oxidation state of +1.5 in the oxidized form. This site exhibits a characteristic seven-line pattern in the EPR spectra and is purple colored. Both copper ions have a tetrahedral geometry and are bridged by two sulfur atoms of two cysteinyl residues. Each copper ion is also coordinated by a nitrogen atom from a histidine residue. The function of this site is long-range electron transfer, and it can be found, for example, in cytochrome *c* oxidase [12–14], and nitrous oxide reductase (Figure 5.1e).

5.2.6
The Cu$_B$ Active Site

This site has been introduced as a class to indicate a mononuclear Cu ion coordinated by three nitrogen atoms from three histidine residues in a trigonal pyramidal geometry, occurring in cytochrome *c* oxidase (Figure 5.1f) [15]. No fourth ligand coordinated to the metal ion has been detected. The vacant position in the copper coordination sphere is directed toward the vacant position in the coordination sphere of the heme iron ion, and the two metal ions are strongly antiferromagnetically coupled when oxidized. The function of the Cu$_B$ site is the four-electron reduction of dioxygen to water.

5.2.7
The Cu$_Z$ Active Site

The Cu$_Z$ active site consists of four copper ions, arranged in a distorted tetrahedron and coordinated by seven histidine residues and one hydroxide anion. This site was detected in nitrous oxide reductase [16, 17] (Figure 5.1g) and is involved in the reduction of N$_2$O to N$_2$. The copper ions in the tetranuclear cluster are bridged by an inorganic sulfur ion [18], which until recently was believed to be a hydroxide anion. Three copper ions are coordinated by two histidine residues, whereas the fourth is coordinated by only one, thus leaving a binding site for the substrate. The oxidation states of the copper ions in the resting state are still unclear, as the EPR spectra of this active site can be explained by two different oxidation schemes: Cu$^I_3$Cu^{II} and Cu^{I}Cu$^{II}_3$.

5.3
Catechol Oxidase: Structure and Function

As stated above, catechol oxidase (COx) is an enzyme containing the type 3 active site. It catalyzes the oxidation of a wide range of *o*-diphenols (catechols) to the corresponding quinones in a process known as a catecholase activity. The latter highly reactive compounds undergo autopolymerization, resulting in the formation of a brown pigment (melanin), a process thought to protect damaged tissues against pathogens or insects. Catechol oxidases are usually found in plant tissues and in crustaceans.

In 1998, Krebs and co-authors reported the crystal structures of the catechol oxidase isolated from sweet potatoes (*Ipomoea batatas*) in three catalytic states: the native *met* (Cu^{II}Cu^{II}) state (Figure 5.2a), the reduced *deoxy* (Cu^{I}Cu^I) form, and the complex with the inhibitor phenylthiourea (Figure 5.2b) [19]. Typically for the type 3 active site, each copper ion is coordinated by three histidine residues from the protein backbone. In the native *met* state, the two copper ions are 2.9 Å apart and, in addition to six histidine residues, a bridging solvent molecule, most likely a hydroxide anion, has been refined in close proximity to the two metal centers

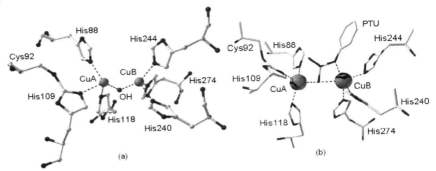

Figure 5.2 (a) Coordination sphere of the dinuclear copper(II) center in the *met* state. (b) Crystal structure of the inhibitor complex of catechol oxidase with phenylthiourea. Redrawn after Krebs and co-workers [60].

(CuA—O 1.9 Å, CuB—O 1.8 Å). The coordination spheres of both copper ions in the *met* state are therefore trigonal pyramids, with the apical positions occupied by one of the histidine residues. EPR data reveal a strong antiferromagnetic coupling between the copper ions; therefore, the presence of a bridging OH⁻ ligand between the copper(II) ions has been proposed for the *met* form of the enzyme.

The reduction of both copper(II) ions to the oxidation state +1 leads to the *deoxy* form of the enzyme, in which the distance between metal ions increases to 4.4 Å [19]. Instead of a hydroxide anion bridging the two copper centers, a water molecule is now coordinated to one of the metal ions (the distance CuA—O(H$_2$O) is 2.2 Å) in the *deoxy* state. Thus, the coordination sphere around the CuA ion is a distorted trigonal pyramid, with three nitrogen atoms from the histidine residues forming a basal plane, while the coordination sphere around the CuB ion can be best described as square planar with one missing coordination site. In the complex with the inhibitor phenylthiourea (Figure 5.2b), the latter compound binds to the active site by replacing the hydroxido bridge, present in the *met* form. The sulfur atom of phenylthiourea is coordinated to both copper(II) centers, increasing the distance between them to 4.2 Å. The amide nitrogen interacts weakly with the CuB center (Cu—N distance is 2.6 Å), completing its square-pyramidal geometry.

Upon treating the *deoxy* form of the enzyme with dihydrogen peroxide, two absorption bands at 343 nm ($\varepsilon = 6500\,M^{-1}cm^{-1}$) and 580 nm ($\varepsilon = 450\,M^{-1}cm^{-1}$) develop in the UV-Vis spectrum. These bands reflect the formation of another form of the enzyme – the *oxy* state. In this highly unstable state, dioxygen is bound to two Cu^{II} ions as a peroxide anion in a μ-η^2:η^2 fashion (Figure 5.3a). The first strong absorption in the spectrum corresponds to a peroxido $O_2^{2-}(\pi_\sigma^*) \rightarrow Cu^{II}(d_{x^2-y^2})$ charge transfer, whereas the second weak band around 580 nm corresponds to a peroxido $O_2^{2-}(\pi_v^*) \rightarrow Cu^{II}(d_{x^2-y^2})$ charge transfer transition [4, 20].

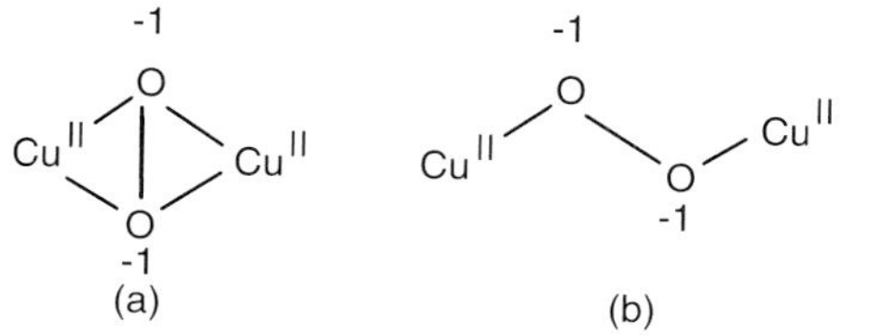

Figure 5.3 Schematic representations of two different dicopper–dioxygen cores: "side-on" $\mu\text{-}\eta^2:\eta^2$ (a) and "end-on" trans-μ-1,2 (b).

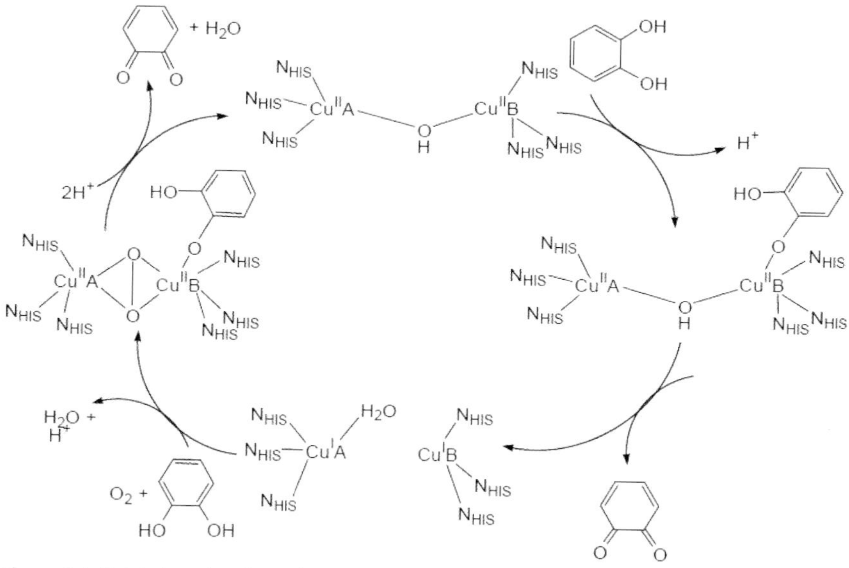

Figure 5.4 Catalytic cycle of catechol oxidase from *Ipomoea batatas*, as proposed on the basis of structural, spectroscopic, and biochemical data. Two molecules of catechol (or derivatives thereof) are oxidized, coupled with the reduction of molecular oxygen to water. The ternary COx—O_2^{2-}–catechol complex was modeled, guided by the binding mode observed for the inhibitor phenylthiourea. Redrawn after Krebs and co-workers [44].

5.3.1
Catalytic Reaction Mechanism

Catechol oxidase catalyzes the oxidation of catechols to the respective quinones through a four-electron reduction of dioxygen to water. Whereas the exact mechanism of the enzymatic conversion remains uncertain, the commonly accepted mechanism is that proposed by Krebs and co-workers [3, 21] (Figure 5.4). The catalytic cycle begins with the *met* form of catechol oxidase, which is the resting form of the enzyme. The dicopper(II) center of the *met* form reacts with one equivalent of catechol, leading to the formation of quinone and the reduced *deoxy* dicopper(I) state. Based on the structure of COx with the bound inhibitor phenylthiourea, the monodentate binding of the substrate to the CuB center has been proposed. Afterwards, dioxygen binds to the dicopper(I) active site, replacing the solvent molecule bonded to CuA in the *deoxy* form and resulting in the formation of the highly unstable *oxy* state. The binding of a second molecule of catechol to this form results in the oxidation of the latter compound and the reduction of the peroxide moiety to water. This reaction restores the native *met* form, completing the catalytic cycle.

5.4
Model Systems of Catechol Oxidase: Historic Overview

The interest in catechol oxidase, as well as in other copper proteins with the type 3 active site, is to a large extent due to their ability to process dioxygen from air at ambient conditions. While hemocyanin is an oxygen carrier in the hemolymph of some arthropods and mollusks, catechol oxidase and tyrosinase utilize it to perform the selective oxidation of organic substrates, for example, phenols and catechols. Therefore, establishment of structure–activity relationships for these enzymes and a complete elucidation of the mechanisms of enzymatic conversions through the development of synthetic models are expected to contribute greatly to the design of oxidation catalysts for potential industrial applications.

The ability of copper complexes to oxidize phenols and catechols has been well known for at least 40 years. For example, in 1964, Grinstead reported the oxidation of 3,5-di-*tert*-butylcatechol (DTBCH$_2$) to the respective 3,5-di-*tert*-butyl-o-benzoquinone (DTBQ) with 55% yield in 75% aqueous methanol, in the presence of 1% copper(II) chloride [22]. In 1978, after quite a few reports of copper-mediated catechol oxidations [23, 24], one of the pioneering studies on the mechanism of this reaction was presented by Lintvedt and Thuruya [25]. These authors studied the kinetics of the reaction of DTBCH$_2$ with dioxygen catalyzed by the complex bis(1-phenyl-1,3,5-hexanetrionato)dicopper(II), and showed that the overall reaction was first-order in substrate and second-order in Cu^{II}. These findings allowed them to conclude that the reaction proceeds via the formation of a dicopper–catecholate species, which is apparently an active reaction intermediate involved in the rate-determining step. Another interesting example of early mechanistic studies is the work of Demmin, Swerdloff, and Rogić [26], who recognized the main steps in the catalytic process: (i) formation of a dicopper(II)–catecholate intermediate; (ii) electron transfer from the aromatic ring to the two copper(II) centers, resulting in the formation of o-benzoquinone and two copper(I) centers; (iii) irreversible reaction of the generated copper(I) species with dioxygen, resulting in a copper(II)–dioxygen adduct; and (iv) the reaction of this adduct with catechol, leading to regeneration of the dicopper(II)–catecholate intermediate and formation of water as by-product.

The hypothesis of Lintvedt and Thuruya about the formation of the dicopper–catecholate intermediate in the catalytic process [25] was soon after confirmed by findings from Oishi *et al.*, who reported higher activities of dinuclear copper(II) complexes in the oxidation of DTBCH$_2$ in comparison to their mononuclear analogues [27]. These findings were consistent with the hypothesis that catechol was binding to two copper centers at some point during the course of the catalytic reaction. The same authors also reported a stoichiometric oxidation of DTBCH$_2$ in anaerobic conditions to the respective quinone by a number of mononuclear and dinuclear copper(II) complexes, in agreement with the first step of the mechanism proposed by Demmin, Swerdloff, and Rogić [26]. Another interesting fact was that the complexes for which the copper–copper separation was estimated to be more than 5 Å showed very low catalytic activities in comparison to the complexes in which the copper–copper distance was estimated to be short (about 3 Å). This difference allowed the authors to propose that the catecholase activity of dinuclear copper(II) complexes depends on the metal–metal distance, and is regulated by a steric match between the dicopper(II) center and the substrate. The higher activity of dinuclear copper(II) complexes towards catechol oxidation in comparison to mononuclear copper(II) complexes has also been pointed out by some other authors, for example, Malachowski [28] and Casellato *et al.* [29].

In 1985, the hypothesis about the formation of a dicopper–catecholate intermediate at the first stage of the catalytic reaction was further supported by Karlin and co-workers [30], who crystallized an adduct between tetrachlorocatechol (TCC) and a dicopper(II) complex with a phenol-based dinucleating ligand (Figure 5.5). This

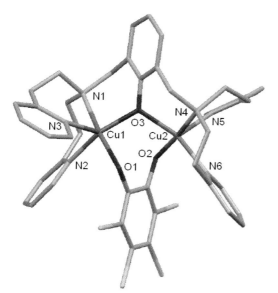

Figure 5.5 Crystal structure of the complex cation of [Cu$_2$(L-O$^-$)(TCC)]$^+$. LOH: 2,6-bis(N,N-bis(2-methylpyridyl)aminomethyl)phenol. The Cu$\cdots$Cu distance is 3.248(2) Å. Redrawn after Karlin and co-workers [30].

compound was prepared through the reaction of tetrachloro-1,2-benzoquinone with the dicopper(I) precursor complex in dichloromethane. The catecholate anion was found to bind as a bridging ligand in a *syn-syn* fashion to both copper(II) ions, resulting in a metal-metal separation of 3.248(2) Å. Both copper(II) ions in the complex adopt a square-pyramidal geometry, with the oxygen atoms of the catecholate anion occupying the basal plane.

However, the findings of Thompson and Calabrese [31], published only one year after the work of Karlin *et al.* [30], called the mechanistic hypothesis of Demmin, Swerdloff, and Rogić into question [26]. Whereas earlier studies all agreed on the two-electron stoichiometric reduction of catechol by the dinuclear copper(II) core, these authors proposed that the catalytic reaction proceeds via a one-electron transfer from catechol to the copper(II) ion, resulting in the formation of a semiquinone intermediate species. The authors have prepared and characterized a bis(3,5-di-*tert*-butyl-*o*-semiquinonato)copper(II) complex by reaction of $[Cu_2(py)_4(OCH_3)_2](ClO_4)_2$ with $DTBCH_2$ in anaerobic conditions. Interestingly, the simultaneous two-electron transfer yielding DTBQ and two copper(I) centers was not observed. Finally, the formation of the semiquinone species in the catalytic cycle was later reported by other authors [32–34].

Even more controversy arose when, in the early 1990s, several research groups reported the formation of dihydrogen peroxide instead of water as the product of dioxygen reduction in the catalytic oxidation of $DTBCH_2$ by the copper(II) complexes [35, 36]. In order to explain their experimental results, Chyn and Urbach proposed two different mechanisms for the catalytic cycle, as depicted in Scheme 5.1 [35].

(1)

$Cu^{II}...Cu^{II}$ + 3,5-$DTBCH_2$ ⟶ $Cu^{I}...Cu^{I}$ + 3,5-DTBQ + $2H^+$ (fast)

$Cu^{I}...Cu^{I}$ + O_2 ⟶ $Cu^{II}(O_2)^{2-}Cu^{II}$ (slow)

$Cu^{II}(O_2)^{2-}Cu^{II}$ + $2H^+$ ⟶ $Cu^{II}...Cu^{II}$ + H_2O_2 (fast)

(2)

Initial step

$Cu^{II}...Cu^{II}$ + 3,5-$DTBCH_2$ ⟶ $Cu^{I}...Cu^{I}$ + 3,5-DTBQ + $2H^+$

Redox cycle

$Cu^{I}...Cu^{I}$ + O_2 $\underset{k_2}{\overset{k_1}{\rightleftarrows}}$ $Cu^{II}(O_2)^{2-}Cu^{II}$

$Cu^{II}(O_2)^{2-}Cu^{II}$ + $2H^+$ ⟶ $Cu^{II}...Cu^{II}$ + H_2O_2

Scheme 5.1 Two possible mechanistic pathways resulting in the formation of H_2O_2 as a by-product, as proposed by Chyn and Urbach [35].

5.4 Model Systems of Catechol Oxidase: Historic Overview

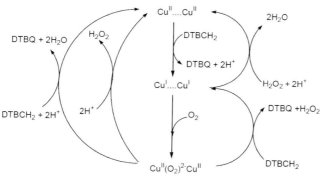

Scheme 5.2 The possible reaction pathways in the catalytic cycle of catechol oxidation by dicopper(II) complexes, as proposed by Casella and co-workers. Redrawn after Casella et al. [38].

These controversial findings inspired numerous subsequent studies on the structure-activity relationship of catalytically active compounds. Very detailed mechanistic studies on the catecholase activity of a series of structurally related dicopper(II) complexes have also been published by Casella and co-workers [37–40], who have grouped together different mechanisms earlier proposed for the catecholase activity of dicopper(II) complexes, as shown in Scheme 5.2.

In 1998, the crystal structure of catechol oxidase from sweet potatoes was solved by Krebs and co-workers [19], who also proposed the mechanism of the enzymatic conversion, as described above. However, certain aspects of this mechanism remained unclear, among them the binding mode of catechol to the dicopper core. The various possible binding modes of catechol to the copper centers are summarized in Figure 5.6. Although Krebs and co-authors proposed an asymmetric binding of catechol based on the structure of the enzyme with the inhibitor thiourea [3, 21], unfortunately no direct proof of this hypothesis could be offered. On the other hand, various other examples of adducts of catechol with model compounds, besides the work of Karlin, were reported in the literature. Thus, Comba and co-authors [41] have reported the crystal structures of four different copper–tetrachlorocatecholate adducts, with three different modes of substrate coordination to the metal centers (Figure 5.7a, c, d): as a monodentate, monoprotonated ligand (**1**), as a bidentate fully deprotonated chelating ligand (**2**, a and **4**, d), and as a bridging deprotonated ligand between the two copper(II) centers (**3**, c, *anti-anti* binding mode). Meyer and co-workers [42] have reported the structures of three dinuclear Cu^{II} complexes, in which the deprotonated tetrachlorocatecholate is bound to only one of the two copper(II) ions in a bidentate chelating fashion (Figure 5.7b). Apparently, the distance between the two copper centers played some role in the latter case, as the copper-copper separation in the precursor dicopper(II) complexes was found to exceed 4 Å, possibly precluding the simultaneous binding of the catecholate to both copper(II) ions. Casella and co-workers

112 5 Mimicking Nature: Bio-inspired Models of Copper Proteins

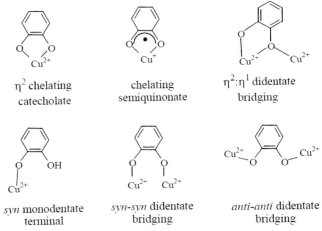

Figure 5.6 Different binding modes of the (deprotonated) catechol substrate to the copper centers.

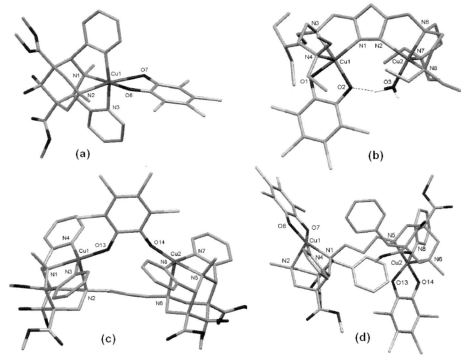

Figure 5.7 X-ray crystal structure projections of copper–catecholate adducts obtained by Comba and co-workers [41] (a, c and d), and x-ray crystal structure of one of the dicopper(II)-catecholate adducts crystallized by Meyer and co-workers (b) [42].

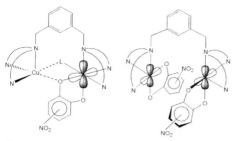

Figure 5.8 Structure proposals for dicopper–catecholate adducts obtained by Casella and co-workers [43].

[43] have used inactive p-nitrocatechol (NCat) to isolate and spectroscopically characterize catecholate adducts of mononuclear and dinuclear copper(II) complexes. IR and resonance Raman spectroscopic studies on these complexes allowed the authors to propose that catechol was bound as a catecholate anion with a chelating η^2 binding mode to only one copper ion. This catecholate may eventually exhibit an additional η^1 bridging coordination to a second copper atom in the dicopper(II) complex, as depicted in Figure 5.8. In addition, the second molecule of catechol could bind to the dicopper complex, forming a biscatecholate adduct, similar to the complex 4 reported by Comba [41]. Thus, owing to the many crystallographically and/spectroscopically characterized examples of possible binding modes reported in the literature, the question of which one is realized in the natural enzyme could not readily be answered.

The first step of the catalytic reaction also raised some questions. Krebs and co-workers observed the formation of stoichiometric amounts of quinone when catechol was added to the natural enzyme, even in the absence of dioxygen [19, 44]. This result allowed them to propose that, on the first stage, a two-electron transfer between catechol and the dicopper(II) core takes place. However, other authors [31–34] suggested that the reaction proceeds via the formation of semiquinone radicals. Another ambiguous point in the mechanism of the catalytic reaction was the reduction mode of dioxygen. Whereas the formation of water as a single by-product was reported for the natural enzyme, as well as for some model systems, the reduction of dioxygen to dihydrogen peroxide was also observed in a few cases [35, 36]. Even more interestingly, some authors reported the formation of H_2O_2 [45], as well as semiquinone radicals [46], during the catalytic oxidation of DOPA by the structurally related enzyme tyrosinase in the hemolymph of some insects, thus raising the question whether the formation of these species could also be possible in case of catechol oxidase. One thing was thus certain: despite over 40 years of research and numerous studies devoted to this topic, the exact mechanism of the functioning of catechol oxidase remained far from clear.

Another fascinating topic which attracted the attention of many researchers is the difference in the enzymatic activities between catechol oxidase and the structurally related enzyme tyrosinase. Although both enzymes contain type 3 active sites, their reactivities differ significantly: in contrast to catechol oxidase, tyrosinase is also able to oxidize selectively an *ortho*-position of the aromatic rings of phenols, besides the oxidation of catechol. As the activation of C—H bonds is a crucial issue for the fine chemicals industry, the elucidation of the structural and/or chemical disparities between catechol oxidase and tyrosinase, which would explain the drastic differences in their behavior, would be of paramount interest for the design of new oxidation catalysts.

5.5
Our Research on Catechol Oxidase Models and Mechanistic Studies

5.5.1
Ligand Design

As discussed above, the active site of catechol oxidase comprises two copper ions, each of which surrounded by three nitrogen donor atoms from histidine residues. To model the active site of this enzyme, we have designed and synthesized new macrocyclic pyrazole-based ligands: [22]py4pz (9,22-bis(2-pyridylmethyl)-1,4,9,14, 17,22,27,28,29,30-decaazapentacyclo-[22.2.1.1^{4,7}.1^{11,14}.1^{17,20}]triacontane-5,7(28),11(2 9),12,18,20(30),24(27),25-octaene) and [22]pr4pz. (9,22-bispropyl-1,4,9,14,17,22,27, 28,29,30-decaazapentacyclo-[22.2.1.1^{4,7}.1^{11,14}.1^{17,20}]triacontane-5,7(28),11(29),12,18, 20(30),24(27),25-octaene) [47–49]. The reaction scheme for the syntheses of these ligands is presented in Figure 5.9. These macrocyclic ligands are able to bind two copper ions, keeping them in close proximity and providing each of them with either four ([22]py4pz) or three ([22]pr4pz) nitrogen donor atoms. Pyrazole moieties were chosen as isostructural analogues of imidazole rings, present at the active site, as they mimic their electronic properties most closely. The flexibility of the macrocyclic cavities also allows significant variations in the metal–metal distance, depending on the bridging ligand between the metal ions, which can be used to study the influence of the metal–metal distance on the catalytic properties of the respective complexes.

A number of copper(I) and copper(II) complexes with [22]py4pz and [22]pr4pz have been isolated and structurally characterized [47–49]. Their structural and catalytic properties, as well as studies on the mechanism of the catalytic oxidation of catechol performed by some of these compounds, are discussed below.

5.5.2
Copper(I) and Copper(II) Complexes with [22]py4pz: Structural Properties and Mechanism of the Catalytic Reaction

The interaction of [22]py4pz with copper(II) and copper(I) salts led to the isolation of a dinuclear copper(II) complex of composition [Cu$_2$([22]py4pz)(μ-OH)] (ClO$_4$)$_3$·H$_2$O, and its reduced dicopper(I) analog [Cu$_2$([22]py4pz)](ClO$_4$)$_2$·2CH$_3$OH. The dicopper(II) complex was obtained by the reaction of copper(II) perchlorate

5.5 Our Research on Catechol Oxidase Models and Mechanistic Studies | 115

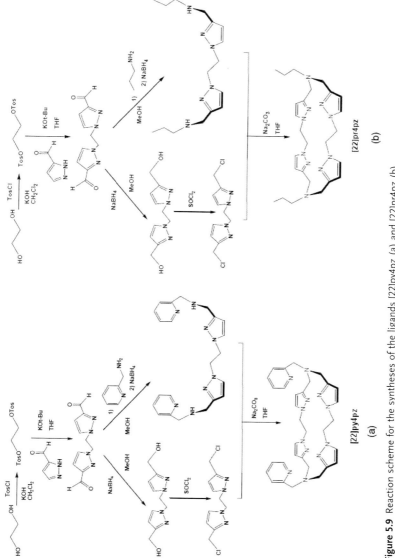

Figure 5.9 Reaction scheme for the syntheses of the ligands [22]py4pz (a) and [22]pr4pz (b).

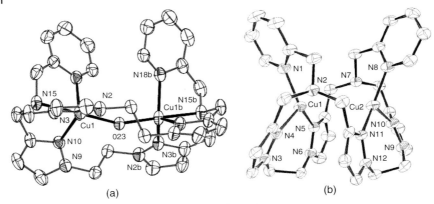

Figure 5.10 ORTEP representations of the complex cations [Cu$_2$([22]py4pz)(μ-OH)]$^{3+}$ (a) and [Cu$_2$([22]py4pz)]$^{2+}$ (b). Hydrogen atoms are omitted for clarity.

with the ligand in the presence of base, whereas the dicopper(I) complex was obtained by reacting two equivalents of [Cu(CH$_3$CN)$_4$](ClO$_4$) with [22]py4pz in methanol in a dry glove box atmosphere.

The molecular plot of the complex cation [Cu$_2$([22]py4pz)(μ-OH)]$^{3+}$ is depicted in Figure 5.10a. The Cu^{II} ions are in an N$_4$O environment, which can best be described as a distorted trigonal bipyramid. The equatorial positions are occupied by the two pyrazole nitrogen atoms and the pyridine nitrogen atom, whereas the bridging oxygen atom of the hydroxide moiety and the tertiary amine N15 atom occupy the axial positions. The bridging oxygen atom O23 connects the two central copper atoms Cu1 and Cu1b, resulting in a Cu—Cu distance of 3.7587(11) Å and a Cu—O—Cu angle of 156.0(3)°. The macrocycle adopts a *cis*-(boat)-conformation, with the two pyridine groups being located at the same side above the macrocyclic ring.

The molecular plot of the complex cation [Cu$_2$([22]py4pz)]$^{2+}$ is shown in Figure 5.10b. The macrocyclic ligand adopts a saddle-shaped structure with a roof closed by the two pyridines. The coordination mode of both copper ions is trigonal planar, as the two tripodal nitrogen atoms are located too far from the metal centers to be considered as coordinating. The complex cation encloses two copper(I) ions at a distance of 3.3922(7) Å, which is shorter than the intermetallic distance in its dicopper(II) analogue. The Cu1 and Cu2 cations are surrounded by the nitrogen atom of the pyridine ring and the nitrogen atoms of two pyrazole rings, with N—Cu—N angles around both metal ions ranging between 76° and 143°.

The complexes obtained can be considered as structural models of two states of catechol oxidase: the native *met* state and the reduced *deoxy* state. In order to investigate whether the dicopper(II) complex also possesses the functionality of the natural enzyme, for example, if it can perform the catalytic oxidation of cate-

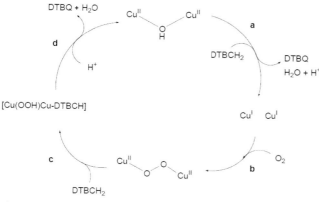

Figure 5.11 Proposed mechanism for the oxidation of 3,5-di-tert-butylcatechol by $[Cu_2([22]py4pz)(\mu\text{-}OH)](ClO_4)_3 \cdot H_2O$. The peroxido-dicopper(II) intermediate was characterized by UV-Vis and resonance Raman spectroscopy [47].

chol, its catalytic behavior in acetonitrile solution in the presence of an excess of catechol has been studied. The complex was found to oxidize catalytically the model substrate 3,5-di-*tert*-butylcatechol (DTBCH$_2$) to the corresponding quinone, with a turnover number of 36, determined after one hour. The reaction follows a Michaelis–Menten behavior, with $V_{max} = 1.3 \times 10^{-6}\,M\,s^{-1}$ and $K_M = 4.9\,mM$ [47]. Thus, the complex can be considered as a satisfactory functional model of the natural enzyme as well.

In order to investigate the mechanism of the catalytic reactions, we have studied separately all steps of the catalytic cycle, which led to the mechanistic proposal presented in Figure 5.11. In the first stage of the reaction, a stoichiometric oxidation of catechol by the dicopper(II) complex takes place (step **a**, Figure 5.11). This step does not require the presence of dioxygen and is thus similar to the first step of the mechanism proposed by Krebs and co-workers for natural catechol oxidase. In the second stage, the formed dicopper(I) complex binds dioxygen to form a highly reactive peroxido-dicopper intermediate (Figure 5.11, step **b**). The latter species is highly reactive, and is only stable for a few hours in acetonitrile solution at −40 °C. Spectroscopic studies on this species clearly showed, however, that dioxygen is not bound in $\mu\text{-}\eta^2:\eta^2$ "side-on" fashion, proposed for catechol oxidase, but forms a *trans*-μ-1,2-peroxido-dicopper(II) species (see also Figure 5.3b). Copper–dioxygen adducts of this type were first reported by Karlin and co-workers [50]. This species oxidizes a second equivalent of catechol in a stoichiometric reaction through a two-electron transfer from the catechol to the peroxide moiety, resulting in the formation of one equivalent of water as a by-product, along with quinone formation. The latter reaction proceeds in two steps: as the *trans*-μ-1,2-peroxido-dicopper(II) complex is essentially nucleophilic in nature, it first abstracts a proton from the catechol molecule (Figure 5.11, step **c**), and subsequently oxidizes the bound catecholate (Figure 5.11, step **d**). After the quinone molecule is released, the dicopper(II) complex is regenerated, and the catalytic cycle can continue. Two equivalents of quinone are thus generated per one catalytic cycle.

This mechanism is in fact very similar to the mechanism proposed by Krebs and co-workers for catechol oxidase [19, 44], the major difference between the two proposals being the structure of the peroxido-dicopper intermediate. Whereas the formation of a μ-η^2:η^2 peroxido-dicopper intermediate was proposed for catechol oxidase, in the present case dioxygen is clearly bound in a trans-μ-1,2-peroxido ("end-on") fashion. Interestingly, there is a significant difference regarding the chemical behavior of these copper–dioxygen adducts. While a trans-μ-1,2-peroxido-dicopper(II) species, as stated above, is a nucleophile, a μ-η^2:η^2 peroxido-dicopper intermediate is electrophilic in nature. [51] This in turn results in their different reactivities: whereas the former type of copper–dioxygen species is able to oxidize catechols (catecholase activity), the latter type of copper–dioxygen adducts can oxidize not only catechols but phenols as well (tyrosinase activity). This observation generates an interesting speculation: whether the difference in behavior toward phenol and catechol substrates might be dependent on the structure of the copper–dioxygen adduct. Although it should be pointed out that trans-μ-1,2-peroxido intermediates have never been observed in natural systems, an earlier report from Karlin and co-workers showed that a trans-μ-1,2-peroxido-dicopper intermediate could interconvert very rapidly into a μ-η^2:η^2 final species [52]. These findings allowed the authors to suggest that the "end-on" species may initially form upon dioxygen binding by the type 3 copper proteins, due to the long distance between the metal ions in the reduced dicopper(I) core, and may afterwards rapidly interconvert into "side-on" species [52]. It is thus fascinating to note the possible existence of "end-on" peroxido-dicopper moieties in living systems and embrace the possibility that the reactivity toward different substrates might be able to be tuned by the type of dicopper–dioxygen adduct. Although these assumptions are largely speculative at the moment, future studies on dicopper-dioxygen model systems and their reactivity would be crucial to interpretation of the mechanisms of monophenolase and diphenolase activity of type 3 copper proteins.

5.5.3
Copper(I) and Copper(II) Complexes with [22]pr4pz: Unraveling Catalytic Mechanisms

Two copper complexes, namely $[Cu^{II}_2([22]pr4pz)(CO_3)(H_2O)]_2(CF_3SO_3)_4 \cdot 2CH_3CN \cdot 4H_2O$ and its reduced analog $[Cu^{I}_2([22]pr4pz)(CH_3CN)_2](ClO_4)_2$, have been obtained with the ligand [22]pr4pz and structurally characterized [48]. The former complex was obtained by diffusion of diethyl ether in an acetonitrile solution containing two molar equivalents of copper(II) triflate, one molar equivalent of the ligand, and one molar equivalent of sodium carbonate. The copper(I) complex was isolated by reacting two molar equivalents of Cu^I as a tetrakis(acetonitrile) complex, with a solution of the ligand in methanol, followed by precipitation with diethyl ether.

5.5 Our Research on Catechol Oxidase Models and Mechanistic Studies

The molecular structure of the isolated solid dicopper(II) complex consists of a tetracopper complex cation $[Cu_2([22]pr4pz)(CO_3)(H_2O)]_2^{4+}$ (Figure 5.12a), four counterions $CF_3SO_3^-$, two noncoordinated acetonitrile molecules, and four noncoordinated water molecules. The dimeric cation contains four copper(II) centers, two macrocyclic ligands, two coordinated carbonates and two coordinated water molecules. Each macrocyclic unit encloses two copper(II) ions, which are bridged by a carbonate anion, with an intramacrocyclic Cu···Cu distance of 4.5427(18) Å.

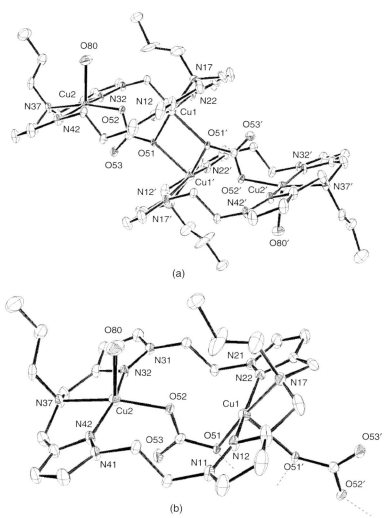

Figure 5.12 (a) ORTEP projection of the tetranuclear cation $[Cu_2([22]pr4pz)(CO_3)(H_2O)]_2^{4+}$ and (b) ORTEP projection of half of the tetranuclear cation (1^{2+}). Hydrogen atoms and solvent molecules are omitted for clarity.

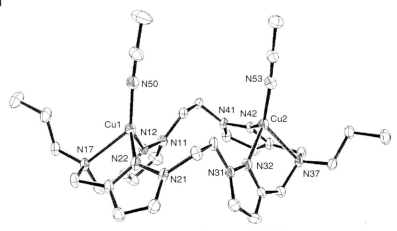

Figure 5.13 ORTEP projection of the complex cation [Cu$_2$([22] pr4pz)(CH$_3$CN)$_2$]$^{2+}$. Hydrogen atoms are omitted for clarity.

Two bridging oxygen atoms from two different carbonate anions further connect the two central copper atoms, resulting in a Cu···Cu intermacrocyclic distance of 3.281(2) Å. Each carbonate ion is thus bound in a *syn, syn-anti* fashion; for example, one of the oxygen atoms of the carbonate anion bridges the copper ions of two different macrocyclic rings, whereas another oxygen atom binds to another copper ion of the same unit. The coordination spheres around both copper ions in the macrocyclic unit can be described as weakly distorted square pyramids, with either water or carbonate anions occupying the apical positions (Figure 5.12b).

An ORTEP projection of the dicopper(I) complex cation is shown in Figure 5.13. Similarly to the structure of the dicopper(I) complex with [22]py4pz, the macrocyclic ligand adopts a saddle-shaped conformation [47]. However, in the [22]py4pz complex, the tertiary nitrogen atoms of the ligand fail to bind to the metal ions, resulting in distorted trigonal surroundings for both copper(I) ions. In the present case, in contrast, both copper ions are tetracoordinated and have distorted tetrahedral surroundings, with three positions in the coordination sphere occupied by the nitrogen atoms from the ligand and one by the nitrogen atom of an acetonitrile molecule. The N—Cu—N angles for both copper(I) ions vary in quite a large range, from 78° to 132°, indicating a significant distortion of the coordination sphere from a regular tetrahedral geometry. The distance between the two copper(I) ions is very large: 5.547(2) Å.

Despite its tetranuclear structure in the solid state, the dicopper(II) complex was found to dissociate in solution into dinuclear units at the concentration levels used for catecholase activity studies. Similarly to the copper(II) complex with the ligand [22]py4pz, the present complex also catalyzes the oxidation of the model substrate DTBCH$_2$ in methanol. However, several unexpected observations have been made in the present case. First, the rate-determining step in the catalytic reaction was found to change with the substrate-to-complex ratio. Thus, at low substrate-to-

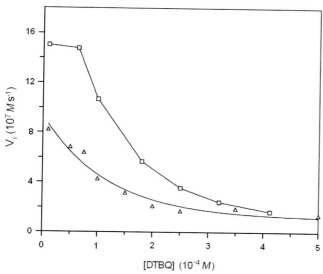

Figure 5.14 Dependence of the initial reaction rates on the concentration of di-*tert*-butylquinone at low (△, 10:1) and high (□, 50:1) DTBCH$_2$-to-complex ratios. The concentration of the dinuclear cation [Cu$_2$([22]pr4pz)(CO$_3$)(H$_2$O)]$^{2+}$ was 2 × 10^{-5} M.

complex ratios (<12:1, [[Cu$_2$([22]pr4pz)(CO$_3$)(H$_2$O)]$^{2+}$] = 2 × 10^{-5} M) the reaction shows Michaelis–Menten behavior, with $K_M = 0.176$ mM and $V_{max} = 2.47 \times 10^{-6}$ M s^{-1}, while at high substrate-to-complex ratios (up to 200:1) the reaction rate was found to depend linearly on the DTBCH$_2$ concentration, with a first-order rate constant $k_1 = 2 \times 10^{-4}$ s^{-1}. Second, while the catalytic reaction was finished within a few minutes at low catechol-to-complex ratios, at high catechol to complex ratios no full conversion of DTBCH$_2$ could be achieved, not even after 24 hours. Surprisingly, the reaction was found to be inhibited by the product of catechol oxidation, 3,5-di-*tert*-butylquinone (Figure 5.14). In addition, dihydrogen peroxide was found to form as a by-product at an early stage of the catalytic reaction, although its concentration in the reaction mixture ceased to increase after a few minutes. These unprecedented findings prompted us to study the mechanism of the catalytic reaction in more detail.

The mechanism of the catalytic reaction proved indeed to be very different from that found for [Cu$_2$([22]py4pz)(µ-OH)](ClO$_4$)$_3$·H$_2$O. Thus, in the first step of the reaction, a stoichiometric oxidation of catechol by the dicopper(II) complex takes place; however, only one electron is transferred in this stoichiometric reaction, resulting in the formation of a semiquinone radical and a mixed-valence Cu^{II}Cu^{I} species. Interestingly, the dicopper(II) complex was found to be essentially dinuclear in solution; nevertheless, only one of the two copper(II) ions was found to participate in the redox process, whereas the second one played a purely structural

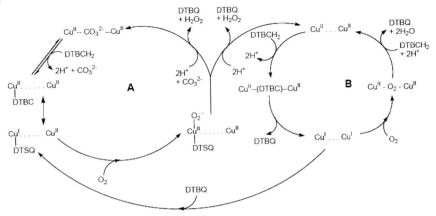

Figure 5.15 The proposed mechanism for the catalytic catechol oxidation by [Cu$_2$([22]pr4pz)(CO$_3$)(H$_2$O)]$^{2+}$ [48].

role. The mixed-valence Cu^{II}Cu^{I}-semiquinone species was also found to form readily when treating the solution of the dicopper(I) complex with one molar equivalent of DTBQ under anaerobic conditions. Whereas the oxidation of catechol by the dicopper(II) complex was found to be slow (rate constant $k = 1.8 \times 10^{-3}\,M^{-1}\,s^{-1}$), the oxidation of the dicopper(I) complex by DTBQ occurred virtually immediately. The re-oxidation of the obtained mixed-valence species and semiquinone by dioxygen subsequently led to the formation of one equivalent of quinone and one equivalent of dihydrogen peroxide as a by-product, completing the cycle (Figure 5.15, cycle A).

However, this simple reaction scheme does not explain the fact that the formation of dihydrogen peroxide stops at early stages of the reaction, suggesting that a different catalytic mechanism may take place in later stages. A proposal of H$_2$O$_2$ being consumed in the course of the reaction, as suggested by some authors [53], seems unlikely in this case, as kinetic measurements performed in the presence of variable amounts of H$_2$O$_2$ indicated that its presence has virtually no influence on the catalytic cycle [48], except at unrealistically high concentrations levels, which were never reached during the reaction. Furthermore, the inhibiting effect of DTBQ also suggested a different mechanistic pathway operating at later stages of the catalytic oxidation, as these results indicated that the formed quinone does not simply accumulate in the reaction mixture but obviously also participates in the catalytic process.

In order to explain these apparent contradictions, we have proposed that a different reaction mechanism takes place at later stages of the catalytic reaction, namely the oxidation of DTBCH$_2$ by a "classic" mechanism, proposed by Krebs *et al.* [19] for the natural enzyme, involving a stoichiometric reaction between the dicopper(II) species and the substrate, leading to the reduced dicopper(I) species. The oxidation of the second equivalent of substrate by a peroxido-dicopper(II) adduct, formed upon dioxygen binding to the dicopper(I) intermediate, results in the formation the second molecule of quinone and water as a by-product (Figure 5.15, cycle B). The inhibiting effect of DTBQ on the catalytic cycle can then be explained by its very fast reaction with the reduced dicopper(I) species (which is the only intermediate species able to react with the quinone), leading to semiquinone formation (Figure 5.15). Thus, at high concentrations, DTBQ competes with dioxygen in the re-oxidation of the reduced dicopper(I) species, resulting in the mixed-valence $Cu^{II}Cu^{I}$-semiquinone species, which is then subsequently oxidized in the less efficient cycle A. Consequently, the concentration of quinone increases more slowly, as only one molecule of DTBQ is produced in cycle A, in contrast to cycle B.

On the other hand, when present in high concentration, H_2O_2 can compete with dioxygen in the re-oxidation of the dicopper(I) complex to the dicopper(II) state; therefore, the increase in its concentration results in its progressive involvement as a copper(I) oxidant. The increase of the reaction rates in this case can be explained by a change in the rate-determining step of the reaction, as was previously proposed by Casella and co-workers [38].

The change of the catalytic mechanism in the course of the reaction can then be explained considering that the different mechanistic pathways are directly related to the binding mode of the substrate to the dicopper(II) core. As stated before, the latter has often been debated in the literature in the past [30, 41–43, 54] and different examples of substrate coordination modes to dicopper(II) complexes have been reported [30, 42, 43]. Our results suggest that the binding mode of the substrate is to a large extent determined by the distance between the two copper centers and their accessibility. Thus, the long Cu1$\cdots$Cu2 separation (4.5427(18) Å) observed for two copper ions in the macrocyclic unit in the dicopper(II) complex prohibits the substrate binding in a bridging bidentate fashion, leading to the binding of catechol to only one metal center. Consequently, only one electron can be transferred from the substrate to the metal center, resulting in semiquinone formation. On the other hand, when the distance between the two copper(II) ions is sufficiently short, for example, in $[Cu_2([22]py4pz)(\mu\text{-}OH)](ClO_4)_3 \cdot H_2O$, binding of the catecholate in a bidentate bridging fashion occurs, leading to both copper ions being reduced in the stoichiometric reaction with the substrate and resulting in the formation of quinone and the reduced dicopper(I) species.

The difference in the first step of the catalytic cycle in turn results in two different mechanisms of the catalytic reaction: one suggested by Krebs *et al.* [19] for the natural enzyme, during which two equivalents of the substrate are oxidized and water is formed as a single by-product, and a less efficient mechanism resulting in the oxidation of only one equivalent of substrate and with subsequent formation of H_2O_2.

It is possible to imagine that in the case of $[Cu^{II}_2([22]pr4pz)(CO_3)(H_2O)]_2(CF_3SO_3)_4 \cdot 2CH_3CN \cdot 4H_2O$, the distance between the two copper(II) ions may change in the course of the catalytic reaction, as the carbonate bridge is likely to be cleaved by the incoming catecholate, as proposed in Figure 5.15. As the macrocyclic cavity possesses sufficient flexibility to bring two copper ions to a short distance, required for a bidentate bridging coordination mode of catechol, it can be imagined that in the absence of the rigid carbonate bridge the substrate can bind to both copper(II) ions in a bridging fashion, pushing the catalytic reaction toward cycle B, and thus explaining a change in the reaction mechanism as the reaction proceeds.

5.6
Concluding Remarks

The studies on the copper(II) complexes with the macrocyclic ligands led to several important conclusions concerning the mechanism of the catalytic oxidation of catechol by model compounds. First, the studies on the copper(II) complex with the ligand [22]pr4pz settled an apparent contradiction concerning the binding mode of the substrate to the dicopper(II) core. The results proved that in fact both binding modes are possible, depending on the distance between the two metal ions. It is also clear that different binding modes of the substrate result in completely different mechanisms of catechol oxidation by dicopper(II) complexes; furthermore, in case of sufficiently flexible ligands, both mechanistic pathways can be realized.

In addition, a mechanism for the formation of dihydrogen peroxide has now been established. The results obtained strongly suggest that it is formed as a by-product during the oxidation of the semiquinone intermediate with dioxygen, when the metal–metal distance within a dicopper(II) complex is too long to allow the binding of the substrate in a bidentate bridging fashion. This hypothesis is also consistent with the findings of other authors; indeed, Meyer and co-workers reported a recovery of dihydrogen peroxide of 58–71% for a series of dicopper(II) complexes [42], for which an asymmetric coordination mode of a nonreactive catechol to only one of the copper(II) centers has been clearly established. In addition, two other crystallographically characterized dicopper(II) complexes with essentially dinucleating ligands, for which the reduction mode of dioxygen to dihydrogen peroxide has been definitely established, also possess a large metal–metal separation (3.7 Å and 7.8 Å) [55].

It is also interesting to consider the role of copper–dioxygen species in the catalytic cycle. The studies on the copper(II) complex with the ligand [22]py4pz showed that the catalytic oxidation of catechol may also proceed via the formation of trans-μ-1,2-peroxido-dicopper(II) species; however, the latter species is incapable of oxidizing the ortho-position of phenol rings, for example, of exhibiting a tyrosinase activity. This in turn opens possible speculation about the difference in the reactivity of the two enzymes being related to the dioxygen binding mode.

In conclusion, the mechanism of catechol oxidation by the model compounds is very intricate, which obviously explains often contradictory literature reports on the catalytic behavior of copper(II) complexes. However, despite being sometimes controversial, studies on model compounds offer stimulating results, which improve our knowledge of the structure–activity relationships in natural systems. There is little doubt that the combination of distinct but complementary disciplines of contemporary chemistry–biochemistry, synthetic, and inorganic chemistry, and spectroscopy–will be of paramount importance in order to elucidate the ingenious ways in which Nature operates. This understanding is essential for a sustainable development of industrial chemistry, since effective, selective, and ecologically friendly catalysts may be produced via a biomimetic approach.

5.7
Acknowledgments

Much of the work summarized in this chapter was performed in close collaboration with Dr. Catherine Belle, Dr. Katalin Selmeczi and Prof. Dr. Eric Saint-Aman from the J. Fourier University (Grenoble, France), and with the group of Prof. B. Krebs from the Westfalische Wilhelms University Münster (Germany) and with the department of molecular crystallography at Utrecht University (The Netherlands). I.A.K. acknowledges the organizers of the European Young Chemist Award at the first European Congress on Chemistry (Budapest, Hungary, 2006) for their invitation to write this chapter.

References

1 Gray, H.B., Malmström, B.G. and Williams, R.J.P. (2000) *Journal of Biological Inorganic Chemistry: JBIC: A Publication of the Society of Biological Inorganic Chemistry*, **5**, 551–9.
2 Guckert, J.A., Lowery, M.D. and Solomon, E.I. (1995) *Journal of the American Chemical Society*, **117**, 2817–44.
3 Solomon, E.I., Sundaram, U.M. and Machonkin, T.E. (1996) *Chemical Reviews*, **96**, 2563–605.
4 Solomon, E.I., Baldwin, M.J. and Lowery, M.D. (1992) *Chemical Reviews*, **92**, 521–42.
5 Ettinger, M.J. (1974) *Biochemistry*, **13**, 1242–7.
6 Solomon, E.I., Hemming, B.L. and Root, D.E. (1993) *Bioinorganic Chemistry of Copper*, (Eds.: Karlin, K.D. and Tyeklar, Z.), Kluwer Academic Publishers Group, Dordrecht, 3–20.
7 Piontec, K., Antorini, M. and Choinowski, T. (2002) *The Journal of Biological Chemistry*, **277**, 37663–9.
8 Claus, H. (2004) *Micron*, **35**, 93–6.
9 Bertrand, T., Jolivalt, C., Briozzo, P., Caminade, E., Joly, N., Madzak, C. and Mougin, C. (2002) *Biochemistry*, **41**, 7325–33.
10 Messerschmidt, A., Rossi, A., Ladenstein, R., Huber, R., Bolognesi, M., Gatti, G., Machesini, A., Petruzzelli, R. and Finazzi-Agró, A. (1989) *Journal of Molecular Biology*, **206**, 513–29.
11 Musci, G. (2001) *Protein and Peptide Letters*, **8**, 156–69.
12 Itawa, S., Ostermeier, C., Ludwig, B. and Michel, H. (1995) *Nature*, **376**, 660–9.
13 Tsukihara, T., Aoyama, H., Yamashita, E., Tomizaki, T., Yamaguchi, H., Shinzawa-Itoh, R., Nakashima, R., Yaono, R. and Yoshikawa, S. (1995) *Science*, **269**, 1069–74.
14 Wilmanns, M., Lappalainen, P., Kelly, M., Sauer-Eriksson, E. and Saraste, M. (1995) *Proceedings of the National Academy of Sciences of the United States of America*, **92**, 11955–9.
15 Kadenbach, B. (1995) *Angewandte Chemie (International Ed. in English)*, **34**, 2635–7.
16 Brown, K., Djinovic-Carugo, K., Haltia, T., Cabrito, I., Saraste, M., Moura, J.J.G., Moura, I., Tegoni, M. and Cambillau, C.

(2000) *The Journal of Biological Chemistry*, **275**, 41133–6.

17 Haltia, T., Brown, K., Tegoni, M., Cambillau, C., Saraste, M., Mattila, K. and Djinovic-Carugo, K. (2003) *The Biochemical Journal*, **369**, 77–88.

18 Brown, K., Dijnovic-Carugo, K., Haltia, T., Cabrito, I., Saraste, M., Moura, J.J., Moura, I., Tegoni, M. and Cambillau, C. (2000) *The Journal of Biological Chemistry*, **275**, 41133–6.

19 Klabunde, T., Eicken, C., Sacchettini, J.C. and Krebs, B. (1998) *Nature Structural Biology*, **5**, 1084–90.

20 Solomon, E.I., Tuczek, F., Root, D.E. and Brown, C.A. (1994) *Chemical Reviews*, **94**, 827–56.

21 Wilcox, D.E., Porras, A.G., Hwang, Y.T., Lerch, K., Winkler, M.E. and Solomon, E.I. (1985) *Journal of the American Chemical Society*, **107**, 4015–27.

22 Grinstead, R.R. (1964) *Biochemistry*, **3**, 1308–14.

23 Thuji, J. and Takayanagi, H. (1974) *Journal of the American Chemical Society*, **96**, 7349–50.

24 Rogic, M.M. and Demmin, T.R. (1978) *Journal of the American Chemical Society*, **98**, 7441–3.

25 Thuruya, S. and Lintvedt, R.L. (1978) *176th National Meeting of the American Chemical Society* (Miami, Sept. 1978).

26 Demmin, T.R., Swerdloff, M.D. and Rogic, M.M. (1991) *Journal of the American Chemical Society*, **103**, 5795–804.

27 Oishi, N., Nishida, Y., Ida, K. and Kida, S. (1980) *Bulletin of the Chemical Society of Japan*, **53**, 2847–50.

28 Malachowski, M.R. (1989) *Inorganica Chimica Acta*, **162**, 199–204.

29 Casellato, U., Tamburini, S., Vigato, P.A., de Stefani, A., Vidali, M. and Fenton, D.E. (1983) *Inorganica Chimica Acta*, **69**, 45–51.

30 Karlin, K.D., Gultneh, Y., Nicholson, T. and Zubieta, J. (1985) *Inorganic Chemistry*, **24**, 3725–7.

31 Thompson, J.S. and Calabrese, J.C. (1986) *Journal of the American Chemical Society*, **108**, 1903–7.

32 Speier, G. (1994) *New Journal of Chemistry*, **18**, 143–7.

33 Kodera, M., Kawata, T., Kano, K., Tachi, Y., Itoh, S. and Kojo, S. (2003) *Bulletin of the Chemical Society of Japan*, **76**, 1957–64.

34 Kaizer, J., Pap, J., Speier, G., Parkanyi, L., Korecz, L. and Rockenbauer, A. (2002) *Journal of Inorganic Biochemistry*, **91**, 190–8.

35 Chyn, J.-P. and Urbach, F.L. (1991) *Inorganica Chimica Acta*, **189**, 157–63.

36 Balla, J., Kiss, T. and Jameson, R.F. (1992) *Inorganic Chemistry*, **31**, 58–62.

37 Monzani, E., Battaini, G., Perotti, A., Casella, L., Gullotti, M., Santagostini, L., Nardin, G., Randaccio, L., Geremia, S., Zanello, P. and Opromolla, G. (1999) *Inorganic Chemistry*, **38**, 5359–69.

38 Monzani, E., Quinti, L., Perotti, A., Casella, L., Gulotti, M., Randaccio, L., Geremia, S., Nardin, G., Faleschini, P. and Tabbi, G. (1998) *Inorganic Chemistry*, **37**, 553–62.

39 Monzani, E., Casella, L., Zoppellaro, G., Gullotti, M., Pagliarin, R., Bonomo, R., Tabbi, G., Nardin, G. and Randaccio, L. (1998) *Inorganica Chimica Acta*, **282**, 180–92.

40 Granata, A., Monzani, E. and Casella, L. (2004) *Journal of Biological Inorganic Chemistry: JBIC: A Publication of the Society of Biological Inorganic Chemistry*, **9**, 903–13.

41 Börzel, H., Comba, P. and Pritzkow, H. (2001) *Chemical Communications*, 97–8.

42 Ackermann, J., Meyer, F., Kaifer, E. and Pritzkow, H. (2002) *Chemistry – A European Journal*, **8**, 247–58.

43 Plenge, T., Dillinger, R., Santagostini, L., Casella, L. and Tuczek, F. (2003) *Zeitschrift für anorganische und allgemeine Chemie*, **629**, 2258–65.

44 Eicken, C., Krebs, B. and Sacchettini, J.C. (1999) *Current Opinion in Structural Biology*, **9**, 677–83.

45 Komarov, D.A., Slepneva, I.A., Glupov, V.V. and Khramtsov, V.V. (2005) *Free Radical Research*, **39**, 853–8.

46 Slepneva, I.A., Komarov, D.A., Glupov, V.V., Serebrov, V.V. and Khramtsov, V.V. (2003) *Biochemical and Biophysical Research Communications*, **300**, 188–91.

47 Koval, I.A., Belle, C., Selmeczi, K., Philouze, C., Saint-Aman, E., Schuitema,

A.M., Gamez, P., Pierre, J.-L. and Reedijk, J. (2005) *Journal of Biological Inorganic Chemistry: JBIC: A Publication of the Society of Biological Inorganic Chemistry*, **10**, 739–50.

48 Koval, I.A., Selmeczi, K., Belle, C., Philouse, C., Saint-Aman, E., Schuitema, A.M., van Vliet, M., Gamez, P., Roubeau, O., Lüken, M., Krebs, B., Lutz, M., Spek, A.L., Pierre, J.-L. and Reedijk, J. (2006) *Chemistry–A European Journal*, **12**, 6138–50.

49 Koval, I.A., van der Schilden, K., Schuitema, A.M., Gamez, P., Belle, C., Pierre, J.-L., Luken, M., Krebs, B., Roubeau, O. and Reedijk, J. (2005) *Inorganic Chemistry*, **44**, 4372–82.

50 Jacobson, R.R., Tyeklar, Z., Karlin, K.D., Liu, S. and Zubieta, J. (1988) *Journal of the American Chemical Society*, **110**, 3690–2.

51 Paul, P. P., Tyeklár, Z., Jacobson, R.R. and Karlin, K.D. (1991) *Journal of the American Chemical Society*, **113**, 5322–32.

52 Jung, B., Karlin, K.D. and Zuberbühler, A.D. (1996) *Journal of the American Chemical Society*, **118**, 3763–4.

53 Neves, A., Rossi, L.M., Bortoluzzi, A.J., Szpoganicz, B., Wiezbicki, C., Schwingel, E., Haase, W. and Ostrovsky, S. (2002) *Inorganic Chemistry*, **41**, 1788–94.

54 Torelli, S., Belle, C., Hamman, S., Pierre, J.L. and Saint-Aman, E. (2002) *Inorganic Chemistry*, **41**, 3983–9.

55 Selmeczi, K., Reglier, M., Giorgi, M. and Speier, G. (2003) *Coordination Chemistry Reviews*, **245**, 191–201.

56 Colman, P. M., Freeman, H.C., Guss, J.M., Norris, V.A., Ramshaw, J.A.M. and Venkatappa, M.P. (1978) *Nature*, **272**, 319–24.

57 Ito, N., Phillips, S., Stevens, C., Ogel, Z.B., McPherson, M.J., Keen, J.N., Yadav, K.D.S. and Knowles, P.F. (1991) *Nature*, **350**, 87–90.

58 Volbeda, A. and Hol, W.G. (1989) *Journal of Molecular Biology*, **209**, 249–79.

59 Prudêncio, M., Pereira, A.S., Tavares, P., Besson, S., Cabrito, I., Brown, K., Samyn, B., Devreese, B., Cambillau, C. and Moura, I. (2000) *Biochemistry*, **39**, 3899–907.

60 Gerdemann, C., Eicken, C. and Krebs, B. (2002) *Accounts of Chemical Research*, **35**, 183–91.

6
From the Past to the Future of Rotaxanes

Andreea R. Schmitzer

6.1
Introduction

The construction of useful devices is the essence of technology, and has always been a key issue for human development. In general, a device is an assembly of components designed to achieve a specific function, resulting from the cooperation of the acts performed by each component. Another distinctive feature of a device that has grown in importance is its size. In the last fifty years, many fields of technology, in particular information processing, have benefited from progressive miniaturization of the components of devices. A common prediction is that further progress in miniaturization will not only decrease the size and increase the power of computers, but could also open the way to new technologies in the fields of medicine, environment, energy, and materials.

Research in supramolecular chemistry has shown that molecules are convenient nanometer-scale building blocks that can be used, in a bottom-up approach, to construct ultraminiaturized devices and machines. Chemists are in an ideal position to develop such a molecular approach to functional nanostructures because they are able to design, synthesize, investigate, and organize molecules – that is, make them react or bring them together into larger assemblies.

Much of the inspiration to construct molecular devices and machines comes from the outstanding progress in molecular biology that has begun to reveal the secrets of the natural nanodevices that constitute the material base of life [1]. Surely, the supramolecular architectures of the biological world are themselves the premier, proven examples of the feasibility and utility of nanotechnology, and constitute a sound rationale for attempting the realization of artificial molecular devices [2,3]. Chemists have tried to construct much simpler systems, without mimicking the complexity of the biological structures. The aim is to investigate the challenging problems posed by interfacing artificial molecular devices with the macroscopic world, particularly as far as energy supply and information exchange are concerned. In the last few years the development of powerful synthetic methodologies, combined with a device-driven ingenuity evolved from the attention to

---

Tomorrow's Chemistry Today. Concepts in Nanoscience, Organic Materials and Environmental Chemistry.
Edited by Bruno Pignataro
Copyright © 2008 WILEY-VCH Verlag GmbH & Co. KGaA, Weinheim
ISBN: 978-3-527-31918-3

functions and reactivity, has led to remarkable achievements in the field of molecular devices [4].

A thorough knowledge and understanding of molecular recognition processes – particularly those based upon hydrogen bonding [5], metal–ligand coordination [6], π–π interactions [7, 8], hydrophobic interaction [9], ion pairing [10] and van der Waals [11] interactions – facilitates the rational design of small, relatively simple building blocks that are capable of assembling into larger superstructures. In effect, as a consequence of the synthetic chemist's own judicious design, the components do all of the hard work by themselves, self-assembling [12] into extended arrays by virtue of complementary recognition features. Application of the chemistry of the noncovalent bond to the challenge of fabricating larger and larger "structures," has proved to be very effective, resulting in the creation of many elaborate, and intricate, supramolecular architectures [10–13]. Supramolecular chemistry has also assisted in the realization of novel molecular topologies [14, 15] held together by mechanical bonds.[1]

The serendipitous discovery of crown ethers by Pedersen [16] just over thirty years ago is arguably the origin of many aspects of supramolecular chemistry [17] as we know it today. Although not the first to synthesize macrocyclic polyethers, it was Pedersen who realized their importance in the context of host–guest chemistry and went on to investigate the binding properties of crown ether hosts with a wide variety of cationic guests [18] . Among the many guests studied [19, 20], the ammonium (NH_4^+) and primary alkylammonium (RNH_3^+) ions were shown [21] by Cram and others to bind dibenzo[18]crown-6 in a face-to face manner (Figure 6.1).

The binding of secondary dialkylammonium ions ($R_2NH_2^+$) was largely unexplored [22] until recent times when it was discovered that, if a [24]crown-8-containing macrocyclic polyether is employed, the $R_2NH_2^+$ ion can interpenetrate fully the macrocyclic cavity, generating a supramolecular complex in which the cation is threaded through the center of the crown ether [23]. This breakthrough heralded the arrival of a new paradigm for the construction of discrete interlocked molecules such as rotaxanes, pseudorotaxanes, catenanes [24], and extended interwoven supramolecular arrays [25].

Interlocked molecules such as rotaxanes initially gained interest due to their interesting topology and associated synthetic challenge, but recent efforts have showed that they can be used in many important applications that will be discussed in this chapter (Scheme 6.1).

The common representation of a rotaxane is that of a thin axle threaded through the cavity of a macrocycle, which is hindered from dethreading by two bulky

---

1) Unlike classical molecular structures, interlocked molecules consist of two or more separate components which are not connected by chemical (i.e. covalent) bonds. These structures are true molecules and not a supramolecular species, as each component is intrinsically linked to the other, resulting in a mechanical bond which prevents dissociation without cleavage of one or more covalent bonds. It should be noted that the "mechanical bond" is a relatively new terminology and at this point has limited usage in chemical literature relative to more well-established bonds, such as covalent, hydrogen, or ionic bonds.

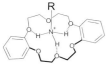

Figure 6.1 The face-to-face interaction between [18]crown-6-containing macrocyclic polyether DB18C6 and a primary ammonium ion.

Scheme 6.1 Schematic representation of interlocked molecules: (a) pseudorotaxane.; (b) rotaxane; and (c) catenane.

stopper groups at the axle ends. The word rotaxane is derived from the Latin "rota" meaning wheel and "axis" meaning axle. In chemistry, [2]rotaxanes (the prefix indicates the number of interlocked components) are a group of compounds in which a dumbbell-shaped molecule is encircled by a macrocycle. No covalent bonds hold the components together, but rather mechanical bonds are responsible for the linking of the components. A pseudorotaxane is a supramolecular system composed only of a thread-like species inserted through the cavity of a macrocycle. Since there are no stoppers at the ends of the thread, dissociation of the complex can occur and the pseudorotaxane is always equilibrated with the "free" molecular components. A similar concept is used in [2]catenanes where two cyclic molecules are mechanically linked with each other. The disruption of a rotaxane or catenane into its separate components requires the breaking of one or more covalent bonds in the mechanically linked molecule. Thus, rotaxanes and catenanes behave as well-defined molecular compounds with properties significantly different from those of their individual components.

6.2 Synthesis of Rotaxanes

For the preparation of rotaxanes, three different routes can be followed (Scheme 6.2A). In the "threading" procedure, the macrocycle will first encircle the thread to form a so-called pseudorotaxane. By end-capping the thread with bulky groups that prevent de-threading, a [2]rotaxane is formed. The "clipping" method that is used for the preparation of catenanes can also be applied for the synthesis of rotaxanes; hence the macrocycle is assembled in the presence of the end-capped

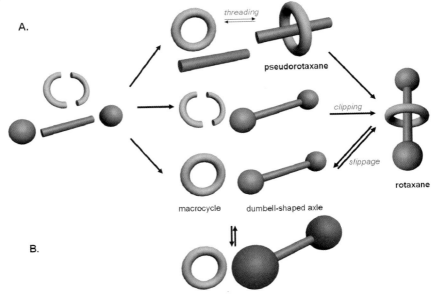

Scheme 6.2 Schematic representation of different strategies employed for the synthesis of [2]rotaxanes.

thread or dumbbell. As a third classical possible route, careful selection of the size of the macrocycle can allow threading of the cyclic molecule over the blocking groups at elevated temperatures. In addition, a new fourth route giving access to [2]- and [3]rotaxanes has been developed recently, in which the photoinduced conformational change of the axle's bulky end groups leads to a pseudorotaxane-like complex and allows the threading of the macrocycle over the irradiated, less-hindered end groups. UV-controlled assembly of rotaxanes in this way has been achieved by means of photocyclizable fulgide end groups (Scheme 6.2B) [26].

The choice of one of these four routes for the preparation of a rotaxane depends mainly on the chemical nature of the different components and the chemistry required to establish the interlocked molecule. An interaction between the two individual components is very often the driving force in the synthesis of rotaxanes.

6.2.1
Van der Waals Interactions in the Synthesis of Rotaxanes

The first rotaxanes were prepared by a statistical synthetic approach. The threading of a macrocycle around a linear chain is based purely on chance; no specific

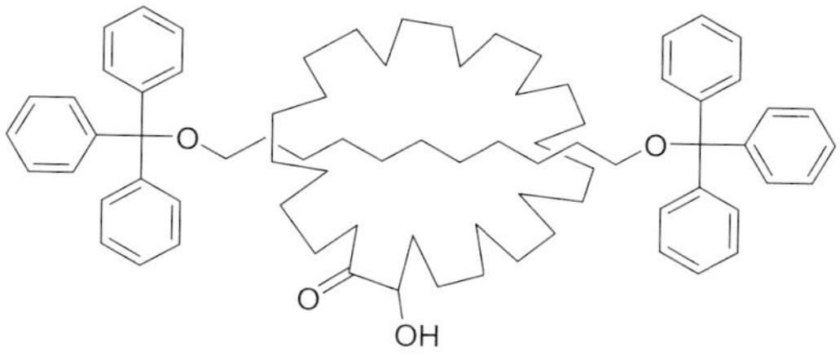

Figure 6.2 Example of a [2]rotaxane based on van der Waals interactions.

interactions are involved to prepare the interlocked molecules. Since the threading of the macrocycle is entropically unfavorable, very low yields were obtained. For example the [2]rotaxane [27] **1** was prepared in a 0.1% yield by the threading method (Figure 6.2).

Fortunately, more efficient methods for the complexation of macrocyclic hosts with acyclic guest molecules have become available with the advent of supramolecular chemistry, resulting in higher yields in rotaxane and catenane synthesis. In the following sections, the preparation of different types of interlocked molecules, with the use of host–guest recognition, is discussed. It should be noted that these template-directed methods differ significantly from the above-mentioned stochastic approach [28].

6.2.2
Hydrophobic Interactions in the Synthesis of Rotaxanes

The most extensively studied type of rotaxanes, which possess hydrophobic interactions between the axle and the wheel, are cyclodextrin-based structures. Cyclodextrins (CDs) are cyclic oligosaccharides consisting of six or more α-1,4-linked D-glucopyranose rings [29]. The conformation of CDs provides a rigid, well-defined cavity with a conical shape. The wide side of the cavity is encircled by secondary hydroxyl groups (two per glucopyranose unit), the narrow side is encircled with primary hydroxyl groups (one per glucopyranose unit). These groups give the molecule a hydrophilic exterior, which makes it water-soluble. In contrast, the inside cavity is relatively hydrophobic, giving the molecule the ability to complex a wide variety of molecular guests in water [30]. The binding interaction in a CD–guest complex is based on a summation of weak effects, namely van der Waals interactions, hydrophobic binding [31], and the release of "high-energy water" from the cavity [29a].

Examples of cyclodextrin-based [2]rotaxanes, which reflect the diversity of possible structures, are depicted in Figure 6.3. Aliphatic as well as aromatic compounds can complex with CDs. Different end groups can be used in the rotaxanes to prevent the dethreading of the CD, even within one rotaxane molecule as in **2** [32].

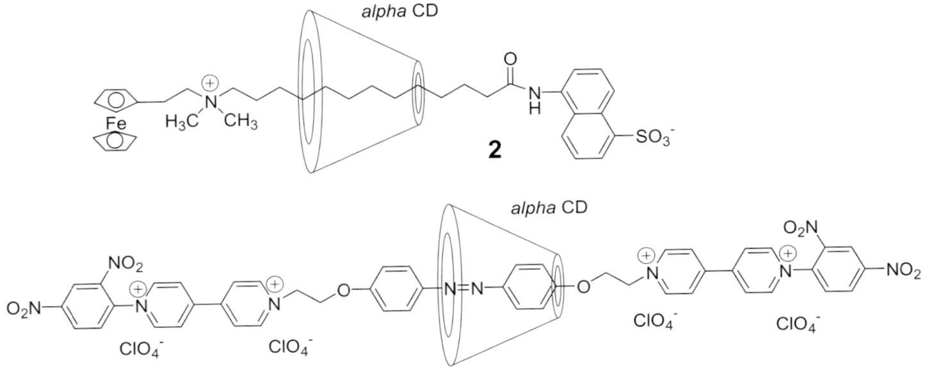

Figure 6.3 Examples of two cyclodextrin-based [2]rotaxanes possessing mainly hydrophobic interactions.

6.2.3
Hydrogen Bonding in Rotaxane Synthesis

One of the best-known examples of molecules held together by hydrogen bonds is DNA. Unnatural DNA in the form of catenanes has already been synthesized [33] and is seen as the forerunner to new types of drug delivery systems and nanoscale mechanical devices. Recently, rotaxanes [34] have also been found in polypeptides and proteins. Although these structures might be expected to be observed only in complex biological systems, examples of synthetic rotaxanes based on hydrogen bonds have also been described. For example, Leigh and co-workers [35] have synthesized tertiary amide-based rotaxanes, as depicted in Figure 6.4.

The rotaxane is composed of a glycylglycine chain, a simple dipeptide, bearing two diphenylmethane blocking groups, which prevent the dethreading of the benzylic amide macrocycle. The synthesis of the rotaxane structure is based on

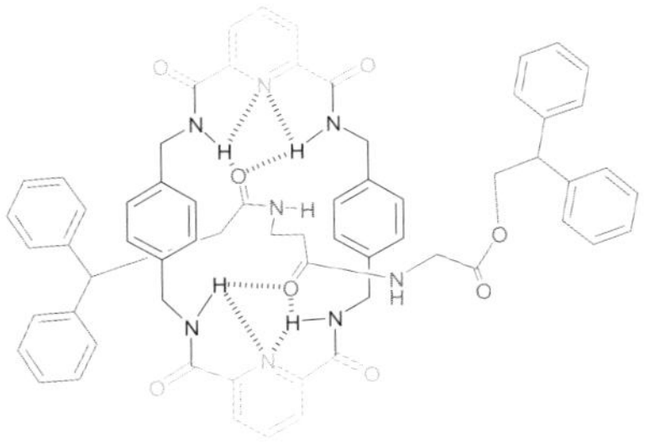

Figure 6.4 A [2]rotaxane based on hydrogen bonding interactions.

the "clipping" procedure. Due to hydrogen bonding, four small molecules (represented in different colors in Figure 6.4) are complexed around the glycylglycine template and subsequent reaction with each other gives a [2]rotaxane in ~60% yield. In complex **4**, four hydrogen bonds are formed between the cyclic unit and the carbonyl groups of the dipeptide thread. The structure of this rotaxane is rigid in apolar solvents because of the strong hydrogen bonds in the molecule. However, when they are dissolved in a polar solvent such as DMSO, which can break the hydrogen bonding between the peptide and the macrocycle, different conformations are observed [36].

6.2.4
Donor–Acceptor Interactions in the Synthesis of Rotaxanes

Rotaxanes possessing donor–acceptor interactions have been studied extensively by Stoddart and co-workers [37]. Their work is mainly based on the combination of π-electron-deficient bipyridinium and π-electron-rich hydroquinone moieties. The donor–acceptor interactions between these π-systems have already been studied for over 40 years [38] and have been applied in the template-directed synthesis of interlocked molecules after the discovery of the complexes **5** [39] and **6** [40] (Figure 6.5). In [2]pseudorotaxane **5**, the 1,1′-dimethyl-4,4′-bipyridinium dication (the bipyridinium herbicide paraquat) is threaded inside the crown ether. A combination of two types of interactions is present, namely [C—H···O] hydrogen bonding between some of the polyether oxygen atoms and the α-bipyridinium protons and [π···π] stacking between the complementary aromatic units. In the [2]pseudorotaxane **6**, the 1,4-dioxybenzene derivative is inserted through the center of the rigid cavity of the cyclobis(paraquat-p-phenylene) tetracation. The same noncovalent bonds, supplemented by [C—H···π] interactions between the 1,4-dioxybenzene protons and the p-phenylene protons in the tetracationic cyclophane component, assist in the complexation in the [2]pseudorotaxane.

The observation that it was possible to obtain inclusion complexes of π-electron-deficient guests in a π-electron-rich cavity, and vice versa, suggested that

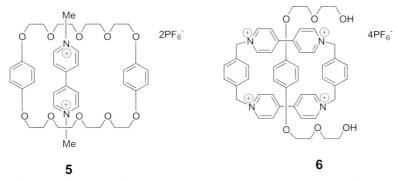

Figure 6.5 Examples of [2]pseudorotaxanes based on donor–acceptor interactions.

a combination of these two complexes within one molecular assembly should be feasible and has led to the synthesis of a [2]catenane [41], in which hydrogen bonding and [π···π] stacking are again present.

6.2.5
Transition-Metal Coordination in the Synthesis of Rotaxanes

Among the strongest types of interactions used in the synthesis of interlocked molecules is the metal coordination of organic ligands. For example Cu^+ ions coordinate tetrahedrally with phenanthroline groups (Figure 6.6). The synthesis and studies of rotaxanes and catenanes, based on these interactions, have been extensively studied since 1983 by Sauvage's group [42]. Due to the strength of the interactions involved, high yields up to 92% can be obtained in the preparation of this type of rotaxanes [43]. After the interlocked molecule has been synthesized, it can be demetallated and leave the two interlocked parts free to move relative to each other. This metal coordination is exploited as a tool to increase the yields in the synthesis of these molecules. The [2]rotaxane 7 [44], in which the Cu^{2+} ion is still present, is interesting since two nonequivalent locations are present in the thread at which the Cu complex can be stationed. By an external stimulus it is possible to alter the position of the macrocycle on the thread in a controlled way. This is the basis of controllable molecular shuttles and switches, which will be discussed later.

The development of good synthetic methods for the preparation of rotaxanes in moderate to good yields has opened new ways to explore the physical properties

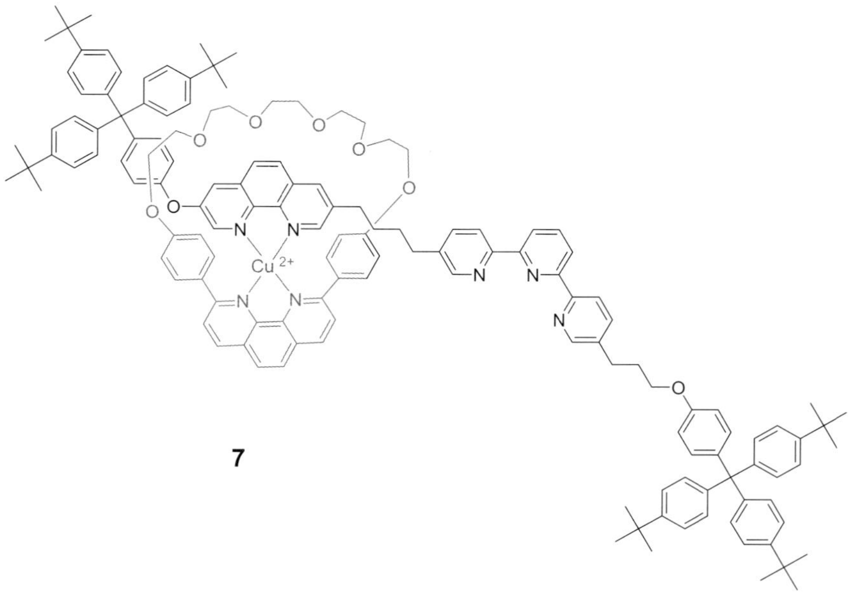

Figure 6.6 Example of a [2]rotaxane based on transition metal coordination.

of these compounds. Although devices based on rotaxanes have not been commercialized yet, much research is done on applications for these interlocked molecules. A promising application is in the field of nanotechnology. The miniaturization of semiconductor technology by the "top-down" approach is limited by the wavelength of the radiation used in the lithography to prepare electronic components. Therefore, the "bottom-up" approach may be more suitable to create rotaxane-based devices with sizes down to the molecular level. Molecular or supramolecular systems, in which the relative position of the component parts can be altered by an external stimulus (i.e. a molecular switch) can act as the basic unit for the creation of new molecular devices and are discussed in the next sections.

6.3
Applications of Rotaxanes

6.3.1
Rotaxanes as Molecular Shuttles

The conformational freedom in most rotaxanes leads to random motion. The motion of the ring can be controlled if the rotaxane has two potential binding sites, referred to as stations (Scheme 6.3). Rotaxanes are appealing systems for the construction of molecular machines because (i) the mechanical bond allows a large variety of mutual arrangements of the molecular components, while conferring stability to the system; (ii) the interlocked architecture limits the amplitude of the intercomponent motion in the three directions; (iii) the stability of a specific arrangement (co-conformation) is determined by the strength of the intercomponent interactions; and (iv) such interactions can be modulated by external stimulation. In particular, two interesting molecular motions can be envisaged in rotaxanes, namely (i) translation, that is, shuttling, of the ring along the axle, and (ii) rotation

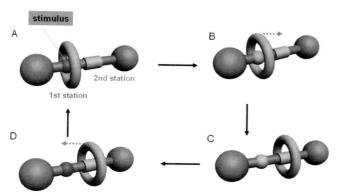

Scheme 6.3 Principle of shuttling in a [2]rotaxane.

of the ring around the axle. Hence, rotaxanes are good prototypes for the construction of both linear and rotary molecular motors. Systems of type (i), termed molecular shuttles (Scheme 6.3), constitute the most common implementation of the molecular machine concept with rotaxanes.

In the initial state the ring binds to the recognition site (first station) (Scheme 6.3A). The macrocycle can move from the first station to the second when the relative binding energy of one of the stations is changed. The second station is changed by applying an external stimulus, either chemical, electrochemical or photochemical, in such a way that it becomes a better binding site for the macrocycle (Scheme 6.3B). Alternatively, the first station can be changed in such a way that it becomes a worse binding site. This causes the macrocycle to move from the first to the second station (Scheme 6.3C). The macrocycle moves back to station 1 when the axle reverses to its original state (Scheme 6.3D), either by another stimulus or by a spontaneous process.

The words motor and machine are often used interchangeably when referring to molecular systems. It should be noted that a motor converts energy into mechanical work, while a machine is a device, usually containing a motor component, designed to accomplish a function. Molecular machines and motors operate via electronic and/or nuclear rearrangements and, like the macroscopic ones, are characterized by (i) the kind of energy input supplied to make them work; (ii) the type of motion (linear, rotatory, oscillatory, . . .) performed by their components; (iii) the way in which their operation can be monitored; (iv) the possibility of repeating the operation at will (cyclic process); and (v) the timescale needed to complete a cycle. According to the view described above, an additional and very important distinctive feature of a molecular machine with respect to a molecular motor is (vi) the function performed [45].

As far as point (i) is concerned, a chemical reaction can be used, at least in principle, as an energy input. In such a case, however, if the machine has to work cyclically [point (iv)], it will require the addition of reactants at any step of the working cycle, and the accumulation of by-products resulting from the repeated addition of matter can compromise the operation of the device. On the basis of this consideration, the best energy inputs to make a molecular device work are photons [46] and electrons [47]. It is indeed possible to design very interesting molecular devices based on appropriately chosen photochemically and electrochemically driven reactions. In order to control and monitor the device operation [point (iii)], the electronic and/or nuclear rearrangements of the component parts should cause readable changes in some chemical or physical property of the system. In this regard, photochemical and electrochemical techniques are very useful since both photons and electrons can play the dual role of "writing" (i.e. causing a change in the system) and "reading" (i.e. reporting the state of the system). The operation timescale of molecular devices [point (v)] can range from less than picoseconds to seconds, depending on the type of rearrangement (electronic or nuclear) and the nature of the components involved. Finally, as far as point (vi) is concerned, the functions that can be performed by exploiting the movements of the component parts in molecular machines are various and, to a large extent, still unpredictable. It is worth noting that the mechanical movements taking place in molecular-level machines, and the related changes in the spectroscopic and electrochemical properties, usually obey binary logic and can thus be taken as a basis for information processing at the molecular level. Artificial molecular machines capable of performing logic operations have been reported [48].

6.3.1.1 Acid–Base-controlled Molecular Shuttle

In rotaxanes containing two different recognition sites or stations along the dumbbell-shaped axle, it is possible to switch the position of the ring between the two "stations" by an external stimulus. A system which behaves as a chemically controllable molecular shuttle is compound 8^{3+} shown in Figure 6.7 [49]. It is made of a dibenzo[24]crown-8 (DB24C8) macrocycle and a dumbbell-shaped component containing a dialkylammonium center and a 4,4'-bipyridinium unit. An anthracene moiety is used as a stopper because its absorption, luminescence, and redox properties are useful to monitor the state of the system. Since the $N^+\!\!-\!\!H\cdots O$ hydrogen-bonding interactions between the DB24C8 macrocycle and the ammonium center are much stronger than the electron donor–acceptor interactions of the macrocycle with the bipyridinium unit, the rotaxane exists as only one of the two possible translational isomers. Deprotonation of the ammonium center with a base (a tertiary amine) causes 100% displacement of the macrocycle to the bipyridinium unit; reprotonation directs the macrocycle back onto the ammonium center. Such a switching process has been investigated in solution by ^{1}H NMR spectroscopy and by electrochemical and photophysical measurements. The full chemical reversibility of the energy supplying acid–base reactions guarantees the reversibility of the mechanical movement, in spite of the formation of waste products. Note that this system may be useful for information processing since it exhibits a binary logic behavior. It should also be noted that, in the deprotonated rotaxane, it is possible to displace the crown ring from the bipyridinium station by destroying the donor–acceptor interaction through reduction of the bipyridinium station or oxidation of the dioxybenzene units of the macrocyclic

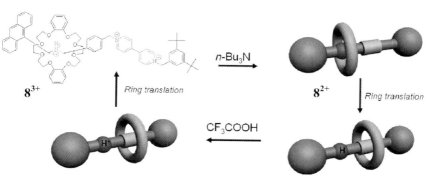

Figure 6.7 A chemically controllable molecular shuttle. The macrocyclic ring can be switched between the two stations of the dumbbell-shaped component by acid–base inputs.

ring. Therefore, in this system, mechanical movements can be induced by two different types of stimuli (acid–base and electron–hole).

6.3.1.2 A Light-driven Molecular Shuttle

For a number of reasons, light is the most convenient form of energy to make artificial molecular machines work [50]. In order to achieve photoinduced ring shuttling in rotaxanes containing two different recognition sites in the dumbbell-shaped component, the carefully designed compound 9^{6+} (Figure 6.8) was synthesized [51]. This compound is made of the electron-donor macrocycle, and a dumbbell-shaped component which contains (i) [Ru(bpy)3]$^{2+}$ (P) as one of its stoppers; (ii) a 4,4'-bipyridinium unit (A1) and a 3,3'-dimethyl-4,4'-bipyridinium unit (A2) as electron-accepting stations; (iii) a *p*-terphenyl-type ring system as a rigid spacer (S); and (iv) a tetraarylmethane group as the second stopper (T). The structure of rotaxane 9^{6+} was characterized by mass spectrometry, ^{1}H NMR spectroscopy and cyclic voltammetry. These methods established that the stable translational isomer is the one in which the macrocycle encircles the A1 unit. This is expected as the A1 station is a better electron acceptor than the A2 station. The electrochemical, photophysical and photochemical (under continuous and pulsed excitation) properties of the rotaxane, its dumbbell-shaped component, and some model compounds have been investigated and two strategies have been devised in order to obtain the photoinduced abacus-like movement of the macrocycle between the two stations A1 and A2: one was based on a processes involving only the rotaxane components (intramolecular mechanism), while the other required the help of external reactants (sacrificial mechanism).

The intramolecular mechanism, illustrated on the left-hand side of Figure 6.8, is based on four separate operations [52]. (a) *Destabilization of the stable translational isomer:* light excitation of the photoactive unit P (step 1) is followed by the transfer of an electron from the excited state to the A1 station, which is encircled by the macrocycle (step 2); with the consequent "deactivation" of this station; such a photoinduced electron-transfer process has to compete with the intrinsic decay of P (step 3). (b) *Ring displacement:* the ring moves from the reduced station A1$^-$ to A2 (step 4), a step that has to compete with the back electron-transfer process from A1$^-$ (still encircled by the macrocycle) to the oxidized photoactive unit P$^+$ (step 5). This is the most difficult requirement to meet in the intramolecular mechanism. (c) *Electronic reset:* a back electron-transfer process from the "free" reduced station A1$^-$ to P$^+$ (step 6) restores the electron-acceptor power to the A1 station. (d) *Nuclear reset:* as a consequence of the electronic reset, back movement of the ring from A2 to A1 takes place (step 7).

It is worth noting that in a system which behaves according to the intramolecular mechanism shown in Figure 6.8, each light input causes the occurrence of a forward and back ring movement (i.e. a full cycle) without generation of any waste product. In some way, it can be considered as a "four-stroke" cyclic linear motor powered by light.

A less demanding mechanism is based on the use of external sacrificial reactants (a reductant like triethanolamine and an oxidant like dioxygen) that operate as

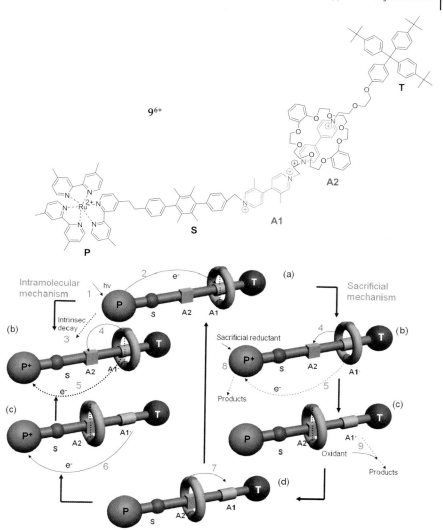

Figure 6.8 Structural formula of the rotaxane 9^{6+} and schematic representation of the intramolecular (below left) and sacrificial (below right) mechanisms for the photoinduced shuttling movement of macrocycle between the two stations A1 and A2.

illustrated in the right-hand side of Figure 6.8. (a) *Destabilization of the stable translational isomer*, as in the previous mechanism. (b) *Ring displacement after scavenging of the oxidized photoactive unit*: since the solution contains a suitable sacrificial reductant, a fast reaction of such species with P$^+$ (step 8) competes successfully with the back electron-transfer reaction (step 5); therefore, the originally occupied station remains in its reduced state A1⁻, and the displacement of the macrocycle to A2 (step 4), even if it is slow, does take place. (c) *Electronic reset:* after an appropriate time, restoration of the electron-acceptor power of the A1 station is obtained by oxidizing A1⁻ with a suitable oxidant, such as O_2 (step 9). (d) *Nuclear reset*, as in the previous mechanism (step 7). The results obtained show that such a sacrificial mechanism is fully successful. Of course, this mechanism is less appealing than the intramolecular one because it causes the formation of waste products. An alternative strategy is to use a nonsacrificial (reversible) reductant species that is regenerated after the back electron-transfer process [52].

6.3.2
Molecular Lifts

By using an incrementally staged strategy, the architectural features of the acid-controllable molecular shuttle 8^{3+} (Figure 6.9) were integrated with those of a supramolecular bundle. This bundle is obtained by self-assembly of a trifurcated compound and a tris-crown ether to afford the two-component molecular device 10_3^{9+} (Figure 6.9a) that behaves as a nanoscale lift [53]. This interlocked compound, which is ~2.5 nm in height and has a diameter of ~3.5 nm, consists of a tripod component containing two different notches – one ammonium center and one 4,4′-bipyridinium unit – at different levels in each of their three legs. The latter are interlocked by a tritopic host, which plays the role of a platform that can be made to stop at the two different levels. The three legs of the tripod carry bulky feet that prevent the loss of the platform. Initially, the platform resides exclusively on the "upper" level, that is, with the three rings surrounding the ammonium centers (Figure 6.9b). This preference results from strong N⁺–HO hydrogen bonding and weak stabilizing π–π stacking forces between the aromatic cores of the platform and tripod components. Upon addition of a strong, nonnucleophilic phosphazene base to an acetonitrile solution of $8H_3^{9+}$, deprotonation of the ammonium center occurs and, as a result, the platform moves to the lower level, that is, with the three macrocyclic rings surrounding the bipyridinium units (Figure 6.9c).

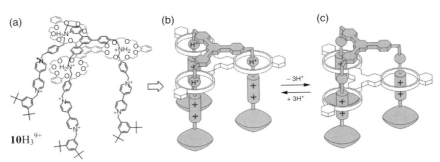

Figure 6.9 Chemical formula (a) and operation scheme in solution (b, c) of the molecular lift $10H_3^{9+}$.

This co-conformation is stabilized mainly by electron donor–acceptor interactions between the electron-rich aromatic units of the platform and the electron-deficient bipyridinium units of the tripod component. Subsequent addition of acid to 10^{6+} restores the ammonium centers, and the platform moves back to the upper level. The "up and down" liftlike motion corresponds to a quantitative switching. This motion can be repeated many times, and can be monitored by NMR spectroscopy, electrochemistry, and absorption and fluorescence spectroscopy [54]. Interestingly, the experimental results show also that the platform operates by taking three distinct steps associated with each of the three deprotonation/reprotonation processes. Hence, this molecular machine is more reminiscent of a legged animal than it is of a lift. It can also be noted that the acid–base-controlled mechanical motion in $10H_3^{9+}$ is associated with remarkable structural modifications, such as the opening and closing of a large cavity and the control of the positions and properties of the bipyridinium legs. This behavior can in principle be used to control the uptake and release of a guest molecule, a function of interest for the development of drug delivery systems.

6.3.3
Artificial Molecular Muscles

The construction of molecular machines that exhibit controlled extension–contraction motions, reminiscent of the operation of sarcomere units in skeletal muscles, is a topic of great interest. The molecular muscle concept was first implemented with artificial molecules by designing a doubly threaded rotaxane dimer [55]. Recently, mechanical actuation in a submillimeter-scale device was observed by means of a self-assembled monolayer of cleverly designed rotaxane molecules exhibiting redox-switchable musclelike motions [56]. The motor molecule is the palindromic three-component rotaxane 11^{8+} (Figure 6.10), consisting of two mechanically mobile rings encircling the same dumbbell. The two electron-deficient rings are initially located on the electron-rich tetrathiafulvalene (TTF) stations, and the inter-ring distance is approximately 4.2 nm. Chemical or electrochemical oxidation of the TTF units leads to the displacement of the two rings onto the naphthalene (NP) stations, and to an inter-ring distance of 1.4 nm. Reduction of the TTF cationic units restores the original mechanical state (Figure 6.10). Owing to the disulfide tether covalently linked to each ring, a monolayer of 11^{8+} could be assembled onto a gold surface. An array of flexible silicon microcantilever beams (500 × 100 × 1 μm), coated on one side with a monolayer of rotaxane molecules, was shown to undergo controllable and reversible bending up and down when exposed to the successive addition of an aqueous chemical oxidant $(Fe(ClO_4)_3)$ and reductant (ascorbic acid) in a transparent fluid cell. The position of each cantilever beam was monitored by an optical lever. Since a monolayer of the dumbbell component alone does not bend the cantilevers under the same conditions, the beam bending can be correlated with flexing of the surface-bound molecular muscles. Hence, the rotaxane molecules employ the free energy of a chemical reaction to do mechanical work against the spring force of the cantilever.

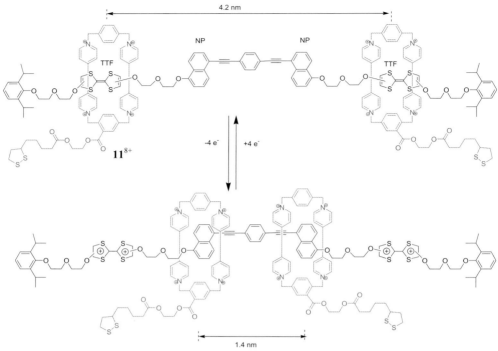

Figure 6.10 Palindromic rotaxane **11**[8+] and redox-controlled switching between its contracted and extended forms.

It is worth noting that no alignment of the rotaxane molecules with respect to the cantilevers is required to observe a bending effect because only the component of the contraction that is parallel to the long axis of the cantilever contributes effectively to the bending.

6.3.4
Redox-activated Switches for Dynamic Memory Storage

Amphiphilic rotaxane **12** comprising one monopyrrolotetrathiafulvalene (MPTTF) unit and one 1,5-dioxynaphthalene (DNP) site on the rod along with a cyclobis(paraquat-p-phenylene) (CBPQT^{4+}) ring was constructed using the clipping method by Stoddart and co-workers (Figure 6.11) [57]. As observed by cyclic voltammetry, UV-vis spectroscopy, and ^{1}H NMR spectroscopy, switching occurred when MPTTF was oxidized while the ring was encircling it, which is designated as the ground state. Upon oxidation, a MPTTF radical cation was formed, and the ring, repelled by Coulombic forces, moved over to the DNP site. Once the MPTTF unit was reduced back to its neutral starting state, the molecule existed as a metastable isomer isoelectronic with its ground state, the only difference being the location of the ring (Figure 6.11). This isomer presumably has a different conductivity than the ground state, which thus makes these molecules useful in molecular

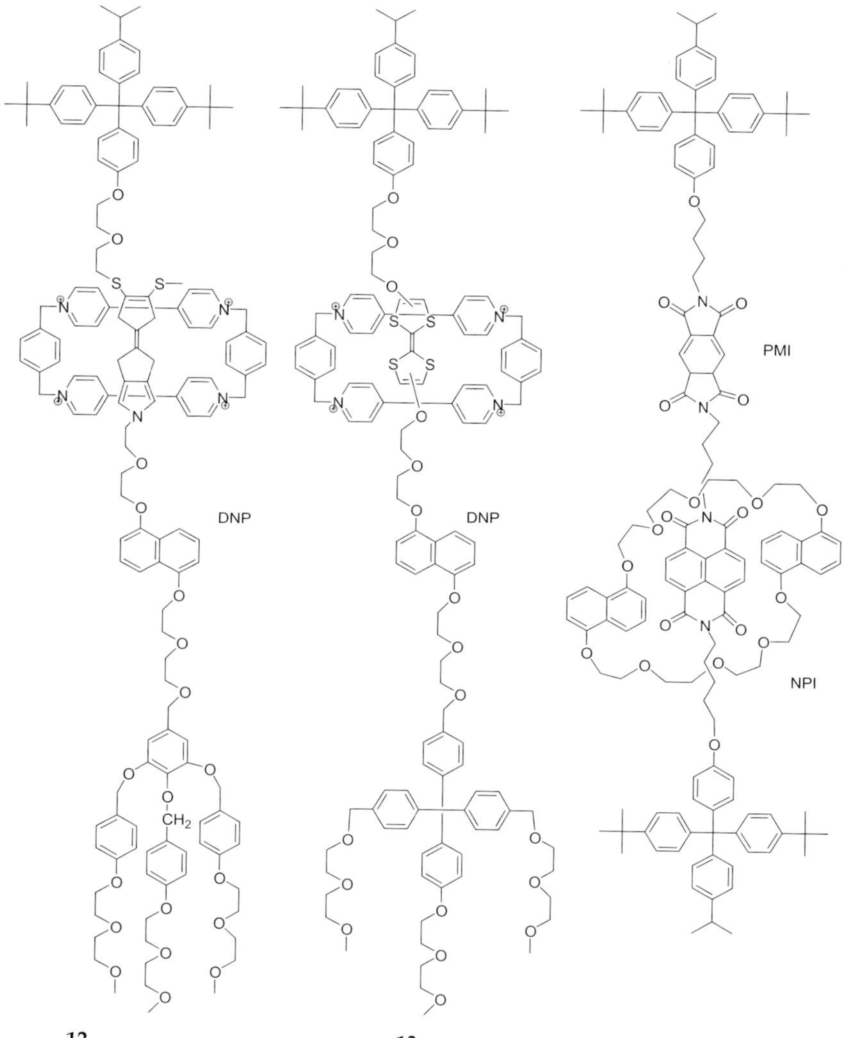

Figure 6.11 Rotaxanes used in molecular electronics.

electronics. The molecule can equilibrate back to the ground state either by thermal equilibration or by reduction and oxidation of the CBPQT^{4+} ring. Unfortunately, it was discovered that in acetonitrile at room temperature only half of the ground-state material existed in the conformation having the ring encircling the MPTTF site. Therefore, only half of the rotaxanes are working as switches. Attempting to improve the ground-state co-conformation ratio, Stoddart and co-workers synthesized rotaxane **13** having a tetrathiafulvalene unit instead of the MPTTF unit (Figure 6.11) [58].

In addition to finding that the mechanism for switching was the same, they also observed that **12** existed solely as the co-conformation in which the ring encircled the TTF unit because it forms a tighter donor–acceptor complex than with MPTTF. Both **12** and **13** have been incorporated into devices. Although TTF-containing switches have worked reasonably well, they still have problems associated with them. First, any rotaxane with a CBPQT^{4+} ring contains counterions. Since it is unknown how counterions affect the switching mechanism, their elimination would remove a level of uncertainty. Secondly, electrochemical and photophysical investigations provide evidence that in solution, and on monolayers at the air–water interface, there can be folded conformations of the rotaxanes in which the hydrophilic chains on the stopper group, or the other recognition site, fold over to interact with the positively charged macrocyle (Figure 6.11). A neutral switch would presumably lessen these conformations. Stoddart, Sanders, and co-workers have recently synthesized, via slippage, a neutral rotaxane **14** containing one pyromellitic diimide (PMI) unit and one 1,4,5,8-naphthalene-tetracarboxylate diimide (NPI) site along with a 1,5-dinaphtho[38]crown-10 macrocycle. ^{1}H NMR analysis of rotaxane **14** shows that the crown ether resides primarily over the NP unit. The co-conformation having the crown ether ring surrounding the PM unit could not be detected using variable-temperature ^{1}H NMR spectroscopy, and free energy difference data acquired from studying degenerate versions of rotaxane **14** indicated that fewer than 1% of the population of molecules reside in the conformation containing the PMI unit surrounded by the ring. This was to be expected because NPI is known to be a much better electron acceptor unit than PMI. After one-electron reduction, the NPI unit is deactivated and the ring moves to the PMI site. Further reduction of the NPI and PMI site results in a situation where there are no donor–acceptor interactions with the macrocycle. The ability to destabilize the metastable state is favorable because it provides a mechanism to "reset" the switch without reduction or thermal relaxation. This switch has only been studied in solution, and efforts are currently under way to make the molecule amphiphilic and incorporate it into half devices and full devices. After studies were done on self-assembled monolayers on the air–water interface (Figure 6.12) [59] and on gold [60], rotaxanes **13** and **14** were incorporated into a crossbar device [61]. A layer of rotaxanes was sandwiched between a silicon electrode and another of titanium and aluminum. When tested against controls, such as eicosanoic acid and a version of **12** missing the ring component, the devices with rotaxanes showed switching behavior while the controls did not. The device stored and read out words with ASCII characters when voltages were applied to the electrodes by assigning the low-conductance ground state to 0 and the high conductance metastable state to **1**. One obstacle in the creation of such solid stable devices is the lack of any direct characterization techniques available to study the molecules themselves. This was demonstrated when the crossbar was constructed using two metal electrodes instead of one metal and one silicon electrode. This device showed switching that is independent of the molecule that is contained between the wires. Williams and co-workers showed that this behavior is the result of metal filaments forming between the electrodes [62].

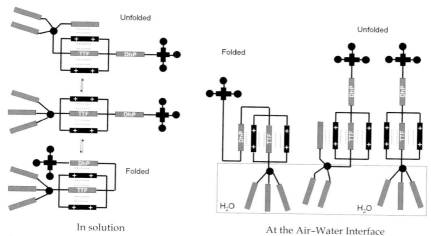

Figure 6.12 Folded conformations of amphiphilic rotaxane.

6.3.5
Bioelectronics

Redox enzymes not only have high specificity for their substrates but also have very high turnover rates, thereby offering an efficient and potent means for energy acquisition and integration into biofuel cells. However, a major problem arises because, in many redox enzymes, the active centers are located deep inside the protein's tertiary or quaternary structure and are insulated from the electrode. Charge carriers have been used to transfer electrons from the enzyme to the submerged electrode in attempts to solve this problem. Ferrocene derivatives have been used that are free in solution, covalently linked to the enzyme, and covalently linked to a cofactor that binds to the enzyme [63]. In all these cases the goal was to allow the carrier molecule to capture electrons from inside the enzyme and, because the molecule would have freedom to leave the active site, transfer current to the electrode surface. Additionally, conducting polymers have been used to entrap the proteins and mediate the transfer of electrons. Although these techniques did achieve electron transfer, it was to a lesser extent than for the natural enzyme with its natural substrates. Presumably the attachments of the charge carrier groups on the enzyme are nonoptimal and alignment with the electrode was poor. Attaching the enzymes or the substrates to the electrode can alleviate these problems because of the consistency in the attachment. The challenge arises in creating attachments that are capable of transferring electrons to the electrode. Willner and co-workers have devised several methods to harness the redox capabilities of glucose oxidase (GOx) by creating monolayers of its cofactor, flavin–adenine dinucleotide (FAD), on gold surfaces. In solution this enzyme catalyzes the oxidation of glucose in the presence of oxygen when bound to FAD. Gluconic acid and hydrogen peroxide are products of the reaction, which has an overall turnover rate of approximately $700\,s^{-1}$. When GOx binds to the monolayer assembled FAD, electrons from the reaction can be transferred to the gold surface provided there is a transport mechanism. Several attachments have achieved high levels of electron transfer. When a pyrroloquinoline quinone (PQQ) unit is present in the chain, the electrons can transfer from FAD to PQQ and then to the surface [64]. The turnover rate for this process was found to be equivalent to that of the natural enzyme. FAD has also been attached to gold surfaces through connections to gold nanoparticles and single-walled carbon nanotubes [65]. The turnover rates for these half-cells are $5000\,s^{-1}$ and $4100\,s^{-1}$, respectively. The most recent development in creating electrical communication to GOx has been through the use of rotaxanes [66]. The incorporation of rotaxanes into this strategy increases the intricacy of electron transfer due to the freedom of motion that the ring component has between the enzyme and the surface. A rotaxane containing a $CBPQT^{4+}$ ring encircling a diiminobenzene was synthesized by the threading method (Figure 6.13). Synthesized by attachment to a gold surface, the other end was capped with FAD. Due to the strength of the donor–acceptor complex, the ring preferentially remains around the diiminobenzene unit. Once FAD is reduced to $FADH_2$, electrons are transferred to the $CBPQT^{4+}$ ring, thereby reducing it. The reduced form of the ring no longer has an affinity for the diiminobenzene unit. As a result it is free to move along the rod. The positively charged ring is attracted to the negatively charged gold surface and shuttles toward it, leading to fast electron transfer. It was found, though, that the electron transfer turnover rate of this reaction is only approximately $400\,s^{-1}$, which is lower than the rate of the natural process. However, of crucial importance in creating efficient half-cells for biofuel devices is the potential at which reduction occurs. The thermodynamic redox potential of FAD at pH 8.0 is $-0.51\,V$. Oxidation occurs here at $-4.0\,V$, a value close to the thermodynamic potential. Maximum power extraction occurs when these values are as close as possible. The control electrode that was missing the ring component exhibited no electron transfer to the electrode. As a result, rotaxanes act as efficient electron transport agents in gold assembled monolayers of GOx, creating an anode well suited for incorporation into glucose-based biofuel cells.

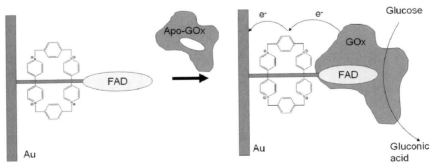

Figure 6.13 Mechanism of construction between GOx and gold electrode via a rotaxane ring.

6.3.6
Membrane Transport

Membrane permeability can be a significant impediment in drug effectiveness. The pharmaceutical industry invests much time and effort into developing drugs that bind tightly and selectively to their targets; yet if the compound is unable to access its target, it will have little medicinal value. Although several synthetic membrane transporters have been developed, there is a need for new transporters, especially considering the structural diversity of synthetic drugs. Since polar or charged molecules have trouble crossing lipid bilayers, it is no surprise that many small-molecule transporters function by a mechanism that compensates by shielding the functionality from the hydrophobic interior of the membrane. For example, glutathione has been transported when attached to a cholate–spermidine–Ellman's reagent derivative [67]. This molecule folded upon itself to encompass the polar groups and expose the nonpolar substituents. This umbrella mechanism allowed passive passage through the membrane. Smithrud and co-workers [68], noticing the mechanistic freedom of the components of the rotaxane, have devised a new cellular transport agent. Rotaxane **15** consisting of a diarginine derivatized dibenzo-24-crown-8 ring was synthesized using the threading strategy (Figure 6.14). The stopper groups were composed of a bulky aromatic group and a cyclophane pocket. Fluorescein and fluorescein-protein kinase C inhibitors were held in place by the diarginine units and bound in the cyclophane pocket. In the absence of the transport agent, fluorescein did not permeate the membrane and was not detected in the cell. However, when bound to **15**, the fluorophore was noticeably transported into COS-7 cells. The complex was judged to be nontoxic to the cell line because mitotic behavior was observed in their presence. At this time no current investigation has been done on the mechanism of transport, but it has been proposed that transport occurs through the ability of the complex to alter conformation based

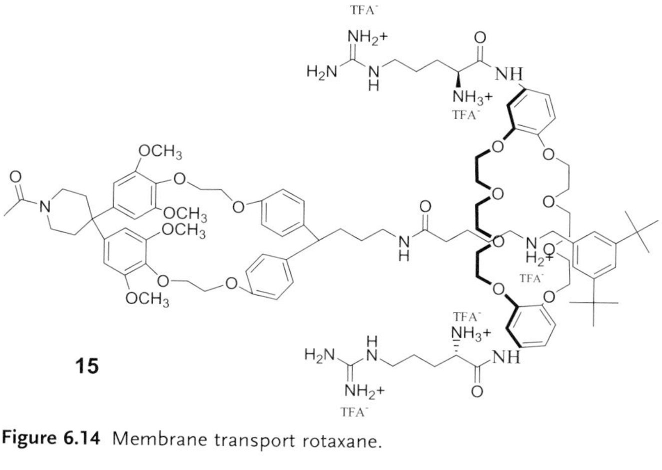

Figure 6.14 Membrane transport rotaxane.

on its environment. In aqueous conditions the aromatic components on the ring would move closer to the encapsulated guest. In the apolar environment of the membrane, the ring can move, allowing optimal salt bridging and hydrogen bonding between the diarginine components on the ring and the bound guest. The rotaxane bound to a variety of other molecules such as small peptides, but no cell transport studies have been done. The effectiveness of this agent will be evaluated once a larger selection of substrates has been investigated. For instance, the peptides mentioned are only two or three amino acids in length. It would be interesting to determine the size limit for transport. Expanding the scope to include the transport of oligonucleotides would also be of interest.

Inspired by this work, we developed a new concept of umbrella-rotaxanes in an effort to create a way of promoting the passive transport of polar, biologically active agents (which represent the rotaxane's wheel) across lipid membranes. We exploited the facial amphiphilicity of the cholate derivatives (which form one of the bulky end groups of the rotaxane) by allowing the rotaxane's axle and wheel components to be shielded from the hydrophobic core as it traverses the membrane. As a first step toward this goal, we have shown that molecular umbrella-rotaxanes display "molecular amphomorphism," the ability to form a shielded or exposed conformation in hydrophobic and hydrophilic environments, respectively. In essence, our "molecular umbrella-rotaxane" concept may be summarized as follows: two amphiphilic walls (i.e., rigid hydrocarbon units that maintain a hydrophobic and a hydrophilic face) are coupled to a suitable linear molecule that bears a threaded biologically active agent (Figure 6.15, where the "darkened face" is hydrophobic and the agent is hydrophilic); the second end group of the rotaxane is a hydrophilic photoisomerizable group (in our case the fulgide) [26], which conformation can be modified by UV irradiation. When the rotaxane is immersed in an aqueous environment, a fully exposed conformation is favored such that intramolecular hydrophobic interactions are maximized and the external face of each wall is hydrated. When the rotaxane is immersed in a hydrocarbon environment, the umbrella then favors a shielded conformation such that intramolecular

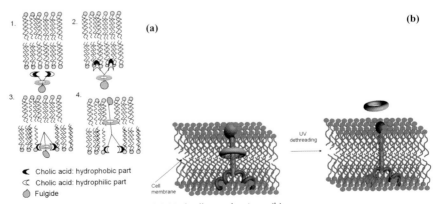

Figure 6.15 Umbrella-rotaxane. (a) Umbrella mechanism. (b) Rotaxane's cell membrane cell insertion and UV dethreading of the bioactive macrocycle.

dipole–dipole and hydrogen–bonding interactions are maximized and the hydrophobic faces are effectively solvated. Our working hypothesis is that such a complex will permeate across a lipid membrane via the following sequence of events: (i) diffusion of the conjugate to the biomembrane surface in a fully exposed state; (ii) insertion into the outer monolayer leaflet by flipping into a shielded state; (iii) diffusion to the inner monolayer leaflet; and (iv) release of the biologically active wheel by photoisomerization.

6.3.7
Catalytically Active Rotaxanes as Processive Enzyme Mimics

The rotaxane assembly is adopted by many enzymes that operate on nucleic acids and proteins. In the case of processive enzymes, the catalytic reaction drives the sequential motion of the enzyme on its polymeric substrate. Therefore, these enzymes can be viewed as molecular motors powered by chemical reactions and moving one-dimensionally on a track, in which fuel is provided by the track itself. An initial attempt to carry out processive catalysis with a synthetic rotaxane has been described [69].

The catalyst **16⁺** is a macrocycle consisting of a substrate-binding cavity that incorporates a Mn(III) porphyrin complex able to oxidize alkenes to the corresponding epoxides (Figure 6.16). It was shown that oxidation occurs within the cavity, provided that a ligand (such as **17** in Figure 6.16) is complexed by the outer face of the porphyrin for both activating the catalytic complex and preventing the oxidation reaction from taking place outside the cavity. In the presence of an oxygen donor and the activating axial ligand, a Mn(V)O species is formed, which

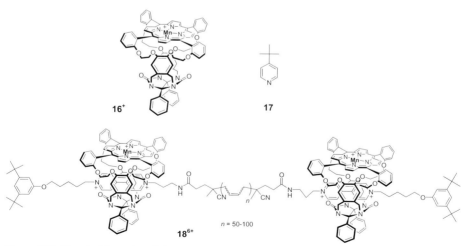

Figure 6.16 Chemical formulae of the macrocyclic catalyst **16⁺**, ligand **17** and three-component rotaxane **18⁶⁺** which mimics the operation of topologically linked enzymes.

transfers its oxygen atom to the alkene substrate. Macrocycle **16**$^+$ was employed as the ring to make the three-component rotaxane **16**$^{6+}$, in which two **16**$^+$ rings surround an axle containing a polybutadiene chain. It was shown that the macrocyclic components of **18**$^{6+}$ are able to catalyze the conversion of the polybutadiene chain incorporated in the axle into the corresponding polyepoxide. This result indicates that the macrocyclic catalysts can surround each diene unit of the polybutadiene chain by moving along the axle. The rotaxane **18**$^{6+}$ can be described as a simple mimic of topologically linked enzymes. A feature that is not mimicked is the sequential nature of the reaction, which gives rise to processive behavior. It is not straightforward to determine whether the macrocyclic units of **18**$^{6+}$ convert each butadiene unit into the corresponding epoxide in a sequential manner or randomly hop along the chain. However, calculations based on the catalytic rate suggested that the rate of displacement of the macrocycle by the effect of the reaction would be much slower than the thermal shuttling rate, pointing to a random hopping mechanism. To make the system sequentially processive, one has to precisely balance the speed of the movement of the catalyst and the rate of the catalytic reaction. In such a case the unidirectional linear motion of the macrocycle along the axle would be powered by the exergonic chemical reaction between the "fuels" PhIO and the diene units.

6.4
Conclusion and Perspectives

In the last few years, several examples of molecular machines and motors have been designed and constructed. It should be noted, however, that the molecular-level machines described in this chapter operate almost in solution. Although the solution studies of chemical systems as complex as molecular machines are of fundamental importance, it seems reasonable that, before functional supramolecular assemblies can find applications as machines at the molecular level, they have to be interfaced with the macroscopic world by ordering them in some way.

The next generation of molecular machines and motors will need to be organized at interfaces, deposited on surfaces, or immobilized into membranes or porous materials so that they can behave coherently. Indeed, the preparation of modified electrodes represent one of the most promising ways to achieve this goal. Solid-state electronic devices based on functional rotaxanes have already been developed. Furthermore, addressing a single molecular-scale device by instruments working at the nanometer level is no longer a dream. Apart from more or less futuristic applications, the extension of the concept of a machine to the molecular level is of interest not only for the development of nanotechnology but also for the advance of basic research. Looking at supramolecular chemistry from the viewpoint of functions with reference to devices of the macroscopic world is indeed a very interesting exercise which introduces novel concepts into chemistry as a scientific discipline. The recent outstanding results achieved in this research field, some of which have been reviewed here, let us optimistically hope that useful devices based on artificial nanomachines will see the light in the not too distant future. Apart from applications, the study of motion at the molecular level and the extension of the concept of motor and machine to the nanoscale are indeed fascinating topics for basic research.

References

1 Schliwa, M. (ed.) (2003) *Molecular Motors*, Wiley-VCH Verlag GmbH, Weinheim.
2 Goodsell, D.S. (2004) *Bionanotechnology – Lessons from Nature*, Wiley, Hoboken.
3 Jones, R.A.L. (2005) *Soft Machines – Nanotechnology and Life*, OUP, Oxford.
4 a. Balzani, V., Credi, A., Raymo, F.M. and Stoddart, J.F. (2000) *Angewandte Chemie (International Ed. in English).*, **39**, 3349–91. Balzani, V., Credi, A. and Venturi, M. (2003) *Molecular Devices and Machines – A Journey into the Nano World*, Wiley-VCH Verlag GmbH, Weinheim. b. Flood, A.H., Ramirez, R.J.A., Deng, W.-Q., Muller, R.P., Goddard, W.A., III and Stoddart, J.F. (2004) *Australian Journal of Chemistry.*, **57**, 301–22. c. van Delden, R.A., ter Wiel, M.K.J., Pollard, M.M., Vicario, J., Koumura, N. and Feringa, B.L. (2005) *Nature*, **437**, 1337–40.
5 a. Seto, C.T. and Whitesides, G.M. (1993) *Journal of the American Chemical Society*, **115**, 1330–40. b. Kotera, M., Lehn, J.-M. and Vigneron, J.-P. (1994) *Chemical Communications*, **2**, 197–199. c. Subramanian, S. and Zaworotko, M. (1994) *Coordination Chemistry Reviews*, **137**, 357–401. d. Leigh, D.A., Murphy, A., Smart, J.P. and Slawin A.M.Z. (1997) *Angewandte Chemie (International Ed. in English)*, **36**, 728–32. e. Jeffrey, G.A. (1997) *An Introduction to Hydrogen Bonding*, Oxford University Press, New York. f. Philp, D. and Robinson, J.M.A. (1998) *Journal of the Chemical Society. Perkin Transactions*, **2**, 1643–50. g. Martin, T., Obst, U. and Rebek, J., Jr. (1998) *Science*, **281**, 1842–45. h. Melendez, R.E. and Hamilton, A.D. (1998) *Topics in Current Chemistry*, **198**, 97–129. i. Sijbesma, R.P. and Meijer, E.W. (1999) *Current Opinion in Colloid and Interface Science*, **4**, 24–32. j. Seel, C., Parham, A.H., Safarowsky, O., Hübner, G.M. and Vögtle F. (1999) *The Journal of Organic Chemistry*, **64**, 7236–42.
6 a. Sauvage, J.-P. (1998) *Accounts of Chemical Research*, **31**, 611–9. b. Fujita, M. (1998) *Chemical Society Reviews*, **27**, 417–25. c. Saalfrank, R.W. and Bernt, I. (1998) *Current Opinion in Solid State and Materials Science*, **3**, 407–13. d. Raymo, F.M. and Stoddart, J.F. (1998) *Current Opinion in Colloid and Interface Science*, **3**, 150–9. e. Batten, S.R. and Robson, R. (1998) *Angewandte Chemie (International Ed. in English)*, **37**, 1461–94. f. Fujita, M. 1999, *Accounts of Chemical Research*, **32**, 53–61. g. Caulder, D.L. and Raymond, K.N. (1999) *Journal of the Chemical Society. Dalton Transactions*, **8**, 1185–1200. h. Constable, E.C. and Haverson, P. (1999) *Polyhedron*, **18**, 3093–106. i. Caulder, D.L. and Raymond, K.N. (1999) *Accounts of Chemical Research*, **32**, 975–82. j. Leininger, S., Olenyuk, B. and Stang, P.J. (2000) *Chemical Reviews*, **100**, 853–907.
7 a. Rebek, J., Jr and Nemeth, D. (1986) *Journal of the American Chemical Society*, **108**, 5637–8. b. Burley, S.K. and Petsko, G.A. (1986) *Journal of the American Chemical Society*, *108*, 7995–8001. c. Hamilton, A.D. and Van Engen, D. (1987) *Journal of the American Chemical Society*, **109**, 5035–6. d. Zimmerman, S.C. and Vanzyl, C.M. (1987) *Journal of the American Chemical Society*, **109**, 7894–6. e. Hunter, C.A. and Sanders, J.K.M. (1990) *Journal of the American Chemical Society*, **112**, 5525–34. f. Cozzi, F., Cinquini, M., Annunziata, R., Dwyer, T. and Siegel, J.S. (1992) *Journal of the American Chemical Society*, **114**, 5729–33. g. Hunter, C.A. (1993) *Angewandte Chemie (International Ed. in English)*, **32**, 1584–6. h. Hunter, C.A. (1994) *Chemical Society Reviews*, **23**, 101–9. i. Claessens, C.G. and Stoddart, J.F.

(1997) *Journal of Physical Organic Chemistry*, **10**, 254–72. j. Nelson, J.C., Saven, J.G., Moore, J.S. and Wolynes, P.G. (1997) *Science*, **277**, 1793–6. k. Lokey, R.S., Kwok, Y., Guelev, V., Pursell, C.J., Hurley, L.H. and Iverson, B.L. (1997) *Journal of the American Chemical Society*, **119**, 7202–10. l. Whitten, D.G., Chen, L.H., Geiger, H.C., Perlstein, J. and Song, X.D. (1998) *The Journal of Physical Chemistry B*, **102**, 10098–111.

8 An edge-to-face interaction between two aromatic rings is simply an example of a C—H···π interaction. In general terms, the C—H donor does not necessarily have to be part of an aromatic ring. For literature on the C—H···π interaction, see: a. Ferguson, S.B. and Diederich, F. (1986) *Angewandte Chemie (International Ed. in English)*, **25**, 1127–9. b. Nishio, M. and Hirota, M. (1989) *Tetrahedron*, **45**, 7201–45. c. Oki, M. (1990) *Accounts of Chemical Research*, **23**, 351–6. d. Etter, M.C. (1991) *The Journal of Physical Chemistry*, **95**, 4601–10. e. Zaworotko, M.J. (1994) *Chemical Society Reviews*, **23**, 283–8. f. Boyd, D.R., Evans, T.A., Jennings, W.B., Malone, J.F., O'Sullivan, W. and Smith, A. (1996) *Chemical Communications*, **19**, 2269–70. g. Ashton, P.R., Hörner, B., Kocian, O., Menzer, S., White, A.J.P., Stoddart, J.F. and Williams, D.J. (1996) *Synthesis*, **8**, 930–40. h. Adams, H., Carver, F.J., Hunter, C.A., Morales, J.C. and Seward, E.M. (1996) *Angewandte Chemie (International Ed. in English)*, **35**, 1542–4. i. Carver, F.J., Hunter, C.A. and Seward, E.M. (1998) *Chemical Communications*, **7**, 775–6. j. Desiraju, G.R. and Steiner, T. (1999) *The Weak Hydrogen Bond in Structural Chemistry and Biology*, Oxford University Press, Oxford.

9 a. Ben-Naim, A. (1980) *Hydrophobic Interactions*, Plenum Press, New York and London. b. Tanford, C. (1980) *The Hydrophobic Effect: Formation of Micelles and Biological Membranes*, 2nd edn, John Wiley & Sons, Ltd, Chichester. c. Privalov, P.L. and Gill, S.J. (1989) *Pure and Applied Chemistry Chimie Pure Et Appliquee*, **61**, 1097–104. d. Blokzijl, W. and Engberts, J.B.F.N. (1993) *Angewandte Chemie (International Ed. in English)*, **32**, 1545–79. e. Silverstein, K.A.T., Hayment, A.D.J. and Dill, K.A. (1998) *Journal of the American Chemical Society*, **120**, 3166–75. f. Rekharsky, M.V. and Inoue, Y. (1998) *Chemical Reviews*, **98**, 1875–917. g. Marmur, A. (2000) *Journal of the American Chemical Society*, **122**, 2120–1.

10 a. Msayib, K.J. and Watt, C.I.F. (1992) *Chemical Society Reviews*, **21**, 237–43. b. Aida, M. (1994) *Journal of Molecular Structure*, **117**, 45–53. c. Hamelin, B., Jullien, L., Derouet, C., du Penhoat, C.H. and Berthault, P. (1998) *Journal of the American Chemical Society*, **120**, 8438–47. d. Mascal, M., Hansen, J., Fallon, P.S., Blake, A.J., Heywood, B.R., Moore, M.H. and Turkenburg, J.P. (1999) *Chemistry—A European Journal*, **5**, 381–4. e. du Mont, W.W. and Ruthe, F. (1999) *Coordination Chemistry Reviews*, **189**, 101–33. f. Müller-Dethlefs, K. and Hobza, P. (2000) *Chemical Reviews*, **100**, 143–67. g. Sheinerman, F.B., Norel, R. and Honig, B. (2000) *Current Opinion in Structural Biology*, **10**, 153–9.

11 a. Nevinskii, G.A. (1995) *Molecular Biology*, **29**, 6–19. b. Ohta, K., Ikejima, M., Moriya, M., Hasebe, H. and Yamamoto, I. (1998) *Journal of Materials Chemistry*, **8**, 1971–7. c. Novoa, J.J., Lafuente, P. and Mota, F. (1998) *Chemical Physics Letters*, **290**, 519–25. d. Street, A.G. and Mayo, S.L. (1999) *Proceedings of the National Academy of Sciences of the United States of America*, **96**, 9074–6. e. Nakamura, K. and Houk, K.N. (1999) *Organic Letters*, **1**, 2049–51. f. Issaenko, S.A. and Harris, A.B. (2000) *Physical Review. E*, **61**, 2777–91. g. Berezovsky, I.N., Esipova, N.G., Tumanyan, V.G. and Namiot, V.A. (2000) *Journal of Biomolecular Structure and Dynamics*, **17**, 799–809.

12 a. Lindsey, J.S. (1991) *New Journal of Chemistry*, **15**, 153–180. b. Philp, D. and Stoddart, J.F. (1991) *Synlett*, 445–458. c. Whitesides, G.M., Mathias, J.P. and Seto, C.T. (1991) *Science*, **154**, 1312–19. d. Lawrence, D.S., Jiang, T. and Levett, M. (1995) *Chemical Reviews*, **95**, 2229–60. e. Philp, D. and Stoddart, J.F. (1996) *Angewandte Chemie (International Ed. in*

English), **35**, 1154–96. f. Stang, P.J. and Olenyuk, B. (1997) *Accounts of Chemical Research*, **30**, 502–18. g. Cusack, L., Rao, S.N., Wenger, J. and Fitzmaurice, D. (1997) *Chemistry of Materials: A Publication of the American Chemical Society*, **9**, 624–31. h. Conn, M.M. and Rebek, J., Jr. (1997) *Chemical Reviews*, **97**, 1647–68. i. Linton, B. and Hamilton, A. D. (1997) *Chemical Reviews*, **97**, 1669–80. j. Breen, T.L., Tien, J., Oliver, S.R.J., Hadzic, T. and Whitesides, G.M. (1999) *Science*, **284**, 948–51. k. Tomalia, D.A., Wang, Z.G. and Tirrel, M. (1999) *Current Opinion in Colloid and Interface Science*, **4**, 3–5.

13. Fyfe, M.C.T. and Stoddart, J.F. (1999) *Coordination Chemistry Reviews*, **183**, 139–55.

14. a. Schill, G. (1971) *Catenanes, Rotaxanes, and Knots*, Academic Press, New York. b. Breault, G.A., Hunter, C.A. and Mayers, P.C. (1999) *Tetrahedron*, **55**, 5265–93. c. Sauvage, J.-P. and Dietrich-Buchecker, C. (1999) *Molecular Catenanes, Rotaxanes and Knots*, Wiley-VCH Verlag GmbH, Weinhem. d. Hubin, T.J., Kolchinski, A. G., Vance, A.L. and Busch, D.H. (1999) *Advances in Supramolecular Chemistry*, **5**, 237–357.

15. Amabilino, D.B. and Stoddart, J.F. (1995) *Chemical Reviews*, **95**, 2725–828.

16. Pedersen, C.J. (1988) *Angewandte Chemie (International Ed. in English)*, **27**, 1021–7.

17. a. Lehn, J.-M. (1995) *Supramolecular Chemistry*, Wiley-VCH Verlag GmbH, Weinheim.
    b. Atwood, J.L., Davies, J.E.D., MacNicol, D.D. and Vögtle, F. (1996) *Comprehensive Supramolecular Chemistry*, Vol. 11, Pergamon, Oxford.

18. Pedersen, C.J. (1967) *Journal of the American Chemical Society*, **89**, 7017–36.

19. For in-depth and historical analyses of cation/crown ether complexation, see: a. Bradshaw, J.S., Izatt, R.M., Bordunov, A. V., Zhu, C.Y. and Hathaway, J.K. (1996) *Comprehensive Supramolecular Chemistry*, Vol. 1 (eds J.L. Atwood, J.E.D. Davies, D. D. MacNicol, F. Vögtle and G.W. Gokel), Pergamon, Oxford, pp. 35–95. b. Izatt, R. M., Pawluk, K., Bradshaw, J.S. and Bruening, R.L. (1995) *Chemical Reviews*, **95**, 2529–86. c. Izatt, R.M., Pawluk, K., Bradshaw, J.S. and Bruening, R.L. (1991) *Chemical Reviews*, **91**, 1721–2085. d. Gokel, G.W. (1991) *Crown Ethers and Cryptands*, The Royal Society of Chemistry, Cambridge, pp. 64–98. e. Inoue, Y. and Gokel, G.W. (eds) (1990) *Cation Binding by Macrocycles*, Marcel Dekker, New York. f. Vögtle, F. and Weber, E. (1989) *Crown Ethers and Analogs* (eds S. Patai and Z. Rappoport), John Wiley & Sons, Inc., New York, pp. 207–304. g. Goldberg, I. (1984) *Inclusion Compounds*, Vol. 2 (eds J.L. Atwood, J.E.D. Davies, and D.D. MacNicol), Academic Press, London, pp. 261–335.

20. For some examples, see: a. Drljaca, A., Hardie, M.J., Raston, C.L. and Spiccia, L. (1999) *Chemistry – A European Journal*, **5**, 2295–99. b. Marchand, A.P., Chong, H.-S., Alihodzic, S., Watson, W.H. and Bodige, S.G. (1999) *Tetrahedron*, **55**, 9687–96. c. Liu, H., Liu, S. and Echegoyen, L. (1999) *Chemical Communications*, **16**, 1493–4. d. Flink, S., van Veggel, F.C.J. and Reinhoudt, D.N. (1999) *The Journal of Physical Chemistry. B*, **103**, 6515–20. e. Bartsch, R.A., Hwang, H.-S., Talanov, V.S., Talanova, G.G., Purkiss, D.W. and Rogers, R.D. (1999) *The Journal of Organic Chemistry*, **64**, 5341–9.

21. In Pedersen's seminal paper (ref. 18), he noted that primary alkylammonium ions form 1:1 complexes with dibenzo[18]crown-6 (DB18C6). This ground-breaking discovery subsequently spawned much research in the areas of host–guest and supramolecular chemistry. For examples spanning from Pedersen's day until the present, see: a. Kyba, E.B., Koga, K., Sousa, L.R., Siegel, M.G. and Cram, D.J. (1973) *Journal of the American Chemical Society*, **95**, 2692–93. b. Cram, D.J. and Cram, J.M. (1974) *Science*, **183**, 803–9. c. Tarnowski, T.L. and Cram, D.J. (1976) *Journal of the Chemical Society. Chemical Communications*, **16**, 661–3. d. Timko, J.M., Moore, S.S., Walba, D.M., Hiberty, P.C. and Cram, D.J. (1977) *Journal of the American Chemical Society*, **99**, 4207–19. e. Kyba, E.P., Timko, J.M., Kaplan, L.J., de Jong, F., Gokel, G.W. and Cram, D.J. (1978) *Journal of the American*

Chemical Society, **100**, 4558–68. f. Cram, D.J. and Cram, J.M. (1978) *Accounts of Chemical Research*, **11**, 8–14. g. Stoddart, J.F. (1979) *Chemical Society Reviews*, **8**, 85–142. h. Bovill, M.J., Chadwick, D.J., Johnson, M.R., Jones, N.F., Sutherland, I.O. and Newton, R.F. (1979) *Journal of the Chemical Society. Chemical Communications*, **23**, 1065–6. i. Lehn, J.-M. (1982) *IUPAC Frontiers of Chemistry* (ed. K.J. Laidler), Pergamon, Oxford, pp. 265–72. j. Goldberg, I. (1980) *Journal of the American Chemical Society*, **102**, 4106–13. k. de Jong, F. and Reinhoudt, D.N. (1980) *Advances in Physical Organic Chemistry*, **17**, 279–433. l. Cram, D.J. and Trueblood, K.N. (1981) *Topics in Current Chemistry*, **98**, 43–106. m. Bradshaw, J.S., Baxter, S.L., Lamb, J.D., Izatt, R.M. and Christensen, J.J. (1981) *Journal of the American Chemical Society*, **103**, 1821–7. n. Trueblood, K.N., Knobler, C.B., Lawrence, D.S. and Stevens, R.V. (1982) *Journal of the American Chemical Society*, **104**, 1355–62. o. Aldag, R. and Schröder, G. (1984) *Justus Liebigs Annalen Der Chemie*, **5**, 1036–45. p. Sutherland, I.O. (1986) *Chemical Society Reviews*, **15**, 63–91. q. Stoddart, J.F. (1987) *Top Stereochem*, **17**, 205–88. r. Cram, D.J. (1988) *Angewandte Chemie (International Ed. in English)*, **27**, 1009–20. s. Misumi, S. (1990) *Pure And Applied Chemistry Chimie Pure Et Appliquee*, **62**, 493–8. t. Sutherland, I.O. (2000) *Pure And Applied Chemistry Chimie Pure Et Appliquee.*, **3**, 499–504. u. Izatt, R.M., Wang, T., Hathaway, J.K., Zhang, X.X., Curtis, J.C., Bradshaw, J.S., Zhu, C.Y. and Huszthy, P. (1994) *Journal of Inclusion Phenomena and Molecular Recognition in Chemistry*, **17**, 157–75. v. Reetz, M.T., Huff, J., Rudolph, J., Töllner, K., Deege, A. and Goddard, R. (1994) *Journal of the American Chemical Society*, **116**, 11588–9. w. Williamson, B.L. and Creaser, C.S. (1998) *International Journal of Mass Spectrometry*, **188**, 53–61. x. Hansson, A.P., Norrby, P.-O. and Wärnmark, K. (1998) *Tetrahedron Letters*, **39**, 4565–8. y. Tsukube, H., Wada, M., Shinoda, S. and Tamiaki H. (1999) *Chemical Communications*, **11**, 1007–8.

22 Complexes formed between secondary dialkylammonium ions and crown ethers having fewer than 24 atoms in their macrocycles have been observed to occur in a *face-to-face* manner. See: a. Metcalfe, J.C., Stoddart, J.F. and Jones, G. (1977) *Journal of the American Chemical Society*, **99**, 8317–9. b. Krane, J. and Aune, O. (1980) *Acta Chemica Scandinavica*, **34B**, 397–401. c. Metcalfe, J.C., Stoddart, J.F., Jones, G., Atkinson, A., Kerr, I.S. and Williams, D.J. (1980) *Journal of the Chemical Society Chemical Communications*, **12**, 540–3. d. Abed-Ali, S.S., Brisdon, B.J. and England, R. (1987) *Journal of the Chemical Society Chemical Communications*, **20**, 1565–6.

23 a. Ashton, P.R., Campbell, P.J., Chrystal, E.J.T., Glink, P.T., Menzer, S., Philp, D., Spencer, N., Stoddart, J.F., Tasker, P.A. and Williams, D.J. (1995) *Angewandte Chemie (International Ed. in English)*, **34**, 1865–9. b. Ashton, P.R., Chrystal, E.J.T., Glink, P.T., Menzer, S., Schiavo, C., Spencer, N., Stoddart, J.F., Tasker, P.A., White, A.J.P. and Williams, D.J. (1996) *Chemistry–A European Journal*, **2**, 709–28.

24 a. Kolchinski, A.G., Busch, D.H. and Alcock, N.W. (1995) *Journal of the Chemical Society. Chemical Communications*, **12**, 1289–91. b. Ashton, P.R., Glink, P.T., Stoddart, J.F., Tasker, P.A., White, A.J.P. and Williams, D.J. (1996) *Chemistry–A European Journal*, **2**, 729–36. c. Martínez-Díaz, M.-V., Spencer, N. and Stoddart, J.F. (1997) *Angewandte Chemie (International Ed. in English)*, **36**, 1904–7. d. Ashton, P.R., Baxter, I., Fyfe, M.C.T., Raymo, F.M., Spencer, N., Stoddart, J.F., White, A.J.P. and Williams D.J. (1998) *Journal of the American Chemical Society*, **120**, 2297–307. e. Ashton, P.R., Ballardini, R., Balzani, V., Baxter, I., Credi, A., Fyfe, M.C.T., Gandolfi, M.T., Gómez-López, M., Martínez-Díaz, M.-V., Piersanti, A., Spencer, N., Stoddart, J.F., Venturi, M., White, A.J.P. and Williams, D.J. (1998) *Journal of the American Chemical Society*, **120**, 11932–42. f. Kolchinski, A.G., Alcock, N.W., Roesner, R.A. and Busch, D.H. (1998) *Chemical Communications*, **14**, 1437–8. g. Cantrill, S.J., Fulton, D.A., Fyfe, M.C.T., Stoddart, J.

F., White, A.J.P. and Williams, D.J. (1999) *Tetrahedron Letters*, **40**, 3669–72. h. Rowan, S.J., Cantrill, S.J. and Stoddart, J.F. (1999) *Organic Letters*, **1**, 129–32. i. Takata, T., Kawasaki, H., Asai, S., Furusho, Y. and Kihara, N. (1999) *Chemistry Letters*, **3**, 223–4. j. Cantrill, S.J., Rowan, S.J. and Stoddart, J.F. (1999) *Organic Letters*, **1**, 1363–6. k. Cao, J., Fyfe, M.C.T., Stoddart, J.F., Cousins, G.R.L. and Glink, P.T. (2000) *The Journal of Organic Chemistry*, **65**, 1937–46. l. Rowan, S.J. and Stoddart, J.F. (2000) *Journal of the American Chemical Society*, **122**, 164–5. m. Furusho, Y., Hasegawa, T., Tsuboi, A., Kihara, N. and Takata T. (2000) *Chemistry Letters*, **1**, 18–9. n. Kihara, N., Yuya, T., Kawasaki, H. and Takata T. (2000) *Chemistry Letters*, **5**, 506–7.

25 a. Ashton, P.R., Glink, P.T., Martínez-Díaz, M.-V., Stoddart, J.F., White, A.J.P. and Williams, D.J. (1996) *Angewandte Chemie (International Ed. in English)*, **35**, 1930–3. b. Ashton, P.R., Collins, A.N., Fyfe, M.C.T., Menzer, S., Stoddart, J.F. and Williams D.J. (1997) *Angewandte Chemie (International Ed. in English)*, **36**, 735–9. c. Ashton, P.R., Fyfe, M.C.T., Hickingbottom, S.K., Menzer, S., Stoddart, J.F., White, A.J.P. and Williams, D.J. (1998) *Chemistry – A European Journal*, **4**, 577–89. d. Fyfe, M.C.T. and Stoddart, J.F. (1999) *Coordination Chemistry Reviews*, **183**, 139–55. e. Yamaguchi, N. and Gibson, H.W. (1999) *Angewandte Chemie (International Ed. in English)*, **38**, 143–7. f. Yamaguchi, N. and Gibson, H.W. (1999) *Chemical Communications*, **9**, 789–90.

26 Noujeim, N., Faure, D. and Schmitzer, A.R. (2007) **22**, 5723–4.

27 Harrison, I.T. and Harrison, S. (1967) *Journal of the American Chemical Society*, **89**, 5723.

28 Vickers, M.S. and Beer, P.D. (2007) *Chemical Society Reviews*, **36**, 211–25.

29 a. Saenger, W. (1980) *Angewandte Chemie*, **82**, 343–7. b. Saenger, W. (1980) *Angewandte Chemie (International Ed. in English)*, **19**, 344–7.

30 a. Cramer, F. (1951) *Chemische Berichte*, **84**, 851–7. b. Cramer, F. (1952) *Angewandte Chemie*, **64**, 136–42. c. Cramer, F. (1953) *Chemische Berichte*, **86**, 1576. d. Cramer, F. (1954) *Einschussverbindungen*, Springer, Berlin, in (1987) *Cyclodextrins and Their Industrial Uses* (ed. D. Duchêne), Editions de Santé, Paris.

31 a. Muller, N. (1990) *Accounts of Chemical Research*, **23**, 23–8. b. Blokzijl, W. and Engberts, J.B.F.N. (1993) *Angewandte Chemie*, **105**, 1610–44. c. Blokzijl, W. and Engberts, J.B.F.N. (1993) *Angewandte Chemie (International Ed. in English)*, **32**, 1545–79.

32 Isnin, R. and Kaifer, A.E. (1991) *Journal of the American Chemical Society*, **113**, 8188–90.

33 For example, see: Chen, J. and Seeman, N.C. (1991) *Nature*, **350**, 631–34.

34 For example, see: Liang, C. and Mislow, K. (1995) *Journal of the American Chemical Society*, **117**, 4201–13.

35 Leigh, D.A., Murphy, A., Smart, J.P. and Slawin, A.M.Z. (1997) *Angewandte Chemie (International Ed. in English)*, **36**, 752–6.

36 Clegg, W., Gimenez-Saiz, C., Leigh, D.A., Murphy, A., Slawin, A.M.Z. and Teat, S.J. (1999) *Journal of the American Chemical Society*, **121**, 4124–9.

37 For a review, see: Philp, D. and Stoddart, J.F. (1996) *Angewandte Chemie (International Ed. in English)*, **35**, 1154–96.

38 For reviews, see: a. Mulliken, R.S. and Person, W.B. (1962) *Annual Review of Physical Chemistry*, **8**, 107. b. Foster, R. (1969) *Organic Charge-Transfer Complexes*, Academic Press, New York. c. Mulliken, R.S. and Person W.B. (1980) *The Journal of Physical Chemistry*, **84**, 2135–41.

39 a. Allwood, B.L., Spencer, N., Shahriari-Zavareh, H., Stoddart, J.F. and Williams, D.J. (1987) *Journal of the Chemical Society Chemical Communications*, **14**, 1064–6. b. Ashton, P.R., Slawin, A.M.Z., Spencer, N., Stoddart, J.F. and Williams, D.J. (1987) *Chemical Communications*, **14**, 1066–70. c. Slawin, A.M.Z., Spencer, N., Stoddart, J.F. and Williams, D.J. (1987) *Chemical Communications*, **14**, 1070–2.

40 a. Ashton, P.R., Odell, B., Reddington, M.V., Slawin, A.M.Z., Stoddart, J.F. and Williams D.J. (1988) *Angewandte Chemie*, **100**, 1608–11. b. Ashton, P.R., Odell, B.,

Reddington, M.V., Slawin, A.M.Z., Stoddart, J.F. and Williams, D.J. (1988) *Angewandte Chemie (International Ed. in English)*, **27**, 1550–3.

41 a. Ashton, P.R., Goodnow, T.T., Kaifer, A.E., Reddington, M.V., Slawin, A.M.Z., Spencer, N., Stoddart, J.F., Vicent, C. and Williams, D.J. (1989) *Angewandte Chemie (International Ed. in English)*, **28**, 1396–9. b. Anelli, P.-L., Ashton, P.R., Ballardini, R., Balzani, V., Delgado, M., Gandolfi, M.T., Goodnow, T.T., Kaifer, A.E., Philp, D., Pietraszkiewicz, M., Prodi, L., Reddington, M.V., Slawin, A.M.Z., Spencer, N., Stoddart, J.F., Vicent, C. and Williams, D.J. (1992) *Journal of the American Chemical Society*, **114**, 193–218.

42 a. Dietrich-Buchecker, C.O., Sauvage, J.-P. Kintzinger, J.-P. (1983) *Tetrahedron Letters*, **24**, 5095–8. b. Sauvage, J.-P. (1990) *Accounts of Chemical Research*, 23, 319–27.

43 a. Mohr, B., Weck, M., Sauvage, J.-P. and Grubbs, R.H. (1997) *Angewandte Chemie (International Ed. in English)*, **36**, 1308–10. b. Weck, M., Mohr, B., Sauvage, J.-P. and Grubbs, R.H. (1999) *The Journal of Organic Chemistry*, **64**, 5463–71.

44 Metzger, R.M. (2003) *Chemical Reviews*, **103**, 3803–34.

45 Stoddart, J.F. (2001) *Accounts of Chemical Research*, **34**, no. 6, Special Issue on Molecular Machines; Guest Editor: J.F. Stoddart.

46 Ballardini, R., Balzani, V., Credi, A., Gandolfi, M.T. and Venturi, M. (2001) *Accounts of Chemical Research*. **34**, 445. b. Ballardini, R. Balzani, V. Credi, A. Gandolfi, M.T. and Venturi M. (2001) *International Journal of Photoenergy*, **3**, 63–77.

47 Kaifer, A.E. and Gómez-Kaifer, M. (1999) *Supramolecular Electrochemistry*, Wiley-VCH Verlag GmbH, Weinheim.

48 Balzani, V., Credi, A. and Venturi, M. (2003) *A European Journal of Chemical Physics and Physical Chemistry*, **4**, 49–59.

49 Ashton, P.R., Ballardini, R., Balzani, V., Baxter, I., Credi, A., Fyfe, M.C.T., Gandolfi, M.T., Gomez-Lopez, M., Martinez-Diaz, M.V., Piersanti, A., Spencer, N., Stoddart, J.F., Venturi, M., White, A.J.P. and Williams, D.J. (1998) *Journal of the American Chemical Society*, **120**, 11932–42.

50 a. Ballardini, R., Balzani, V., Credi, A., Gandolfi, M.T. and Venturi M. (2001) *Accounts of Chemical Research*, **34**, 445–55.

51 Ashton, P.R., Ballardini, R., Balzani, V., Credi, A., Dress, K.R., Ishow, E., Kleverlaan, C.J., Kocian, O., Preece, J.A., Spencer, N., Stoddart, J.F., Venturi, M. and Wenger, S. (2000) *Chemistry – A European Journal*, **6**, 3558–74.

52 Brouwer, A.M., Frochot, C., Gatti, F.G., Leigh, D.A., Mottier, L., Paolucci, F., Roffia, S. and Wurpel, G.W.H. (2001) *Science*, **291**, 2124–8.

53 Badjic, J.D., Balzani, V., Credi, A., Silvi, S. and Stoddart, J.F. (2004) *Science*, **303**, 1845.

54 Badjic, J.D., Ronconi, C.M., Stoddart, J.F., Balzani, V., Silvi, S. and Credi, A. (2006) *Journal of the American Chemical Society*, **128**, 1489–99.

55 Sauvage, J.-P (2005) *Chemical Communications*, **12**, 1507–11.

56 Liu, Y., Flood, A.H., Bonvallett, P.A., Vignon, S.A., Northrop, B.H., Tseng, H.-R., Jeppesen, J.O., Huang, T.J., Brough, B., Baller, M., Magonov, S., Solares, S.D., Goddard, W.A., Ho, C.M. and Stoddart, J.F. (2005) *Journal of the American Chemical Society*, **127**, 9745–59.

57 Jeppesen, J.O., Nielsen, K.A., Perkins, J., Vignon, S.A., Fabio, A.D., Ballardini, R., Gandolfi, M.T., Venturi, M., Balzani, V., Becher, J. and Stoddart, J.F. (2003) *Chemistry – A European Journal*, **9**, 2982–3007.

58 Tseng, H.-R., Vignon, S.A., Celestre, P.C., Perkins, J., Jeppesen, J.O., Fabio, A.D., Ballardini, R., Gandolfi, M.T., Venturi, M., Balzani, V. and Stoddart, J.F. (2004) *Chemistry – A European Journal*, **10**, 155–72.

59 a. Huang, T.J., Tseng, H.-R., Sha, L., Lu, W., Brough, B., Flood, A.H., Yu, B.-D., Celestre, P.C., Chang, J.P., Stoddart, J.F. and Ho, C.-M. (2004) *Nano Letters*, **4**, 2065–71. b. Lee, I.C., Frank, C.W., Yamamoto, T., Tseng, H.-R., Flood, A.H., Stoddart, J.F. and Jeppesen, J.O. (2004)

Langmuir: The ACS Journal of Surfaces and Colloids, **20**, 5809–28. c. Norgaard, K., Jeppesen, J.O., Laursen, B.W., Simonsen, J.B., Weygand, M.J., Kjaer, K., Stoddart, J.F. and Bjornholm, T. (2005) *The Journal of Physical Chemistry B*, **109**, 1063–6.

60 Tseng, H.-R., Fang, D., Wu, N.X., Zhang, X. and Stoddart, J.F. (2004) *A European Journal of Chemical Physics and Physical Chemistry*, **5**, 111–6.

61 Luo, Y., Collier, P., Jeppesen, J.O., Nielsen, K.A., Delonno, E., Perkins, G., Ho, J., Tseng, H.-R., Yamamoto, T., Stoddart, J.F. and Heath, J.R. (2002) *A European Journal of Chemical Physics and Physical Chemistry*, **3**, 519–5.

62 Lau, C.N., Stewart, D.R., Williams, R.S. and Bockrath, M. (2004) *Nano Letters.*, **4**, 569–72.

63 Willner, I. and Katz, E. (2000) *Angewandte Chemie (International Ed. in English)*, **39**, 1180–218 [and the references therein].

64 Zayats, M., Katz, E. and Willner, I. (2002) *Journal of the American Chemical Society*, **124**, 2120–1.

65 Patolsky, F., Weizmann, Y. and Willner, I. (2004) *Angewandte Chemie (International Ed. in English).*, **43**, 2113–7.

66 Katz, E., Sheeney-Haj-Ichia, L. and Willner, I. (2004) *Angewandte Chemie (International Ed. in English)*, **43**, 3292–300.

67 a. Janout, V., Staina, I.V., Bandyopadhyay, P. and Regen, S.L. (2001) *Journal of the American Chemical Society* **123**, 9926–7. b. Janout, V., Jing, B. and Regen, S.L. (2002) *Bioconjugate Chemistry*, **13**, 351–6.

68 Dvornikovs, V., House, B.E., Kaetzel, M., Dedman, J.R. and Smithrud, D.B. (2003) *Journal of the American Chemical Society*, **125**, 8290–301.

69 Thordarson, P., Bijsterveld, E.J.A., Rowan, A.E. and Nolte, R.J.M. (2003) *Nature*, **424**, 915–8.

7
Multiphoton Processes and Nonlinear Harmonic Generations in Lanthanide Complexes
Ga-Lai Law

7.1
Introduction

In recent years, considerable effort has been devoted to materials that will emit in the region of short wavelengths such as the visible region when excited by radiation from another region such as the infrared [1–3]. The production of high-energy output when pumped with a longer wavelength input leads to many advantages, especially in biological studies, telecommunications, and three-dimensional optical storage, with emerging interest in the possibility of producing contrast agents for two-photon excited fluorescence microscopy [4–8]. Thus there has been burgeoning interest in the design of efficient luminescent organic–lanthanide complexes for application in these areas [9].

The use of lanthanides are common for optical purposes because of their narrow and sharp bands, and distinguishable long lifetimes, accompanied by low transition probabilities due to the forbidden nature of the transitions [10–13]. Thus chromophoric sensitization of ligand to metal has been subjected to numerous theoretical and experimental investigations [14–16]. However, only limited classes of organic–lanthanide complexes have been developed and shown to display nonlinear processes [17–19]. Common nonlinear processes from lanthanide complexes include harmonic generation, photon up-conversion and multiphoton absorption induced emission.

This chapter reviews the important aspects of multiphoton absorption sensitization of lanthanide complexes and their nonlinear behavior; some typical nonlinear processes from the conversion of long-wavelength excitations to give short-wavelength emissions will be presented in Section 7.2. Their basic features and their differences will be described.

Examples of studies on multiphoton absorption processes and nonlinear second- and third-harmonic generation processes will be discussed along with some possible radiative and nonradiative processes. The selection rules for multiphoton absorption will be mentioned in Section 7.3, and molecular examples will be shown along with their correlating photophysical properties in Section 7.4. The effect of some parameters relating to second-order activity along the lanthanide series and some recent investigations on the role of 4f electrons will be discussed in Section 7.5. Section 7.6 gives a personal view on the future research perspective of optical properties of lanthanide complexes.

7.2
Types of Nonlinear Processes

In nonlinear processes, the emission/luminescence intensity (I) does not increase proportionally with increase in excitation power density. Thus, it is nonlinear in intensity of applied light.

Linear: Signal $\propto I$

Nonlinear: Signal $\propto I^2, I^3$ and so forth.

The effect that the light itself induces as it propagates through the medium determines the different types of nonlinear processes and optical phenomena. These phenomena are usually only observed at very high light intensities and such nonlinearity requires the use of high-power pulsed lasers [20].

The conversions of long to short and short to long wavelengths are nonlinear processes such as *up-conversion* and *down-conversion*. These are very "general" terms normally used to differentiate the difference occurring in the conversion of the wavelength. Down-conversion refers to excitation at high energy which is converted to lower frequency, such as in quantum cutting where excitation is in the vacuum ultraviolet with emission in the visible or from the visible to the infrared region [21, 22]. Up-conversion is where low-frequency photoexcitation energy is converted to a higher frequency, such as excitation in the near infrared region to produce emission in the visible [23, 24]. The focus of this chapter is on the "up-conversion" of nonlinear processes.

Three major approaches to achieve short wavelength generation from longer wavelengths in photophysics are illustrated in Figure 7.1. The first approach is nonlinear second and/or third harmonic generation (SHG/THG) [25]. This is a laser-dependent system where upon change of the pumping wavelength the emission output will be changed accordingly. In the other two processes these are system dependent, as changing the pumping energy will not change the output. The second approach is up-conversion, based on sequential absorption such as in excited state absorption (ESA) or energy transfer involving two centers, mostly found in rare-earth-doped materials, and is normally known as energy transfer up-conversion (ETU) [26, 27]. The last approach is based on (one-center) direct multiphoton absorption (MPA), in which absorption of multiple photons is simultaneous; this phenomenon is mostly observed in organic dyes [28]. This has only been realized from the 1970s, with interest in two-photon absorption leading to increased interest in other, higher-order processes [28–30]. However, it is impor-

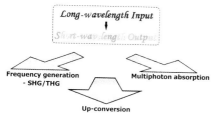

Figure 7.1 Illustration of three mechanisms proposed for the energy conversion of long-wavelength excitation to short wavelength emission (SHG/THG: second/third harmonic generation).

tant to distinguish carefully the difference between these processes and to understand the principles involved.

In general, nonlinear harmonic generation, as in SHG/THG is when light is generated at a frequency that is two or three times, respectively, that of the incident light. The criteria for these phenomena are a stable, intense, and coherent excitation source. Crystals with noncentrosymmetric space groups are required for SHG but not for THG. SHG is not possible in isotropic materials, but THG is allowed in all materials. This is because these processes require phase matching and thus are coherent processes and also exhibit wavelength-dependent hyperpolarizability [31–33]. The intermediate states are all "virtual" states as the absorption and emission are simultaneous; one such example is given by lithium niobate [34].

Up-conversion relies on sequential absorption and luminescence with intermediate steps to generate shorter wavelengths. Hence, the presence of *more than one metastable excited state* is required; the intermediate metastable states act as excitation reservoirs. One typical example is ground-state absorption followed by intermediate-state excitation, excited-state absorption, and final-state excitation to give the up-conversion (the intermediate states and final states are real states) [1, 35]. There are many types of up-conversion mechanisms such as excited-state absorption, energy transfer up-conversion and cooperative up-conversion. All these up-conversion processes can be differentiated by studying the energy dependence, lifetime decay curve, power dependence, and concentration dependence by experimental measurements [36–39].

In contrast, multiphoton absorption requires only *one real excited state* and the accumulation of the photons is via virtual states; which need not correspond to any real electronic or vibrational energy eigenstate. The absorption of the photons is simultaneous, with the extra energy of the excited states corresponding to the sum of the energies of the incident photons.

Usually luminescence follows multiphoton absorption. If an accepting metal is involved, the energy is sometimes transferred via intersystem crossing to give sensitized emission if there is appropriate matching of the energy levels [40–42]. Such antenna sensitization has been demonstrated from both singlet and triplet states of chromophoric ligands [43, 44]. Selection rules are important in these processes as they impose certain parity restrictions on the transitions. Thus some transitions are forbidden – that is, transitions for odd or even numbers of photon absorptions are different, such as in one-photon absorption and two-photon absorption, respectively.

7.3
Selection Rules for Multiphoton Absorption

The selection rules will be mentioned briefly here. In general, the process of multiphoton absorption is similar to that of single-photon absorption. The multiple photons are absorbed simultaneously to a real excited state in the same quantum event, where the energy of the transition corresponds to the sum of the energies of the incident photons. Thus selection rules for these transitions may be derived from the selection rules for one-photon transitions as they can be considered *multiple* one-photon transitions [20].

The selection rules governing transitions between electronic energy levels are the spin rule ($\Delta S = 0$), according to which allowed transitions must involve the promotion of electrons without a change in their spin, and the Laporte rule ($\Delta L = \pm 1$ for one photon). This parity selection rule specifies whether or not a change in parity occurs during a given type of transition. It states that one-photon electric dipole transitions are only allowed between states of different parity [45].

Relaxation of the rules can occur, especially since the selection rules apply strongly only to atoms that have "pure" Russell–Saunders (L–S) coupling. In heavy atoms such as lanthanides, the Russell–Saunders coupling is not entirely valid as there is the effect of the spin–orbit interactions, or so called j mixing, which will cause a breakdown of the spin selection rule. In lanthanides, the f–f transitions, which are parity-forbidden, can become weakly allowed as electric dipole transitions by admixture of configurations of opposite parity, for example d states, or charge transfer. These f–f transitions become parity-allowed in two-photon absorptions that are g ↔ g and u ↔ u. These even-parity transitions are forbidden for one photon but not for two photons, and vice versa for g ↔ u transitions [46].

According to the selection rules, one-photon absorption occurs only if the change in angular momentum (change in L) is +1 or −1 ($\Delta l = \pm 1$, $\Delta J = 0, \pm 1$ (0 ↔ 0 not allowed), $\Delta L = 0, \pm 1$, $\Delta S = 0$) (Δl is according to the hydrogenic atom model, whereas ΔL is for multielectron atoms). The selection rules allow transition in one-photon absorption only to the p states from the s ground state; as a result only even-to-odd parity is allowed.

Since photons have angular momentum of +1 or −1, an electronic state absorbing two photons simultaneously may change angular momentum by +2, 0. Two $L = +1$ photons cause a change of +2; a photon of $L = +1$ and one of $L = -1$ cause a change of 0 ($\Delta l = 0, \pm 2$, $\Delta J = 0, \pm 2$, $\Delta L = 0, \pm 2$, $\Delta S = 0$). Thus the selection rules for two-photon absorption allow the excited electron to be either in an s or a d state, states which are of even-to-even parity or odd-to-odd parity such as f–f transitions, which now become allowed. An electron therefore cannot go from an s state to a p state by two-photon absorption, but transition is allowed for states of the same parity [47, 48].

This rule is analogous for any number of photons; thus, in a three-photon absorption process the allowed states are the p and f states, which allows even-to-odd or odd-to-even parity. This means that transitions can be divided between states of either the same or different parity corresponding to absorption of either even or odd numbers of photons. Hence the selection rules are the same for one and three photons and for two and four photons, respectively [49].

7.4
Multiphoton Absorption Induced Emission

Multiphoton absorption for organic–lanthanide complexes can be achieved by a careful choice of sensitizing ligand [19, 44, 50, 51]. It has been shown that the tripodal amide ligands can give rise to two different types of coordinated complexes [34, 35]. They were found to exhibit rather different nonlinear processes and could be distinguished by their spectral properties. Some observed coordination modes of these complexes are shown in Figure 7.2. Several similar types of ligands have previously been studied by Raymond and co-workers in terms of their magnetic properties with transition metals for relaxation agents in magnetic resonance imaging applications. It has been shown that the metal ion is encapsulated in a coordination mode similar to type B, with bonding from the metal to the substituent groups of the ligands as well [52].

The organic–lanthanide complexes described here are illustrated in Figure 7.3 with their respective tripodal amide ligands, **1–3**. These tripodal ligands were designed by the incorporation of a suitable sensitizing group, usually an aromatic moiety, into the amide backbone. The chromophores selected are a benzoate group for **1**, a 2-methoxybenzoate for **2**, and 3-methoxybenzoate for **3**, which allowed for simple synthesis. Complexation of these ligands with their respective lanthanide nitrate salts at different ratios afforded the desired lanthanide complexes [44, 52, 53].

These complexes were found to display two different polymeric architectures with tripodal ligand coordination of types A and C as illustrated in Figure 7.2. Both types of polymeric complexes have a coordination number of 9, which is typical of the lanthanide ions in the middle of the series, namely Eu^{3+}, Gd^{3+} and Tb^{3+}. Type A was found to form as a metallodendrimer, whereas type C formed as a one-dimensional linear polymeric chain. The metal in the metallodendrimer is coordinated to three carbonyl oxygens of three independent tripodal amide ligands. Thus the metal can be described as being in a 1:3 ratio with respect to the ligand. This, along with the coordinated nitrates, forms an architecture that is higher in symmetry than that of type C in the linear chain. Crystals obtained for these structures such as **1**, with a noncentrosymmetric space group, allowed the observation of second- and third-harmonic generation in addition to the multiphoton absorption induced emissions [50]. The linear polymeric chain of type C such as in **2** and **3** crystallized in a centrosymmetric space group [44]. The metal and ligand

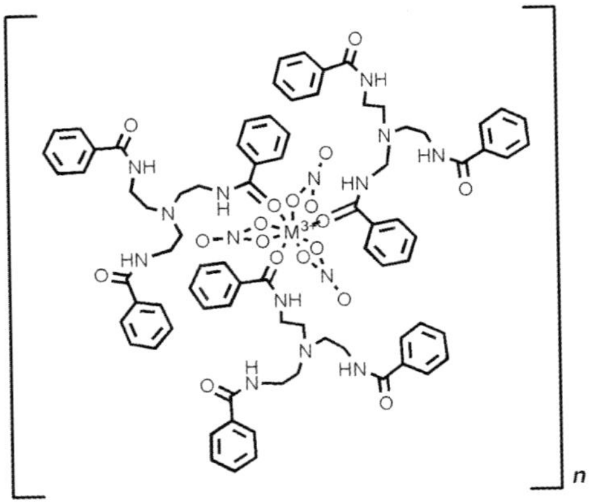

Figure 7.2 Different types of coordination complexes with the tripodal amide ligand. Type A and type C are adapted from [29] and [30].

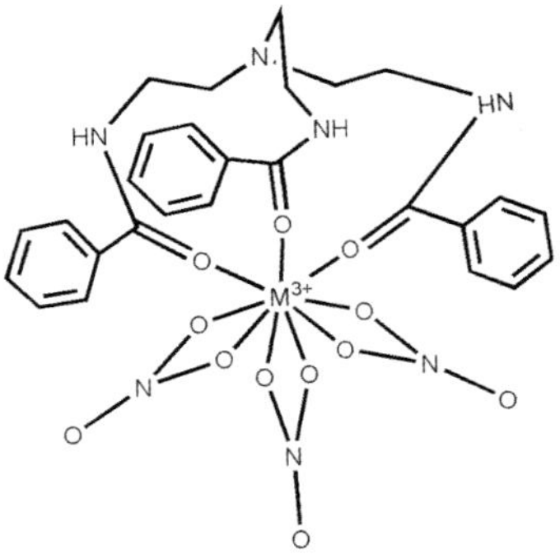

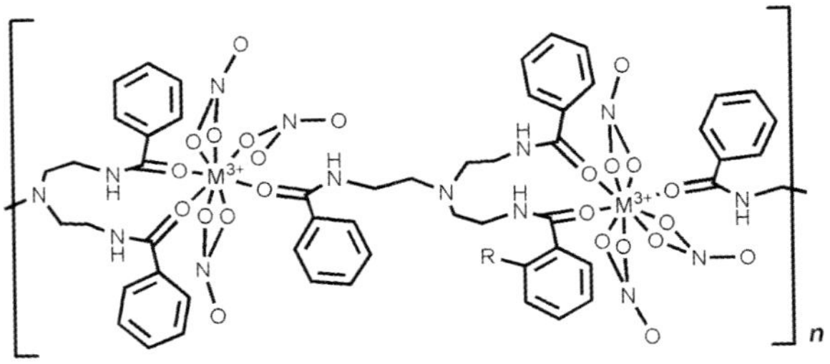

here are in alternating fashion where two tripodal amide ligands are coordinated to one metal center. The coordination of the metal to ligand is also via three carbonyl oxygen groups with six coordinations associated with the nitrates. However, one ligand always has two participating arms of two coordinating carbonyl oxygens, whereas the other ligand will have only one coordinating carbonyl group.

The metallodendritic type A complexes and the linear polymeric complexes of type C both show similar absorption band shapes in their spectra that are not affected by the differing coordination as the ligands are the same. However, the absorption of the tripodal amide ligands can vary due to the presence of different substituent groups on the aromatic rings of the ligand [54]. There are slight displacements of the absorption peaks upon complexation with the metal ions, in this case the studied lanthanides are Eu, Gd, and Tb, but spectral discussions will be based on Tb. The excitation spectra of **1** and **2** in the infrared region show resemblance to the single-photon UV-vis absorption spectra, except that the wavelength is tripled; see Figure 7.4. These results with power dependence

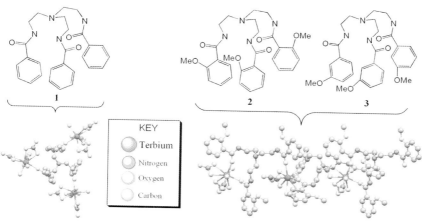

Figure 7.3 Terbium complexes with their ligands and coordination type. **1** is dendritic, whereas **2** and **3** both contain a methoxy substituent group in the ligand and form one-dimensional chains.

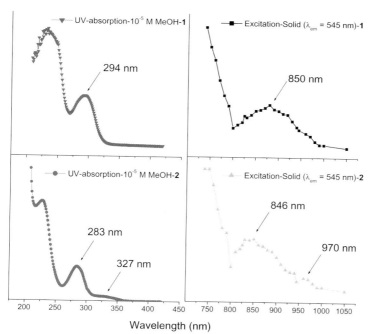

Figure 7.4 Linear UV-absorption and three-photon excitation spectra of complexes **1** and **2**.

experiments confirm that the transition symmetries for one- and three-photon absorption processes are the same [55]. As previously discussed, the selection rules governing the electronic transitions are different for two-and three-photon absorption due to the different symmetry requirements. Thus the intraconfigurational f–f transitions are parity-allowed for two-photon absorption whereas they are forbidden for a three-photon absorption process. However, as the f–f transitions are sensitized by a chromophoric ligand, it is of no significance here. In three-photon absorption, ligand excitation is possible where there is low symmetry and no parity [20, 56].

Photoluminescence spectra of the terbium complexes **1** and **2** are displayed in Figure 7.5; which compares the emission properties from the two different coordination types A and type C, respectively. It is observed that both type A and type C display multiphoton induced f–f emission from the terbium, but only second and third harmonic generations are perceived in type A, which are located in their respective wavelength regions as the harmonic generations are wavelength (excitation)-dependent. The emissions are shown at different excitation wavelengths, progressing from near to mid infrared with a power source of 1.26 μm. At such excitations, power dependence studies have shown absorption of up to four photons. Their respective second, third, and fourth power dependence shown by plotting power dependence on excitation power on a log scale indicates the absorption of two, three, and four photons in the process [57]. The proposed mechanisms for type A and type C complexes are very similar, with A having an extra mechanism from the second- and third-harmonic generation, and are illustrated in Figure 7.6 for **1** and **2**. It is proposed that in type A complexes the nonlinear harmonic generations occur simultaneously and do not play a role in the multi-

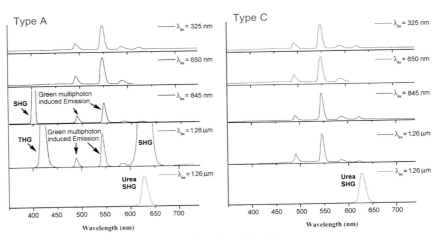

Figure 7.5 Photoluminescence spectra of type A and type C under excitation at different wavelengths. (Urea is used as the standard for SHG. SHG and THG are second- and third-harmonic generation, respectively.)

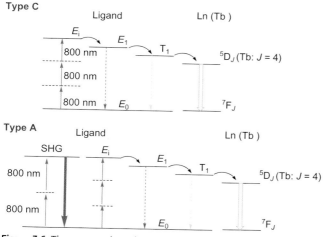

Figure 7.6 The proposed mechanisms for type C complexes and type A complexes (adapted from [29] and [30]).

photon absorption process as they are very different processes. This is because light intensity of SHG is approximately <10^{-6} of the excitation source and thus it is unlikely that it can reexcite the f–f absorption band. The simultaneous occurrence of harmonic generations and the multiphoton absorption induced emissions was shown by lifetime measurements and power dependence experiments. Comparing the intensity and the integrated areas of the emission bands of the f–f transitions shows that the pathways for types A and C are similar and that the SHG/THG in type A does not affect the mechanism for the f–f transition. From the power dependence plots it was also shown that the ligand was not reexcited by the SHG to induce the f–f transitions, which would have given one photon in the power dependence plot [50].

The studies of the respective Gd complexes at low temperatures of both types of complexes allowed the conclusion that the energy transfer is via the triplet state of the ligand [58, 59]. When the ligand absorbs three photons via the multiple virtual states (see Figure 7.6), it goes from the ground state to the excited singlet state. The energy then relaxes to the lowest singlet state, S_1, and then further by intersystem crossing to the lowest triplet excited state, T_1. The triplet state lifetimes here were measured in microseconds when compared to the milliseconds of the lanthanide metal ion and to the nanoseconds of the fluorescence of the singlet state of the ligand. The energy is then transferred by the antenna effect to the lanthanide center, in this case Tb, to give the characteristic green emission from the $^5D_4$–^7F_J bands. The mechanisms of these systems allows one to relate the significance of these different structural motifs, with the additional bonus of tunable nonlinear harmonic generation to give all three additives' colored emissions of blue, green, and red at the same time, which is only possible in type A.

The structure–property relationships are affected by three main structural elements – coordination and hence packing, which determines some of the nonlinear harmonics; the degree of conjugation along the backbone; and the effects of the substituents – all of which have previously been reported to affect two-photon absorption processes; thus it is proposed that the same applies for other higher-order systems. However, it is rare for reports of these high-order multiphoton absorption processes displayed by organic metal complexes to be accompanied with detailed molecular structure to show the correlation between the photoproperties.

Different substituents have been shown to affect and contribute to the properties of the ligand, and thus of the metal as well since the weaker the absorption the weaker is the energy transfer or so-called sensitization of the metal. The substituent groups on the aromatic rings not only affect the multiphoton absorption and thus the S_n, S_1 and T_1 states of the tripodal ligand but will also dictate the compatibility of the f–f transitions of the lanthanide acceptors. By measuring the triplet state of the ligand, one can choose suitable lanthanides for luminescence, which is sometimes inappropriate due to the energy gap matching. If it is too close, quenching by back transfer can affect the emission; if too far, the energy can be lost by greater nonradiative processes [58]. The tripodal ligands studied here mostly absorb in the ultraviolet region, thus mostly favoring the $^5D_4$–^7F_J to the ^7F_J transitions of the terbium and sometimes the $^5D_0$–^7F_J of the europium, which tends to have weaker luminescence and lower quantum efficiency. This has been found to be true in both the linear and the nonlinear processes.

In the aromatic rings of the tripodal ligand, electron-donating groups can expected to be and are found to be more favorable than electron-withdrawing groups, which reduce the π-electron density in the ring, as well as halide groups which promote the inductive effect [60]. The size and position of the substituents affects the chain and molecular packing and thus the luminescent efficiency [61]. Such examples have been found with bulkier side-chains and functional groups which can reduce the efficiency of nonradiative decay pathways due to interchain interactions and so will improve the photoluminescence [59, 62]. The position of the substituents has been shown to strongly affect the linear and multiphoton absorption properties in the tripodal ligands and thus in the complexes. This has been illustrated with two terbium complexes of the same space group with the same type of coordination and substituent (**2** and **3**); Figure 7.3 shows the ligands and molecular structures. Here the terbium complexes studied are the linear polymeric chain type with one methoxy group as the substituent on the aromatic ring of the tripodal ligand. In **2** the methoxy group is placed in the *ortho* position, whereas in **3** it is in the *meta* position of the aromatic ring. Due to *para* and *ortho* electron donating groups having resonance effects, the *ortho* position of the methoxy group is better at directing the electron density to the carbon on the ring that is connected to the outer conjugated backbone. For the *meta*-positioned methoxy group, it has no charge transfer resonant structures, thus less electron density is directed to the conjugated backbone system. This factor has been proposed as the main contributor to the difference between the emission properties as the backbone conjugation is important in the formation of a donor–π-acceptor

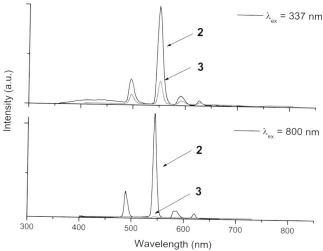

Figure 7.7 Linear (λ_{ex} = 337 nm) and three-photon absorption induced f–f emission (λ_{ex} = 800 nm) of complexes **2** and **3**.

system. The emission spectra of the terbium complexes ($^5D_4 \rightarrow {}^7F_J$ transitions; J = 6, 5 are stronger than $J < 5$) at λ_{ex} = 337 nm and multiphoton absorption induced emission spectra at λ_{ex} = 800 nm can be observed in Figure 7.7. These spectra reveal a much more intense emission for **2** than for **3** at both excitations, but at λ_{ex} = 800 nm excitation the emission of **3** is virtually negligible compared with that of **2**, as the multiphoton process is much weaker. Such a dramatic difference in the photoluminescence from the f–f transition is regarded as the result of the different substituent positioning of the methoxy group in the ligand.

The unit cell packing diagrams of the terbium complexes **2** and **3** with their coordination geometries are depicted in Figure 7.8, which shows some slight packing differences. If viewed from the Tb center in an octahedral arrangement, these differences are more prominent. In **2**, the ligand shows facile arrangement to the Tb center with the nitrates also all in facile arrangement and orthogonal to each other. In contrast, in **3**, the ligands are in a meridional arrangement with two nitrates almost parallel but tangent to the third nitrate. These small differences in arrangement can also be taken to make a small contribution to the differences in luminescence.

Another example of a different type of correlation of structural to photophysical properties is shown in a study of a unique terbium compound [63]. This compound will be briefly discussed and is depicted in Figure 7.9 with its nonlinear emission properties with excitation at 800 nm. The photophysical properties are atypical and rather extraordinary due to the unusual molecular structure of the co-crystallization compound (**4**) of the organic chromophore and the terbium salt. This compound shows both multiphoton absorption induced green f–f emission from the terbium ion as well as second-harmonic generation. However, unlike previously

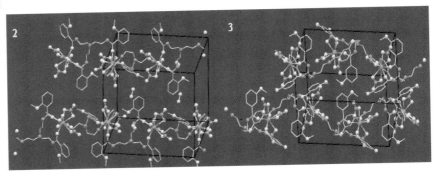

Figure 7.8 Packing of complex **2** with *ortho* methoxy substituent and **3** with *meta* methoxy substituent on their tripodal ligands.

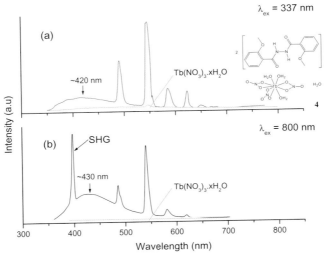

Figure 7.9 The solid-state linear emission (a) and multiphoton absorption induced emission/SHG (b) spectra of **4** at room temperature.

described structures, the organic chromophore and metal here have no direct coordination, which is shown by crystal x-ray determination of the structure. The structure shows the closest Tb to carbonyl oxygen at around 4.4 Å with the Tb center approximately 6–9 Å away from the centroid of the π rings in the ligand, but it is emphasized again that no contact between the terbium and the ligand is observed. However, the terbium is sensitized by an antenna effect: in a control experiment with the terbium nitrate salt, virtually negligible emissions are seen in comparison; very weak at λ_{ex} = 337 nm and unobservable at λ_{ex} = 800 nm [46, 64, 65]. In another comparative study of **2** and **4**, to examine the induced

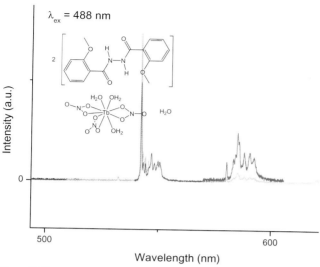

Figure 7.10 Emission spectra of **4** between 500 and 620 nm at 10 K.

$^5D_4 \rightarrow {}^7F_J$ transitions, the emission intensities were of the same order and thus the antenna sensitization was also comparably efficient.

Although the terbium salts contain more quenching aqua ligands due to the OH oscillators, these are relatively unimportant for the $^5D_4$ state because of the large energy gap below. Also, compound **4** itself contains three aqua ligands in its coordination sphere as well as three bidentate nitrate groups to give Tb^{3+} with C_1 site symmetry. This is shown by low-temperature experiments at 10 K in which the fine structures can be obtained. The fine structures of the $^5D_4 \rightarrow {}^7F_5$ transition are shown in Figure 7.10 for the nondegenerate terminal crystal field state which is typical of the C_1 site.

The emission intensity–power dependences on a log-scale for the second-harmonic generation and three-photon absorption are linear with slopes of 2 and 3, respectively which confirms the nonlinearity of the processes. The similarity in lifetimes within experimental errors of the Tb^{3+} emission in **4** (4.11 ms and 4.23 ms) for both linear and multiphoton excitation allows for the proposed energy transfer mechanism. In this the pathway is from $S_1 \rightarrow T_1 \rightarrow {}^5D_4$, which is the same as for complex **1** which exhibits the same nonlinear processes. Furthermore, the emission spectra of **4** at 337 nm and 800 nm excitation show emission from the ligand which arises from the triplet of the ligand and also shows the involvement of the antenna; whereas the second-harmonic generation arises from the compound being in a dipolar group with the correct phase matching in the solid state.

Multiphoton absorption induced emission from lanthanides emitting in the visible region is by far most studied with Eu and Tb, which are the most widely used metals as they have favorable emission windows and larger separations of the *J* levels and thus energy gaps [66]. However, interest in other lanthanides, especially those that can emit in the infrared window, is increasing due to their attraction as near-infrared probes, in telecommunications, and in optical power control. In biological applications, tissues are transparent at such wavelengths and thus deeper penetration into the skin can be obtained because these radiations are not absorbed by cellular and blood constituents [67, 68]. The beneficial properties of lanthanides such as sharp emissions and long lifetimes allow time-gated elimination of the usual biological autofluorescence background. Absorption in the infrared region will also allow applications as sensitizers for near-infrared photorefractive composites [69–72].

The lower energy or emission states of lanthanides such as Nd, Er, and Yb can allow excitation at considerably longer wavelengths such as those used for multiphoton processes. However, most problems encountered with these lanthanides are in the choice of appropriate weakly energy-absorbing sensitizers which are often derived from azide dyes and their derivatives [68, 69]. Advances in molecular engineering suggest that such problems will be overcome.

Organic–lanthanide complexes with infrared emission by multiphoton absorption have been reported only rarely; an example is shown in Figure 7.11, unpublished work which illustrates the power dependence and emission spectrum that are characteristic of a Nd complex, **5** [73–74]. Excitation of the complex is shown at 800 nm, with three-photon absorption induced emission, which is an advantage over two-photon absorption especially for potential as probes because the cubic dependence on the input light intensity provides stronger spatial confinement, which allows higher contrast in imaging. This is confirmed again by power dependence experiments, which also substantiate that the emission is not from direct excitation of the Nd^{3+} ion (there is an appropriate energy transition at 800 nm that can cause dispute whether the emission is via a nonlinear process). The problem with such a Nd complex is that its emission window is between 850 nm and 1400 nm, which makes excitation in this region impossible. Use of far-infrared

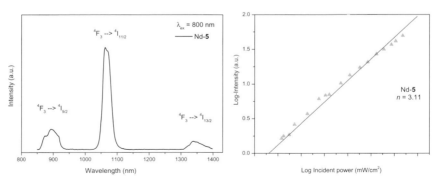

Figure 7.11 Three-photon absorption induced f–f emission spectra (left) of **5** in the infrared region with its corresponding power dependence plot (right).

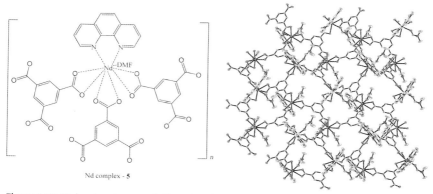

Figure 7.12 Molecular structure (left) and packing (right) of the Nd complex **5**.

excitation (such as at 1.6 μm) will have to address the problems of thermal stability of the complex as the excitation power is increased. Also, with the increase in the order of photon absorption, there will be a decrease in the probability of the required number of photons reacting at the same time. In other words, for two-photon absorption the photon density must be approximately one million times that required to generate the same rate of one-photon absorption. This has to be accommodated by increase in the power of excitation, which can be enhanced by focusing the light into a small spot; which in turn will affect the sample's thermal stability due to the high laser power [57].

Structural changes and modifications of this molecular complex are also important in relation to its emission properties. From x-ray crystallographic determination, **5** crystallized in an orthorhombic space group *Pbca* with the Nd situated in a coordination geometry that forms a two-dimensional coordination network which is portrayed in Figure 7.12. The Nd ion is in a nine-coordination environment with coordination to six carboxyl groups from three bidentate benzene 1,3,5 tricarboxylic acids; with two coordinations to a bidentate phenanthroline and a dimethylformamide (DMF) solvent molecule. This results in a distorted monocapped square antiprism geometry of the Nd complex, which has the above-described multiphoton absorption induced emission properties. However, carrying out the synthetic procedure without the addition of phenanthroline, the polymeric complex obtained was found to show only linear properties. Hence it is suggested that the structural modification plays a role in the nonlinear process. Unfortunately, it was not possible to obtain crystals of these intermediates for better characterization.

To summarize, as shown by the work discussed in this section, it is obvious that studies in this area are necessary for an understanding of the different possible mechanisms in nonlinear processes. Apart from its intrinsic interest, it will open the door for more advanced developments in photonic imaging and biosystems involving antenna and targeted sensitization.

7.5
Nonlinear Harmonic Generation

Nonlinear harmonic generation properties, especially in thin films and inorganic crystals, have been intensively and in some cases well explored, such as in potassium titanyl phosphate and lithium niobate [75–78]. This has enabled the design of materials for use in such applications laser systems and photonic devices, areas of still growing interest. There has also been emerging interest in organic materials as an important class of nonlinear optical materials due to their structural flexibility at both the molecular and bulk levels [24]. However, investigations with organic metal complexes have been comparatively rare, being first reported in 1980 with the use of transition metals, and lanthanide complexes have only really been looked at during the last two decades or so [79–82]. There are thus many properties still unknown and to be unexplored.

LeBozec and co-workers have reported nonlinear behavior in a series of terpyridyl and dipicolinic acid complexes, with further studies on these complexes by Maury and co-workers [83, 84]. Their research was on new molecular materials for optoelectronics, with studies based on octupolar nonlinear optical molecules showing that molecular quadratic hyperpolarizability values were strongly influenced by the symmetry of the complexes [85]. Other studies on organic–lanthanide complexes with nonlinear optics have also reported second- and third-harmonic generation behavior with simultaneous multiphoton absorption properties [50]. Such studies have shown the importance of coordination chemistry as a versatile tool in the design of nonlinear materials.

The primary objective of this section is to show the importance of lanthanides to nonlinear harmonic generation. This work, which is unpublished, evolved from earlier studies that have shown the correlation of nonlinear optical effects in dipolar and octupolar compounds which involve the use of one or more electron-donating groups connected via a conjugated π-linker to an electron acceptor group.

Discussion will be based on as series of lanthanide complexes covering the whole row of the 4f elements in the periodic table. These complexes were generated with cinnamic acid used as the coordinating ligand and can be subdivided into two types of complexes due to different metal center coordination and thus space group generation. All these lanthanide complexes were found to show second-order harmonic generation at excitation of 800 nm with the use of a Ti:sapphire femtosecond laser as the power source, as shown in Figure 7.13. The space groups are noncentrosymmetric and thus noncentrosymmetry is a universal requirement for materials to have bulk quadratic nonlinear optics, offering a basic explanation for the occurrence of the nonlinear behavior [31, 32, 86].

The first half or more of the series, known as the early and mid-lanthanides (La–Tb), are referred to here as type I; the later or second half of the lanthanide period (Dy–Lu), and also for Y complex, are known as type II. The type I complexes all have larger ionic radii (1.10–1.22 Å) than type II, which are typically between 0.92 and 0.97 Å. The size of the metals was found to govern their coordination to the cinnamic acid, thus giving rise to the two different coordinations and space

7.5 Nonlinear Harmonic Generation

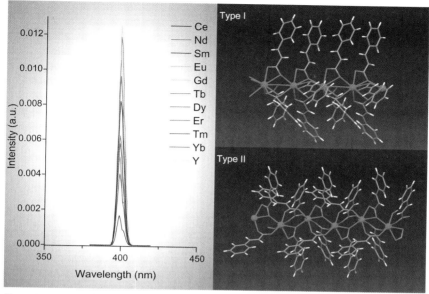

Figure 7.13 SHG of the lanthanide and Y complexes (left), with packing of type I and type II (right; type I = Ce, Nd, Sm, Eu, Gd, and Tb; type II. Dy, Er, Tm, Yb, and Y).

groups. This was confirmed by the additional Y complex, which has a small ionic radius and fits in with the trend by displaying type II coordination. Y is often classed with the lanthanides as it has strong similarities in chemical properties, with the same preference for forming in the 3+ oxidation state with ionic radius of 0.96 Å, similar to Dy and Ho. These different types of coordinated complexes all form polymeric chains in which types I and II show nine- and seven-coordination, respectively, as illustrated in Figure 7.13, which shows a Ce and an Er complex; the difference in the ligand arrangements of the respective space groups $R3_c$ and $P2_1$ are also portrayed.

Past studies have shown that increase in ionic radius increases the nonlinear activity of lanthanides, for example studies by Khamaganova [87]. However, newer work by LeBozec and co-workers, in which they determined the first hyperpolarizability experimentally, demonstrated the first direct contribution from f electrons to nonlinear activity by investigating the contribution of the f orbitals to the hyperpolarizability tensor. Their molecular first hyperpolarizabilities were measured in water solution using the harmonic light scattering (HLS) technique at 1.06 mm. However work by other groups such as Bogari and Cavigli, have shown no obvious trends relating to the effect of the metal centre in their family of lanthanide complexes; which was studied by the Kurtz-Perry method in the solid state [89].

The lanthanide complexes measured here were measured in the solid state due to the effect of the polymeric nature of the complexes, which reduced their solubility in virtually all solvents, with limited solubility shown in dimethyl sulfoxide.

Given that concentration and solubility are important factors in the measurement of hyperpolarizability, β, the conventional harmonic light scattering technique could not provide accurate results. For this reason, results based on solid-state measurements are discussed here. X-ray diffraction has been performed to calculate the crystallite size, and scanning electron microscope (SEM) pictures were taken of the surface that were used in the photo measurements to confirm that there was not a large deviation in the surface which would affect the reported results. The calculation of the particle size was derived from Schemer's equation. The crystallite size was estimated as an average for the two different series, which gives a rough estimate where **type I** is 93.5 ± 4.5 nm and **type II** is 90.2 ± 5.1 nm. Averages were also obtained for the SHG intensity and plotted by the statistical method based on the Student's t-distribution. Error bars on the intensity were also calculated using the uncertainty propagation method and given a confidence level of 95% [90–92].

For both type I and type II complexes, the second-harmonic generation intensities were found to vary significantly from one metal complex to another. In general, the second-harmonic generation was found to surpass that of the common standard urea. Results from preliminary studies illustrated in Figure 7.14 suggest a trend which is different to those previously reported, a trend that relates to unpaired electrons. This is shown by the corresponding increase in nonlinear activity with an increase in the number of unpaired electrons in the metal ion, this trend being observed in both type I and type II complexes. As expected, the preeminent second-harmonic generation is observed for Gd, which contains the maximum number of unpaired electrons in the f orbitals. It is also observed that for type I and type II, the nonlinear activities are almost the same (by intensity and integrated area of their emissions) with respect to the number of unpaired electrons. This can be discerned from the plot as the intensity scales are the same for the two types of complexes; for example, the lowest intensity arises from Ce and Yb with one unpaired electron, compared to the higher intensities for Eu and Tb, both with six unpaired electrons. This shows that nonlinear optics has little effect dependence on the space group. Nevertheless, the two different types of coordination should be taken in account as they will affect the packing of the bulk molecule and thus contribute to the intensity difference, though in this case the effect is not found to be striking. The plots for type I and II complexes against ionic radii also shown in Figure 7.14 further confirm that the metal size has no major role in the nonlinear activity as there is no general trend. Anomalies with La and Tb are observed for type I complexes and with Y and Lu for type II.

It is also apparent from Figure 7.14 that other factors besides electronic parameters concurrently make their respective contributions, though more weakly. This is shown by a trend found for La, Y, and Lu. It is observed that, when no unpaired electrons are present, the nonlinear activity follows the typical ionic radius trend of increase with metal size from Lu to La. Any association of the contribution from f electrons can be ignored here because La and Y have $4f^0$ electronic configuration although Lu has a full f-shell. Again the difference in space group seems to have little effect in disrupting the associated trends.

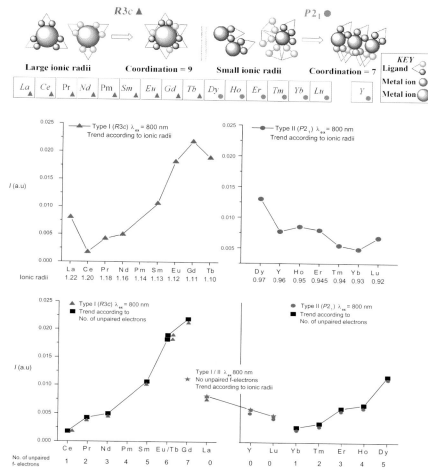

Figure 7.14 (Upper) Different space groups and the types of coordination along the lanthanide series. (Middle) Plots of the SHG intensity along the type I and type II series as according to ionic radius and also (bottom) to the number of unpaired f electrons. Complexes with no unpaired f electrons are shown to follow a different trend (La, Y, Lu).

To conclude, it is recognized that electronic and geometric parameters both affect the nonlinear activity; however, for lanthanide complexes the electronic parameters have been to date been found to be more significant for the nonlinear activity in second-harmonic generation. Such a hypothesis is outlined by previous studies by other groups on direct f electron contributions and by this work, which suggests that the number of unpaired electron can act as an additional factor

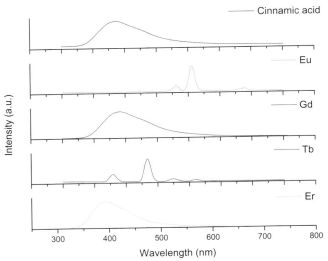

Figure 7.15 Linear emission spectrum of selected lanthanide complexes (Eu, Gd, Tb, Er) and cinnamic acid (λ_{ex} = 337 nm).

contributing to nonlinear activity. However, further studied are still required as other groups have shown that this is not necessarily true for all types of lanthanide complexes with SHG activity. It is hard to draw conclusive propensity at the molecular/electronic level from measurements at the macroscopic level from such prelimary investigations.

In addition to special nonlinear properties, other optical properties will also be mentioned in brief. The optical absorption of these complexes is at approximately 328 nm with a bathochromic shift of 5 nm upon complexation with the metal. Here the linear photoproperties will be discussed as these polymeric structures display no multiphoton absorption unlike those previously reviewed. The photoluminescence spectrum of the complexes shows a broad ligand peak from the cinnamic acid with a maximum at ~430 nm which is very weak and unobservable if looked at with the f–f emission of the lanthanides, which dominates the spectral intensity when allowed matching of the energy levels is obtained. This is illustrated in Figure 7.15, which shows the characteristic $^5D_4 \rightarrow {}^7F_J$ transition from europium and $^5D_0 \rightarrow {}^7F_J$ transition from terbium at λ_{ex} = 337 nm. Otherwise, as with most of the lanthanide complexes, nonradiative emission occurs, radiative emission being possible only by fluorescence and phosphorescence, which will dominate due to the presence of a heavy atom as it will lead to higher intersystem crossing due to stronger spin–orbit coupling.

The spectra of a few arbitrarily selected emissions from other lanthanides, such as Gd and Er, are also illustrated in Figure 7.15. These results elucidate the incompatible f–f transitions to the ligand triplet state [93–95]. These photoemission studies were performed using a Ti:sapphire femtosecond laser with all measurements carried out in air and at room temperature.

7.6
Conclusion and Future Perspectives

In this chapter, different properties of nonlinear behavior in lanthanide complexes have been overviewed, including brief examination of their optical emission and excitation spectra. A perception of the importance of photophysical relationships of molecular complexes to these phenomena has also been conveyed.

These studies provide interesting developments for optical materials and perhaps possibilities for bio-imaging systems. This is only achievable if a good understanding of the structures and their optical behavior is available. Their potential is highlighted by the rarity and novelty of some of these systems, for which studies are only in their infancy, indicating that there is great scope for the optimization of multiphoton absorption and nonlinear properties of metallochromophore complexes by molecular engineering. A major future objective is to investigate further the parameters governing these optical properties, which hold promise for switching behavior, as knowledge in this areas is still lacking or in its preliminary stages. Intriguing and challenging possibilities await to be uncovered.

7.7
Acknowledgments

Special thanks go to Prof. Pignataro for the opportunity of writing this article as well as to the committee members of the first European Chemistry Congress and to the Royal Society of Chemistry for the conference award.

The author wishes to acknowledge Prof. W.T. Wong for his invaluable comments. Gratitude also goes to Prof. P. A. Tanner, Prof. D. L. Phillips, Dr. W. M. Kwok, and Prof. K. W. Cheah for their excellent collaboration as well as all research members who contributed to this work as co-authors in some of the publications that are reviewed here. This work was supported by The University of Hong Kong.

References

1. Gamelin, D.R. and Güdel, H.U. (2000) *Accounts of Chemical Research*, **33**, 235–42.
2. Maury, O. and Le Bozec, H. (2005) *Accounts of Chemical Research*, **38**, 691–704.
3. He, G.S., Markowicz, P.P., Lin, T.-C. and Prasad, P.N. (2002) *Nature*, **415**, 767–70.
4. Gao, D., Agayan, R.R., Philbert, H., Xu, M.A. and Kopelman, R. (2006) *Nano Letters*, **6**, 2383–6.
5. Zijlmans, H.J.M.A.A., Bonnet, J., Burton, J., Kardos, K., Vail, T., Niedbala, R.S. and Tanke, H.J. (1999) *Analytical Biochemistry*, **267**, 30–6.
6. Wang, L. and Li, Y. (2006) *Chemical Communications*, 2557–9.
7. Reinhardt, B.A., Brott, L.L., Clarson, S.J., Dillard, A.G., Bhatt, J.C., Kannan, R., Yuan, L., He, G.S. and Prasad, P.N. (1998) *Chemistry of Materials: A Publication of the American Chemical Society*, **10**, 1863–74.

8 Rapaport, A., Milliez, J., Szipocs, F., Bass, M., Cassanho, A. and Jenssen, H. (2004) *Applied Optics*, **43**, 6477–80.
9 Ostroverkhova, O. and Moerner, W.E. (2004) *Chemical Reviews*, **104**, 3267–314.
10 Lemmetyinen, H., Vuorimaa, E., Jutila, A., Mukkala, V-M., Takalo, H. and Kankare, J. (2000) *Luminescence: The Journal of Biological and Chemical Luminescence*, **15**, 341–50.
11 Johansson, M.K., Cook, R.M., Xu, J. and Raymond, K.N. (2004) *Journal of the American Chemical Society*, **126**, 16451–5.
12 Parker, D. (2004) *Chemical Society Reviews*, **33**, 156–65.
13 de Sá, G.F., Malta, O.L., de Mello Donegá, C., Simas, A.M., Longo, R.L., Santa-Cruz, P.A. and da Silva, E.F., Jr (2000) *Coordination Chemistry Reviews*, **196**, 165–95.
14 Shi, M., Li, F., Yi, T., Zhang, D., Hu, H. and Huang, C.-H. (2005) *Inorganic Chemistry*, **44**, 8929–36.
15 Hebbink, G.A., Klink, S.I., Grave, L., Alink, P.G.B.O. and van Veggel, F.C.J.M. (2002) *A European Journal of Chemical Physics and Physical Chemistry*, **3**, 1014–8.
16 Law, G.-L., Wong, K.-L., Zhou, X.-J., Wong, W.-T. and Tanner, P.A. (2005) *Inorganic Chemistry*, **44**, 4142–4.
17 Shi, J.-M., Xu, W., Liu, Q.-Y., Liu, F.-L., Huang, Z.-L., Lei, H., Yu, W.-T. and Fang, Q. (2002) *Chemical Communications*, 756–7.
18 Yang, J., Yue, Q., Li, G.-D., Cao, J.-J., Li, G.-H. and Chen, J.-S. (2006) *Inorganic Chemistry*, **45**, 2857–65.
19 Lai, W.P.-W. Wong, W.-T. Li, B.K.-F. and Cheah, K.-W. (2002) *New Journal of Chemistry*, **26**, 576–81.
20 Di Bartolo, B. (1992) *Optical Properties of Excited States in Solids*, Plenum.
21 Andrews, D.L. and Jenkins, R.D. (2001) *The Journal of Chemical Physics*, **114**, 1089–100.
22 Wegh, R.T., Donker, H., Oskam, K.D. and Meijerink, A. (1999) *Science*, **283**, 663–6.
23 Lau, A.M.F. (1979) *Physical Review. A*, **19**, 1117–31.
24 Messier, J., Kajzar, F. and Prasad, P. (1990) *Organic Molecules for Nonlinear Optics and Photonics*, Kluwer Academic Publishers, p. 194.
25 de la Torre, G., Vázquez, P., Agulló-López, F. and Torres, T. (2004) *Chemical Reviews*, **104**, 3723–50.
26 Auzel, F. (2004) *Chemical Reviews*, **104**, 139–74.
27 Scheps, R. (1996) *Prog Quantum Electron*, **20**, 271–358.
28 Chung, S.-J., Rumi, M., Alain, V., Barlow, S., Perry, J.W. and Marder, S.R. (2005) *Journal of the American Chemical Society*, **127**, 10844–5.
29 Yoshino, F., Polyakov, S., Liu, M. and Stegeman, G. (2003) *Physical Review Letters*, **91**, 063902-1–063902-4.
30 Hernandez, F.E., Belfield, K.D., Cohanoschi, I., Balu, M. and Schafer, K.J. (2004) *Applied Optics*, **43**, 5394–8.
31 Dong, W., Zhang, H., Lin, Q., Su, Y., Wang, S. and Zhu, C. (1999) *Journal of Solid State Chemistry*, **148**, 302–7.
32 Fateley, W.G., Devitt, N.T. and Mc Bentley, F.F. (1971) *Applied Spectroscopy*, **25**, 155–73.
33 Coe, B.J. (2006) *Accounts of Chemical Research*, **39**, 383–93.
34 Tejer, M.M., Mabel, G.A., Jundt, D.H. and Byer, R.L. (1992) *IEEE Journal of Quantum Electronics*, **28**, 2631–54.
35 Heer, S., Kömpe, K., Güdel, H.-U. and Haase, M. (2004) *Advanced Materials (Deerfield Beach, Fla.)*, **16**, 2102–5.
36 Chen, X.B., Zhang, G.Y., Mao, Y.H., Hou, Y.B., Feng, Y. and Hao, Z. (1996) *Journal of Luminescence*, **69**, 151–60.
37 Riedener, T. and Güdel, H.U. (2001) *The Journal of Chemical Physics*, **107**, 2169–74.
38 Rico, M., Volkov, V. and Zaldo, C. (2001) *Journal of Alloys and Compounds*, **806**, 323–4.
39 Courrol, L.C., de Lima, B.L.S., Kassab, L.R.P., Del Cacho, V.D., Tatumi, S.H., Gomes, L. and Wetter, N.U. (2004) *Journal of Non-Crystalline Solids*, **348**, 98–102.
40 Maciel, G.S., Kim, K.-S., Chung, S.-J., Swiatkiewicz, J., He, G.S. and Prasad, P.N. (2001) *The Journal of Physical Chemistry. B*, **105**, 3155–7.
41 Piszczek, G., Maliwal, B.P., Gryczynski, I., Dattelbaum, J. and Lakowicz, J.R. (2001) *Journal of Fluorescence*, **11**, 101–7.

42. Piszczek, G., Gryczynski, I., Maliwal, B.P. and Lakowicz, J.R. (2002) *Journal of Fluorescence*, **12**, 15–7.
43. Yang, C., Wang, L.-M., Fu, Y., Zhang, J.-P., Wong, W.-T., Qiao, X.-C., Ai, Y.-F., Zou, B.-S. and Gui, L.-L. (2004) *Angewandte Chemie (International Ed. in English)*, **43**, 5010–3.
44. Wong, K.-L., Kwok, W.-M., Wong, W.-T., Phillips, D.L. and Cheah, K.-W. (2004) *Angewandte Chemie (International Ed. in English)*, **43**, 4659–62.
45. Shirver, D.F. and Atkins, P.W. (1999) *Inorganic Chemistry*, 3rd edn, Oxford University Press, New York.
46. Sastri, V.S., Bünzli, J.-C., Ramachandra Rao, V., Rayudu, G.V.S. and Perumareddi, J.R. (2000) *Modern Aspects of Rare Earths and Their Complexes*, Elsevier.
47. Birch, D.J.S. (2001) *Spectrosc Acta. Part A*, **57**, 2313–36.
48. Fröhlich, D., Itoh, M. and Pahlke-Lerch, C. (1994) *Physical Review Letters*, **72**, 1001–3.
49. Kajzar, F. and Agranovich, M.V. (1999) *Multiphoton and Light Driven Multielectron Processes in Organics: New Phenomena, Materials and Applications*, Kluwer Academic Publishers, p. 79.
50. Wong, K.-L., Law, G.-L., Kwok, W.-M., Wong, W.-T. and Phillips, D.L. (2005) *Angewandte Chemie (International Ed. in English)*, **44**, 3436–9.
51. Luo, L., Lai, W.P.-W., Wong, K.-L., Wong, W.-T., Cheah, K.-F. and Li, K.-W. (2004) *Chemical Physics Letters*, **398**, 372–6.
52. Cohen, S.M., Meyer, M. and Raymond, K.N. (1998) *Journal of the American Chemical Society*, **120**, 6277–86.
53. Hajela, S., Botta, M., Giraudo, S., Raymond, J., Xu, K.N. and Aime, S. (2000) *Journal of the American Chemical Society*, **122**, 11228–9.
54. Sabbatini, N., Guardigli, M. and Manet, I. (1996) *Handbook on the Physics and Chemistry of Rare Earths*, Vol. 23, Elsevier Science Publishers BV, The Netherlands, pp. 69–119.
55. Werts, M.H.V., Verhoeven, J.W. and Hofstraat, J.W. (2000) *Journal of the Chemical Society. Perkin Transactions 2*, **3**, 433–9.
56. Lin, S.H. (1984) *Advances in Multiphoton Processes and Spectroscopy*, World Scientific, p. 15.
57. Tanabe, S., Tamai, K., Hirao, K. and Soga, N. (1993) *Physical Review. B, Condensed Matter*, **47**, 2507–14.
58. Latva, M., Takalo, H., Mukkala, V.-M., Matachescu, C., Rodriguez-Ubis, J.C. and Kankare, J. (1997) *Journal of Luminescence*, **75**, 149–69.
59. Beeby, A., Bushby, L.M., Maffeo, D. and Williams, J.A.G. (2002) *Journal of the Chemical Society. Dalton Transactions*, 48–54.
60. Baur, J.W., Jr, M. Banach, M.D.A., Denny, L.R., Reinhardt, B.A., Vaia, R.A., Fleitz, P.A. and Kirkpatrick, S.M. (1999) *Chemistry of Materials: A Publication of the American Chemical Society*, **11**, 2899–906.
61. Brousmiche, D.W., Serin, J.M., Frechet, J.M.J., He, G.S., Lin, T.-C., Chung, S.J. and Prasad, P.N. (2003) *Journal of the American Chemical Society*, **125**, 1448–9.
62. Dahiya, P., Kumbhakar, M., Mukherjee, T. and Pal, H. (2006) *Journal of Molecular Structure*, **798**, 40–8.
63. Law, G.-L., Kwok, W.-T., Wong, K.-L., and Tanner, P.A. (2007) *The Journal of Physical Chemistry B*, **111**, 10858–61.
64. Güdel, H.U. and Pollnau, M. (2000) *Journal of Alloys and Compounds*, **303–304**, 307–15.
65. Lakowicz, J.R., Piszczek, G., Maliwal, B.P. and Gryczynski, I. (2001) *A European Journal of Chemical Physics and Physical Chemistry*, **2**, 247–52.
66. Bünzli, J.-C. and Piguet, C. (2002) *Chemical Reviews*, **102**, 1897–928.
67. Frangioni, J.V. (2003) *Current Opinion in Chemical Biology*, **7**, 626–34.
68. Uzunbajakava, N. and Otto, C. (2003) *Optics Letters*, **28**, 2073–5.
69. Hemmilá, I. and Mukkala, V.-M. (2001) *Critical Reviews in Clinical Laboratory Sciences*, **38**, 441–519.
70. Yu, J., Parker, D., Pal, R., Poole, R.A. and Cann, M.J. (2006) *Journal of the American Chemical Society*, **128**, 2294–9.
71. Beverina, L., Leclercq, J., Fu, A., Zojer, E., Pacher, P., Barlow, S., Van Stryland, E.W., Hagan, D.J., Brédas, J.-L. and Marder, S.R.

(2005) *Journal of the American Chemical Society*, **127**, 7282–3.
72 Schaller, R.D., Petruska, M.A. and Klimov, V.I. (2003) *The Journal of Physical Chemistry. B*, **107**, 13765–8.
73 Imbert, D., Comby, S., Chauvin, A.S. and Bünzli, J.-C.G. (2005) *Chemical Communications*, 1432–4.
74 Coldwell, J.B., Felton, C.E., Harding, L.P., Moon, R., Pope, S.J.A. and Rice, C.R. (2006) *Chemical Communications*, 5048–50.
75 Nagayasu, T., Matsumoto, K., Morino, S., Tagawa, T., Nakamura, A., Abo, T., Yamasaki, N. and Hayashi, T. (2006) *Lasers in Surgery and Medicine*, **38**, 290–5.
76 Hertle, R.W., Mollet, L. and Prayer, W.C. (1994) *Lasers in Surgery and Medicine*, **15**, 83–90.
77 Tur i ová, H., Zemek, J., Vacík, J., Cervená, J., Perina, V., Polcarová, M. and Brádler, J. (2000) *Surface and Interface Analysis*, **29**, 260–4.
78 Arizmendi, L. (2004) *Physica Status Solidi*, **201**, 253–83.a
79 Zyss, J. and Ledoux, I. (1994) *Chemical Reviews*, **94**, 77–105.
80 Cho, M., Lee, S.-Y., An, H., Ledoux, I. and Zyss, J. (2002) *The Journal of Chemical Physics*, **116**, 9165–73.
81 Le Bozec, H. and Renouard, T. (2000) *European Journal of Inorganic Chemistry*, **2**, 229–39.
82 Houbrechts, S., Clays, K., Persoons, A., Cadierno, V., Gamasa, M.P. and Gimeno, J. (1996) *Organometallics*, **15**, 5266–8.
83 Sénéchal, K., Toupet, L., Ledoux, I., Zyss, J., Le Bozec, H. and Maury, O. (2004) *Chemical Communications*, 2180–1.
84 Tancrez, N., Feuvrie, C., Ledoux, I., Zyss, J., Toupet, L., Le Bozec, H., and Maury, O. (2005) *Journal of the American Chemical Society*, **127**, 13474–5.
85 Maury, O. and Le Bozec, H. (2005) *Accounts of Chemical Research*, **38**, 691–704.
86 Li, Z.H., Zeng, J.H., Zhang, G.C. and Li, Y.D. (2005) *Journal of Solid State Chemistry*, **178**, 3624–30.
87 Khamagunova, T.N., Kuperman, N.M. and Bazarova, Z.G. (1999) *Journal of Solid State Chemistry*, **145**, 33–6.
88 Sénéchal-David, K., Hemeryck, A., Tancrez, N., Toupet, L., Williams, J.A.G., Ledoux, I., Zyss, J., Boucekkine, A., Guégan, J.-P., Le Bozec, H. and Maury, O. (2006) *Journal of the American Chemical Society*, **128**, 12243–55.
89 Bogari, L., Cavigli, L., Bernot, K., Sessoli, R., Gurioli, M. and Gatteschi, D. (2006) *Journal of Materials Chemistry*, **16**, 2587–92.
90 Le Bouder, T., Maury, O., Bondon, A., Costuas, K., Amouyal, E., Ledoux, I., Zyss, J. and Le Bozec, H. (2003) *Journal of the American Chemical Society*, **125**, 12284–99.
91 Kurtz, S.K. and Perry, T.T. (1968) *Journal of Applied Physics*, **39**, 3798–813.
92 Burland, D.M., Miller, R.D. and Walsh, C.A. (1994) *Chemical Reviews*, **94**, 31–75.
93 Mürner, H.-R., Chassat, E., Thummel, R.P. and Bünzli, J.-C.G. (2000) *Journal of the Chemical Society. Dalton Transactions*, **2**, 2809–16.
94 Dadabhoy, A., Faulkner, S. and Sammes, P.G. (2002) *Journal of the Chemical Society. Perkin Transactions*, **2**, 348–57.
95 Gúillaumont, D., Bazin, H., Benech, J.-M., Boyer, M. and Mathis, G. (2007) *A European Journal of Chemical Physics and Physical Chemistry*, **8**, 1–10.

8
Light-emitting Organic Nanoaggregates from Functionalized *para*-Quaterphenylenes

Manuela Schiek

8.1
Introduction to *para*-Phenylene Organic Nanofibers

Nanofibers or nanoneedles in the context of this chapter are highly crystalline, light-emitting nanoaggregates from organic molecules, such as *para*-hexaphenylene (p6P). A p6P oligomer consists of six phenyl rings connected at the *para*-position to give a rodlike molecule. The nanofibers possess widths and height on the nanometer scale, whereas the length can be up to 1 millimeter. Accordingly, the fibers bridge the gap between microscopic and macroscopic dimensions. They can be imaged by, for example, fluorescence microscopy, Figure 8.1a, or atomic force microscopy (AFM), Figure 8.1b.

Since p6P is insoluble in common solvents, deposition on substrates has to be carried out by vacuum sublimation processes such as molecular-beam epitaxy [1] or hot-wall epitaxy [2]. Sublimation processes are advantageous compared with precipitation from solution because the organic molecules deposited from the gas phase in high vacuum possess superior purity suitable for high performance applications. In this way the nanofiber growth is a bottom-up nano-engineering approach [3]. Here the organic molecules are evaporated in high vacuum (10^{-7}–10^{-8} mbar) from a home-built Knudsen cell and deposited on a growth substrate where they self-assemble into well-defined nanofibers. The substrate temperature is controlled during deposition, since only a small temperature range is suitable for nanofiber growth [4], and the flux of the evaporating molecules is monitored by a water-cooled quartz-microbalance. An appropriate growth substrate is required: A freshly cleaved muscovite mica sheet emerged as the ideal match for p6P for generating long, mutually parallel aligned nanofibers, whereas fiberlike aggregates but without parallel alignment are observed on alkali halides [5], titanium dioxide [6] and even on phlogopite [7, 8], a different mica polymorph. The driving forces for the aligned growth on muscovite mica are dipole–induced-dipole interactions together with epitaxy.

Their outstanding optical properties, which are a result of their high crystallinity, make the fibers extraordinary. They show blue fluorescence in high quantum

Tomorrow's Chemistry Today. Concepts in Nanoscience, Organic Materials and Environmental Chemistry.
Edited by Bruno Pignataro
Copyright © 2008 WILEY-VCH Verlag GmbH & Co. KGaA, Weinheim
ISBN: 978-3-527-31918-3

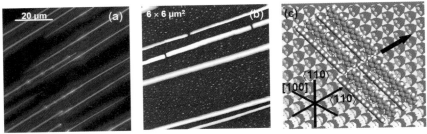

Figure 8.1 Nanofibers from p6P on muscovite mica. (a) Fluorescence microscopy image. (b) Atomic force microscopy image; height scale 70 nm. p6P clusters are visible between the fibers. (c) Needles consist of lying molecules with the (1-1-1) plane facing the substrate; the long needle axis (arrow) is parallel to the grooved mica direction. Dr. Frank Balzer is thanked for providing the images.

yields [9] that is strongly polarized after excitation with unpolarized UV light. The fibers serve as waveguides [10–12], and even exhibit lasing [13–16]. Beside these optical properties they are thermally, mechanically, and chemically stable and can easily be transferred from the growth substrate to other template surfaces [17, 18].

The combination of p6P as a molecular building block with muscovite turned out to be a unique combination for nanofiber growth. All previous attempts to realize organic nanofibers from molecules other than the phenylenes have resulted in only short needlelike structures with no or only weak mutual alignment or more complex needlelike structures [19, 20]. The *para*-phenylene basis gives the highest probability for growth of well-shaped nanofibers on muscovite mica [21]. Comparing the growth of commercially available *para*-phenylenes consisting of four, five and six phenyl rings, respectively, the nanoaggregates become shorter and less oriented with decreasing chain length of the molecular building block. An important step in exploiting the full potential of organic nanoaggregates is to functionalize the molecular building block. This possibility is a great advantage of organic molecules, in contrast to inorganic compounds. Such primary functionalization has advantages over modification of existing nanoaggregates. Although the aggregates can be modified by adjusting the growth conditions such as substrate temperature or deposition rate, or varying and/or modifying the growth substrate [22] before deposition of the organic material, this is only possible within rather narrow limits. But chemical functionalization of the molecular basis widens the scope of possibilities for modifications. This not only allows tailoring and improvement of morphology and optical properties, but also allows the creation of new properties. It is challenging to explore the prospects and limits of this bottom-up nanoengineering approach: Which chemically functionalized oligomers would still undergo a similar self-assembly process and allow creation of quantitative amounts of crystalline nanofibers with tailored morphologies and optical, electrical and mechanical properties and even novel properties?

8.2
General Aspects of Nanofiber Growth

Attention will be focused here on the growth process of p6P and the growth substrate muscovite mica, but no detailed growth model will be given.

Freshly cleaved muscovite mica with the (ideal) formula ($K_2Al_4[Si_6Al_2O_{20}](OH)_4$) was used as growth substrate. It is reasonably cheap and easy to cleave, giving a clean and atomically flat surface. This allows preparation of well-defined samples by deposition of organic molecules in high vacuum without too much disturbance from such substrate defects as steps, as is the case for example with the alkali halides KCl and NaCl. Muscovite mica is a sheet silicate (monoclinic lattice, $a = 5.2$ Å, $b = 9.0$ Å, $c = 20.1$ Å, $\gamma = 95.8°$) [23] consisting of octahedral Al—O layers sandwiched between two tetrahedral Si—O layers. One out of four Si atoms in the tetrahedral layers is replaced by an Al atom. The charge resulting from this substitution is compensated by intercalation of potassium ions in between two tetrahedral sheets. The cleavage of the substrate occurs along these interlayers and is almost perfect, resulting in large atomically flat areas on the surface. Each cleavage face possesses half of the potassium cations, which leads to the formation of surface dipoles. Their existence has been postulated from low-energy electron diffraction (LEED) patterns of vacuum- and air-cleaved muscovite mica [24, 25]. A freshly cleaved mica surface is positively charged and hydrophilic.

The orientation of the high-symmetry directions can be obtained *in situ* by LEED or *ex situ* by a so-called *Schlagfigur*. A pin is punched through a mica sample, leading to cracks along [100], [110] and [1–10] [26]. Usually the crack along [100] is the longest, but looking through two crossed polarizers at the *Schlagfigur* makes it distinctive: if the light is polarized either parallel or perpendicular to [100], no light passes the polarizers. In case the pin is not sharp enough or is pressed with too much force onto the mica sample before punching, a *Druckfigur* is obtained, which is rotated by 30° with respect to the *Schlagfigur* [27].

Muscovite mica is a dioctahedral mica [28], which means that only two out of three octahedral sites are occupied by an anion. This leads to a tilt of the Al/Si oxide tetrahedra and to the formation of grooves along a single [110] direction of the surface. They have been observed experimentally by atomic force microscopy [29]. These grooves alternate with an angle of 120° between consecutive cleavage layers in the case of the most common $2M_1$ polytype (Figure 8.2) together with the oriented electric fields of the surface, leading to grooves along the two [110] directions, but not along the optical axis [100]. In this way the surface exhibits two domains with a onefold symmetry instead of a threefold symmetry.

Accordingly, on every dipolar domain there is only a single needle direction. This dipolar domain rotates by an angle of 120° between consecutive cleavage planes. The fibers change their orientation when they cross an odd number of mica cleavage steps, and maintain their orientation on crossing an even number of steps. This results in the formation of two domains on a single sample, rotated by 120° with respect to each other. The growth direction of the *para*-phenylene nanofibers is always along a muscovite [110] direction, and for symmetry

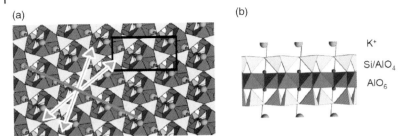

Figure 8.2 Growth substrate muscovite mica. (a) Top view. Red line indicates the grooved muscovite direction. Arrows show surface electric fields, which are connected to the groove direction. Rectangle indicates the surface unit cell. (b) Side view along the line in (a). Dr. Frank Balzer is thanked for providing the images.

reasons this is supposed to be the *grooved* [110] direction. Surface electric fields are assumed to be along two directions at 90° ± 15° with respect to the groove direction. These electric fields are supposed to align molecules on the surface, as was demonstrated many years ago [30, 31]. For the nonpolar molecules this alignment is caused by dipole–induced-dipole interaction, that is, a dipole-assisted self-assembly [1].

Whereas the long fiber axis is parallel to one of the high-symmetry directions, the molecule's long axis is approximately perpendicular to this direction with an angle of about 15° (Figure 8.1c). In this way the molecules orientate with their long axes parallel to the surface dipoles. The important role of the grooves and the corresponding electrical fields becomes obvious if a mica substrate without these features is used: On phlogopite, a trioctahedral mica without pronounced grooves [28, 29], a single growth direction is no longer preferred and needles grow along all the three epitaxially favored high-symmetry directions simultaneously [8, 7].

Epitaxy leads to formation of fibers along the high-symmetry directions. In general, epitaxy reflects the crystallographic relationship between the organic overlayer (the nanofibers) and the growth substrate. The nanofibers face the substrate with their close-packed (1-1-1) face of the herringbone bulk crystal structure [32], and the unit cell's short axis is oriented along the high-symmetry directions [33]; that is, molecules are lying on the surface and their stacking is along muscovite [110] (Figure 8.1c). Since epitaxy alone would result in three growth directions along the high symmetry direction, the dipole–induced-dipole interactions are responsible for choosing the grooved muscovite directions as the only growth direction. According to this, the driving force behind the self-assembled growth process is a combination of epitaxy and dipole-assisted alignment.

In a fluorescence microscope the nanofibers emit polarized blue fluorescence after excitation with unpolarized UV light (Figure 8.1a) owing to their high degree

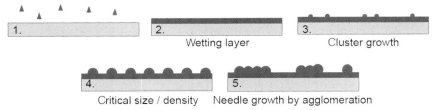

Figure 8.3 Sketch of the growth model for nanofibers via sublimation in high vacuum.

of crystallinity. The plane of polarization is almost perpendicular to the long fiber axis, with an angle of approximately 15°. Because of the molecule's symmetry, the transition dipole for absorption of UV light and emission of fluorescence light is parallel to the long molecular axis [34]. Therefore, analysis of the nanofiber's polarization properties gives information about the molecule's orientation within the fibers. This is confirmed by x-ray diffraction (XRD) experiments, which reveal that the nanofibers with a herringbone bulk crystal structure are lying with their (1-1-1) face parallel to the substrate [32].

Atomic force microscopy (AFM) shows morphological details (Figure 8.1b), for example, width and height, which are not obtainable from optical images. AFM additionally finds small clusters between the fibers, which are believed to be needle precursors [35, 36] reflecting the bottom-up fashion of the growth process (Figure 8.1b). It has been postulated on the basis of LEED patterns [33] and thermal desorption spectroscopy [37] that a wetting layer of organic molecules is initially formed on the surface. This reduces surface energy, which leads to formation of clusters (i.e., Stransky–Krastanov growth) with the same crystallographic alignment of oligomers as in the wetting layer [32]. The nanofibers grow by agglomeration of these clusters after a critical size/density is reached (Figure 8.3).

8.3
Synthesis of Functionalized *para*-Quaterphenylenes

Comparing the growth of commercially available *para*-phenylenes consisting of four, five, and six phenyl rings, respectively, the nanoaggregates become shorter and less oriented with decreasing chain length of the molecular building block. This is to some extent attributed to the decrease in polarizability of the oligomers [1, 33]. Accordingly, it is inviting to use a *para*-phenylene basis and to maintain approximately the polarizability of the p6P while changing optical and electronic properties of the molecule via functionalization. A coplanar arrangement of the molecule's phenylene rings is important for ensuring optimum CH—π interactions between the individual molecules within the crystalline nanoaggregates so as to maintain the fiberlike structure. As typical for many organic molecular crystals from aromatic molecules, the p6P crystallizes in a herringbone packing [38] because this allows the most dense packing with minimum repulsion [39]. The crystals are held together by weak van der Waals interactions; the identity of the individual molecule is retained in the crystal.

Since *para*-phenylenes are twisted in the gas phase, the surface mobility of the vapor-deposited molecules is high enough for them to migrate on the surface with a long diffusion length, which promotes growth. It has been demonstrated for different phenyl-thiophene co-oligomers with varying degrees of torque in the gas phase, vapor-deposited onto KCl, that the rather coplanar molecules favor hit-and-stick growth, which results in inferior fiber growth [40]. The coplanar arrangement within the bulk structure is important for the nanofibers' optical properties because of better conjugation of the π-electron systems, but functionalization can alter the packing of the molecules considerably. Exchange of a hydrogen atom at the *meta*-position of a phenyl ring by only a small fluorine atom leads to deviation from coplanarity of the phenyl rings within the crystal structure [41, 42], whereas symmetric substitution with fluorine atoms in the *para*-positions does not basically change the crystal structure. In the case of 1,4'''-difluoro-quaterphenylene (FP4) the typical herringbone structure is retained [43] and the packing is not changed (Table 8.1).

Therefore, the only positions suitable for carrying functional groups seem to be the two *para*-positions, because any other position would lead to a significant out-of-plane orientation of the phenylene rings and probably also prevent the self-assembly for steric reasons.

One could think of two different classes of compounds, symmetrically and nonsymmetrically (i.e., differently di- or mono-) substituted oligomers. The latter are more interesting because they offer more possibilities to fine tune the desired molecular properties and also for new properties to be exhibited. Owing to the non-centrosymmetry of the molecules, it is expected from theory that they will possess nonzero second-order susceptibility [44–46]; therefore the respective nano-aggregates could feature nonlinear optical activity. Molecular engineering has already enabled the design of molecules with specific optical properties such as a large two-photon absorption cross-section [47, 48] or large second-harmonic (SH) response [49, 50]. The challenge is now to obtain not only tailor-made oligomers but also morphologically well-defined nanofibers.

Unfortunately, it is rather difficult to modify *para*-phenylenes because of their low solubility, which decreases further with increasing chain length of the

Table 8.1 Lattice constants of p4P [38], FP4 [43] and p6P [38]

Molecule	Symmetry	$a(Å)$	$b(Å)$	$c(Å)$	$\beta(°)$
p4P	$P2_1/a$	8.11	5.61	17.91	95.8
FP4	$P2_1/a$	7.91	5.69	18.39	96.59
p6P	$P2_1/c$	26.24	5.57	8.09	98.2

molecule. It is therefore reasonable to try functionalization of the rather short *para*-quaterphenylene oligomers. Since synthesis of the nonsymmetrically functionalized oligomers is complex, the symmetrically functionalized oligomers are ideal candidates to test whether the introduction of functionalities is possible and whether the functional groups will be accepted within the growth process.

In the following the development is presented of a general approach for the synthesis of symmetrically and differently 1,4'''-substituted *p*-quaterphenylenes by application of a reliable Suzuki cross-coupling strategy. Experimental details are not given here but can be found elsewhere [51–53]. Note that organic synthesis is a key step for the bottom-up nanotechnology approach.

Although the nonsubstituted *p*-quaterphenylene has been known for more than 125 years [54], and like its higher oligomers is even commercially available, 1,4'''-disubstituted derivatives and especially nonsymmetrically functionalized compounds are still rare. This is mainly due to the notoriously low solubility of these compounds, which virtually prevents a (regio-)selective functionalization [55, 56]. In order to get access to substances with a defined substitution pattern, it is therefore mandatory to introduce the desired functional groups into smaller building blocks and use these precursors to establish the synthesis of the *p*-quaterphenylene scaffold. In the past this was achieved by cyclotrimerizations of acetylenes [57–61], Diels–Alder reactions of cyclopentadienones and subsequent aromatization [62–68], Wittig reactions of cinnamaldehydes followed by Diels–Alder reactions with acetylenic dicarboxylates and subsequent aromatization [69], addition of Grignard reagents to arines [70–73], Grignard reactions with *p*-quinones and subsequent dehydratization [74, 75], or Ullmann-type coupling reactions [76].

Modern transition-metal-catalysed homo- and cross-coupling reactions have become increasingly popular over the last 30 years [77–80] and they dominate the synthesis of oligophenylenes today. Kharash- and Suzuki-type couplings using Grignard reagents or arylboronic acids or esters have in particular been used very successfully in this context [81–111]. Although most of the molecules prepared in this way carry long alkyl or alkoxy groups that ensure solubility in common organic solvents, a few examples of only 1,4'''-disubstituted *p*-quaterphenylenes could also be prepared [80, 81, 99, 112–116].

Thus, for our present purposes a similar approach was followed using Suzuki cross-coupling reactions as the key steps in the synthesis of our target compounds. Symmetrically substituted compounds were synthesized in a twofold Suzuki cross-coupling reaction from commercially available *p*-substituted phenylboronic acids or esters and 4,4'-dibromobiphenyl or 4,4'-biphenyl-bis-boronic acid ester and a *p*-substituted arylhalide, respectively, using tetrakis (triphenylphosphino) palladium as catalyst together with cesium fluoride as base in dry tetrahydrofuran as shown in Scheme 8.1. The desired products were obtained in respectable yields after heating at reflux for 50 h.

The preparation of nonsymmetrically disubstituted derivatives was achieved in a multistep synthesis involving three Suzuki cross-coupling reactions and an iodination reaction as the key steps. It starts with a building block consisting of a

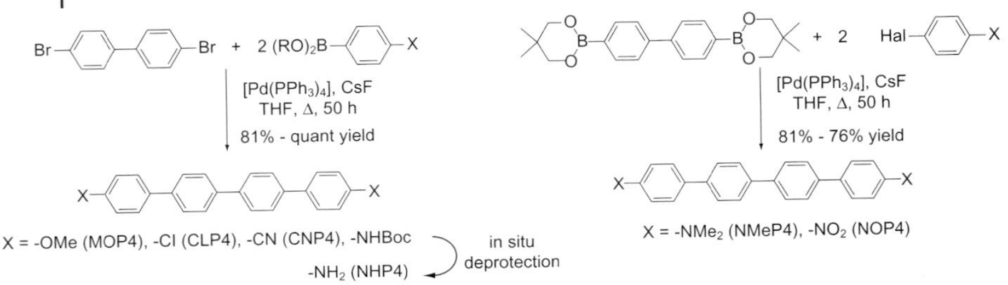

Scheme 8.1 Synthesis of symmetrically 1,4‴-disubstituted p-quaterphenylenes MOP4, CLP4, CNP4, NHP4, NMeP4, and NOP4 via Suzuki cross-coupling reactions (R = H or alkyl).

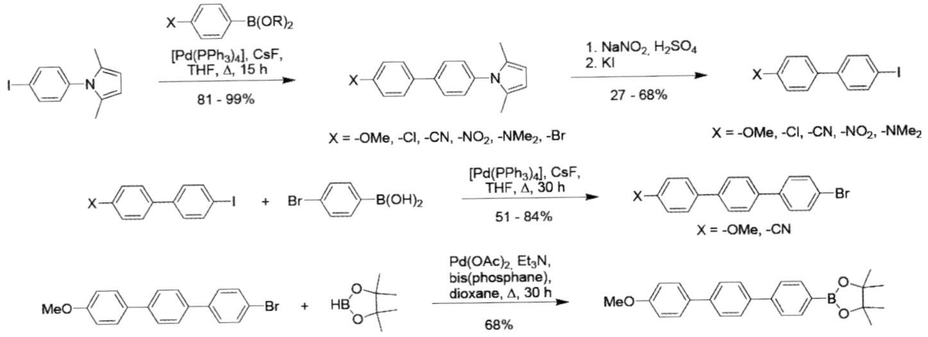

Scheme 8.2 Synthesis of nonsymmetrically 1,4″-disubstituted biphenyl building blocks and 1,4″-disubstituted p-terphenylene building blocks (R = H or alkyl).

single phenyl ring substituted with a protective group and a reactive group in para-positions. Further phenyl rings are added stepwise at reactive groups using Suzuki cross-coupling reactions to finally give the p-quaterphenylene core bearing functional groups at the 1,4‴-positions (Schemes 8.2–8.4).

This strategy has the advantage of allowing access to a variety of p-quaterphenylenes with different combinations of functional groups from the same precursors. Furthermore, it is flexible in the sense that the sequence of Suzuki coupling and iodination reactions can be changed or additional functional group manipulations like palladium-catalyzed borylations of halogenated compounds can be performed to synthesize other functionalized oligo-p-phenylenes (Scheme 8.2).

Because previous growth studies by the author gave good results for oligomers bearing methoxy groups [117, 118], and differently functionalized oligomers with a methoxy group on one side should be synthetically accessible in good yields,

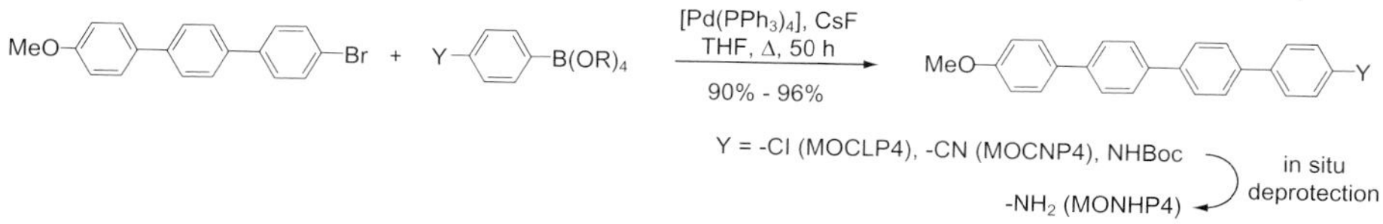

Scheme 8.3 Synthesis of nonsymmetrically 1,4'''-disubstituted p-quaterphenylenes MOCLP4, MOCNP4, and MONHP4 from a methoxy functionalized p-terphenylene (R = H or alkyl).

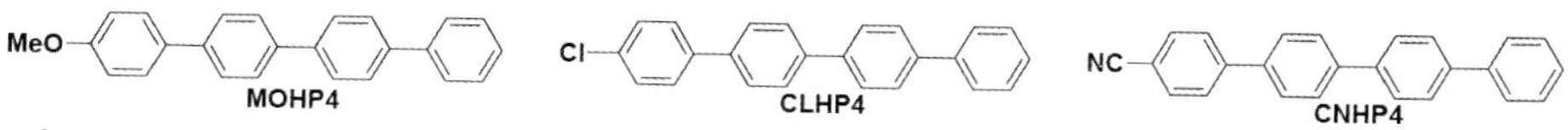

Y = -Cl (MOCLP4), -CN (MOCNP4), NHBoc → -NH$_2$ (MONHP4) → in situ deprotection

Scheme 8.4 Molecular formulas of 1-mono-functionalized p-quaterphenylenes MOHP4, CLHP4, and CNHP4.

an assortment of these has been synthesized. The three different compounds MOCLP4, MOCNP4, and MONHP4, also carrying a methoxy group at the 1-position and a chloro, cyano, or amino substituent in the 4'''-position, respectively, have been prepared as a first set of nonsymmetrically (i.e., differently di-) substituted p-quaterphenylenes in respectable overall yields of around 30% (Scheme 8.3).

As in the case of the symmetrical analogues, the final products precipitated from the reaction mixture and were washed repeatedly with water and organic solvents for purification. Residual water and organic solvents were then removed by outgassing *in vacuo* to give the desired functionalized compounds in high purity for the vapor deposition experiments.

Finally, a set of different 1-mono functionalized *para*-quaterphenylenes was obtained by means of a similar Suzuki cross-coupling strategy [119]. Three of these oligomers were chosen here: CLHP4, CNHP4, and MOHP4, bearing a chloro, cyano, or methoxy group, respectively, at one of the *para*-positions of the *para*-quaterphenylene core (Scheme 8.4).

A new practical approach is presented leading to versatile symmetrically, nonsymmetrically 1,4'''-disubstituted and 1-mono substituted p-quaterphenylene employing Suzuki cross-coupling reactions. These are promising molecular building blocks for the formation of defined nanoaggregates via vapor deposition techniques. Exploration of their acceptability for crystalline nanofiber growth is the next step and is presented in the following section.

8.4
Variety of Organic Nanoaggregates from Functionalized *para*-Quaterphenylenes

The functionalized oligomers were vapor-deposited onto muscovite mica via organic molecular beam epitaxy (OMBE) as for the p6P in the previous growth studies. Typical deposition rates of the organic molecules were 0.1–0.2 Å/s. Since there is a rather narrow temperature window for generating nanofibers [120], the investigated substrate temperatures range from room temperature to about 400 K. The optimal temperatures for fiber growth vary for the individual oligomers, being usually in an individual range of ±15 K around the respective optimum temperature.

Similar to p6P the nanostructures form two domains on a single sample rotated by 120°, and growth is usually along muscovite if not otherwise indicated. The morphology and the optical properties of the resulting nanostructures are determined by the functional groups attached to the molecular building block. Most of the nanostructures show strong, polarized blue luminescence after excitation with unpolarized UV light at normal incidence, which indicates molecules lying on the surface as light emitters [121, 122] ordered with a high degree of crystallinity. The plane of polarization is always approximately perpendicular to muscovite [110], that is, almost perpendicular to the long fiber axes. Because of the assumed herringbone structure with parallel packing of the molecules, and because the main transition dipole is approximately parallel to the long molecular axis [34], analysis of the polarized emission gives a clue about the molecules' orientations within the nanofibers.

In the following an overview of the individual structural features of the nanoaggregates is given, aiming at qualitative and comparative impressions and focusing on representative features.

8.5
Symmetrically Functionalized p-Quaterphenylenes

The brightly blue fluorescent nanofibers from MOP4 [117, 118, 123, 124] are aligned almost parallel with a mean width of several hundred nanometers and mean height of several tens nanometers, and a length of several hundred micrometers (Figure 8.4a). For the optimum substrate temperature of about $T_S = 340$ K,

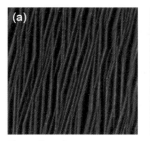

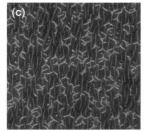

Figure 8.4 Fluorescence microscopy images ($85 \times 85\,\mu m^2$) of nanoaggregates from (a) MOP4, $T_S = 340$ K, (b) CLP4, $T_S = 370$ K, and (c) CNP4, $T_S = 300$ K. Nominal film thicknesses are around 8 nm.

fibers with lengths up to 800 μm are nearly as long as the prominent p6P fibers. The fibers are in general wider and flatter than p6P fibers for similar growth conditions. The top surface of MOP4 aggregates is exceedingly flat; the cross-section is slablike.

A special feature of MOP4 is that within each domain are visible two orientations with an angle of about 14° between them: the nanofibers are regularly bent every few tens of micrometers to form kinks. They are oriented at ±7° (experimental error ±3°) with respect to muscovite. A possible explanation for the two orientations within a single domain considers an epitaxial relationship with the substrate: Needles with the same crystallographic orientation can be mirrored along the high-symmetry direction. Or two types of needles exist with different crystal faces being parallel to the substrate.

The nanofibers from CLP4 [125] exhibit widths and heights comparable to those of the fibers from MOP4 but they are significantly shorter, with a length up to 30 μm at maximum. Figure 8.4b shows parallel, well-separated nanofibers, which emit polarized blue light in a fluorescence microscope. Again length, width, and height depend on the substrate temperature during deposition, as found for p6P and MOP4, but the temperature additionally has a pronounced influence on the aggregates' shape. Here the case of high surface temperature is shown, which is favorable if one is aiming for straight, parallel-aligned nanoneedles.

CNP4 self-assembles into diverse nanostructures within a single domain (Figure 8.4c). Besides mutually aligned fiberlike structures there are visible aggregates with increased height and with a shape reminiscent of swallows' wings. The fibers have a mean length of about 8 μm as well as a typical width of several 100 nm, and height up to 100 nm. A single swallow-wing exhibits a mean length of 4 μm and a width in the same dimension as the fiberlike structures, but the height is up to 200 nm, that is, about twice that of the fibers. The fibers grow along muscovite, whereas the swallow-wings grow along (but rotated by 120° with respect to the fibers) as well as along [99]. CNP4 is the only one among the functionalized *p*-quaterphenylenes tested so far that shows growth not only along both muscovite directions within a single domain but also along muscovite [99].

Both nanofiber types emit polarized blue light in the fluorescence microscope with the polarization vector pointing in the same direction, that is, almost perpendicular to the long axes of the fibers. Since the molecular orientation determines the polarization properties, the oligomers are orientated in the same direction for the fibers and the swallow-wings. The different aggregates thus reflect different crystallographic properties, either varying crystal structures or varying crystal faces parallel to the substrate. However, the fluorescent light is always polarized almost perpendicularly to the grooved muscovite, which means that the oligomers orientate with their long molecular axes along the surface electric fields.

The generation of well-defined nanostructures from NMeP4 (Figure 8.5) demonstrates that even the comparatively bulky and highly polar *N,N*-dimethyl amino group is accepted within the growth process. Again their shape is strongly

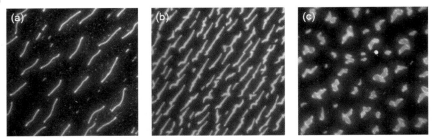

Figure 8.5 Fluorescence microscopy images ($40 \times 40\,\mu m^2$) of nanoaggregates from NMeP4 for increasing substrate temperatures (a) $T_S = 340\,K$, (b) $T_S = 380\,K$, and (c) $T_S = 400\,K$ and similar nominal film thicknesses, namely 10 nm.

dependent on the substrate temperature during deposition. In this case, however, low temperature (Figure 8.5a) leads to the needlelike aggregates in contrast to the case with CLP4, which needs high substrate temperature for generating parallel-orientated nanofibers. The low-temperature nanofibers from NMeP4 look slightly bent, like worms. They have a mean length of about 10 μm as well as a typical width of a few 100 nm and height of several tens of nanometers. The growth direction is along muscovite as usual. Between the "nanoworms," small poorly fluorescent aggregates are visible. Atomic force microscopy images (not shown here) reveal that the space between the nanoworms is densely filled with small, elongated clusters. These hardly fluoresce owing to their small size compared with the nanoworms, and they are believed to be needle precursors as already described for the case of p6P above. For higher substrate temperatures, these clusters seem to be consumed by the nanoaggregates owing to increased surface mobility at elevated temperatures.

For somewhat higher substrate temperatures, the medium-temperature case, the nanoaggregates look similar to tadpoles (Figure 8.5b). The mean length increases to about 8 μm for higher film thicknesses as shown here. The width can be more than 1.5 μm. For nanostructures wider than about 800 nm, the top surface is flat and even.

The high surface temperature case leads to nanoaggregates which look like flakes (Figure 8.5c). The flakes are well separated and show hardly any preference for growth direction, though they show the same polarization properties as the fiberlike nanostructures, indicating a highly crystalline structure. They reach size of about $3 \times 3\,\mu m^2$ and about 200 nm in height. The top surface is exceedingly flat, exhibiting a few well-shaped steps with a height of a few tens of nanometers.

Interestingly, the low-temperature case leads to formation of elongated nanofibers (i.e., nanoworms) with NMeP4 as opposed to CLP4, with the high temperature being favorable for generating parallel-aligned nanofibers. Since p6P and MOP4 form longer nanofibers with increasing substrate temperatures (up to a certain limit), the behavior of NMeP4 may be described as inverted temperature dependence. Such behavior is contrary to basic nucleation theory [126], which predicts larger and more separated crystallites for higher substrate temperatures, so long as no degradation of the aggregates as a whole takes place, as described in detail for p6P on muscovite mica [7, 33]. However, the possibility of growing different structures from a single molecule just by changing the substrate temperature adds another degree of freedom for the design of nanoscale structures.

8.6 Differently Di-functionalized p-Quaterphenylenes

The whole variety of nanoaggregates shown so far was generated from comparatively simple, symmetrically functionalized oligomers, but nonsymmetrically functionalized oligomers also self-assemble into well-shaped nanostructures.

Figure 8.6a shows a dark-field microscopy image of MONHP4 nanofibers [127, 128]. These do not fluoresce in the fluorescence microscope, in contrast to all other nanostructures. This fluorescence quenching is attributed to the formation of intermolecular hydrogen bonding between the hydrogen of the amino groups and the lone pairs of the oxygen of the methoxy group. That the formation of hydrogen bonding is a reasonable explanation is confirmed by the fact that nanostructures of NMeP4 show bright blue fluorescence.

This problem of fluorescence quenching is also well known for naphthalene bisimide dyes core-substituted with arylamines. Formation of intramolecular hydrogen bonding to a neighboring carbonyl group is considered to be the pathway for nonradiative deactivation of the excited state. To avoid this problem, alkyl amines [129] and amino-linked benzyl groups [130] have been attached successfully to the naphthalene core, leading to a strong, tunable fluorescence.

Vapor deposition of MOCLP4 results in well-defined nanostructures [128] that show a strong dependence on the growth temperature: clearly a high surface

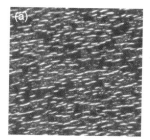
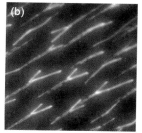
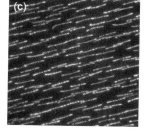

Figure 8.6 (a) Dark-field microscopy image (85 × 85 μm^2) of nanofibers from MONHP4, T_S = 380 K, 7 nm. Fluorescence microscopy images (40 × 40 μm^2) of nanoaggregates from MOCLP4: (b) low surface temperature case nanobranches, T_S = 360 K, 5 nm, and (c) high surface temperature case nanofibers, T_S = 400 K, 10 nm.

temperature case and a low surface temperature case can be distinguished. Similar behavior has been found for CLP4, the symmetrically functionalized oligomer with two chloride atoms. This is not shown here, but it seems that this strong morphological temperature dependence is an intrinsic property of functionalization with chloride groups for *p*-quaterphenylenes.

The low surface temperature case (Figure 8.6b) leads to mutually aligned nanobranches. The growth direction is the muscovite direction before the fiber splits. Then the muscovite direction is the bisecting line of the angle between the branches, which is about 30° with an experimental error of ±3°. But they exhibit no bilateral symmetry, for most nanobranches one branch is significantly longer, sometimes more than twice as long. The nanobranches have overall lengths up to 30 μm. With heights up to 150 nm they are almost twice as high as the parallel nanofibers for the high-temperature case (Figure 8.6c). These fibers form kinks with an angle of about 15°, which have already been observed in the case of MOP4, the symmetrically functionalized oligomer with two methoxy groups. Thus it seems that the trend to form kinks is an intrinsic property due to the substitution with methoxy groups.

MOCNP4 forms well-defined mutually aligned nanostructures, reminiscent of walking sticks. The nanoaggregates show bright, polarized blue fluorescence after excitation with UV light from a high-pressure mercury lamp (Figure 8.7a). That the "handle" of the walking stick consists of a bent fiber piece was shown by polarization measurements of the fluorescence: the polarization vector follows the bending. This in turn means that the molecules within the bend change their alignment with respect to the substrate and its surface electric fields, similarly to the p6P micro-rings [131]. This contrasts with the CNP4 swallow-wings, whose molecular alignment is strongly determined by the substrate and unaffected by the nanostructures' shape.

A cause for the bending can be found by means of AFM imaging (Figure 8.7b and c). The sticks invariably bend when large dendritic islands occur on the

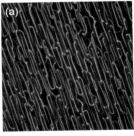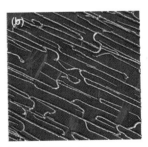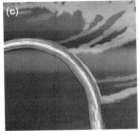

Figure 8.7 Nano "walking sticks" of MOCNP4: (a) fluorescence microscopy image (100 × 100 μm²), and (b, c) atomic force microscopy images: (b) 42 × 42 μm², height scale 260 nm, (c) 5 × 5 μm², height scale 110 nm. T_S = 380 K, film thickness 8 nm.

surface. The handle ends in a knob, which is about twice as high as the stick, growing up to 300 nm. The islands cover several square micrometers and have heights of several tens of nanometers. Step heights of 2.5 nm indicate that the islands are formed by upright standing oligomers on the surface. This is in accordance with the fact that these islands do not fluoresce in the fluorescence microscope. However, this new structure motif of bent fibers with a long straight part opens up more possibilities for implementing such organic nanoaggregates in future integrated optical circuits.

8.7
Monofunctionalized *p*-Quaterphenylenes

Since the methoxy group emerged as a well-suited functional group in the previous growth studies, MOHP4 was also considered as a promising oligomer; in fact the resulting nanoaggregates turned out to be comparatively short and less well aligned, but they nevertheless show blue luminescence (Figure 8.8a). CLHP4 nanofibers look very similar to the di-functionalized analogue (Figure 8.8b).

Since MOHP4 shows clearly inferior fiber growth compared with the prominent di-functionalized oligomer MOP4, it is clear that not only the functional group itself decides on improved growth: the counterpart functionality also has to be taken into account. In the case of the cyano group, the mono-functionalized oligomer CNHP4 shows superior growth properties to the di-functionalized CNP4. The brightly blue fluorescent fibers have lengths of several hundred micrometers (Figure 8.8c). They are mutually aligned but not strictly parallel. Within each domain, three growth directions can be observed: along muscovite ⟨110⟩ and along ±7° with an experimental error of ±3°. The nanofibers branch out, forming kinks and switch between growth directions.

The superior fiber growth of CNHP4 compared to CNP4 is attributed to enhanced intermolecular interactions, that is, hydrogen bonding between the lone pair of

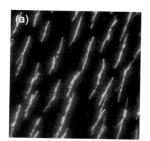
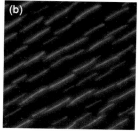
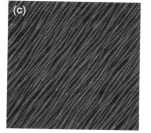

Figure 8.8 Fluorescence microscopy images (85 × 85 μm^2) of nanofibers from mono functionalized p4Ps: (a) MOHP4, T_S = 340 K, 6 nm; (b) CLHP4, T_S = 360 K, 8 nm; (c) CNHP4, T_S = 360 K, 12 nm.

the cyano group's nitrogen and a hydrogen atom of the adjacent molecule. These lead to improved packing of the molecules within the nanofibers.

It can be concluded that not only the functional group itself determines the growth behavior but also the counterpart functionalization has to be taken into account, leading to adjusted possibilities for intermolecular interactions. That only two positions are considered to be functionalized demonstrates the complexity of effects caused by functionalization.

8.8
Tailoring Morphology: Nanoshaping

The detailed morphology–dimensions, cross-sectional shape, aggregate density, and so on–is determined by the functional groups as well. Atomic force microscopy has been used to visualize the diversity in morphology. The cross-sectional shape varies from rounded (p6P) through slablike (MOP4) to peaked like a pitched roof (CLP4). The new nanofibers from MOP4 and CLP4 are the most striking candidates for nanoshaping. Of course, other functionalities generate similar shape motifs as shown in Figure 8.9. Introduction of amino groups to the *p*-quaterphenylene core (NHP4) results in rather short, nonfluorescent nanoworms, which possess a rounded cross-sectional shape, while N,N-dimethylamino groups (NMeP4) can lead to exceedingly flat nanoflakes that emit bright blue fluorescence light. Functionalization with cyano groups (CNP4) creates rather short and straight nanofibers with a prismatic cross-section, among other structural types. The change in shape occurs for nanofibers that are still parallel-aligned, emit intense

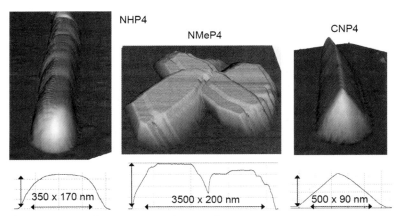

Figure 8.9 Atomic force microscopy images of nanostructures from the symmetrically substituted p4Ps NHP4, NMeP4, and CNP4. Height scales 150 nm, 250 nm, 100 nm, respectively. Cross-sections (width × height) show the respective dimensions.

blue fluorescence light, and show waveguiding. Hence, nanoshaping opens up various new application possibilities such as slab-based nanophotonic architectures or surface enhanced Raman scattering (SERS) nanosensors, after coating with ultrathin metal films.

Additionally, dimensions and aggregate density can be varied according to the growth parameters; for example, different structure types are generated depending on the substrate temperature during deposition.

8.9
Tailoring Optical Properties: Linear Optics

Most nanoaggregates exhibit polarized fluorescence after excitation with unpolarized UV light ($\lambda = 365$ nm) at normal incidence. Fluorescence spectra of the *para*-phenylenes are all within the blue, but the fine-tuning of the emission color is determined by the functional groups attached to the molecular building block. Functionalization allows a shift of the peak emission frequency of the nanoaggregates from 383 nm (CLP4) to 452 nm (NMeP4). This is due to inductive and mesomeric effects of the functional groups on the conjugated π-electron system, which is responsible for the fluorescence [132]. While a change in the peak emission frequency via substitution of polymeric or molecular compounds is a well known in organic thin films, it is demonstrated here that this is possible within the concept of crystalline nanofibers. The spectra have been recorded after continuous wave (CW) UV excitation of the nanoaggregates on the mica substrate at 325 nm. The effect of substituents on the fluorescence properties of aromatic hydrocarbons is complex and additional solid-state effects in general are hard to predict. Molecular crystals so named because the molecules with their individual properties are retained in the crystal phase [133]. However, explanation of the optical properties of nanofibers must start with consideration of the properties of the oligomers, but should be viewed cautiously.

Fluorescence spectra of the symmetrically functionalized p4Ps' nanoaggregates, all within the blue, can be seen in Figure 8.10. Most of the spectra, apart from those of CLP4, are red-shifted compared to p4P. The spectra of CLP4 and MOP4 show well-resolved vibronic structures, whereas the spectrum of CNP4 is broadened but still shows vibronic structures, and the NMeP4 spectrum is broad as well as structureless. The vertical line, representing the maximum emission (that is, the (0–1) transition of the nonfunctionalized *para*-quaterphenylene (p4P)), is given for comparison.

The chlorine atom is an electron-pulling substituent with a strong −I-effect overcoming a weak +M-effect; therefore the electron density of the aromatic system is reduced. This results in a blue-shifted fluorescence compared to the parent p4P. The fact that the spectrum of CLP4 shows well-resolved vibronic structures indicates that the lone pairs of the chlorine atoms are not directly involved in the aromatic system; that is, the fluorescence relevant transition has no charge transfer character. The blueshift is clearly due to an inductive effect, since the

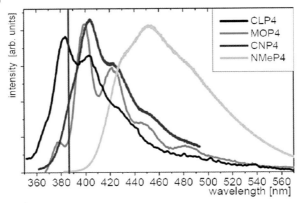

Figure 8.10 Fluorescence spectra of nanoaggregates from the symmetrically functionalized oligomers CLP4, MOP4, CNP4, and NMeP4 after continuous-wave excitation at 325 nm. The vertical line represents the (0–1) transition of the parent p4P.

positive mesomeric effect of the chlorine substituents would result in an opposite shift.

The methoxy group is an electron-pushing functional group; that is, it has a strong +M-effect as well as a moderate −I-effect, so the electron density of the conjugated system is increased. This leads to the red-shifted fluorescence of the nanofibers of MOP4 compared with those of p4P. The spectrum shows well-resolved vibronic structures, although spectra of phenols are often broad and structureless compared with the parent aromatic hydrocarbon [132]. This loss of structure happens when the lone pairs of the oxygen are directly involved in the π-bonding of the aromatic system due to significant intramolecular charge transfer character of the fluorescence-relevant transitions. This is only possible for the nearly coplanar arrangement of the aromatic ring and the —OH substituent. However, replacing the hydrogen by alkyl chains, for example, an —OMe substituent, leads to departure from coplanarity of the aromatic ring and the functional group. Thus, the lone pairs of the oxygen atoms of MOP4 cannot conjugate with the π-electron system to the full extent. The degree of conjugation of the lone pairs is decreased, which reduces the charge 'transfer character of the fluorescence-relevant transitions, so that the vibronic structures are retained on the one hand. But on the other hand, the positive mesomeric effect, that is, the electron pushing effect, is still pronounced enough to cause a red shift in the fluorescence.

The cyano group is an electron-pulling substituent with a strong −M- and −I-effect, but it contains a triple bond, which elongates the π-conjugated system. This overcomes the electron-reducing effect, no matter whether it is based on negative inductive or on negative mesomeric effects. Therefore, a red-shifted fluorescence can be observed compared with the parent p4P. Since a heteroatoms involved in the fluorescence-relevant transitions, they possess a certain charge transfer character, because the triple bond of the cyano group is directly involved in the π-electron system. This results in a broadened spectrum, which still shows vibronic structures. Owing to the symmetric functionalization, the charge transfer character is less pronounced.

The spectrum of the nanostructures of the N,N-dimethylamino functionalized oligomer NMeP4 exhibits the longest wavelength among all investigated functionalized p-quaterphenylenes. The amino group possesses a strong +M- and a weak –I-effect: the lone pair of the nitrogen conjugates into the aromatic system, since the N—Me bondings are arranged coplanar to the phenyl ring. This increases the electron density as well as the conjugation length, which is reflected in an obvious red shift in the fluorescence. This strong positive mesomeric effect greatly overpowers the negative inductive effect. Since the lone pairs of the heteroatom are directly involved in the aromatic system, the spectrum is broadened and structureless owing to the pronounced charge transfer character of the transition momentum.

8.10
Creating New Properties: Nonlinear Optics

Nonsymmetric substitution of oligomers should not only allow tuning of the luminescence behavior but also should give access to systems that show nonlinear optical (NLO) properties. The differently functionalized oligomers MONHP4, MOCLP4, and MOCNP4 are (to different extents) push–pull substituted systems as well as the mono-functionalized oligomers MOHP4, CLHP4, NMeHP4, and CNHP4. They can act as NLO prototypes, since they are expected from theory to possess nonzero second-order susceptibility owing to the nonsymmetric functionalization [44, 45]. If this anisotropy is retained in the highly crystalline nanoaggregates, which should be the case for head-to-tail orientation of the molecules within the nanofibers' assumed herringbone packing, the bulk oligomers should also exhibit NLO activity. Rodlike conjugated oligomers with push–pull substituents often crystallize in centrosymmetric space groups [46], which would prevent NLO activity for crystallites. But this is not the case for the nanofibers presented here: Upon irradiation with femtosecond laser pulses the nanofibers act as frequency doublers and emit a strong second-harmonic signal. Such a property is of obvious importance for future integrated optical circuits, but optical second-harmonic generation (SHG) is also a powerful technique for understanding the correlation between morphology and optoelectronic response of nanoaggregates and for characterization.

It has been reported before that nanofibers from *para*-hexaphenylene exhibit nonlinear optical activity, but this can mainly be assigned to surface second-harmonic generation from a continuous organic film (wetting layer) on the growth substrate and from the nanofiber surfaces [134], according to recent investigations on transferred nanofibers [127, 128]. The amount of true second-harmonic

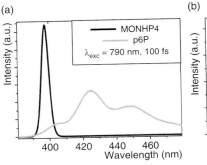 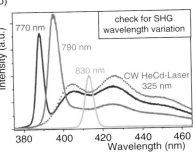

Figure 8.11 (a) Emission of second-harmonic signal from transferred MONHP4 nanofibers and from p6P nanofibers on muscovite mica after excitation with femtosecond near-infrared laser pulses (100 fs, 4.5 mW) at 790 nm. (b) Emission of second-harmonic signal from MOCLP4 nanostructures after excitation with femtosecond near-infrared laser pulses (100 fs) at different wavelengths (770 nm, 790 nm, 830 nm). The fluorescence spectrum after continuous-wave (CW) UV excitation is shown for comparison.

generation from the nanofibers of p-hexaphenylene is reported to be on the order of a small percentage of the two-photon luminescence intensity at infrared femtosecond excitation around 800 nm.

The nonsymmetrically functionalized p-quaterphenylenes, however, behave differently [135].

For the observation of the fibers' intrinsic properties, nanofibers obtained from MONHP4 and MOCLP4 have been transferred from mica to glass (silicon oxide) using a standard procedure [17, 18] in order to avoid second-harmonic generation from the underlying substrate and a wetting layer.

Upon irradiation with femtosecond pulses, a strong second-harmonic peak was observed, clearly confirming true second harmonic generation from the bulk of the fiber [127]. Figure 8.11 shows the emission of a second-harmonic signal from transferred MONHP4 fibers and for comparison the emission from p6P nanofibers on muscovite mica. Obviously the MONHP4 nanofibers emit a strong second-harmonic signal, whereas the frequency-doubled signal originating from the p6P nanofibers cannot even be seen in this graph. Only the fluorescence signal of the p6P fibers based on two-photon absorption is clearly visible.

The basic difference between MOCLP4 and MONHP4 nanofibers is that the latter can be considered as nonfluorescent, whereas the nanostructures from MOCLP4 emit intense blue fluorescence light. Upon excitation of MOCLP4 nanofibers with femtosecond near-infrared laser pulses (100 fs) at different wavelengths, they show a pronounced second-harmonic signal (Figure 8.11). As expected the second-harmonic signal shifts with the wavelength of the exciting laser beam. An accompanying fluorescence signal appears for excitation wavelengths shorter than 800 nm. This fluorescence spectrum is the same as obtained after CW excitation with UV light.

8.11
Summary

To return to the initial question: Is the combination of *p*-hexaphenylene as a molecular building block and muscovite mica as a growth substrate truly unique for generating nanofibers? No, it is not. In chapter the following results have been presented:

- A reliable synthesis strategy to obtain new either symmetrically or nonsymmetrically *para*-functionalized *para*-quaterphenylenes.
- The possibility of well-defined nanofiber growth on muscovite starting with substituted oligomers; that is, fiber growth is not prevented by the functional groups but is even improved in some cases.
- The distinctive properties of the nanostructures are determined by the respective functional groups attached to the molecular building block:
  - The exact wavelength of the fluorescence color shifts within the blue depending on the functional group.
  - The individual morphology – that is, cross-sectional shape – is defined by the functional group.
- New properties such as nonlinear optical activity (frequency doubling) of the nanofibers are created due to intrinsic nonzero hyperpolarizability of the nonsymmetrically substituted molecular building blocks.

Thus, it is possible to obtain tailored nanofibers from appropriately functionalized *p*-quaterphenylenes. A requirement for generating aligned nanofibers still seems to be muscovite mica as growth substrate, but the molecular basis is not restricted to p6P; a *para*-phenylene molecular basis can be employed which tolerates functional groups at the *para*-positions.

8.12
Acknowledgments

Research presented here was conducted in the context of the author's PhD thesis under supervision of her doctoral advisor Prof. Dr. K. Al-Shamery (University of Oldenburg). Organic synthesis was done under supervision of Prof. Dr. Arne Lützen (University of Bonn), and growth and measurements of optical properties under supervision of Prof. Dr. H.-G. Rubahn (University of Southern Denmark, Sonderborg www.nanosyd.sdu.dk).

Dipl.-Chem. Ivonne Wallmann (University of Bonn) provided additional organic molecules for nanofiber growth in the context of her diploma thesis.

Jonathan R. Brewer (University of Southern Denmark, Odense) carried out the measurement of nonlinear optical properties in the context of his PhD research.

Additionally the author thanks Dr. Frank Balzer for continuous (scientific) support during the time of her PhD research.

References

1 Balzer, F. and Rubahn, H.-G. (2001) Dipole-assisted self-assembly of light-emitting p-np needles on mica. *Applied Physics Letters*, **79**, 3860–2.

2 Andreev, A., Matt, G., Brabec, C.J., Sitter, H., Badt, D., Seyringer, H. and Sariftci, N.S. (2000) Highly anisotropically self-assembled structures of *para*-sexiphenyl grown by hot-wall epitaxy. *Advanced Materials (Deerfield Beach, Fla.)*, **12**, 629–33.

3 General aspects of nanotechnology: see for instance, Rubahn, H.-G. (2004) *Nanophysik und Nanotechnologie*, 2nd edn, Teubner, Stuttgart.

4 Balzer, F. and Rubahn, H.-G. (2002) Laser-controlled growth of needle-shaped organic nanoaggregates. *Nano Letters*, **2**, 747–50.

5 Yanagi, H. and Morikawa, T. (1999) Self-waveguided blue light emission in *p*-sexiphenyl crystals epitaxially grown by mask-shadowing vapor deposition. *Applied Physics Letters*, **75**, 187–9.

6 Koller, G., Berkebile, S., Krenn, J.R., Tzvetkov, G., Hlawacek, G., Lengyel, O., Netzer, F.P., Teichert, C., Resel, R. and Ramsey, M.G. (2004) Oriented sexiphenyl single crystal nanoneedles on TiO_2 (110). *Advanced Materials (Deerfield Beach, Fla.)*, **16**, 2159–62.

7 Balzer, F. (2006) *Organic Nanoaggregates*, Habilitationsschrift, Humboldt-Universität zu Berlin.

8 Balzer, F., Schiek, M., Lützen, A., Al-Shamery, K. and Rubahn, H.-G. (2007) Growth of nanofibers from thiophenes, thiophene-phenylenes and phenylenes: A systematic study. *Proceedings of SPIE*, **6470**, -06.

9 Stampfl, J., Tasch, S., Leising, G. and Scherf, U. (1995) Quantum efficiencies of electroluminescent poly(para)-phenylenes. *Synthetic Metals*, **71**, 2125–8.

10 Balzer, F., Bordo, V.G., Simonsen, A.C. and Rubahn, H.-G. (2003) Isolated hexaphenyl nanofibers as optical waveguides. *Applied Physics Letters*, **82**, 10–3.

11 Balzer, F., Bordo, V.G., Simonsen, A.C. and Rubahn, H.-G. (2003) Optical waveguiding in individual nanometer-scale organic fibers. *Physical Review. B, Condensed Matter*, **67**, 1154081–8.

12 Yanagi, H. and Morikawa, T. (1999) Self-waveguided blue light emission in *p*-sexiphenyl crystals epitaxially grown by mask-shadowing vapor deposition. *Applied Physics Letters*, **75**, 187–9.

13 Quochi, F., Cordella, F., Orru, R., Communal, J.E., Verzeroli, P., Mura, A., Bongiovanni, G., Andreev, A., Sitter, H. and Sariftci, N.S. (2004) Random laser action in self-organized *para*-sexiphenyl nanofibers grown by hot-wall epitaxy. *Applied Physics Letters*, **84**, 4454–6.

14 Quochi, F., Cordella, F., Mura, A., Bongiovanni, G., Balzer, F. and Rubahn, H.-G. (2005) One dimensional random lasing in a single organic nanofiber. *The Journal of Physical Chemistry. B*, **109**, 21690–3.

15 Quochi, F., Cordella, F., Mura, A., Bongiovanni, G., Balzer, F. and Rubahn, H.-G. (2006) Gain amplification and lasing properties of individual organic nanofibers. *Applied Physics Letters*, **88**, 0411061–3.

16 Quochi, F., Andreev, A., Cordella, F., Orru, R., Mura, A., Bongiovanni, G., Hoppe, H., Sitter, H. and Sariftci, N.S. (2005) Low-threshold blue lasing in epitaxially grown *para*-sexiphenyl nanofibers. *Journal of Luminescence*, **112**, 321–4.

17 Brewer, J., Henrichsen, H.H., Balzer, F., Bagatolli, L., Simonsen, A.C. and Rubahn, H.-G. (2005) Nanofibers made to order: Free floating, transferred and gel-packed organic nanoaggregates. *Proceedings of SPIE*, **5931**, 250–7.

18 Brewer, J., Maibohm, C., Jozefowski, L., Bagatolli, L. and Rubahn, H.-G. (2005) A 3D view on free-floating, space-fixed and surface-bound *para*-phenylene nanofibers. *Nanotechnology*, **16**, 2396–401.

19 Balzer, F., Kankate, L., Niehus, H. and Rubahn, H.-G. (2005) Nanoaggregates from oligothiophenes and oligophenylenes – a systematic survey. *Proceedings of SPIE*, **5724**, 285–94.

20 Balzer, F., Schiek, M., Lützen, A., Al-Shamery, K. and Rubahn, H.-G. (2007) Growth of nanofibers from thiophenes, thiophene-phenylenes and phenylenes: A systematic study. *Proceedings of SPIE*, **6470**, -06.

21 Balzer, F. and Rubahn, H.-G. (2005) Growth control and optics of organic nanoaggregates. *Advanced Functional Materials*, **14**, 17–24.

22 Balzer, F., Beermann, J., Bozhevolnyi, S.I., Simonsen, A.C. and Rubahn, H.-G. (2003) Optically active organic microrings. *Nano Letters*, **3**, 1311–4.

23 Knurr, R.A. and Bailey, S.W. (1986) Refinement of Mn-substituted muscovite and phlogophite. *Clays and Clay Minerals*, **34**, 7–16.

24 Müller, K. and Chang, C.C. (1969) Electric dipoles on clean mica surfaces. *Surface Science*, **14**, 39–51.

25 Müller, K. and Chang, C.C. (1968) Low energy electron diffraction observations of electric dipoles on mica surfaces. *Surface Science*, **9**, 455–8.

26 Tröger, W.E. (1971) *Optische Bestimmung der Gesteinsbildenden Minerale*, 4th edn, Schweizerbartsche Verlagsbuchhandlung, Stuttgart.

27 Tschermak, G. (1894) *Lehrbuch der Mineralogie*. Alfred Hölder, Wien, p. 135.

28 Griffen, D.T. (1992) *Silicate Crystal Chemistry*, Oxford University Press, New York.

29 Kuwahara, Y. (1999) Muscovite surface structure imaged by fluid contact mode AFM. *Physics and Chemistry of Minerals*, **26**, 198–205.

30 Uyeda, N., Ashida, M. and Suito, E. (1965) Orientation overgrowth of condensed polycyclic aromatic compounds vacuum-evaporated onto cleaved face of mica. *Journal of Applied Physics*, **36**, 1453–60.

31 Ashida, M. (1966) The orientation overgrowth of metal-phthalocyanines on the surface of single crystals. I. vacuum condensed films on muscovite. *Bulletin of the Chemical Society of Japan*, **39**, 2625–31.

32 Plank, H., Resel, R., Purger, S., Keckes, J., Thierry, A., Lotz, B., Andreev, A., Sariciftci, N.S. and Sitter, H. (2001) Heteroepitaxial growth of self-assembled highly ordered *para*-sexiphenyl films: A crystallographic study. *Physical Review. B, Condensed Matter*, **64**, 235423.

33 Balzer, F. and Rubahn, H.-G. (2004) Chain-length dependent *para*-phenylene film- and needle-growth on dielectrics. *Surface Science*, **548**, 170–82.

34 Ambrosch-Draxl, C., Majewski, J.A., Vogl, P. and Leising, G. (1995) First-principles studies of the structural and optical properties of crystalline poly(*para*-phenylene). *Physical Review. B, Condensed Matter*, **51**, 9668–76.

35 Andreev, A.Y., Teichert, C., Hlawacek, G., Hoppe, H., Resel, R., Smilgies, D.-M., Sitter, H. and Sariciftci, N.S. (2004) Morphology and growth kinetics of organic thin films deposited by hot wall epitaxy. *Electronic Advocacy*, **5**, 23–7.

36 Balzer, F. (2007) Growth of oriented organic nanoaggregates via organic molecular beam deposition, in *Organic Nanofibers for Next Generation Devices* (eds K. Al-Shamery, H.-G. Rubahn and H. Sitter), Springer Series in Material Science, Springer, Berlin.

37 Teichert, C., Hlawacek, G., Andreev, A.Y., Sitter, H., Frank, P., Winkler, A. and Sariciftci, N.S. (2006) Spontaneous rearrangement of *para*-sexiphenyl crystallites into nanofibers. *Applied Physics. A*, **82**, 665–9.

38 Baker, K.N., Fratini, A.V., Resch, T., Knachel, H.C., Adams, W.W., Socci, E.P. and Farmer, B.L. (1993) Crystal structures, phase transition and energy calculations of poly-*p*-phenylene oligomers. *Polymer*, **34**, 1571–87.
39 Schwoerer, M. and Wolf, H.C. (2005) *Organische Molekulare Festkörper – Einführung in die Physik von π-Systemen*, Wiley-VCH Verlag GMBH, Weinheim.
40 Yanagi, H., Araki, Y., Ohara, T., Hotta, S., Ichikawa, M. and Taniguchi, Y. (2003) Comparative carrier transport characteristics on organic field-effect transistors with vapor deposited thin films and epitaxially grown crystals of biphenyl-capped thiophene oligomers. *Advanced Functional Materials*, **13**, 767–73.
41 Yoon, M.-H., Facchetti, A., Stern, C.E. and Marks, T.J. (2006) Fluorocarbon-modified organic semiconductors: Molecular architecture, electronic, and crystal tuning of arene- versus fluoroarene-thiophene oligomer thin-film properties. *Journal of the American Chemical Society*, **128**, 5792–801.
42 Facchetti, A., Yoon, M.-H., Stern, C.L., Katz, H.E. and Marks, T.J. (2003) Building blocks for n-type organic electronics: Regiochemically modulated inversion of majority of carrier sign in perfluoroarene-modified polythiophene semiconductors. *Angewandte Chemie (International Ed. in English)*, **42**, 3900–3.
43 Blanchard, M.D., Hughes, R.P., Concolino, T.E. and Rheingold, A.L. (2000) π-stacking between pentafluorophenyl and phenyl groups as a controlling feature of intra- and intermolecular crystal structure motifs in substituted ferrocenes: Observation of unexpected facte-to-face stacking between pentafluorophenyl rings. *Chemistry of Materials: A Publication of the American Chemical Society*, **12**, 1604–10.
44 Fanti, M. and Almlöf, J. (1996) Hyperpolarizabilities of substituted polyphenyls. *Journal of Molecular Structure: THEOCHEM*, **388**, 305–13.
45 Matsuzawa, N. and Dixon, D.A. (1992) Semiempirical calculations of hyperpolarizabilities for extended π-systems: Polyenes, polyynes, and polyphenyls. *International Journal of Quantum Chemistry*, **44**, 497–515.
46 Meier, H. (2005) Conjugated oligomers with terminal donor-acceptor substitution. *Angewandte Chemie (International Ed. in English)*, **44**, 2482–506.
47 Albota, M., Beljonne, D., Bredas, J.L., Ehrlich, J.E., Fu, J.Y., Heikal, A.A., Hess, S.E., Kogej, T., Levin, M.D., Marder, S.R., McCord-Maughon, D., Perry, J.W., Roeckel, H., Rumi, M., Subramaniam, G., Web, W.W., Wu, X.L. and Xu, C. (1998) Design of organic molecules with large two-photon absorption cross sections. *Science*, **281**, 1653.
48 Ventelon, L., Charier, S., Moreaux, L., Mertz, J. and Blanchard Desce, M. (2001) Nanoscale push-pull dihydrophenanthrene derivatives as novel fluorophores for two-photonexcited fluorescence. *Angewandte Chemie (International Ed. in English)*, **40**, 2098–101.
49 Thalladi, V.R., Brasselet, S., Weiss, H.-C., Blaser, D., Katz, A.K., Carrell, H.L., Boese, R., Zyss, J., Nangia, A. and Desiraju, G.R. (1998) Octupolar nonlinear materials. *Journal of the American Chemical Society*, **120**, 2563–77.
50 Ashwell, G.J. (1999) Langmuir-blodgett films: Molecular engineering of noncentrosymmetric structures for second order nonlinear optical applications. *Journal of Materials Chemistry*, **9**, 1991–2003.
51 Schiek, M., Al-Shamery, K. and Lützen, A. (2007) Synthesis of symmetrically and nonsymmetrically *para*-functionalised *para*-quaterphenylenes. *Synthesis*, **4**, 613–21.
52 Schiek, M. (2007) Organic Molecular Nanotechnology. Ph.D. thesis, University of Oldenburg.
53 Wallmann, I. (2006) Synthese von monofunktionalisierten para-Tetraphenylenen und Thiophenphenylenen. Diploma thesis, University of Bonn.

54 Schmidt, H. and Schultz, G. (1880) *Justus Liebigs Annalen Der Chemie*, **203**, 129.

55 Scheinbaum, M.L. (1969) Nitration of *p*-quaterphenyl. *Journal of the Chemical Society D*, 1235a.

56 Pavlopoulos, T.G. and Hammond, P.R. (1974) Spectroscopic studies of some laser dyes. *Journal of the American Chemical Society*, **96**, 6568–79.

57 Keegstra, M.A., De Feyter, S., De Schryver, F.C. and Müllen, K. (1996) Hexaterphenylyl- and hexaquaterphenylylbenzene: The behavior of chromophores and electrophores in a restricted space. *Angewandte Chemie (International Ed. in English)*, **35**, 774–6.

58 Iyer, V.S., Wehrmeier, M., Brand, J.D., Keegstra, M.A. and Müllen, K. (1997) From hexaperihexabenzocoronene to 'superacenes'. *Angewandte Chemie (International Ed. in English)*, **36**, 1604–7.

59 Müller, M., Iyer, V.S., Kübel, C., Enkelmann, V. and Müllen, K. (1997) Polycyclic aromatic hydrocarbons by cyclodehydrogenation and skeletal rearrangement of oligophenylenes. *Angewandte Chemie (International Ed. in English)*, **36**, 1607–10.

60 Fechtenkötter, D., Saalwächter, A., Harbison, M.A., Müllen, K. and Spiess, H.W. (1999) Highly ordered columnar structures from hexa-peri-hexabenzocoronenes – synthesis, X-ray diffraction, and solid-state heteronuclear multiple-quantum NMR investigations. *Angewandte Chemie (International Ed. in English)*, **38**, 3039–42.

61 Ito, S., Herwig, P.T., Böhme, T., Rabe, J.P., Rettig, W. and Müllen, K. (2000) Bishexaperihexabenzocoronenyl: A 'superbiphenyl'. *Journal of the American Chemical Society*, **122**, 7698–706.

62 Stille, J.K., Rakutis, R.O., Mukamal, H. and Harris, F.W. (1968) Diels-Alder polymerizations. IV. Polymers containing short phenylene blocks connected by alkylene units. *Macromolecules*, **1**, 431–6.

63 Morgenroth, F., Reuter, E. and Müllen, K. (1997) Polyphenylene dendrimers: From threedimensional to two-dimensional structures. *Angewandte Chemie (International Ed. in English)*, **36**, 631–4.

64 Wiesler, U.-M. and Müllen, K. (1999) Polyphenylene dendrimers via Diels-Alder reactions: The convergent approach. *Chemical Communications*, 2293–4.

65 Dötz, F., Brand, J.D., Ito, S., Gherghel, L. and Müllen, K. (2000) Synthesis of large polycyclic aromatic hydrocarbons: variation of size and periphery. *Journal of the American Chemical Society*, **122**, 7707–17.

66 Wiesler, U.-M., Berresheim, A.J., Morgenroth, F., Lieser, G. and Müllen, K. (2001) Divergent synthesis of polyphenylene dendrimers: the role of core and branching reagents upon size and shape. *Macromolecules*, **34**, 187–99.

67 Weil, T., Wiesler, U.M., Herrmann, A., Bauer, R., Hofkens, J., De Schryver, F.C. and Müllen, K. (2001) Polyphenylene dendrimers with different fluorescent chromophores asymmetrically distributed at the periphery. *Journal of the American Chemical Society*, **123**, 8101–8.

68 Simpson, C.D., Brand, J.D., Berresheim, A.J., Przybilla, L., Räder, H.J. and Müllen, K. (2002) Synthesis of a giant 222 carbon graphite sheet. *Chemistry – A European Journal*, **8**, 1424–9.

69 Subramaniam, G. and Gilpin, R.K. (1992) A convenient synthesis of 2′,3′′′′-dimethyl-*p*-sexiphenyl. *Synthesis*, 1232–4.

70 Hart, H. and Harada, K. (1985) A new synthesis of *para*-terphenyls. *Tetrahedron Letters*, **26**, 29–32.

71 Hart, H., Harada, K. and Frank Du, C.-J. (1985) Synthetically useful aryl-aryl bond formation via Grignard generation and trapping of arynes. A one-step synthesis of *p*-terphenyl and unsymmetrical biaryls. *The Journal of Organic Chemistry*, **50**, 3104–10.

72 Harada, K., Hart, H. and Frank Du, C.-J. (1985) Reaction of aryl Grignard reagents with hexahalobenzenes: Novel arenes via multiple aryne sequence. *The Journal of Organic Chemistry*, **50**, 5524–8.

73 Frank Du, C.-J., Hart, H., and Ng, K.-K.D. (1986) A one-pot synthesis of m-terphenyls, via a two-aryne sequence. *The Journal of Organic Chemistry*, **51**, 3162–5.

74 Rebmann, A., Zhou, J., Schuler, P., Stegmann, H.B. and Rieker, A.J. (1996) *Journal of Chemical Research*, 318–9.S

75 Rebmann, A., Zhou, J., Schuler, P., Rieker, A.J. and Stegmann, H.B. (1997) Synthesis of two novel *para*-extended bisaroxyls and characterization of their triplet spin states. *Journal of the Chemical Society. Perkin Transactions*, **2**, 1615–7.

76 Harley-Mason, J. and Mann, F.G. (1940) *Journal of the Chemical Society*, 1379–85.

77 de Meijere, A. and Diederich, F. (eds) (2004) *Metal-Catalyzed Cross-Coupling Reactions*, 2nd edn, Wiley-VCH Verlag GmbH, Weinheim.

78 Lie, J.J. and Gribble, G.W. (2000) *Palladium in Herterocyclic Chemistry*, Pergamon Press, Elsevier, Amsterdam.

79 Miyaura, N. (ed.) (2002) *Cross-Coupling Reactions*, Springer, Berlin, Heidelberg.

80 Tamao, K., Hiyama, T. and Negishi, E. (eds) (2002) Special issue on cross-coupling reactions. *Journal of Organometallic Chemistry*, **653**, 1–299.

81 Kraft, A., Grimsdale, A.C. and Holmes, A.B. (1998) Electroluminescent conjugated polymers - Seeing polymers in a new light. *Angewandte Chemie (International Ed. in English)*, **37**, 402–28.

82 Saitoh, H., Saito, K., Yamamura, Y., Matsuyama, H., Kikuchi, K., Iyoda, M. and Ikemoto, I. (1993) Crystal structures of the room-temperature phase of 4,4″-difluoro-*p*-terphenyl and 4,4‴-difluoro-*p*-quaterphenyl. *Bulletin of the Chemical Society of Japan*, **66**, 2847–53.

83 Ung, V.A., Bardwell, D.A., Jeffery, J.C., Mahler, J.P., Ward, M.D. and Williamson, A. (1996) Dinuclear oxomolybdenum(V) complexes showing strong interactions across diphenol bridging ligands: Syntheses, structures, electrochemical properties, and EPR spectroscopic properties. *Inorganic Chemistry*, **35**, 5290–9.

84 Abdul-Raman, A., Amoroso, A.A., Branston, T.N., Das, A., Mahler, J.P., McCleverty, J.A., Ward, M.D. and Wlodarczyk, A. (1997) Dinuclear molybdenum complexes derived from diphenols: Electrochemical interactions and reduced species. *Polyhedron*, **16**, 4353–62.

85 Kallitsis, J.K., Kakahli, F. and Gravalos, K.G. (1994) Synthesis and characterization of soluble aromatic polyesters containing oligophenyl moieties in the main chain. *Macromolecules*, **27**, 4509–15.

86 Kallitsis, J.K., Gravalos, K.G., Hilberer, A. and Hadziioannou, G. (1997) Soluble polymers with laterally attached oligophenyl units for potential use as blue luminescent materials. *Macromolecules*, **30**, 2989–96.

87 Kauffmann, J.M. (1999) A really convenient synthesis of 2′,3‴-dimethyl-*p*-sexiphenyl. *Synthesis*, 918–20.

88 Rathore, R., Bruns, C.L. and Deselnicu, M.I. (2001) Multiple-electron transfer in a single step: Design and synthesis of highly charged cation-radical salts. *Organic Letters*, **3**, 2887–90.

89 Liess, P., Hensel, V. and Schlüter, A.-D. (1996) *Justus Liebigs Annalen Der Chemie*, 1037–40.

90 Frahn, J., Karakaya, B., Schäfer, A. and Schlüter, A.-D. (1997) Suzuki polycondensation: On catalyst derived phosphorus incorporation and reproducibility of molecular weights. *Tetrahedron*, **53**, 15459–67.

91 Hensel, V. and Schlüter, A.-D. (1999) A cyclotetraicosaphenylene. *Chemistry – A European Journal*, **5**, 421–9.

92 Sakai, N., Brennan, K.C., Weiss, L.A. and Matile, S. (1997) Toward biomimetic ion channels formed by rigid-rod molecules: Length-dependent ion-transport activity of substituted oligo-(*p*-phenylene)s. *Journal of the American Chemical Society*, **119**, 8726–7.

93 Ghebremariam, B. and Matile, S. (1998) Synthesis of asymmetric septi-(*p*-phenylene)s. *Tetrahedron Letters*, **39**, 5335–8.

94 Ghebremariam, B., Sidorov, V. and Matile, S. (1999) Direct evidence for the importance of hydrophobic mismatch for cell membrane recognition. *Tetrahedron Letters*, **40**, 1445–8.

95 Winum, J.-Y. and Matile, S. (1999) Rigid push-pull oligo(p-phenylene) rods: Depolarization of bilayer membranes with negative membrane potential. *Journal of the American Chemical Society*, **121**, 7961–2.

96 Robert, F., Winum, J.-Y., Sakai, N., Gerard, D. and Matile, S. (2000) Synthesis of multiply substituted, ion channel forming octi(p-phenylene)s: Theme and variations. *Organic Letters*, **2**, 37–9.

97 Sakai, N., Gerard, D. and Matile, S. (2001) Electrostatics of cell membrane recognition: Structure and activity of neutral and cationic rigid push-pull rods in isoelectric, anionic, and polarized lipid bilayer membranes. *Journal of the American Chemical Society*, **123**, 2517–24.

98 Sakai, N. and Matile, S. (2002) Recognition of polarized lipid bilayers by p-oligophenyl ion channels: From push-pull rods to push-pull barrels. *Journal of the American Chemical Society*, **124**, 1184–5.

99 Galda, P. and Rehahn, M. (1996) A versatile palladium-catalyzed synthesis of n-alkylsubstituted oligo-p-phenyls. *Synthesis*, 614–20.

100 Kim, S., Jackiw, J., Robinson, E., Schanze, K.S. and Reynolds, J.R. (1998) Water soluble photo- and electroluminescent alkoxy-sulfonated poly(p-phenylenes) synthesised via palladium catalysis. *Macromolecules*, **31**, 964–74.

101 Goldfinger, M.B., Crawford, K.B. and Swager, T.M. (1998) Synthesis of ethynyl-substituted quinquephenyls and conversion to extended fused-ring structures. *The Journal of Organic Chemistry*, **63**, 1676–86.

102 Konstandakopoulou, F.D., Gravalos, K.G. and Kallitsis, J.K. (1998) Synthesis and characterization of processable aromatic-aliphatic polyethers with quinquephenyl segments in the main chain for light-emitting applications. *Macromolecules*, **31**, 5264–71.

103 Morikawa, A. (1998) Preparation and properties of hyperbranched poly(ether ketones) with a various number of phenylene units. *Macromolecules*, **31**, 5999–6009.

104 Schlicke, B., Belser, P., De Cola, L., Sabbioni, E. and Balzani, V. (1999) Photonic wires of nanometric dimensions: electronic energy transfer in rigid rodlike Ru(bpy)$_3^{2+}$-(ph)n-Os(bpy)$_3^{2+}$ compounds (ph = 1,4-phenylene; n = 3, 5, 7). *Journal of the American Chemical Society*, **121**, 4207–14.

105 Taylor, P.N., Connell, M.J., O'McNeill, L.A., Hall, M.J., Alpin, R.T. and Anserson, H.L. (2000) Insulated molecular wires: Synthesis of conjugated polyrotaxanes by Suzuki coupling in water. *Angewandte Chemie (International Ed. in English)*, **39**, 3456–60.

106 Read, M.W., Escobedo, J.O., Willis, D.M., Beck, P.A. and Strongin, R.M. (2000) Convenient iterative synthesis of an octameric tetracarboxylate-functionalized oligophenylene rod with divergent end groups. *Organic Letters*, **2**, 3201–4.

107 Hwang, S.-W. and Chen, Y. (2001) Synthesis and electrochemical and optical properties of novel poly(aryl ether)s with isolated carbazole and p-quaterphenyl chromophores. *Macromolecules*, **34**, 2981–6.

108 Park, J.-W., Ediger, M.D. and Green, M.M. (2001) Chiral studies in amorphous solids: The effect of the polymeric glassy state on the racemization kinetics of bridged paddled binaphthyls. *Journal of the American Chemical Society*, **123**, 49–56.

109 Deng, X., Mayeux, A. and Cai, C. (2002) An efficient convergent synthesis of novel anisotropic adsorbates based on nanometer-sized and tripod-shaped oligophenylenes end-capped with triallylsilyl groups. *The Journal of Organic Chemistry*, **67**, 5279–83.

110 Lightowler, S. and Hird, M. (2004) Palladium-catalyzed cross-coupling reactions in the synthesis of novel aromatic polymers. *Chemistry of Materials*:

A *Publication of the American Chemical Society*, **16**, 3963–71.
111 Lightowler, S. and Hird, M. (2005) Monodisperse aromatic oligomers of defined structure and large size through selective and sequential suzuki palladium-catalyzed crosscoupling reactions. *Chemistry of Materials: A Publication of the American Chemical Society*, **27**, 5538–49.
112 Percec, V. and Okita, S. (1993) Synthesis and Ni(0)-catalyzed oligomerization of isomeric 4,4'''-dichloroquaterphenyls. *Journal of Polymer Science. Part A*, **31**, 877–84.
113 Li, Z.H., Wong, M.S., Tao, Y. and Iorio, M.D (2004) Synthesis and functional properties of strongly luminescent diphenylamino end-capped oligophenylenes. *The Journal of Organic Chemistry*, **69**, 921–7.
114 Lee, M., Jang, C.-J. and Ryu, J.-H. (2004) Supramolecular reactor from self-assembly of rod-coil molecule in aqueous environment. *Journal of the American Chemical Society*, **126**, 8082–3.
115 Ryu, J.-H., Jang, C.J., Yoo, Y.-S., Lim, S.-G. and Lee, M. (2005) Supramolecular reactor in an aqueous environment: Aromatic cross suzuki coupling reaction at room temperature. *The Journal of Organic Chemistry*, **70**, 8956–62.
116 Welter, S., Salluce, N., Benetti, A., Rot, N., Belser, P., Sonar, P., Grimsdale, A. C., Müllen, K., Lutz, M., Spek, A.L. and de Cola, L. (2005) Rodlike bimetallic ruthenium and osmium complexes bridged by phenylene spacers: Synthesis, electrochemistry, and photophysics. *Inorganic Chemistry*, **44**, 4706–18.
117 Schiek, M., Lützen, A., Koch, R., Al-Shamery, K., Balzer, F., Frese, R. and Rubahn, H.-G. (2005) Nanofibers from functionalized *para*-phenylene molecules. *Applied Physics Letters*, **86**, 153107–9.
118 Schiek, M., Lützen, A., Al-Shamery, K., Balzer, F. and Rubahn, H.-G. (2006) Nanofibers from methoxy functionalized *para*-phenylene molecules. *Surface Science*, **600**, 4030–3.
119 Mono functionalised oligomers were synthesized by Ivonne Wallmann, in context of her Diploma thesis, University of Bonn.
120 Balzer, F. and Rubahn, H.-G. (2002) Laser-controlled growth of needle-shaped organic nanoaggregates. *Nano Letters*, **2**, 747–50.
121 Niko, A., Meghdadi, F., Ambrosch-Draxel, C., Vogel, P. and Leising, G. (1996) Optical absorbance of oriented thin films. *Synthetic Metals*, **76**, 177–9.
122 Puschnig, P. and Ambrosch-Draxel, C. (1999) Density-functional study of the oligomers of poly-*para*-phenylene: Band structures and dielectric tensors. *Physical Review. B, Condensed Matter*, **60**, 7891–8.
123 Maibohm, C., Brewer, J., Sturm, H., Balzer, F. and Rubahn, H-G. (2006) Bleaching and coating of organic nanofibers. *Journal of Applied Physics*, **100**, 0543041–4.
124 Thilsing-Hansen, K., Neves-Petersen, M.T., Petersen, S.B., Neuendorf, R., Al-Shamery, K. and Rubahn, H.-G. (2005) Luminescence decay of oriented phenylene nanofibers. *Physical Review. B, Condensed Matter*, **72**, 1152131–7.
125 Schiek, M., Lützen, A., Al-Shamery, K., Balzer, F. and Rubahn, H.-G. (2007) Organic nanofibers from chloride functionalized *para*-quaterphenylenes. *Crystal Growth and Design*, **7**, 229–33.
126 Ohring, M. (1992) *Materials Science of Thin Films*, Academic Press, San Diego.
127 Brewer, J., Schiek, M., Lützen, A., Al-Shamery, K. and Rubahn, H.-G. (2006) Nanofiber frequency doublers. *Nano Letters*, **6**, 2656–9.
128 Schiek, M., Brewer, J., Balzer, F., Lützen, A., Al-Shamery, K. and Rubahn, H.-G. (2007) Tailored nanoaggregates from functionalized organic molecules. *Proceedings of SPIE*, **6475**, -17.
129 Würthner, F., Ahmed, S., Thalacker, C. and Debaerdemaeker, T. (2002) Core-substituted naphthalene bisimides: New fluorophores with tunable emission wavelength for FRET studies. *Chemistry–A European Journal*, **8**, 4742–50.
130 Blaszcyk, A., Fischer, M. and v.Hänisch, C., and Mayor, M. (2006) Synthesis, structure, and optical properties of

terminally sulfur-functionalized core-substituted naphthalenebisimide dyes. *Helvetica Chimica Acta*, **89**, 1986–2005.
131 Balzer, F., Beermann, J., Bozhevolnyi, S.I., Simonsen, A.C. and Rubahn, H.-G. (2003) Optically active organic microrings. *Nano Letters*, **3**, 1311–4.
132 Valeur, B. (2001) *Molecular Fluorescence: Principles and Applications*, Wiley-VCH Verlag GmbH, Weinheim.
133 Schwoerer, M. and Wolf, H.C. (2005) *Organische Molekulare Festkörper – Einführung in die Physik von π-Systemen*, Wiley-VCH Verlag GmbH, Weinheim.
134 Balzer, F., Al-Shamery, K., Neuendorf, R. and Rubahn, H.-G. (2003) Nonlinear optics of hexaphenyl nanofibers. *Chemical Physics Letters*, **368**, 307–12.
135 The nonlinear optical properties of the nanofibers have been investigated by Jonathan R. Brewer, in context of his PhD thesis, University of Southern Denmark.

9
Plant Viral Capsids as Programmable Nanobuilding Blocks
Nicole F. Steinmetz

9.1
Nanobiotechnology – A Definition

Nanotechnology is a collective term for a broad range of relatively novel topics; scale is the main unifying theme with nanotechnology being concerned with matter on the nanometer scale. It is a highly multidisciplinary area that describes a field of applied science and technology focused on the design, synthesis, characterization and application of materials and devices on the nanoscale.

In 1959, the physicist and Nobelist Richard Feynman was the first scientist to describe some distinguishing concepts of nanotechnology in his groundbreaking lecture "There's plenty of room at the bottom" [1]. The term *nanotechnology* was first defined by Norio Taniguchi in his paper "On the basic concept of nano-technology" in 1974 [2]. Then Eric Drexler in 1986 popularized the term *nanotechnology* in his seminal book on molecular nanotechnology with the title "Engines of Creation: The Coming Era of Nanotechnology" [3].

From a practical point of view, nanotechnology and nanosciences started in the early 1980s. Major developments were the birth of cluster science, the development of the scanning tunneling microscope, the discovery of buckminsterfullerene (the C_{60} buckyball) and carbon nanotubes, as well as the synthesis of semiconductor nanocrystals, which led to the development of quantum dots [4].

Nanobiotechnology or bionanotechnology – both terms are used equally in the literature – is a section of the field of nanotechnology; it involves the exploitation of biomaterials, devices, or methodologies in the nanoscale. It has a multidisciplinary character, as it sits at the interface of biology, chemistry, physics, materials science, engineering, and medicine. The subject of nanobiotechnology can be divided into two main areas: the first is defined as the use of nanotechnological devices to probe and understand biological systems. The second is concerned with the exploitation of biomaterials in the fabrication of new nanomaterials and/or nanodevices.

This chapter will focus on the second area of nanobiotechnology: the exploitation of biomolecules for technological applications. A set of biomolecules has

Tomorrow's Chemistry Today. Concepts in Nanoscience, Organic Materials and Environmental Chemistry.
Edited by Bruno Pignataro
Copyright © 2008 WILEY-VCH Verlag GmbH & Co. KGaA, Weinheim
ISBN: 978-3-527-31918-3

been used in recent years and these include: nucleic acid [5–9], crystalline bacterial cell-surface layers [10], whole organisms such as prokaryotic and eukaryotic cells [11–14], and proteins such as ferritin (an iron storage protein) [15–18], heat shock proteins [19–21], and also the capsids of viral particles. The latter have received particular attention; they have been utilized as scaffolds for the assembly and nucleation of organic and inorganic materials, for the selective attachment and presentation of chemical and biological moieties, as well as building blocks for the construction of one-, two-, and three-dimensional arrays [22–26].

9.2
Viral Particles as Tools for Nanobiotechnology

To date, a broad range of viral particles have been studied and used for nanobiotechnological approaches. This includes the full spectrum including bacteriophages (viruses that infect bacteria), plant viruses (infect plants), and animal viruses (infect animals and humans). Viruses are nonliving organisms; they are obligate intracellular parasites, that is, they need the host cell for reproduction. In the simplest form, viral particles consist of nucleic acid, which can be either deoxyribonucleic acid or ribonucleic acid, and a protective protein coat, the capsid. The nucleic acid encodes the genetic information; the function of the capsid is mainly the protection of the encapsidated nucleic acid. To meet these requirements viral capsids are extremely rigid and robust, and their dimensions make them an excellent tool for nanobiotechnological applications. Table 9.1 gives an overview of the range of viral particles studied and used in nanosciences. The majority are plant viruses; the focus of this chapter will be on the plant virus *Cowpea mosaic virus* (CPMV).

9.3
General Introduction to CPMV

CPMV is the type member of the comovirus genera of the family *Comoviridae*; also known as plant picorna-like viruses as they share similarities in structure, genome organization and replication strategy with animal picornaviruses [64]. CPMV has a rather narrow natural host range, it normally infects legumes and was first reported in *Vigna unguiculata*, also known as black-eyed peas. Geographically, CPMV is found in Cuba, Japan, Kenya, Nigeria, Surinam, and Tanzania, where the virus is transmitted by leaf-feeding beetles, thrips, and grasshoppers. In addition to the natural hosts, species in several families – including legumes and *Nicotiana benthamiana* – are known to be susceptible to the virus and transmission can be achieved experimentally by mechanical inoculation. In systemically infected plants, CPMV typically causes mosaic or mottling symptoms (Figure 9.1) [65].

Table 9.1 Overview of the range of viral particles that have been studied and used for nanobiotechnological applications.

	Virus	Shape/dimensions	Selected references
Bacteriophages	M13	rod/20 × 900 nm	[27–33]
	MS2	Icosahedron/25 nm diameter	[34–36]
	T7	Head–tail, head has icosahedral symmetry/ 55 nm diameter	[37–38]
Plant viruses	Cowpea mosaic virus	Icosahedron/28 nm diameter	The reader is referred to references throughout the text and Table 9.2
	Cowpea chlorotic mosaic virus	Icosahedron/28 nm in diameter	[39–46]
	Tobacco mosaic virus	rod/18 × 300 nm	[47–58]
	Brome mosaic virus	Icosahedron/28 nm diameter	[59–61]
	Red clover necrotic mosaic virus	Icosahedron/36 nm diameter	[62]
Animal viruses	Chilo iridescent virus	Sphere/60 nm diameter	[63]

Figure 9.1 Trifoliate leaves from *Vigna unguiculata*. Left: infected with *Cowpea mosaic virus*. Right: noninfected leaf.

CPMV has a bipartite single-stranded positive-sense RNA genome: RNA1 and RNA2, respectively. The RNA segments are encapsidated separately; and both complements are required for infection [66, 67]. By centrifugation on sucrose or cesium chloride density gradients, CPMV particles can be separated into three components, which have identical protein composition but differ in their RNA contents [68–70]. The particles of the top component are devoid of RNA, while the middle and bottom components each contain a single RNA molecule, RNA2 and RNA1, respectively [71].

CPMV particles have an icosahedral symmetry with a diameter of approximately 28 nm (Figure 9.2), the protein shell of the capsid is about 3.9 nm thick [72]. The structure of CPMV is known to near-atomic resolution (Figure 9.3) [73]. The virions are formed by 60 copies of two different types of coat proteins, the small (S) subunit and the large (L) subunit. The S subunit (213 amino acids) folds into one jelly roll β-sandwich, and the L subunit (374 amino acids) folds into two jelly roll β-sandwich domains. The three domains form the asymmetric unit and are arranged in a similar surface lattice to T = 3 viruses, except they have different polypeptide sequences; therefore the particle structure is described as a pseudo T = 3 or P = 3 symmetry [74].

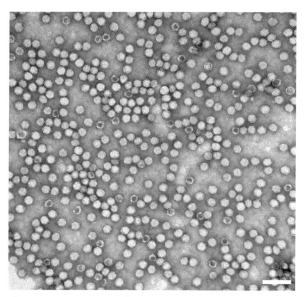

Figure 9.2 Transmission electron micrograph of *Cowpea mosaic virus* particles negatively stained with uranyl acetate. The scale bar is 100 nm.

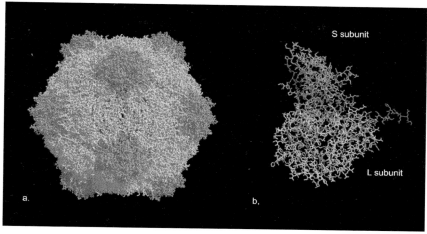

Figure 9.3 The structure of the *Cowpea mosaic virus* (CPMV) capsid (a) and the asymmetric unit (b). The capsid of CPMV is comprised by the small (S) and large (L) subunit.

9.4
Advantages of Plant Viral Particles as Nanoscaffolds

Viruses have been studied extensively and detailed knowledge about the biological, genetic, and physical properties is available. Viral particles – such as CPMV – have appealing features for use in nanobiotechnology. The main advantage is their size, for example CPMV has a diameter of 28 nm (see Table 9.1 for more data). The propensity to self-assemble into discrete, monodisperse nanoparticles of discrete shape and size, with a high degree of symmetry and polyvalency, makes them unique bionanoparticles.

Further, plant viruses are noninfectious toward other organisms, and they do not present a biological hazard; therefore the work with plant viruses can be regarded as safe. The production of the virions is simple and quick. Owing to their autonomous replication and spread through the plant, high expression levels can be reached within a short time (2–4 weeks), and yields in gram scale can be easily obtained from 1 kg of infected leaf material [75]. Plant viral particles are exceptionally stable and robust. CPMV, for example, maintains its integrity at 4 °C or room temperature indefinitely. It can tolerate up to 60 °C for an hour. It remains intact at pH values in the range from 4.0 to 9.0 at room temperature for at least two days. In addition, the virions can tolerate organic solvents such as dimethyl sulfoxide up to 50% by volume for at least two days and they also tolerate an ethanol/buffer mixture at 50% by volume for a few days. Last but not least, in the case of CPMV, chimeric virus technology can be used to modify the capsid surface; cDNA clones of RNA1 and RNA2 are available and allow site-directed and insertional mutagenesis with relative ease [76–78].

Altogether, plant viruses and especially CPMV display a number of features to be exploited for nanobiotechnology and can be regarded as extremely robust, monodisperse nanobuilding blocks.

9.5
Addressable Viral Nanobuilding Block

In 2002, for the first time, CPMV was regarded as an addressable nanobuilding block [79]. First, experiments were conducted in order to test selective chemical derivatization of wild-type virions; later, several mutant particles were made, such as cysteine mutants [80] and histidine mutants [81]. Different bioconjugation strategies were applied and a set of biological and organic and inorganic chemical molecules have been attached to CPMV.

Native CPMV particles display reactive lysines [82], carboxylates derived from aspartic and glutamic acid [83], and tyrosines [84] on their exterior solvent-exposed surface (Figure 9.4). They further display reactive interior cysteines residues [79].

Structural data of the CPMV capsid indicate five exterior lysines to be solvent-exposed (Figure 9.4a). Probing wild-type CPMV [79] and lysine-minus mutants [82] with lysine-selective fluorescent dyes and metals confirmed that all lysines are addressable, but to different degrees. It was found that a maximum labeling of 240 dye molecules per wild-type CPMV particle can be achieved.

In the context of my PhD research with Drs. Dave Evans and George Lomonossoff we have demonstrated that CPMV also displays addressable carboxylates on its solvent-exposed surface (Figure 9.4b) [83]. The structural data from CPMV suggest eight to nine carboxylate groups, derived from aspartic and glutamic acids, to be on the solvent-exposed exterior surface: five on the S subunit, two on the L subunit, and the carboxy-terminus of the solvent exposed terminal domain of S.

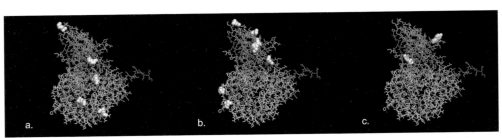

Figure 9.4 The asymmetric unit of the capsid of the *Cowpea mosaic virus*. Depicted are addressable surface residues (highlighted as spheres). (a) Addressable lysines: lys 38 and lys 82 on the small (S) subunit (in dark gray), lys 34, lys 99, and lys 199 on the large (L) subunit (in light gray). (b) Carboxylates: aspartic acid (asp) 26, asp 44, asp 45, asp 85 and glutamic acid (glu) 135 on the S subunit, and asp 273 and glu 319 in the L subunit; c. addressable tyrosines tyr 52 and tyr 103, both located in S.

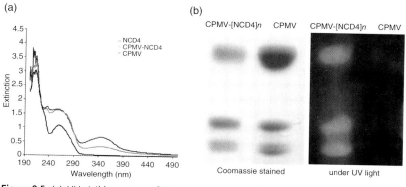

Figure 9.5 (a) UV-visible spectra of N-cyclohexyl-N'-(4-(dimethylamino)naphthyl)carbodiimide (NCD4, dark gray line), native Cowpea mosaic virus (CPMV) particles (black line) and the dye labeled CPMV-[NCD4]$_n$ particles (light gray line). The peak at 345 nm of CPMV-[NCD4]$_n$ indicates that the carboxylate groups are addressable. (b) Wild type CPMV particles and CPMV-[NCD4]$_n$ virions after separation on a denaturating 12% sodium dodecyl sulfate polyacrylamide gel. Detection of the viral capsid subunits— the small (S) and large (L) subunits—was achieved by Coomassie staining (left) or by visualization of the modified fluorescent CPMV-[NCD4]$_n$ proteins under UV light (right).

A further carboxylate is found in the carboxy-terminal domain of S; however, owing to cleavage of 24 amino acids of the S subunit in the plant, the additional carboxylate is present in only a small proportion of the particles [85].

We probed the reactivity of the carboxylates using a fluorescent carboxylate-selective chemical dye, N-cyclohexyl-N'-(4-(dimethylamino)naphthyl)carbodiimide (NCD4). Using UV-visible spectroscopy, native gel electrophoresis, and denaturing gel electrophoresis, covalent modification with the dye was confirmed (Figure 9.5). The latter showed that the dye was attached to both the S and the L subunit, an expected observation, as the structural data suggested carboxylates on both subunits. A quantification of the number of bound dye molecules could not be achieved; owing to the instability of the dye in aqueous solvent a reliable extinction coefficient could not be determined.

Further, native CPMV particles display addressable tyrosine residues. Structural data and the study of tyrosine-minus mutants demonstrated that two tyrosines located in the S subunit (Figure 9.4c) are available for chemical conjugation [84]. The tyrosines can be addressed by treatment with the nickel(II) complex of the tripeptide glycine-glycine-histidine in the presence of magnesium monoperoxyphthalate (Ni/GGH/MMPP), or by visible light irradiation in the presence of tris(bipyridyl)ruthenium(II) dication and ammonium persulfate. Both treatments mediate effective crosslinking of adjacent pentameric subunits, via covalent conjugation of tyrosines residues. Tyrosines also can be functionalized with ligands when a disulfide is present during the Ni/GGH/MMPP treatment, for example an azidoalkyl cystine was selectively attached to tyrosines using this protocol [84].

Besides the reactive exterior, addressable groups have been found on the interior of the capsid as well. CPMV has at least two reactive interior cysteine residues [79], but no cysteines can be found on the solvent-exposed exterior [73]. It was demonstrated that chemical derivatization of the internal cysteines can be achieved using small thiol reactive compounds, such as ethylmercury phosphane, 5-maleimidofluorescein [79], and thiol-selective stilbene derivatives [86].

Thiols are among the most useful functional groups found in proteins; they can react with a large number of organic and inorganic reagents. Since native CPMV particles do not display any cysteines on the exterior solvent-exposed surface, genetic protein modification reactions can be used to introduce cysteines at defined locations. A series of CPMV cysteine mutants have been generated [80]. Previous work designed to employ CPMV as a scaffold for the presentation of antigenic peptide sequences [87–89] established that additional amino acids can be inserted into the highly surface-exposed βE-βF loop on the L subunit, the βB-βC loop and βC'-βC", and the carboxy-terminus of the S subunit without compromising the ability of the resulting mutant virus to propagate in the host plant. Based on this knowledge, cysteine mutants displaying cysteine residues on the surface-exposed βB-βC loop and βE-βF loop were generated, and their addressability was demonstrated using a variety of molecules (see Table 9.2) [79, 80].

A set of five CPMV histidine mutants is also available [81]. Six contiguous histidine residues were genetically inserted at different positions of the capsid, in the βE-βF loop on the L subunit, the βB-βC loop, and at the carboxy-terminus of the S subunit. The affinity to nickel(II) cations was verified for all mutants, although the affinity varies from mutant to mutant. The CPMV histidine mutants can be purified from clarified leaf homogenate added onto nickel-nitrilotriacetic acid (Ni-NTA) columns. The tag also offers a unique and straightforward way for purification in downstream applications. Further, the histidine tag provides a novel attachment site for chemical or biological moieties; the addressability was confirmed by labeling the virus with nanogold derivatized with a Ni-NTA crosslinker [81].

In summary, CPMV wild-type and mutant particles can be regarded as robust and multi-addressable nanobuilding blocks. During the last five years a large number of different biological as well as organic and inorganic chemical moieties have been attached to the virions for different applications. Table 9.2 gives a summary of all molecules that have been reported so far to attach to CPMV.

9.6
From Labeling Studies to Applications

In the early studies fluorescent probes and metals, such as nanogold particles, were mainly attached to the viral nanoscaffold in order to examine the addressability of CPMV wild-type and mutant particles. To date, fluorescent CPMV-derived nanoparticles find applications (i) in biomedicine, (ii) as reporter molecules in sensors or assays, and (iii) as fluorescent tags for understanding biological processes on the nanoscale. To give some examples: fluorescently tagged CPMV

Table 9.2 Overview of chemical moieties that have been attached to wild-type (wt) and mutant *Cowpea mosaic virus* (CPMV) particles and the purpose of the study and/or application.

Chemical/biological moiety conjugated to CPMV	Wild type (wt) and/or mutant/amino acid addressed[a]	Purpose of study	Reference
Fluorescein Nanogold particles	Wt and lysine-minus mutants/lysines	Fluorescent and metal decorated viral nanoparticles for structural analysis	[82, 90]
N-Cyclohexyl-N'-(4-(dimethylamino)naphthyl)carbodiimide	Wt/aspartic and glutamic acids	Probe reactivity of carboxylates present on the solvent-exposed surface	[83]
Treatment with the nickel(II) complex of the tripeptide glycine-glycine-histidine in the presence of magnesium monoperoxyphthalate Visible light irradiation in the presence of tris(bipyridyl)ruthenium(II) dication and ammonium persulfate	Wt and tyrosine minus mutants/tyrosines	Addressability of tyrosines and a protocol for covalently stitching together adjacent subunits of the capsid	[84]
Ethylmercury phosphate Fluorescein Fluorescein and rhodamine (double labeling) Stilbene Nanogold particles Metallic gold particles (0.9 μm diameter) Stilbene and polyethylene glycol (double labeling)	Wt/interior cysteines and cysteine mutants/interior and exterior cysteines Cysteine mutant/lysines and cysteines	Probing the addressability of CPMV and gaining structural information of the viral capsid	[79, 80, 86] [86]
Nanogold Ni-NTA	Histidine mutants/histidines	Addressability of genetically introduced histidine residues	[81]
Azides Alkynes	Wt/lysines	Copper(I)-catalyzed azide-alkyne cycloaddition as a selective bioconjugation technique	[91]
Oregon green Tetramethylrhodamine	Wt/lysines	Study of self-assembly of nanoparticles (fluorescently tagged CPMV) at liquid–liquid interfaces	[92]

Table 9.2 Continued

Chemical/biological moiety conjugated to CPMV	Wild type (wt) and/or mutant/amino acid addressed[a]	Purpose of study	Reference
NeutrAvidin and AlexaFluor 648 (double labeling) Chicken IgG and AlexaFluor 546 (double labeling)	Cysteine mutant/ cysteines	Fluorescent CPMV particles were used as a reporter tag in a standard microtubule gliding assay	[93]
Nanogold	Cysteine mutant/ cysteines	Use of CPMV in order to place and arrange gold nanoparticles at fixed interparticle distances	[94]
Nanogold followed by interconnection of the gold nanoparticles via 1,4-C_6H_4[trans-(4-AcSC_6H_4C≡CPt-(PBu$_3$)$_2$C≡C]$_2$ and/or oligophenylenevinylene	Cysteine mutants/ cysteine	Construction of three-dimensional conducting circuits at the nanometer scale	[95]
Ferrocene	Wt/lyines	Generation of electroactive nanoparticles	[96]
Viologen	Wt/aspartic and glutamic acids	Generation of electroactive nanoparticles	[83]
Quantum dots Carbon nanotubes	Wt/lysines	Covalent network formation of CPMV with quantum dots or carbon nanotubes for the design of new materials	[97]
Oregon green	Wt/lysines	Study of trafficking of fluorescently labeled CPMV in mice after oral or intraveneous injection	[98]
Fluorescent red NIR-782	Wt/lysines	Fluorescent CPMV particles for imaging via diffuse optical tomography for biomedical applications	[99]
AlexaFluor 555 AlexaFluor 488 Fluorescein Polyethylene glycol Fluorescein and polyethylene glycol (double labeling)	Wt/lysines	CPMV as a fluorescent probe for intravital vascular imaging	[100]

Table 9.2 Continued

Chemical/biological moiety conjugated to CPMV	Wild type (wt) and/or mutant/amino acid addressed[a]	Purpose of study	Reference
Fluorescein-polyethyleneglycol-N-hydroxysuccinimide Methoxy-polyethyleneglycol-N-hydroxysuccinimide	Wt/lysines	Pegylation as a strategy for reduced immunogenicity of CPMV virions for diagnostic and therapeutic applications	[101]
Mannose Mannose and fluorescein (double labeling)	Wt/lysines Cysteine mutant/cysteines	Glycolated CPMV as a tool for cellular recognition and signaling events (glucose assay and hemagglutination)	[102]
Side-chain neoglycopolymer	Wt/lysines	CPMV as a platform for covalent clustering of carbohydrates as a tool in the study of carbohydrate-based cellular processes	[103]
T4 lysozyme LRR internalin B Intron 8 of Herstatin	Wt/lysines Cysteine mutant/cysteines Lysine-minus mutant/lysine	CPMV as a platform for the presentation of biologically active proteins with potential for vaccine development, drug delivery, and therapeutic applications	[104]
NeutrAvidin and Cy5 (double labeling)	Cysteine mutant/cysteines	CPMV as a probe for DNA microarray protocols	[105]
Antibodies and AlexaFluor 647 (double labeling) Chicken and/or mouse IgG and AlexaFluor647 (double labeling)	Cysteine mutants/cysteines and lysines	CPMV decorated with fluorescent dyes and either IgG or antibodies as a tracer for direct and indirect sandwich immunoassays	[106]
Oligonucleotides	Wt/lysines and cysteine mutant/cysteines	Self-assembly of CPMV nanoparticles via duplex base-pairing of the attached oligonucleotides in solution	[107]

Table 9.2 Continued

Chemical/biological moiety conjugated to CPMV	Wild type (wt) and/or mutant/amino acid addressed[a]	Purpose of study	Reference
Mannose	Wt/lysines	Controlled aggregation of mannose-labeled CPMV via interaction with concanavalin A	[102]
Biotin	Cysteine mutant/cysteines and wt/lyines	Studying controlled aggregation in solution of CPMV-biotin in interaction with avidin	[80], [90]
Biotin	Wt and histidine mutant/lysines	Immobilization on surfaces and decoration with quantum dots via avidin interaction	[108]
Biotin and AlexaFluor 488 (double labeling) Biotin and AlexaFluor 568, (double labeling) AlexaFluor 488 maleimide Biotin	Wt/lysines Cysteine mutant/cysteines Cysteine mutant/lysines	Construction of arrays of viral particles via the layer-by-layer approach using biotin-streptavidin interaction	[109]

a If not specified, cysteine refers to cysteine residues on the exterior surface of the virus particle.

particles have been used as tools for bioimaging for biomedical applications and fluorescent CPMV particles have been used for near-infrared fluorescence tomography [99]. The main advantage of the viral nanoscaffold is that high local dye concentrations without fluorescent quenching can be achieved; this increases the signal-to-noise ratio and therefore the detection sensitivity. Further, CPMV was labeled with dyes in order to follow CPMV trafficking in mice in vivo after oral or intravenous injection [98] and it has been used as a tool for intravital vascular imaging [100]. In terms of using CPMV as reporter molecules, they have been used as tracers in immunoassays [106]; and improved sensitivity was achieved in DNA microarrays [105]. Other studies have shown the use of CPMV as a fluorescent tag in elucidating nanoparticle assembly at liquid–liquid interfaces [92], and dye-labeled CPMV particles have been used as reporter tags in a standard microtubule gliding assay [93].

In earlier studies CPMV was explored as a scaffold for the presentation of short antigenic peptide sequences for the generation of vaccines. Foreign short peptide sequences can be inserted by means of genetic modifications and displayed in multiple copies on the surface of the particle [89, 110]. However, surface presentation is restricted to 30–40 amino acids by this means. Recently it was demonstrated that full-length biologically active proteins, such as T4 lysozyme (see Table 9.2 for more details) can be covalently attached and presented on the viral nanoscaffold [104]. This complements the genetic modification protocols and offers a route for attaching nearly any peptide or protein to the virus surface. Of course, steric hindrance and conformational restrictions still apply.

In 2006, for the first time, the decoration of viral particles with redox-active moieties was reported [83, 96]. The decoration of CPMV with a redox-active organometallic complex, ferrocene, and also with an organic redox-active compound, viologen, was achieved. Both approaches led to the generation of monodisperse redox-active nanoparticles; the redox centers were presented in multiple copies on the solvent-exposed outer surface.

Ferrocenes are well-characterized molecules noted for their stability and their favorable electrochemical properties. The availability of a large variety of derivatives makes them a popular choice for biological applications such as labels or sensors for electrochemical detection [111]. Ferrocenecarboxylate was used for covalent decoration of CPMV particles [96]. Chemical conjugation of the virions was achieved by activating the ferrocenecarboxylic acid with the coupling reagents N-ethyl-N'-(3-dimethylaminopropyl)carbodiimide hydrochloride (EDC) and N-hydroxysuccinimide (NHS) to accomplish facile coupling of the ferrocenecarboxylic acid to amino groups on the capsid surface.

After chemical derivatization, the integrity of the particles was confirmed by transmission electron microscopy (TEM), dynamic light scattering, and native gel electrophoresis. Measurements of the particle size by dynamic light scattering showed that the ferrocene-decorated CPMV particles (CPMV-Fc_n) had an increase in radius of about 0.7 nm which is in good agreement with the size of the ferrocene moiety.

Electrochemical studies confirmed the presence of redox-active nanoparticles. Differential pulse and cyclic voltammetry studies were conducted. Cyclic voltammetry showed that the complex displays an electrochemically reversible ferrocene/ferrocenium couple (Figure 9.6). The oxidation potential for the hybrid CPMV-Fc_n conjugate and free ferrocenecarboxylic acid in solution was determined; $E_{1/2}$ of CPMV-Fc_n was 0.23 V, and $E_{1/2}$ of free ferrocenecarboxylic acid was 0.32 V versus the Ag/AgCl electrode, respectively. This shift is expected for the conversion of the carboxyl group of ferrocenecarboxylic acid to an amide on coupling to the virus capsid, since the amide is less electron-withdrawing.

Peak currents were measured and the linear plot of i_p versus $v^{1/2}$ ($R = 0.997$) showed that the oxidation process was diffusion-controlled (inset in Figure 9.6). On the basis of the Randles–Sevcik equation [112]. the number of ferrocene molecules was calculated to be around 240 ferrocene moieties per CPMV particle. The appearance of a unique reversible process indicates that the multiple ferrocenyl centers behave as independent, electronically isolated units; therefore the CPMV-Fc_n conjugates are similar to metallodendrimers and could find applications as

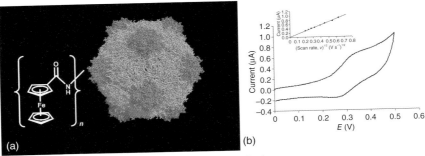

Figure 9.6 (a) Schematic presentation of *Cowpea mosaic virus* (CPMV) particles chemically modified with ferrocenecarboxylic acid (Fc) moieties via the formation of a peptide bond. (b) Cyclic voltammogram of derivatized CPMV-Fc$_n$ at a scan rate v of $0.1\,V\,s^{-1}$, and linear plot of current versus (scan rate)$^{1/2}$ (inset).

multielectron transfer mediators in electrocatalytic processes of biological and industrial importance.

In a second study, addressable carboxylate groups on wild-type CPMV particles were utilized as anchor groups for an organic, redox-active viologen derivative, methyl(aminopropyl)viologen (MAV) [83]. Covalent decoration of CPMV with MAV was achieved by using the coupling reagents EDC and NHS. The particles remained intact after chemical modifications, confirmed by TEM studies. Cyclic voltammetric studies on viologen decorated CPMV-MAV$_n$ nanoparticles showed the characteristic two successive one-electron reversible steps of the methylviologen moieties (Figure 9.7). The two reduction potentials, $E^{0\prime}$ $-0.65\,V$ and $-0.97\,V$ versus the Ag/AgCl electrode, for the virus-bound viologens are comparable to the reduction potentials of free viologen in solution. A large difference in $E^{0\prime}$ is not expected as the methylviologen-N-propylamine and methylviologen-N-propylamide will have similar inductive properties due to the intervening propyl group. Therefore, the attached viologen moieties behave, from an electrochemical point of view, similarly to viologen in solution. Evidently the attached moieties behave as independent, electronically isolated units. Peak currents were measured and the linear plot of i_p versus $v^{1/2}$ ($R = 0.996$) showed that the reduction processes were diffusion controlled (inset in Figure 9.7b). The number of viologen molecules attached to each CPMV virion was estimated by use of the Randles–Sevcik equation, and it was found that around 180 viologens decorated each viral particle.

In the above two independent studies, the feasibility of CPMV as a nanobuilding block for chemical conjugation with redox-active compounds was demonstrated. The resulting robust, and monodisperse particles could serve as a multielectron reservoir that might lead to the development of nanoscale electron transfer mediators in redox catalysis, molecular recognition, and amperometric biosensors and to nanoelectronic devices such as molecular batteries or capacitors.

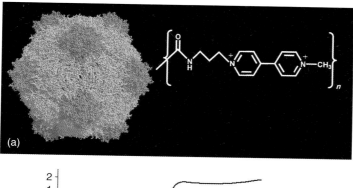

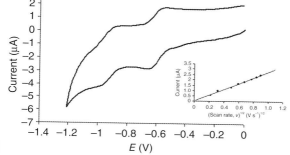

Figure 9.7 (a) Schematic presentation of *Cowpea mosaic virus* (CPMV) particles chemically modified with methylaminoviologen (MAV) via the formation of a peptide bond. (b) Cyclic voltammogram of derivatized CPMV-MAV$_n$ at a scan rate v of 0.15 V s^{-1}, and linear plot of current versus (scan rate)$^{1/2}$ (inset).

9.7
Immobilization of Viral Particles and the Construction of Arrays on Solid Supports

The previous section summarized the available addressable groups on the surface of CPMV and described functionalization studies that have been done in solution. This section will focus on approaches where the particles have been immobilized on surfaces, and highlight a recent study in which CPMV particles were used to construct three-dimensional arrays [109].

Immobilization of biomolecules and the assembly of biomolecules in defined arrays is a desired requirement of nanobiotechnology. Thin films of immobilized proteins on solid supports are of growing interest and find applications in biosensors, information processing, optics, and biomedicine [113–115].

To further extend the utility of CPMV virions as tools in nanobiotechnology, it was shown that CPMV particles can be utilized as building blocks for the construction of monolayer, bilayer, and multilayer arrays on surfaces in a controlled manner [109]. CPMV virions were labeled with two different ligands: fluorescent dyes that enabled differential detection, and biotin molecules that allowed the construction of arrays from the bottom up via a layer-by-layer approach. The construction of the layers was achieved using the high molecular recognition between streptavidin (SAv) and biotinylated CPMV virions (CPMV-biotin$_n$). Owing to the extraordinarily high affinity constant between SAv and biotin ($K_a > 10^{14}\,M^{-1}$) [116], chemical stability of the system is expected. Immobilization of CPMV particles on solid supports was achieved either by direct binding of CPMV cysteine mutant particles onto gold via the gold–thiol interaction, or by indirect immobilization of CPMV-biotin$_n$ mediated via a thiol-modified SAv.

In detail, viral wild-type particles of one set were labeled with the fluorescent dye AlexaFluor (AF) 488 and biotin; both groups were attached to surface available lysines. Another batch was labeled with AF568 and biotin, also at addressable lysines. Both types of building block, in the following referred to as CPMV-biotin-AF488 and CPMV-biotin-AF568, were characterized by TEM, UV-visible spectroscopy, native gel electrophoresis, and dot blot studies. TEM studies verified that the particles remained intact after chemical modification. UV-visible spectroscopy confirmed covalent modification and also allowed quantification of the number of labels per particle; the particles displayed around 40 biotin moieties and around 200 dyes.

Unspecific adsorption on gold surfaces of any of the building blocks (CPMV-biotin-AF488 and CPMV-biotin-AF568 and SAv) was ruled out, indicating that binding of the building blocks occurred in a controlled manner based on a sulfur–gold interaction (SAv–thiol) and on specific interactions between biotin-bound CPMV and SAv.

CPMV bilayers comprising [CPMV-biotin-AF488–streptavidin–CPMV-biotin-AF568], and vice versa, on SAv-functionalized gold surfaces were fabricated and analyzed (Figure 9.8). Fluorescence microscopy imaging of the CPMV arrays was consistent with successful binding of both viral building blocks, CPMV-biotin-AF488 and CPMV-biotin-AF568. The fluorescent viral particles were spread evenly over the whole surface and a dense coverage was achieved. The overlaid image demonstrated that the individual images line up well, indicating that the virions are sitting on top of each other. To further support these observations, a mixed layer was immobilized on the gold surface and analyzed in the same way (Figure 9.8). In this case, the particles compete for the same binding sites, resulting in less dense and less evenly distributed coverage. The merged images do not line up, consistent with the particles occupying the same layer and competing for the same binding sites. Comparison of the overlaid image from the bilayers with that of the mixed layer further supports the successful, controlled fabrication of a bilayer consisting of different fluorescent CPMV particles.

The construction of a trilayer via incorporation of CPMV cysteine mutants has also been reported [109]. In addition, there are a few further examples where CPMV particles have been bound to surfaces using different strategies and templates. For example, CPMV histidine mutants have been immobilized on Neutr-Avidin surfaces bridged with biotin-X-NTA molecules followed by decoration of the viral particles with quantum dots [107]. CPMV cysteine mutants have been successfully immobilized on maleimido-functionalized patterned templates; these

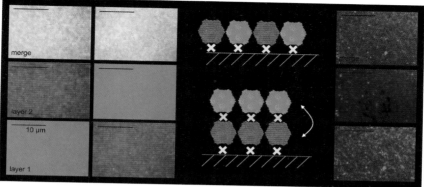

Figure 9.8 Bilayers and a mixed monolayer of biotinylated and fluorescently labeled *Cowpea mosaic virus* (CPMV) particles on gold slides imaged via fluorescence microscopy The bilayers are shown on the left and the mixed monolayer on the right. A schematic representation of the CPMV layer structures is shown in the middle. The green and red viral particles depict CPMV labeled with biotin and AlexaFluor (AF) dye 488 and CPMV-biotin-AF568, respectively. Yellow crosses represent streptavidin; orange crosses represent thiol-modified streptavidin. The scale bar is 10 μm.

templates were prepared by either microcontact patterning or scanning- and dip-pen nanolithography [117, 118]. In a subsequent study the dynamics of virion organization on nanopatterned surfaces was analyzed using *in situ* force microscopy; for this study CPMV histidine particles were immobilized via a Ni-NTA interaction [119].

In summary, the feasibility of CPMV particles as nanobuilding blocks for the controlled fabrication of arrays on solid supports has been demonstrated and different immobilization strategies were found to be applicable. The introduction of functional molecules in the one-, two-, or three-dimensional arrays of uniform nanoparticles may provide potential for the development of novel functional devices on the nanoscale.

9.8
Outlook

Nanotechnology seeks to mimic what nature has achieved, precision at the nanometer level down to the atomic level. In recent years materials scientists have recognized biological structures and biomolecules as promising tools for nanobiotechnological applications, and a set of biomolecules has been evaluated, such as nucleic acid, proteins, and especially viruses. As described above, the plant virus CPMV has been studied extensively, as have other viral particles. Further, the use of the virion as a tool for possible applications such as bioimaging and the use of the particles as reporter molecules for sensors or arrays have been demonstrated. A new field is emerging, a highly interdisciplinary area which involves collaborations between virologists, chemists, physicists, and materials scientists. It is exciting at the virus–chemistry interface.

9.9
Acknowledgments

I would like to thank my PhD studentship supervisors Dr. Dave Evans and Dr. George Lomonossoff. The BBSRC and EU grant MEST-CT-2004-504273 are acknowledged for funding.

References

1 Feynman, R.P. (1960) There's Plenty of Room at the Bottom An Invitation to Enter a New Field of Physics. http://nanoparticles.org/pdf/Feynman.pdf (accessed 28/08/07).
2 Taniguchi, N. (1974) Proceedings of the International Conference on Production Engineering, Part II, Japan Society of Precision Engineering, Tokyo.
3 Drexler, E. (1986) *Engines of Creation The Coming Era of Nanotechnology*, Anchor Books, New York.
4 Ozin, G.A. and Arsenault, A.C. (2005) *Nanochemistry*, The Royal Society of Chemistry, Cambridge.
5 Seeman, N.C. (2005) *Methods in Molecular Biology (Clifton, N.J.)*, **303**, 143–66.
6 Niemeyer, C.M. (2004) *Nanobiotechnology: Concepts, Applications and Perspectives* (eds C.M. Niemeyer and C.A. Mirkin), Wiley-VCH Verlag GmbH, Weinheim, pp. 227–54.
7 Niemeyer, C.M. (2001) *Angewandte Chemie (International Ed. in English)*, **40**, 4128–58.
8 Thaxton, C.S. and Mirkin, C.A. (2004) *Nanobiotechnology: Concepts, Applications and Perspectives* (eds C.M. Niemeyer and C.A. Mirkin), Wiley-VCH Verlag GmbH, Weinheim, pp. 288–307.
9 Mirkin, C.A., Letsinger, R.L., Mucic, R.C. and Storhoff, J.J. (1996) *Nature*, **382**, 607–9.
10 Sleytr, U.B., Egelseer, E.-M., Pum, D. and Schuster, B. (2004) *Nanobiotechnology: Concepts, Applications and Perspectives* (eds C.M. Niemeyer and C.A. Mirkin), Wiley-VCH Verlag GmbH, Weinheim, pp. 77–92.
11 Peelle, B.R., Krauland, E.M., Wittrup, K.D. and Belcher, A.M. (2005) *Acta Biomaterialia*, **1**, 145–54.
12 Flenniken, M., Allen, M. and Douglas, T. (2004) *Chemistry and Biology*, **11**, 1478–80.
13 Kowshik, M., Deshmukh, N., Vogel, W., Urban, J., Kulkarni, S.K. and Paknikar, K.M. (2002) *Biotechnol Bioeng*, **78**, 583–8.
14 Sweeney, R.Y., Mao, C., Gao, X., Burt, J.L., Belcher, A.M., Georgiou, G. and Iverson, B.L. (2004) *Chemistry and Biology*, **11**, 1553–9.
15 Li, M., Wong, K.K.W. and Mann, S. (2003) *Chemistry of Materials: A Publication of the American Chemical Society*, **11**, 23–6.
16 Kramer, R.M., Carter, C., Li, D.C., Stone, M.O. and Naik, R.R. (2004) *Journal of the American Chemical Society*, **126**, 13282–6.
17 Douglas, T. and Stark, V.T. (2000) *Inorganic Chemistry*, **39**, 1828–30.
18 Meldrum, F.C., Heywood, B.R. and Mann, S. (1992) *Science*, **257**, 522–3.
19 Flenniken, M.L., Willits, D.A., Brumfield, S., Young, M.J. and Douglas, T. (2003) *Nano Letters*, **3**, 1573–6.
20 Flenniken, M.L., Liepold, L.O., Crowley, B.E., Willits, D.A., Young, M.J. and Douglas, T. (2005) *Chemical Communications*, 447–9.

21 Klem, M.T., Willits, D., Solis, D.J., Belcher, A.M., Young, M. and Douglas, T. (2005) *Advanced Functional Materials*, **15**, 1489–94.

22 Singh, P., Gonzalez, M.J. and Manchester, M. (2006) *Drug Development Research*, **67**, 23–41. Dev

23 Arora, P.S. and Kirshenbaum, K. (2004) *Chemistry and Biology*, **11**, 418–20.

24 Bittner, A.M. (2005) *Die Naturwissenschaften*, **92**, 51–64.

25 Flynn, C.E., Lee, S.-W., Peelle, B.R. and Belcher, A.M. (2003) *Acta Materialia*, **51**, 5867–80.

26 Douglas, T. and Young, M. (2006) *Science*, **312**, 873–5.

27 Lee, S.W., Mao, C., Flynn, C.E. and Belcher, A.M. (2002) *Science*, **296**, 892–5.

28 Lee, S.W., Lee, S.K. and Belcher, A.M. (2003) *Advanced Materials (Deerfield Beach, Fla.)*, **15**, 689–92.

29 Mao, C., Solis, D.J., Reiss, B.D., Kottmann, S.T., Sweeney, R.Y., Hayhurst, A., Georgiou, G., Iverson, B. and Belcher, A.M. (2004) *Science*, **303**, 213–7.

30 Nam, K.T., Peelle, B.R., Lee, S.W. and Belcher, A.M. (2004) *Nano Letters*, **4**, 23–7.

31 Suh, K.Y., Khademhosseini, A., Jon, S. and Langer, R. (2006) *Nano Letters*, **6**, 1196–201.

32 Lee, S.W., Woods, B.W. and Belcher, A.M. (2003) *Langmuir: The ACS Journal of Surfaces and Colloids*, **18**, 1592–8.

33 Nam, K.T., Kim, D.W., Yoo, P.J., Chiang, C.Y., Meethong, N., Hammond, P.T., Chiang, Y.M. and Belcher, A.M. (2006) *Science*, **312**, 885–8.

34 Anderson, E.A., Isaacman, S., Peabody, D.S., Wang, E.Y., Canary, J.W. and Kirshenbaum, K. (2006) *Nano Letters*, **6**, 1160–4.

35 Hooker, J.M., Kovacs, E.W. and Francis, M.B. (2004) *Journal of the American Chemical Society*, **126**, 3718–9.

36 Peabody, D.S. (2003) *Journal of Nanobiotechnology*, **1**, 5.

37 Liu, C.M., Jin, Q., Sutton, A. and Chen, L. (2005) *Bioconjugate Chemistry*, **16**, 1054–7.

38 Liu, C.M., Chung, S.-H., Jin, Q., Sutton, A., Yan, F., Hoffmann, A., Kay, B.K., Bader, S.D., Makowski, L. and Chen, L. (2006) *J Magn Magn Mater*, **302**, 47–51.

39 Gillitzer, E., Suci, P., Young, M. and Douglas, T. (2006) *Small*, **2**, 962–6.

40 Allen, M., Bulte, J.W., Liepold, L., Basu, G., Zywicke, H.A., Frank, J.A., Young, M. and Douglas, T. (2005) *Magnetic Resonance in Medicine*, **54**, 807–12.

41 Gillitzer, E., Willits, D., Young, M. and Douglas, T. (2002) *Chemical Communications*, 2390–1.

42 Klem, M.T., Willits, D., Young, M. and Douglas, T. (2003) *Journal of the American Chemical Society*, **125**, 10806–7.

43 Douglas, T. and Young, M. (1998) *Nature*, **393**, 152–5.

44 Douglas, T. and Young, M. (1999) *Advanced Materials (Deerfield Beach, Fla.)*, **11**, 679–81.

45 Douglas, T., Strable, E. and Willits, D. (2002) *Advanced Materials (Deerfield Beach, Fla.)*, **14**, 415–8.

46 Slocik, J.M., Stone, M.O. and Naik, R.R. (2005) *Small*, **1**, 1048–52.

47 Kuncicky, D.M., Naik, R.R. and Velev, O.D. (2006) *Small*, **2**, 1462–6.

48 Tseng, R.J., Tsai, C., Ouyang, L., Ma, J., Ozkan, C.S. and Yang, Y. (2006) *National Nanotechnolgy*, **1**, 72–7.

49 Knez, M., Sumser, M.P., Bittner, A.M., Wege, C., Jeske, H., Hoffmann, D.M., Kuhnke, K. and Kern, K. (2004) *Langmuir: The ACS Journal of Surfaces and Colloids*, **20**, 441–7.

50 Knez, M., Sumser, M., Bittner, A.M., Wege, C., Jeske, H., Martin, T.P. and Kern, K. (2004) *Advanced Functional Materials*, **14**, 116–24.

51 Knez, M., Bittner, A.M., Boes, F., Wege, C., Jeske, H., Maiß, E. and Kern, K. (2003) *Nano Letters*, **3**, 1079–82.

52 Demir, M. and Stockwell, M.H.B. (2002) *Journal of Nanobiotechnology*, **13**, 541–4.

53 Fowler, C.E., Shenton, W., Stubbs, G. and Mann, S. (2001) *Advanced Materials (Deerfield Beach, Fla.)*, **13**, 1266–9.

54 Shenton, W., Douglas, T., Young, M., Stubbs, G. and Mann, S. (1999) *Advanced Materials (Deerfield Beach, Fla.)*, **11**, 253–156.

55 Dujardin, E., Peet, C., Stubbs, G., Culver, J.N. and Mann, S. (2003) *Nano Letters*, **3**, 413–7.
56 Fujikawa, S. and Kunitake, T. (2003) *Langmuir: The ACS Journal of Surfaces and Colloids*, **19**, 6545–52.
57 Yi, H., Nisar, S., Lee, S.Y., Powers, M.A., Bentley, W.E., Payne, G.F., Ghodssi, R., Rubloff, G.W., Harris, M.T. and Culver, J.N. (2005) *Nano Letters*, **5**, 1931–6.
58 Schlick, T.L., Ding, Z., Kovacs, E.W. and Francis, M.B. (2005) *Journal of the American Chemical Society*, **127**, 3718–23.
59 Dragnea, B., Chen, C., Kwak, E.S., Stein, B. and Kao, C.C. (2003) *Journal of the American Chemical Society*, **125**, 6374–5.
60 Chen, C., Daniel, M.C., Quinkert, Z.T., Stein, M., De, B., Bowman, V.D., Chipman, P.R., Rotello, V.M., Kao, C.C. and Dragnea, B. (2006) *Nano Letters*, **6**, 611–5.
61 Dixit, S.K., Goicochea, N.L., Daniel, M.C., Murali, A., Bronstein, L., Stein, M., De, B., Rotello, V.M., Kao, C.C. and Dragnea, B. (2006) *Nano Letters*, **6**, 1993–9.
62 Loo, L., Guenther, R.H., Basnayake, V.R., Lommel, S.A. and Franzen, S. (2006) *Journal of the American Chemical Society*, **128**, 4502–3.
63 Radloff, C., Vaia, R.A., Brunton, J., Bouwer, G.T. and Ward, V.K. (2005) *Nano Letters*, **5**, 1187–91.
64 Lin, T. and Johnson, J.E. (2003) *Advances in Virus Research*, **62**, 167–239.
65 Lomonossoff, G.P. and Shanks, M. (1999) Comoviruses (Comoviridae). In *Encyclopaedia of Virology*, 2nd edn, Academic Press.
66 Lomonossoff, G.P. and Shanks, M. (1983) *The EMBO Journal*, **2**, 2253–85.
67 van Wezenbeek, P., Verver, J., Harmsen, J., Vos, P. and Van Kammen, A. (1983) *The EMBO Journal*, **2**, 941–6.
68 Bancroft, J.B. (1962) *Virology*, **16**, 419–27.
69 Bruening, G. and Agrawal, H.O. (1967) *Virology*, **32**, 306–20.
70 Wu, G.J. and Bruening, G. (1971) *Virology*, **46**, 596–612.
71 Lomonossoff, G.P. and Johnson, J.E. (1991) *Progress in Biophysics and Molecular Biology*, **55**, 107–37.
72 Schmidt, T., Johnson, J.E. and Phillips, W.E. (1983) *Virology*, **127**, 65–73.
73 Lin, T., Chen, Z., Usha, R., Stauffacher, C.V., Dai, J.B., Schmidt, T. and Johnson, J.E. (1999) *Virology*, **265**, 20–34.
74 Rossmann, M.G. and Johnson, J.E. (1989) *Annual Review of Biochemistry*, **58**, 533–73.
75 Wellink, J. (1998) *Methods in Molecular Biology (Clifton, N.J.)*, **81**, 205–9.
76 Dessens, J.T. and Lomonossoff, G.P. (1993) *The Journal of General Virology*, **74**, 889–92.
77 Lin, T., Porta, C., Lomonossoff, G.P. and Johnson, J.E. (1996) *Folding and Design*, **1**, 179–87.
78 Lomonossoff, G.P. and Johnson, J.E. (1996) *Current Opinion in Structural Biology*, **6**, 176–82.
79 Wang, Q., Lin, T., Tang, L., Johnson, J.E. and Finn, M.G. (2002) *Angewandte Chemie (International Ed. in English)*, **41**, 459–62.
80 Wang, Q., Lin, T., Johnson, J.E. and Finn, M.G. (2002) *Chemistry and Biology*, **9**, 813–9.
81 Chatterji, A., Ochoa, W.F., Ueno, T., Lin, T. and Johnson, J.E. (2005) *Nano Letters*, **5**, 597–602.
82 Chatterji, A., Ochoa, W.F., Paine, M., Ratna, B.R., Johnson, J.E. and Lin, T. (2004) *Chemistry and Biology*, **11**, 855–63.
83 Steinmetz, N.F., Lomonossoff, G.P. and Evans, D.J. (2006) *Langmuir: The ACS Journal of Surfaces and Colloids*, **22**, 3488–90.
84 Meunier, S., Strable, E. and Finn, M.G. (2004) *Chemistry and Biology*, **11**, 319–26.
85 Taylor, K.M., Spall, V.E., Butler, P.J. and Lomonossoff, G.P. (1999) *Virology*, **255**, 129–37.
86 Wang, Q., Raja, K.S., Janda, K.D., Lin, T. and Finn, M.G. (2003) *Bioconjugate Chemistry*, **14**, 38–43.
87 Porta, C., Spall, V.E., Lin, T., Johnson, J.E. and Lomonossoff, G.P. (1996) *Intervirology*, **39**, 79–84.
88 Lomonossoff, G.P. and Hamilton, W.D. (1999) *Current Topics in Microbiology and Immunology*, **240**, 177–89.

89 Taylor, K.M., Lin, T., Porta, C., Mosser, A.G., Giesing, H.A., Lomonossoff, G.P. and Johnson, J.E. (2000) *Journal of Molecular Recognition: JMR*, **13**, 71–82.

90 Wang, Q., Kaltgrad, E., Lin, T., Johnson, J.E. and Finn, M.G. (2002) *Chemistry and Biology*, **9**, 805–11.

91 Wang, Q., Chan, T.R., Hilgraf, R., Fokin, V.V., Sharpless, K.B. and Finn, M.G. (2003) *Journal of the American Chemical Society*, **125**, 3192–3.

92 Russell, J.T., Lin, Y., Boker, A., Carl, L., Su, P., Zettl, H., Sill, J., He, K., Tangirala, R., Emrick, T., Littrell, K., Thiyagarajan, P., Cookson, D., Fery, A., Wang, Q. and Russell, T.P. (2005) *Angewandte Chemie (International Ed. in English)*, **44**, 2420–6.

93 Martin, B.D., Soto, C.M., Blum, A.S., Sapsford, K.E., Whitley, J.L., Johnson, J.E., Chatterji, A. and Ratna, B.R. (2006) *Journal of Nanoscience and Nanotechnology*, **6**, 2451–60.

94 Blum, A.S., Soto, C.M., Wilson, C.D., Cole, J.D., Kim, M., Gnade, B., Chatterji, A., Ochoa, W.F., Lin, T.W., Johnson, J.E. and Ratna, B.R. (2004) *Nano Letters*, **4**, 867–70.

95 Blum, A.S., Soto, C.M., Wilson, C.D., Brower, T.L., Pollack, S.K., Schull, T.L., Chatterji, A., Lin, T., Johnson, J.E., Amsinck, C., Franzon, P., Shashidhar, R. and Ratna, B.R. (2005) *Small*, **1**, 702–6.

96 Steinmetz, N.F., Lomonossoff, G.P. and Evans, D.J. (2006) *Small*, **2**, 530–3.

97 Portney, N.G., Singh, K., Chaudhary, S., Destito, G., Schneemann, A., Manchester, M. and Ozkan, M. (2005) *Langmuir: The ACS Journal of Surfaces and Colloids*, **21**, 2098–103.

98 Rae, C.S., Khor, I.W., Wang, Q., Destito, G., Gonzalez, M.J., Singh, P., Thomas, D.M., Estrada, M.N., Powell, E., Finn, M.G. and Manchester, M. (2005) *Virology*, **343**, 224–35.

99 Wu, C., Barnhill, H., Liang, X., Wang, Q. and Jiang, H. (2005) *Optics Communications*, **255**, 366–74.

100 Lewis, J.D., Destito, G., Zijlstra, A., Gonzalez, M.J., Quigley, J.P., Manchester, M. and Stuhlmann, H. (2006) *Nature Medicine*, **12**, 354–60.

101 Raja, K.S., Wang, Q., Gonzalez, M.J., Manchester, M., Johnson, J.E. and Finn, M.G. (2003) *Biomacromolecules*, **4**, 472–6.

102 Raja, K.S., Wang, Q. and Finn, M.G. (2003) *Chembiochem: A European Journal of Chemical Biology*, **4**, 1348–51.

103 Sen Gupta, S., Raja, K.S., Kaltgrad, E., Strable, E. and Finn, M.G. (2005) *Chemical Communications*, 4315–7.

104 Chatterji, A., Ochoa, W., Shamieh, L., Salakian, S.P., Wong, S.M., Clinton, G., Ghosh, P., Lin, T. and Johnson, J.E. (2004) *Bioconjugate Chemistry*, **15**, 807–13.

105 Soto, C.M., Blum, A.S., Vora, G.J., Lebedev, N., Meador, C.E., Won, A.P., Chatterji, A., Johnson, J.E. and Ratna, B.R. (2006) *Journal of the American Chemical Society*, **128**, 5184–9.

106 Sapsford, K.E., Soto, C.M., Blum, A.S., Chatterji, A., Lin, T., Johnson, J.E., Ligler, F.S. and Ratna, B.R. (2006) *Biosensors and Bioelectronics*, **21**, 1668–73.

107 Strable, E., Johnson, J.E. and Finn, M.G. (2004) *Nano Letters*, **4**, 1385–9.

108 Medintz, I.L., Sapsford, K.E., Konnert, J.H., Chatterji, A., Lin, T., Johnson, J.E. and Mattoussi, H. (2005) *Langmuir: The ACS Journal of Surfaces and Colloids*, **21**, 5501–10.

109 Steinmetz, N.F., Calder, G., Lomonossoff, G.P. and Evans, D.J. (2006) *Langmuir: The ACS Journal of Surfaces and Colloids*, **22**, 10032–7.

110 Porta, C., Spall, V.E., Loveland, J., Johnson, J.E., Barker, P.J. and Lomonossoff, G.P. (1994) *Virology*, **202**, 949–55.

111 van Staveren, D.R. and Metzler-Nolte, N. (2004) *Chemical Reviews*, **104**, 5931–85.

112 Adams, R.N. (1969) *Electrochemistry at Solid Electrodes*, Marcel Dekker, Inc, New York.

113 Whitesides, G.M. and Grzybowski, B. (2002) *Science*, **295**, 2418–21.

114 Gates, B.D., Stewart, Q., Xu, M., Ryan, D., Willson, C.G. and Whitesides, G.M. (2005) *Chemical Reviews*, **105**, 1171–96.

115 Lowe, C.R. (2000) *Current Opinion in Structural Biology*, **10**, 428–34.
116 Weber, P.C., Ohlendorf, D.H., Wendoloski, J.J. and Salemme, F.R. (1989) *Science*, **243**, 85–8.
117 Cheung, C.L., Camarero, J.A., Woods, B.W., Lin, T., Johnson, J.E. and De Yoreo, J.J. (2003) *Journal of the American Chemical Society*, **125**, 6848–9.
118 Smith, J.C., Lee, K., Wang, Q., Finn, M.G., Johnson, J.E., Mrksich, M. and Mirkin, C.A. (2003) *Nano Letters*, **2**, 883–6.
119 Cheung, C.L., Chung, S.W., Chatterji, A., Lin, T., Johnson, J.E., Hok, S., Perkins, J. and De Yoreo, J.J. (2006) *Journal of the American Chemical Society*, **128**, 10801–7.

10
New Calorimetric Approaches to the Study of Soft Matter 3D Organization

J.M. Nedelec and M. Baba

10.1
Introduction

Nature has always been a source of inspiration for chemists and materials scientists. Among the most striking features of natural materials, the three-dimensional (3D) organization of matter at different length scales has attracted considerable interest during the last 20 years. So-called hierarchical materials are found in many natural systems like bone tissue, wood or nacre and are becoming a major goal for material science. A particular example of 3D organization is the creation of hierarchical porous structures in which matter and voids are organized in a regular and controlled pattern. The siliceous exoskeletons of diatoms provide a magnificent example (Figure 10.1) of such structure occurring in living organisms. In this case, the pores are organized in a beautiful and functional arrangement at different levels exhibiting macro-, meso-, and microporosity. Synthetic approaches are currently being extensively developed to produce so called bio-inspired structures.

The control of 3D organization is also a major challenge in soft matter such as polymers and gels. In this case, the function of the material is controlled by the reticulation level and arrangement of chains. As far as polymers are concerned, the crosslinking level is strongly dependent on the history of the material, from processing to aging. This crosslinking level greatly affects the mechanical properties of the material and knowledge of its evolution is therefore crucial for prediction of long-term durability. Numerous experimental techniques have been used for the study of spatial organization in polymers and gels. In most cases, the information is derived at a molecular or macroscopic level. The distance characteristic of the network is typically found between 1 and 1000 nm (distance between knots) in the mesoscopic range.

The aim of this contribution is to highlight new applications of calorimetric techniques to study soft matter organization directly on this length scale. Two original techniques, namely thermoporosimetry (TPM) and photo-differential scanning calorimetry (photoDSC) will be presented from both the theoretical and experimental points of view. After giving the state of the art of the two techniques,

Figure 10.1 Examples of siliceous exoskeletons of diatoms, showing hierarchical porosity. Photograph courtesy of Prof J. Livage, Collège de France.

several examples taken from the literature will illustrate how these calorimetric techniques can provide a useful tool for the study of polymers and gel networks. Section 10.2 will present the general basis of thermoporosimetry and Section 10.3 will present its applications to soft materials. Section 10.4 describes the principle of photoDSC and its use for the study of photo-initiated reactions, while Section 10.5 gives some examples of the use of photoDSC for the study of polymer aging.

10.2
Transitions in Confined Geometries

Phase transitions are strongly affected by the confinement effect. Among the different techniques using this phenomenon, thermoporosimetry is the most popular. By measuring the shift of the transition temperature of confined liquids, TPM can bring precious information on the morphology of the confining medium including porosity for rigid materials and the network mesh size for soft materials.

In this section, we will first give the theoretical basis for the understanding of transitions in confined geometries and then describe the different techniques based upon this phenomenon, in particular thermoporosimetry will be highlighted.

10.2.1
Theoretical Basis

10.2.1.1 Confinement Effect on Triple-point Temperature

If one considers the thermodynamic equilibrium between gas, liquid, and solid phases in a cylindrical pore, combination of the Clausius–Clapeyron and Kelvin equations yields

$$\ln \frac{T_p}{T_0} = -\frac{2}{\Delta H_m r}(V_l \gamma_l - V_s \gamma_s) \tag{10.1}$$

which relates the melting temperature T_p inside the pore to T_0, the normal melting temperature, and to $1/r$. ΔH_m is the enthalpy of melting, V_l and V_s are the molar volumes of the liquid and solid, respectively, and γ_l, γ_s are the corresponding surface energies. Making the assumption that the change in molar volume can be neglected, this equation can be written

$$\ln \frac{T_p}{T_0} = \frac{2 V_m \gamma_{sl}}{\Delta H_m r} \tag{10.2}$$

Writing $\Delta T = T_p - T_0$, a serial development of Equation 10.2 in the first order yields

$$\Delta T = T_p - T_0 = \frac{2 T_0 V_m \gamma_{sl}}{\Delta H_m r} \tag{10.3}$$

Equation 10.3 is known as the Gibbs–Thomson [1] equation and is classically used experimentally to calculate r from the measurement of ΔT. In the case of a cylindrical pore of radius r_p, Equation 10.4 can be written

$$\Delta T = T_p - T_0 = \frac{2 T_0 V_m \gamma_{sl} \cos\theta}{\Delta H_m r_p} \tag{10.4}$$

where θ is the contact angle between the liquid and the solid, usually taken to be 0°.

A rigorous treatment of melting in confined geometries considering the three interfaces (including the vapor phase) becomes difficult. This has been discussed extensively by Defay *et al.* [2].

In general, the creation of an adsorbed layer of nonfreezing solvent of thickness t is assumed on the surface of the pores, as depicted in Figure 10.2. In this case, the measured r_p corresponds to $(R_p - t)$, R_p being the true radius of the pore. This thickness t is usually found to be less than 2 nm (usually a few molecular layers). This is very important from a practical point of view since t will give the lower limit of measurable pore size. In other words, solvent confined in pores smaller

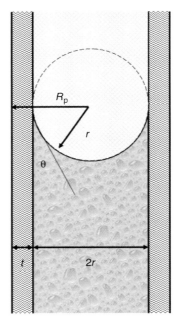

Figure 10.2 Scheme of liquid–solid equilibrium in a cylindrical pore of radius R_p with the creation of an adsorbed layer of thickness t.

than t will not be observable since it will not freeze. This layer must be taken into account for precise derivation of R_p through ΔT measurement.

This treatment of melting in confined geometries is obviously oversimplified and the molecular nature of the phases and interactions between the adsorbent walls and the adsorbate should be taken into account by considering not only the surface energies but also the exact nature of the solid phase (structure, crystalline orientation, crystal defects, and so on).

For a more complete description of freezing and melting in confined geometries, a comprehensive review can be found in the literature [3].

10.2.2
Porosity Measurements via Determination of the Gibbs–Thomson Relation

As shown in Equation 10.4, the depression of the melting point of a given confined solvent is related to the geometry of the pores of the confining material. In principle, measurement of ΔT can give access to the pore size. Three main techniques have been developed to measure porosity in solids via the use of the Gibbs–Thomson equation: thermoporosimetry, NMR cryporometry and surface force apparatus. These techniques are secondary methods since they require preliminary determination of the evolution of ΔT as a function of R_p for a given solvent. Molecular simulations and theoretical methods have also attracted considerable interest for the study of the effect of confinement on melting. This falls beyond the scope of this paper and a comprehensive review can be found in the recent literature [4].

10.2.2.1 Thermoporosimetry

In his seminal work of 1955, Kuhn [5] was the first to propose using the measurement of ΔT by calorimetric techniques to characterize the confining medium; this can be considered the birth of thermoporometry (or thermoporosimetry (TPM)as will be used in the following). TPM has then been further described by Fagerlund and co-workers [6] but Brun and co-workers [7] have gave a real boost to the technique in the late 1970s by giving detailed description of the underlying thermodynamics and proposing various applications. TPM is now recognized as a valuable technique but it is still rarely used compared with gas sorption, which remains the gold standard for porosity measurements in solids. The aim of Section 10.3 is to describe recent developments of thermoporosimetry and its application to soft materials.

10.2.2.2 NMR Cryoporometry

The first nuclear magnetic resonance (NMR) measurement of confined water has to be credited to Derouane *et al.* [8]. In 1993, Strange and co-workers [9] proposed a new technique that they named NMR cryoporometry for the measurement of porous solids through the measurement of ΔT by NMR spectroscopy. This technique has been further developed [10, 11], especially ^{129}Xe NMR spectroscopy [12], which has demonstrated strong potential in particular through the use of hyperpolarized xenon. The use of NMR imaging techniques coupled with the measurement of ΔT of confined liquids [13–15] gives a very unique possibility to measure porosity of solids with a 3D spatial resolution on a macroscopic scale (100 μm–10 mm).

10.2.2.3 Surface Force Apparatus

The surface force apparatus (SFA) has been used extensively over the past 30 years to measure the force directly as a function of separation between surfaces in liquids and vapors. If the force-measuring spring is replaced with a mechanically more rigid support, the two opposing surfaces become an ideal model pore for the study of confinement effects on phase behavior [16]. A detailed review can be found in reference[3]. Briefly, the shift of the melting temperature ΔT can be related to the size h of the condensate measured with SFA according to

$$\Delta T = \frac{4 V_m T_0 \gamma_{SL}}{h \Delta H_m} \tag{10.5}$$

10.2.3
Thermoporosimetry and Pore Size Distribution Measurement

In a thermoporosimetry experiment, the whole DSC curve is indeed representative of the pore size distribution of a given sample. From Equation 10.5 and the DSC thermogram, the pore size distribution (PSD) can be derived as follows:

$$\frac{d(V_p)}{d(R_p)} = \frac{dQ(T)}{dt} \frac{dt}{d(\Delta T)} \frac{d(\Delta T)}{d(R_p)} \frac{1}{W_a(T)} \qquad (10.6)$$

where V_p (cm^3 g^{-1}) is the pore volume, $dQ(T)/dt$ (W g^{-1}) is the heat flow per gram of the dry porous sample given by the ordinate of DSC thermogram, $dt/d(\Delta T)$ (s K^{-1}) is the reverse of the cooling rate, $d(\Delta T)/d(R_p)$ (K nm^{-1}) is derived from the empirical relationship corresponding to Equation 10.4, and $W_a(T)$ (J cm^{-3}) is the apparent energy as defined by Brun. Equation 10.6 is similar to that proposed by Landry [17]. However, the specific energy of crystallization is temperature-dependent and, in addition, not all of the solvent takes part in the thermal transition since, as stated before, a layer of the solvent remains adsorbed on the internal wall of the pores. Only an apparent energy (W_a) can be calculated as the total heat released by the thermal transition divided by the total volume or mass of the confined solvent.

As state before, TPM remains a secondary technique requiring calibration. The extraordinary progress in the preparation of porous materials with well-controlled pore sizes during the last 20 years is now circumventing this problem. Comparison of PSD derived from gas sorption, TPM, and NMR cryoporometry [18, 19] clearly demonstrates the validity of TPM for the study of nanoporous materials.

Various solvents have been calibrated to provide data for TPM measurements. The first studies dealt essentially with water and benzene. In the 1990s Jackson and McKenna published data for *n*-heptane, *cis*- and *trans*-decaline, cyclohexane, naphthalene, and chlorobenzene [20].

In recent years, numerous solvents have been calibrated for thermoporosimetry. In 2004 Wulf published some data on acetonitrile [21] and extensive work has been performed in our group for CCl$_4$ [22, 23], xylenes, various substituted benzenes [24], linear alkanes [25, 26], cyclohexane [27], dioxane [Unpublished results], and acetone [28].

In many papers, an oversimplified expression known as the Kelvin equation $R_p = A/\Delta T + t$, where A is a constant, has been used. In Figure 10.3, two examples clearly demonstrate that this could lead to large errors in pore size measurement. The data are adapted from [29] for water (a) and from work performed in our group [28] for acetone (b).

From Figure 10.3a, an error as high as 300% can be obtained for water for high ΔT (small pores). For acetone (Figure 10.3b) orthogonal distance regression (ODR), which for each point minimizes the distance between the fitting curve and the experimental points, has been performed on the experimental data. The experimental model has proved to be most suitable for other solvents such as substituted benzenes or alkanes [24, 25].

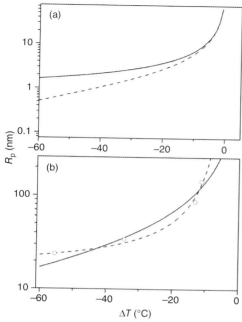

Figure 10.3 Comparison of experimental calibration curves of $R_p = f(\Delta T)$ (dotted lines) with best fit using a simplified Kelvin equation $R_p = A/\Delta T + t$ (solid line). The data are adapted from reference [29] for water (a) and from reference [28] for acetone (b).

Thanks to the calibration curves that we have been proposing for various solvents, thermoporosimetry has become a functional technique. TPM also addresses some major fundamental issues concerning melting/freezing in confined geometries and the underlying energetics and thermodynamics. In the following section we will show how TPM can also be used very efficiently to study less rigid materials like polymers, gels, and hybrids materials. In this sense, TPM appears to be a universal technique directly associated with the physics of confinement.

10.3
Application of Thermoporosimetry to Soft Materials

10.3.1
Analogy and Limitations

In the 1950s, Kuhn had already observed the peculiar behavior of liquids swelling polymeric gels [5]. Indeed, a strong analogy can be proposed between solvent trapped inside the pores of a rigid material and the solvent swelling the chains of a polymeric network. In the case of polymer swelling, the network limits the mobility of the solvent molecules, inducing a confinement effect. As observed for nanoporous materials, the trapped solvent exhibits a shift of its melting/freezing temperature directly related to the size of the confinement domain. The shift ΔT can be associated with the size of the meshes defining the 3D network between successive reticulation knots.

Using TPM as described in Section 10.2 on swollen polymers can give access to the mesh size distribution (MSD) in the swollen state. The ability of the polymer to swell is connected to the reticulation level, which in turn can vary greatly during polymer aging.

From a fundamental point of view, the possible deformation of the network upon freezing of the solvent is an important issue. Possible shrinkage of the material upon swelling liquid crystallization has been reported by Scherer and co-workers [30, 31]. This shrinkage can be particularly important for soft materials like gels and can lead to errors in pore (mesh) size determination. Depending on the bulk modulus of the material K_0 and on the mesh radius R, the error might be very significant (as high as ten times for soft materials and small radius). In any case, even if the swollen gel experiences dimensional fluctuations because of the thermal transition of the liquid-probe, these variations must be small compared to those induced by swelling. Furthermore, mesh size associated with swollen polymeric networks is usually quite large, thus minimizing the error in this case. An example of such a calculation taken from our own work can be found in reference [28], showing negligible effect in the case of polyolefin and silicone. In our opinion, TPM consequently remains an efficient and unique tool with which to compare the relative state of crosslinking of different gel networks. The mesh sizes distributions calculated from the thermoporosimetry formalism reflect the actual state of the sample taking into account the eventual shrinkage and the swelling equilibrium.

Another limitation in applying TPM to gels comes from the necessary extrapolation of calibration curves to high values of R_p not covered by the calibration itself. The use of macroporous reference samples would indeed be very beneficial to ascertain the calibration curves. In the following sections some examples will illustrate the use of TPM to characterize soft materials.

10.3.2
Examples of Use of TPM with Solvent Confined by Polymers and Networks

10.3.2.1 Elastomers
Since the pioneering work of Kuhn, the solvent freezing point depression observed in swollen crosslinked rubbers has been the subject of many works. The observed ΔT can be attributed to two origins. A sizable ΔT is accounted for by the lowering of the thermodynamic potential of solvent molecules in a polymer solution derived from the Flory theory, and the additional ΔT observed for crosslinked rubbers has been attributed to confinement effects. In 1991, Jackson and McKenna [32] studied

both the freezing and melting of solvent crystals in crosslinked and uncrosslinked natural rubber swollen in benzene. This study confirmed that the anomalous ΔT can be accounted for by small crystallite size.

More recently, we proposed a detailed study [33] of natural rubber filled with zinc oxide, sulfur, and carbon black. The sample was especially designed to contain a gradient of sulfur in order to obtain, after curing, a gradient of crosslinking in the material. The sample was then cut into several slices, each of which was of different crosslinking density and submitted to TPM analysis with n-heptane. The originality of this work, which attests the power of TPM, relies in the fact that the experiments were performed on a formulation (polymer + fillers) which correspond to the final application. In this sense, the technique is unique, in particular when compared to Fourier transform infrared (FTIR) spectroscopy. The mesh size distributions obtained are plotted in Figure 10.4.

As expected, the distribution shifts toward low R_p values with respect to the depth, revealing a profile of crosslinking inside the elastomer. An estimate of M_C, the critical mass between knots, can be derived using De Gennes theory [34] from the measured value of R_{max}, the maximum of each distribution curve.

In 2006, McKenna and co-workers reviewed recent work on the use of TPM to study polymer heterogeneity [35]. In this work, uncrosslinked and crosslinked polyisoprene was studied with benzene and hexadecane as swelling solvents. The authors were able to distinguish contributions coming from the confinement as described by the Gibbs–Thomson equation and contributions from polymer–solvent interactions described by the Flory–Huggins (FH) theory [36, 37]. For the first time, it was shown that for an uncrosslinked sample an excess shift ΔT is

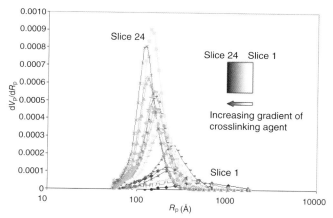

Figure 10.4 Mesh size distributions in a natural rubber sample filled with carbon black and containing a gradient of crosslinking agent. Data were obtained with n-heptane. Reproduced from reference [33], (c) 2004 with permission from Springer.

measured compared to FH theory, and this is interpreted in terms of nanoheterogeneities in the uncrosslinked polymer.

10.3.2.2 Hydrogels

Hydrogels are very important materials for biomedical, pharmaceutical, or membrane applications. Their biocompatibility and low mechanical irritation make them good candidates for applications such as controlled-release pharmaceuticals and implants [38]. These materials are very fragile and can contain large amount of solvent, making their characterization very difficult. Drying the samples can result in large modifications of their architecture. In 2000, Iza et al. [29] described a TPM study of hydrogels prepared from N-(2-hydroxypropyl)methacrylamide (HMPA). Using the swelling water they determined the pore size distribution of the gels. Unexpectedly, two samples originating from the same gel exhibited different PSD with, in particular, a roughly bimodal distribution for one sample. Structural heterogeneity at the nanometric level of the hydrogel could be responsible for such discrepancies. Another possible cause is the alteration of the material during its preparation for TPM measurement. This work demonstrates the interest of the technique for such fragile samples, particularly when compared with gas sorption, which requires drying of the sample that could in turn modify the structure of the material.

10.3.2.3 Polymeric Membranes

Thermoporometry has become a popular tool for measuring pore size distribution in polymer membranes, as illustrated in reviews by Nakao [39] and Kim et al. [40].

In 2000, Hay et al. [41] reported a study of cellulose membranes with water TPM. The authors found a pore size distribution with radii between 5 and 50 nm but noted an important migration of water during DSC measurements.

In 2003, Ksiqzczak et al. [42] used water TPM for the characterization of nitrocellulose prepared by nitration of natural cellulose. The hydrophobic nature of the membrane made the measurements difficult and only partial conclusions were drawn. Despite this, pore size distributions were measured which showed good consistency and confirmed the value of TPM for such studies. Even more recently, Rohman et al. used water TPM to measure pore size distributions in porous polymers networks [43].

10.3.2.4 Crosslinking of Polyolefins

Although they are of major interest for practical applications, polyolefins are usually not studied by TPM because they are difficult to swell. In recent work in our group [44], polyolefin samples presenting various degree of crosslinking were studied by TPM with xylenes. The electron- or γ-irradiated samples presented various swelling ratio (G) measured in p-xylene. Upon increasing the irradiation dose, we observed a decrease of G along with a decrease of the maximum of the MSD (R_{Max}) measured by TPM. These observations are in agreement with the subsequent crosslinking increase upon irradiation. G and R_{Max} decrease correlatively.

The curve $R_{Max} = f(G)$ allows the extrapolation of R_{Max} for $G = 1$, the value corresponding to a dry gel which is usually difficult to measure.

In conclusion, TPM has proved to be a very efficient tool for the characterization of soft materials. The intense development of porous samples with very well controlled pore size distribution, as carried out in our group, allows determination of calibration curves for numerous solvents. In particular, calibrations for solvents able to swell polymeric materials are now available, making TPM a very attractive technique. The simplicity and the low cost associated with TPM are further arguments for extended use of this technique.

10.4
Study of the Kinetics of Photo-initiated Reactions by PhotoDSC

10.4.1
The PhotoDSC Device

photoDSC (differential scanning calorimetry), also named photocalorimetry, combines light irradiation and DSC measurement in a single instrument. Light is brought into the DSC furnace by two symmetrical optical fibers. Arc or electrodeless microwave-powered mercury lamps are commonly used for irradiation. Metal halides and noble gases are frequently added to the mercury vapor in order to shift the spectral output and adapt the source to the requirements of specific applications. The most widely used type of lamp is the medium-pressure mercury lamp enriched with xenon. The possibility of easy control of the temperature (−150 to 500 °C), the nature of the atmosphere (N_2, O_2, air . . .), the light intensity (0–300 mW cm^{-2}), and the irradiation wavelength makes photoDSC a very useful and versatile technique. Furthermore, the monitoring of irradiation time allows photoDSC in pulse mode (short irradiation time, typically 0.6 s) or in prolonged mode (1 minute to several hours). Figure 10.5 shows the principle of this technique and the emission spectrum of a medium-pressure Xe–Hg lamp (Lightingcure LC6).

The light generator is servo-controlled by the DSC software, which permits a choice of both light intensity and duration of the irradiation. The source can be filtered by sapphire disks in order to cut off wavelengths lower than 300 nm and thus to be representative of outdoor aging conditions.

10.4.2
Photocuring and Photopolymerization Investigations

Curing by light irradiation can be applied to a variety of polymers. The acrylic family is a typical example of photosensitive monomers. A well-documented review has recently been dedicated to the curing of composites by ultraviolet radiation [45]. Photocuring time is much shorter than for traditional thermal curing (minutes rather than hours), leading to a significant reduction in the cycle time.

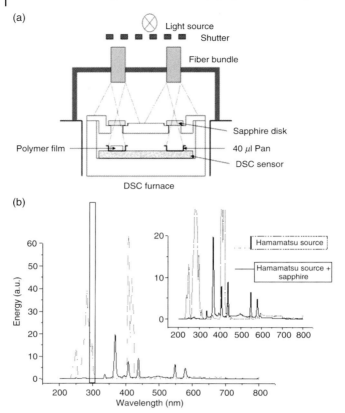

Figure 10.5 Principle of photoDSC. (a) The sample (solid or liquid film) is placed in the DSC pan. Light is introduced by two optical fibers. Two sapphire disks filter the wavelength and complete the thermal insulation of the DSC furnace. (b) Spectrum of the medium-pressure Xe–Hg lamp, not filtered (dashed line) and filtered (solid line) by the sapphire disks.

The most popular field of application of photocuring is the fast drying of varnishes, paints, printing inks, adhesives, printing plates, microcircuits, and optical disks in addition to dental prostheses and rapid prototyping by stereolithography [46, 47]. UV-curing is the most commonly applied method because of such advantages as rapid curing, low curing temperature, possible on-line production, and low energy requirement [48]. Two main curing mechanisms are reported: radical polymerization for acrylic-based resins and cationic polymerization for epoxies and vinyl ether. In both cases the properties of the photocured matrix are determined by its crosslinking density. Khudyakov and co-workers [49, 50] combined photoDSC with curing monitoring and time-resolved electron paramagnetic resonance to follow the course of the photopolymerization of a variety of acrylates. The

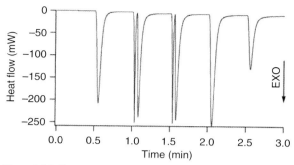

Figure 10.6 PhotoDSC trace obtained during polymerization of VA (10 mg) with BAPO (3%) at 10 °C. The sample was subjected to irradiation by five light pulses, each with a duration of 0.6 s; the pulses started at 0.5, 1.0, 1.5, ... minutes. Reproduced from reference [50], (c) 2001 with permission from the American Chemical Society.

authors described an interesting phenomenon that they named "stumbling polymerization" observed during photopolymerization of vinyl acrylate (VA) in the presence of bis(2,4,6-trimethylbenzoyl)phenylphosphine oxide (BAPO) as photoinitiator. Figure 10.6 illustrates this phenomenon. The fast initial release of heat during photopolymerization is followed by a temporary cessation of the reaction, revealed by the zeroing of the signal.

The authors explained the observed temporary cessation of heat release by the difference between the rate of polymer formation and that of volume relaxation. According to their observations, the difference between the two phenomena leads to a situation in which the free radicals are spatially separated from the unreacted VA molecules. The subsequent spatial relaxation brings them into contact and the reaction starts again.

The same authors used photoDSC to obtain time- or conversion-dependent ratios of the termination and propagation rate coefficients $2k_t/k_p$ for the polymerization of VA. The starting point of this approach is the classical equation derived from formal kinetics:

$$[M]_t/[R]_t - [M]_0/[R]_0 = (2k_t/k_p)t \tag{10.7}$$

where $[M]_0$ and $[R]_0$ are the monomer concentration and the rate of polymerization at $t = 0$ for the dark polymerization. $[M]_t$ and $[R]_t$ are the monomer concentration and the rate of polymerization, respectively, at a given time t.

The rate of polymerization is assumed to be equal to the ordinate of the DSC trace, dQ/dt [51]. At each time t, the concentration of the remaining monomer is proportional to $(Q_{tot} - Q_t)$, with Q_{tot} corresponding to the total heat released when the photopolymerization is completed ($\xi = 100\%$). Equation 10.8 can then be written as follows:

$$\frac{(Q_{\text{tot}} - Q_t)}{dQ/dt} = \left(\frac{2k_t}{k_p}\right) t \qquad (10.8)$$

Q_{tot}, Q_t, and dQ/dt can be derived from the photoDSC trace, which allows the calculation of $2k_t/k_p$ as illustrated in Figure 10.7.

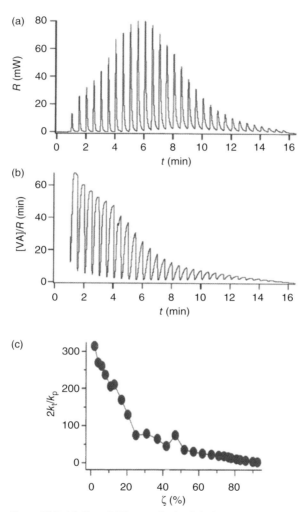

Figure 10.7 (a) PhotoDSC trace obtained during polymerization of VA at 3 °C in the absence of a photoinitiator. (b) Calculated ratio of [VA]/R vs. time. (c) Dependence of $2k_t/k_p$ vs. ξ for the same experiment. Reproduced from reference [50], © 2001 with permission from the American Chemical Society.

PhotoDSC has also been used to study the kinetics of acrylates with functionality of 1 to 6 [52]. The author suggested an autocatalytic or autoaccelerated reaction.

10.5
Accelerated Aging of Polymer Materials

Photo-oxidation is the most common cause of polymer degradation in outdoor conditions. Light–polymer interactions lead to drastic changes in the physical and chemical properties of the material. Many studies have been devoted to the comprehension [53, 54] and the prediction of complex phenomena occurring upon the photo-aging of polymers. Several accelerated photo-oxidation devices have been proposed. SEPAP [55], SunTest [56], Xenotest [57] and WeatherOmeter [58] are the most commonly used. The common characteristic of these devices is that the irradiation stage is disconnected from the aging assessment. Usually, the sample is irradiated, removed from the exposure device, and analyzed by nondestructive methods such as IR spectroscopy.

We recently proposed the use of photoDSC to study polymer photo-aging. photoDSC combines the exposure and the analysis of the polymer sample inside the same DSC furnace. Indeed, the ability of a semicrystalline polymer to crystallize is greatly affected by its photo-aging. Notably, crosslinking and chain scissions, by altering the macromolecular chain mobility, decrease the "crystallizability" of the polymer. PhotoDSC, by following the change of the temperature and the heat of crystallization, is a good tool for studying photo-aging of semicrystalline polymers. In addition, photo-oxidation leads to the formation of macroperoxides inside the polymeric matrix. The subsequent thermal decomposition of these macroperoxides gives an original way to follow the course of photo-oxidation.

10.5.1
Study of Crosslinking of Polycyclooctene

We have published a detailed study of polycyclooctene aging by photoDSC [59]. Figure 10.8 describes the typical temperature and irradiation program that was used in our study.

10.5.1.1 Correlation between Oxidation and Crystallinity

Photo-oxidation leads to the formation of carbonylic products and this is classically monitored by vibrational spectroscopy. To investigate the relation between the accumulation of the oxygenated photoproducts and the change in the crystallinity of polycyclooctene, the decrease of the heat of crystallization was compared with the rise of the concentration of carbonyl function (1721 cm^{-1} band) as displayed in Figure 10.9. The enthalpy of crystallization falls at early stages of irradiation before significant accumulation of the carbonyl. Assuming that the decrease of the

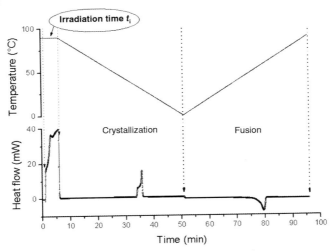

Figure 10.8 Temperature program and thermogram of a photoDSC experiment. The first stage (2 minutes in darkness) allows the completion of melting of the sample. The irradiation period (t_i) follows and then the light is switch off. Crystallization and fusion are recorded in darkness.

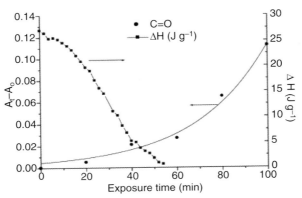

Figure 10.9 Polycyclooctene irradiated in the photoDSC system at 35 °C under air. Left ordinate (●): IR kinetic evolution of carbonyl band (1721 cm^{-1}). Right ordinate (■): change in crystallization heat. Reproduced from reference [59], © 2005 with permission from Elsevier.

polymer crystallizability is related to the network densification, polycyclooctene seems to predominantly undergo the crosslinking reactions before a noticeable formation of oxygenated photoproducts. This behavior has been mentioned previously in other dienic elastomers such as polybutadiene [60] and ethylene-propylene-diene monomer (EPDM) [61].

10.5.1.2 Crosslinking and Crystallizability

The lowering of both the heat and the temperature of crystallization of polycylooctene is related to morphological changes in the polymeric matrix. Instead of "crystallinity," it is more correct to invoke the "crystallizability," which means the ability of the polymer to crystallize from the melt. To clarify the origin of the decrease of the crystallizability of polycyclooctene, the compound has been chemically and progressively crosslinked and the DSC trace corresponding to its crystallization recorded [62]. Figure 10.10 presents the results of this experiment together with the temperature program that was selected. A mixture of elastomer and dicumyl peroxide (DCP, 3w/w %) was prepared by co-dissolution in chloroform and solvent extraction under vacuum. The mixture was then subjected to the DSC program described in Figure 10.10a. The sample was maintained at 130 °C for a certain curing time (10 minutes for the first cycle, 20 minutes for the second and so on). At this temperature, DCP just starts to decompose and crosslinks the elastomer. After 10 minutes, the temperature was quickly lowered (at 50 °C min^{-1}) to 50 °C in order to stop the thermal decomposition of DCP. Between 50 and −20 °C

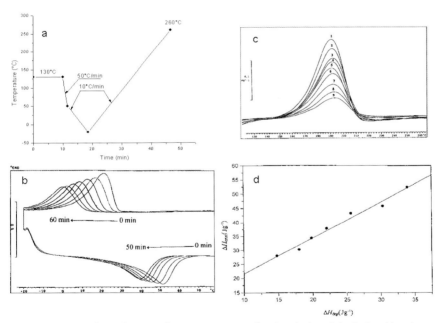

Figure 10.10 Correlation between the crystallizability of polycyclooctene and its crosslinking level. A mixture of polycyclooctene and dicumyl peroxide (DCP; 3% w/w) was subjected to the temperature program shown in (a). The sample is heated at 130° for t_i minutes (10 minutes in this example) for crosslinking. (b) The crystallization (up) and the fusion (down) peaks of the partially crosslinked sample. (c) The peaks of thermal decomposition of the remaining DCP. (d) A representation of the heat of crystallization vs. the heat of the residual DCP. Adapted from reference [62], © 2003 with permission from John Wiley & Sons, Ltd.

the cooling rate was 10 °C min^{-1}, allowing recording of the crystallization trace of the sample. The same rate (10 °C min^{-1}) was adopted to reheat the sample to 260 °C. During this stage two signals were recorded: one associated with the melting of polycyclooctene (around 45 °C, Figure 10.10b) and the second with thermal decomposition of residual DCP (around 170 °C, Figure 10.10c). The same experiment was repeated for several samples. The heat of crystallization and of decomposition of residual DCP can be calculated from the area under the corresponding peaks. Figure 10.10d demonstrates a linear relationship between these two enthalpies, revealing that the amount of decomposed DCP corresponds to the loss of crystallizability of the semicrystalline elastomer. The thermal decomposition of DCP being assumed to provoke the crosslinking, it can be concluded that the lessening of the crystallizability reveals the same phenomenon and thus that crystallizability can be used as a good indicator of the crosslinking level in the material.

Chain-scissions and crosslinking processes provoke a large modification in macromolecular mobility. By NMR experiments, it was shown [63] that the relaxation time (T_2) was drastically modified when the polymer was chemically crosslinked. The polymeric chain mobility decreases because of the increase of crosslinking density. The melting process involves the crystalline part of the polymer. Consequently, when the crosslinked polymer melts, the crystalline part is in its liquid form and it soaks up the amorphous fraction constituting the tridimensional network. When this melted polymer is cooled down, it solidifies but its ability to crystallize depends on the chain mobility, which is reduced when the crosslinking density increases. So, when the exposure time or the light intensity increase, the "crystallizability" decreases. Consequently, the process of germination becomes more and more difficult and the rate of the crystal growth is slowed down. These facts are confirmed by the exotherms of crystallization, which reveal a shift to low temperatures and a noticeable broadening and decrease upon exposure.

10.5.1.3 Photo-aging Study by Macroperoxide Concentration Monitoring

The first stage of polymer oxidation is their combination with a dioxygen biradical. The resulting peroxy macroradical reacts with the surrounding radicular species, leading to the formation of peroxides or hydroperoxides. One of the original applications of photoDSC consists in the combination of the irradiation step and the measurement, by thermal decomposition, of the amount of accumulated peroxides.

A sample of polycyclooctene was irradiated at 60 °C for a time, t_i, selecting 90 mW cm^{-2} as light intensity. After t_i minutes (ranging from 0 to 40 minutes), the light was cut off and the temperature was raised, at 10 °C min^{-1}, to 250 °C in order to decompose the formed peroxides around 170 °C. Figure 10.11a gives some typical thermograms of this decomposition. For clarity, only seven thermograms are presented. It can be noted that the area under the peaks (proportional to the decomposition heat) increases at first time before remaining practically constant as observed in Figure 10.11b.

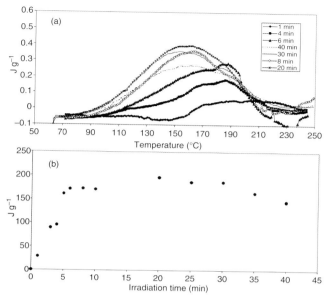

Figure 10.11 Thermal decomposition of macroperoxides accumulated during the photo-oxidation of polycyclooctene irradiated with 90 mW cm^{-2} at 60 °C, under air atmosphere. (a) Thermograms of thermal decomposition of macroperoxides. Each thermogram results from the subtraction of the baseline from the initial signal. The heating rate was 10 °C min^{-1}. (b) Heat of decomposition of peroxides vs. irradiation time.

10.5.2
Kinetics of Chain Scissions during Accelerated Aging of Poly(ethylene oxide)

Poly(ethylene oxide) (PEO) is a semicrystalline water-soluble polymer [64, 65], with a crystallinity that is very sensitive to the thermal history of the sample, making this property interesting as an indicator of degradation. Because it is biodegradable and biocompatible, PEO is a good candidate for environmental and medical applications [66–68]. The mechanisms of thermo- and photo-oxidation of PEO have already been investigated [69, 70] on the basis of IR identification of the oxidation products and are summarized in Scheme 10.1.

The chain scissions predicted by this chemical mechanism have been observed by IR spectroscopy, as shown in Figure 10.12a with the appearance of formate. Rheological investigations give other evidence of the occurrence of the chain scission when PEO is thermo-aged, as can be noted in Figure 10.12b. In our group, we have been recently studying PEO aging by Photo DSC. Figure 10.12c shows the evolution of the thermogram of fusion of various PEO samples as a function of the molecular weight. The lower the molecular weight, the lower is the temperature of fusion of the material. PhotoDSC provides a direct means to observe the transition temperature shift and the change in the heat of crystallization/

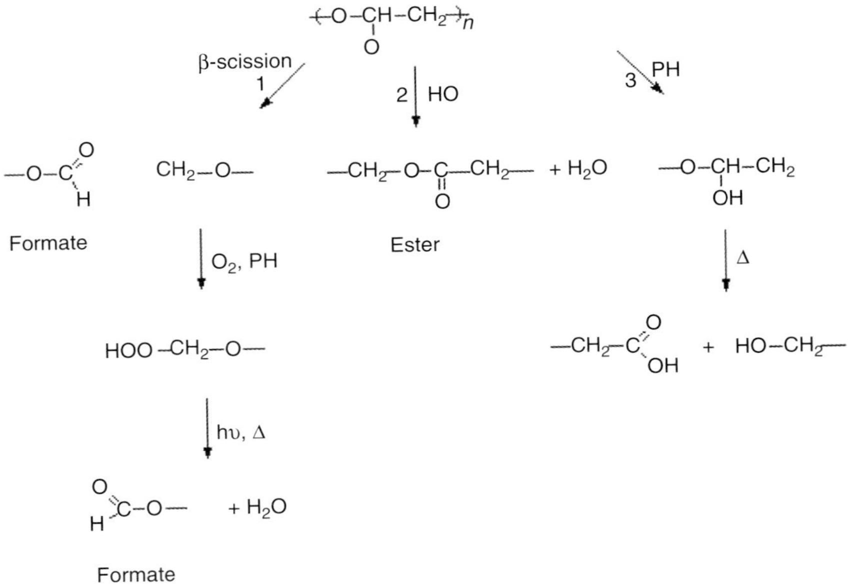

Scheme 10.1 Thermal and photodegradation of PEO.

fusion. It is thus a valuable tool enabling the investigation of chain scission. In the following section are reported the main results obtained in our group concerning chain scissions in PEO matrix upon aging.

10.5.2.1 Chain Scission Kinetics from Melting

We used the melting energy to quantify the extent of degradation of PEO ($M_w = 4 \times 10^6 \, \text{g mol}^{-1}$). Figure 10.13 describes the applied procedure. As expected and as seen in Figure 10.12c, the chain scission leads to a lowering of the temperature of fusion as well as the melting energy. This is clearly demonstrated in Figure 10.13a for the photo-aging and in (b) for the thermo-aging. Figure 10.13c shows the plot of $\ln[H(t)/H(0)]$ as a function of the time of oxidation in the dark or under irradiation at two different intensities. The best fit is a straight line passing through zero.

The linear variation of $\ln[H(t)/H(0)]$ suggests that the rate of the global chain scission reaction follows first-order kinetics that can be characterized by the following relation:

$$\ln \frac{H(t)}{H(0)} = -kt \tag{10.9}$$

where k is the rate constant of chain scission and t is the exposure time.

The theory of the activated complex gives the expression for k:

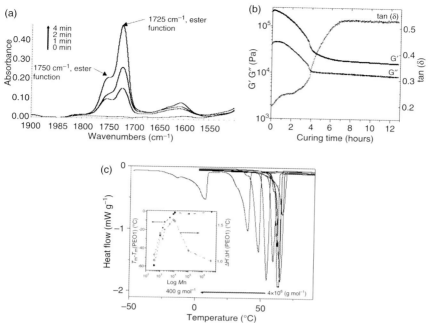

Figure 10.12 Chain scission in poly(ethylene oxide) matrix. (a) Carbonyl region of photo-aged PEO samples irradiated at 100 mW cm^{-2} and 75 °C for various times (0, 1, 2 and 4 minutes). The main band at 1725 cm^{-1} is attributed to formate functions, whereas the shoulder at 1750 cm^{-1} is assigned to esters. (b) Evolution of the storage modulus (G'), the loss modulus (G''), and the tangent of the phase angle tan(δ) versus time (temperature 90 °C). The increase in tan(δ) is evidence of chain scission. (c) Endotherms of the fusion of PEOs samples recorded at 2 °C min^{-1} (PEO1 to PEO8 have average molecular weight (M_n) of 4×10^6, 10^6, 10^4, 8×10^3, 4.6×10^3, 2×10^3, 1.5×10^3, 4×10^2 g mol^{-1}, respectively). In the inset are plotted (empty symbols) the ratio of the heat of fusion of each sample to the heat of fusion of the higher molecular weight sample (PEO1). The full symbols represent the evolution of the shift of melting temperature ($T_m - T_m$(PEO1)) as a function of molecular weight. Reproduced from [71], © 2006 with permission from the American Chemical Society.

$$k = \frac{k_B T}{h} \exp\left(-\frac{\Delta G^{\neq}}{RT}\right) = \frac{k_B T}{h} \exp\left(\frac{\Delta S^{\neq}}{R}\right) \exp\left(-\frac{E_a}{RT}\right) \qquad (10.10)$$

where k_B is Boltzmann's constant, h Planck's constant, $\Delta G^{\neq}$, $\Delta S^{\neq}$, and E_a are the free enthalpy, the entropy, and the energy of activation, respectively, and R is the ideal gas constant. Having performed these experiments at several temperatures, an Eyring plot, representing $\ln(kh/k_B T)$ as a function of the inverse of temperature (T^{-1}), can be drawn. Figure 10.13d shows such a plot. The slopes of these lines represent $-E_a/R$, whereas their intercepts equal $\Delta S^{\neq}/R$. For the

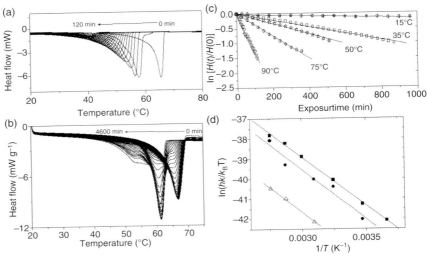

Figure 10.13 Kinetic study of PEO1 (4×10^6 g mol^{-1}) by photoDSC under air. (a) Thermograms of fusion of PEO sample irradiated at 90 °C and under 81 mW cm^{-2} as light intensity. The exposure time increment was of 5 minutes and the rate of temperature change 2 °C min^{-1}. (b) Thermograms of fusion of PEO thermally aged (in darkness) at 50 °C. The time between two DSC traces was 120 minutes. (c) Plots of ln[$H(t)/H(0)$] vs. aging time for PEO1 photo-aged with 81 mW cm^{-2} light intensity. Aging temperature was varied from 15 to 90 °C. $H(t)$ is the heat of fusion derived from the area under the DSC peak. (d) Eyring plot for the photo-oxidation (full symbols) and the thermo-oxidation (empty symbols) of PEO1. The full square symbols correspond to an irradiation intensity of 100 mW cm^{-2} and the full circle symbols to 81 mW cm^{-2}. Reproduced from reference [71], (c) 2006 with permission from the American Chemical Society.

thermo-oxidation (0 mW) and the photo-oxidation experiments (81 and 100 mW cm^{-2}), the derived values of $\Delta S^{\neq}$ and E_a are listed in Table 10.1.

The activation energies, calculated from both thermal and photo-oxidation, were found to be identical. The difference in the rate constant values was attributed to the difference in the activation entropies. A similar study has also been performed in our group for isothermal crystallization of PEO [72].

10.6
Conclusion

Two techniques based upon calorimetric measurements have been presented for the structural characterization of soft matter.

Thermoporosimetry is an old technique but recent developments by our group presented in Section 10.3 have given it a new boost. In particular, TPM application in polymers is an emerging field full of promise. The simplicity and the low cost

Table 10.1 Activation energy and entropy of the photo- and thermo-oxidation of PEO (4×10^6 g mol^{-1}).

Intensity (mW)	$\Delta S^{\neq}$ (J K^{-1} mol^{-1})	E_a (kJ mol^{-1})
0	−222	41.3
500	−204	41.4
750	−203	40.4

Reproduced from [7], (c) 2006 with permission from the American Chemical Society.

of the technique make TPM a very valuable characterization tool for soft materials. The underlying fundamental thermodynamic aspects of transitions in confined geometries are also a very hot topic.

Photo-DSC on the other hand, is a much more recent technique which has been developed thanks to technological developments in thermal analysis and coupled techniques. Until very recently, it has been used mainly to study photopolymerization or photocuring reactions by measuring the heat of reaction. We proposed the use of this powerful technique to study polymer photo-aging, using the photo-DSC as an accelerated aging device and coupled in situ analysis of the modification of the morphology of the materials. In this case, the crystallizability of the polymer is used as an indicator of the structural modifications.

Taken together, these two techniques offer new tools for the study of materials on the nanometer scale and should receive much attention in future.

References

1 Gibbs, J. (1928) *Collected Works*, Yale University Press, New Haven, CT.
2 Defay, R., Prigogine, I., Bellemans, A. and Everett, D.H. (1966) *Surface Tension and Adsorption*, John Wiley & Sons, Inc., New York.
3 Christenson, H.K. (2001) *Journal of Physics. Condensed Matter*, **13**, R95.
4 Alba-Simionesco, C., Coasne, B., Doce, G., Dudziak, G., Gubbins, K.E., Radhakrishnan, R. and Sliwinska-Bartkowiak, M. (2006) *Journal of Physics. Condensed Matter: An Institute of Physics Journal.*, **18**, R15.
5 Kunh, W., Peterli, E. and Majer, H. (1955) *Journal of Polymer Science*, **16**, 539.
6 Fagerlund, G. (1973) *Matériaux et Constructions*, **6** (33), 215.
7 Brun, M., Lallemand, A., Quinson, J.-F. and Eyraud, C. (1977) *Thermochim Acta*, **21**, 59.
8 Derouane, E. (1969) *Bulletin de la Societe Belge*, **78**, 111.
9 Strange, J., Rahman, M. and Smith, E. (1993) *Physical Review Letters*, **71** (21), 3589.
10 Dore, J.C., Webber, J.B.W. and Strange, J.H. (2004) *Colloids and Surfaces A: Physicochemical and Engineering Aspects*, **241**, 191.
11 Valckenborg, R.M.E., Pel, L. and Kopinga, K. (2002) *Journal of Physics D: Applied Physics*, **35**, 249.
12 Ville-veikko, T., Juhani, L. and Jukka, J. (2005) *The Journal of Physical Chemistry. B*, **109**, 757.
13 Strange, J.H. and Webber, J.B.W. (1997) *Measurement Science and Technology*, **8**, 555.

14 Strange, J.H., Webber, J.B.W. and Schmidt, S.D. (1996) *Magn Reson Imaging*, **14** (7/8), 803.
15 Strange, J.H. and Webber, J.B.W. (1997) *Applied Magnetic Resonance*, **12** (2–3), 231.
16 Christenson, H.K. (1997) *Colloids and Surfaces. B, Biointerfaces*, **123**, 355.
17 Landry, R. (2005) *Thermochim Acta*, **433** (1–2), 27.
18 Gane, P., Ridgway, C., Lehtinen, E., Valiullin, R., Furo, I., Schoelkopf, J., Paulapuro, H. and Daicic, J. (2004) *Industrial and Engineering Chemistry Research*, **43**, 7920.
19 Robens, E., Benzler, B. and Unger, K.K. (1999) *Journal of Thermal Analysis and Calorimetry*, **56**, 323.
20 Jackson, C.L. and McKenna, G.B. (1990) *The Journal of Chemical Physics*, **93** (12), 9002.
21 Wulff, M. (2004) *Thermochim Acta*, **419**, 291.
22 Takei, T., Ooda, Y., Fuji, M., Watanabe, T. and Chikazawa, M. (2000) *Thermochim Acta*, **352–353**, 199.
23 Husár, B., Commereuc, S., Lukáč, L., Chmela, S., Nedelec, J.M. and Baba, M. (2006) *The Journal of Physical Chemistry. B*, **110**, 5315.
24 Billamboz, N., Baba, M., Grivet, M. and Nedelec, J.M. (2004) *The Journal of Physical Chemistry. B*, **108**, 12032.
25 Bahloul, N., Baba, M. and Nedelec, J.M. (2005) *The Journal of Physical Chemistry. B*, **109**, 16227.
26 Baba, M., Nedelec, J.M., Lacoste, J., Gardette, J.L. and Morel, M. (2003) *Polymer Degradation and Stability*, **80** (2), 305.
27 Baba, M., Nedelec, J.M., Lacoste, J. and Gardette, J.L. (2003) *Journal of Non-Crystalline Solids*, **315**, 228.
28 Nedelec, J.M., Grolier, J.P.E. and Baba, M. (2006) *Journal of Sol-Gel Science and Technology*, **40**, 191.
29 Iza, M., Woerly, S., Damnumah, C., Kaliaguine, S. and Bousmina, M. (2000) *Polymer*, **41**, 5885.
30 Scherer, G.W. (1993) *Journal of Non-Crystalline Solids*, **155**, 1.
31 Scherer, G.W., Smith, D.M. and Stein, D. (1995) *Journal of Non-Crystalline Solids*, **186**, 309.
32 Jackson, C.L. and McKenna, G.B. (1991) *Rubber Chemistry and Technology*, **64** (5), 760.
33 Nedelec, J.M. and Baba, M. (2004) *Journal of Sol-Gel Science and Technology*, **31**, 169.
34 De, P.G. (1979) *Scaling Concepts in Polymer Physics*, Cornell University Press, Gennes.
35 Qin, Q. and McKenna, G.B. (2006) *Journal of Polymer Science. Part B*, **44**, 3475.
36 Flory, P.J. (1942) *The Journal of Chemical Physics*, **10**, 51.
37 Huggins, M.L. (1942) *Journal of the American Chemical Society*, **64**, 1712.
38 Korsmeyer, R.W. (1990) "Diffusion controlled systems hydrogels" in *Polymers for Controlled Drug Delivery* (ed. P.J. Tarcha), CRC Press, Boca Raton, p. 15.
39 Nakao, S.I. (1994) *Journal of Membrane Science*, **96**, 131.
40 Kim, K.J., Fane, A.G., Aim, R.B., Liu, M.G., Jonsson, G., Tessaro, I.C., Broek, A.P. and Bargeman, D. (1994) *Journal of Membrane Science*, **87**, 35.
41 Hay, J.N. and Laity, P.R. (2000) *Polymer*, **41**, 6171.
42 Ksiqzczak, A., Radomski, A. and Zielenkiewicz, T. (2003) *Journal of Thermal Analysis and Calorimetry*, **74**, 559.
43 Rohman, G., Grande, D., Lauprêtre, F., Boileau, S. and Guérin, P. (2005) *Macromolecules*, **38**, 7274.
44 Billamboz, N., Nedelec, J.M., Grivet, M. and Baba, M. (2005) *Chemphyschem: A European Journal of Chemical Physics and Physical Chemistry*, **6** (6), 1126.
45 Endruweit, A., Johnson, M.S., Long, A.C. (2006) *Polymer Composites*, **27** (2), 119.
46 Decker, C. (2001) *Pigment and Resin Technology*, **30**, 278.
47 Narayanan, V. and Scranton, A.B. (1997) *Trends in Polymer Science*, **5**, 415.
48 Li, L. and Lee, J. (2005) *Polymer*, **46**, 11540.
49 Williams, R.M., Khudyakov, I.V., Purvis, M.B., Overton, B.J. and Turro, N.J. (2000) *The Journal of Physical Chemistry. B*, **104**, 10437.
50 Khudyakov, I.V., Fox, W.S. and Purvis, M.B. (2001) *Industrial and Engineering Chemistry Research*, **40**, 3092.
51 Tryson, G.R. and Shultz, A.R. (1979) *Journal of Polymer Science Part B: Polymer Physics*, **17**, 2059.

52 Khudyakov, I.V., Legg, J.C., Purvis, M.B. and Overton, B.J. (1999) *Industrial and Engineering Chemistry Research*, **38**, 3353.

53 Grassie, N. and Scott, G. (1985) *Polymer Degradation and Stability*, Cambridge University Press.

54 Mattson, B. and Stenberg, B. (1985) *Progress in Rubber and Plastic Technology*, **9**, 1.

55 Lemaire, J., Arnaud, R., Gardette, J.L., Lacoste, J. and Kunststoffe, S. (1986), *German Plastics*, **76**, 149.

56 Chiantore, O. and Lazzari, M. (2001) *Polymer*, **42**, 17.

57 Anna, P., Betalan, G., Marosi, G., Ravadits, I. and Maatoug, M.A. (2001) *Polymer Degradation and Stability*, **73**, 463.

58 Allen, N.S., Edge, M., Ortega, A., Sandoval, G., Liauw, C.M., Verran, J., Stratton, J. and McIntyre, R.B. (2004) *Polymer Degradation and Stability*, **85**, 927.

59 Morel, M., Lacoste, J. and Baba, M. (2005) *Polymer*, **46**, 9274.

60 Bussiere, P.-O., Baba, M., Gardette, J.-L. and Lacoste, J. (2005) *Polymer Degradation and Stability*, **88**, 182.

61 Baba, M., Gardette, J.-L. and Lacoste, J. (1999) *Polymer Degradation and Stability*, **65**, 421.

62 Baba, M., George, S., Gardette, J.-L. and Lacoste, J. (2003) *Polymer International*, **52**, 863.

63 Pilichowski, J.-F., Liptaj, T., Morel, M., Terriac, E. and Baba, M. (2003) *Polymer International*, **52**, 1913.

64 Shieh, Y.-T., Lui, G.-L., Hwang, K.C. and Chen, C.-C. (2005) *Polymer*, **46**, 10945.

65 Sun, L., Zhu, L., Ge, Q., Quirk, R.P., Xue, C., Cheng, S.Z.D., Hsiao, B.S., Avila-Orta, C.A., Sics, I. and Cantino, M.E. (2004) *Polymer*, **45**, 2931.

66 Harris, J.M. (1992) *Poly(Ethylene Glycol) Chemistry: Biotechnical an Biomedical Application*, Plenum Press, New York.

67 Blin, J.-M., Léonard, A., Yuan, Z-Y., Gigot, L., Vantomme, A., Cheetham, A.K. and Su, B.L. (2003) *Angewandte Chemie (International Ed. in English)*, **42**, 2872.

68 Rele, S.M., Cui, W., Wang, L., Hou, S., Barr-Zarse, G., Tatton, D., Gnanou, Y., Esko, J.D. and Chaikof, E.L. (2005) *Journal of the American Chemical Society*, **127** (29), 10132.

69 Morlat, S. and Gardette, J.-L. (2004) *Polymer*, **42**, 6071.

70 Wilhem, C. and Gardette, J.-L. (1998) *Polymer*, **39**, 5973.

71 Fraisse, F., Morlat-Thérias, S., Gardette, J.-L., Nedelec, J.-M. and Baba, M. (2006) *The Journal of Physical Chemistry. B*, **110**, 14678.

72 Fraisse, F., Nedelec, J.-M., Grolier, J.-P.E. and Baba, M. (2007) *Physical Chemistry Chemical Physics: PCCP*, **9**, 2137.

Part Two
Organic Synthesis, Catalysis and Materials

11
Naphthalenediimides as Photoactive and Electroactive Components in Supramolecular Chemistry

Sheshanath Vishwanath Bhosale

11.1
Introduction

Aromatic-based diimides are redox-active molecules whose structural and electronic properties have implications in areas as diverse as chemotherapeutic agents [1] and photosynthetic mimics [2]. They are chemically robust molecules, and a major component in the production of thermally stable polymers [3], yet they are chemically reactive enough to be reduced to the corresponding tertiary amines. Furthermore, aromatic-based diimides such as pyromellitic tetracarboxyl diimides (PMIs) **1**, naphthalenetetracarboxyl diimides (NDIs) **2**, and perylenetetracarboxyl diimides (PDIs) **3**, shown in Figure 11.1, have been used as electron acceptors in many fundamental studies of photoinduced electron transfer, including models for photosynthesis [2, 4], solar energy conversion [2, 3], molecular electronics [5, 6], electrochromic devices [6], and photorefractive materials [7]. Diimides have previously been used to prepare liquid crystalline materials as well being incorporated with porphyrins to form photoconductive polymers.

Of the array of aromatic diimides available, PMIs (**1**), NDIs (**2**), and PDIs (**3**) (Figure 11.1) have been the most widely used, owing to the commercial availability of the starting dianhydride precursors. Although the PDIs are the most attractive in terms of their optical properties, they suffer tremendously from solubility problems – that is, aggregation. In particular, substituted NDIs form stable radical anion and dianion species [1–7], so from this aspect NDIs have become more attractive as supramolecular components. In recent years, NDIs have been shown to be a powerful tool in organic supramolecular chemistry to separate charges from photoexcited donor molecules. In this respect, NDIs are utilized as the electron acceptors [2–5].

Among the aromatic molecules that have found utility in the design of conducting materials, NDI derivatives have attracted much attention owing to their tendency to form *n*-type semiconductor materials, as opposed to most other organic molecules, which are used to fabricate *p*-type semiconductors. NDI

Tomorrow's Chemistry Today. Concepts in Nanoscience, Organic Materials and Environmental Chemistry.
Edited by Bruno Pignataro
Copyright © 2008 WILEY-VCH Verlag GmbH & Co. KGaA, Weinheim
ISBN: 978-3-527-31918-3

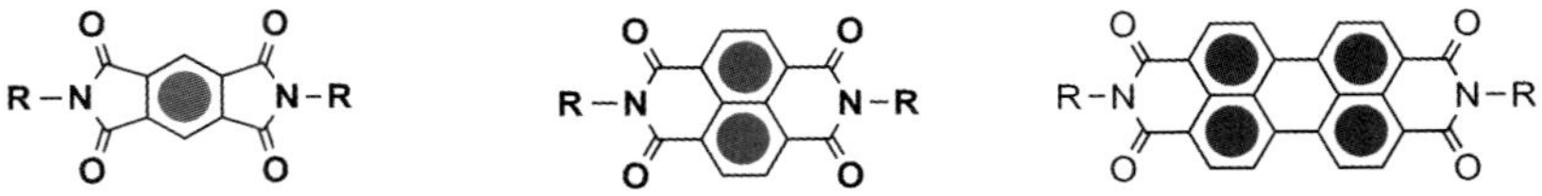

Pyromellitic Diimide (PMI)
1

Naphthalene Diimide (NDI)
2

Perylene Diimide (PDI)
3

Figure 11.1 General structure of pyromellitic diimides, naphthalenediimide, and perylenediimide compounds (R = aryl, alkyl).

is a compact, organizable [8], colorizable [9], and functionalizable organic *n*-semiconductor [10], and was therefore considered as an ideal module for the creation of supramolecular functional materials (neutral catenanes and rotaxanes) [11]. These molecules have been widely used because of their ability to undergo reversible one-electron reduction processes to form stable radical anions [12]. These properties make naphthalene diimides a useful tool in the generation of supramolecular assemblies. As NDIs are a novel class of fluorophore, the interest of their physical and electronic properties extends to the biological (as intercalators) [13] and medical areas (as antibiotics) [14] as well as in material sciences (as molecular wires) [15]. Unsubstituted NDIs are colorless nonfluorescent electron traps and they have been used as electron acceptor models. Core-substituted NDIs, however, are rapidly emerging as effective building blocks for creating conducting functional materials [9]. Despite being colorless and nonfluorescent, unsubstituted NDIs require only minor change on the structural level in order to introduce all of the characteristics essential for photosynthetic activity (color, fluorescence, and redox potential). This chapter summarizes the fundamental principles and methods of NDI in the field of supramolecular chemistry, offering a more subjective primer on new sources of inspiration from related fields. We conclude with our views on the prospects of NDIs for future research endeavors.

11.2
General Syntheses and Reactivity

Three approaches have been to the synthesis of diimides have been implemented (Scheme 11.1). In the symmetric case, 1,4,5,8-naphthalenetetracarboxylic dianhydride was condensed with the appropriate amine in a high-boiling solvent (dimethyl formamide (DMF)).The NDI is easily isolated by pouring the cooled reaction mixture into water and filtering the precipitate [16c]. This method is clean and high-yielding for a variety of amines. In some cases, a stoichiometric amount of K_2CO_3 is added in order to drive the reaction to completion at low temperature. Generally the reaction time is commensurate with the temperature of the reaction mixture; generally for the synthesis of NDI derivatives, reaction time varies from

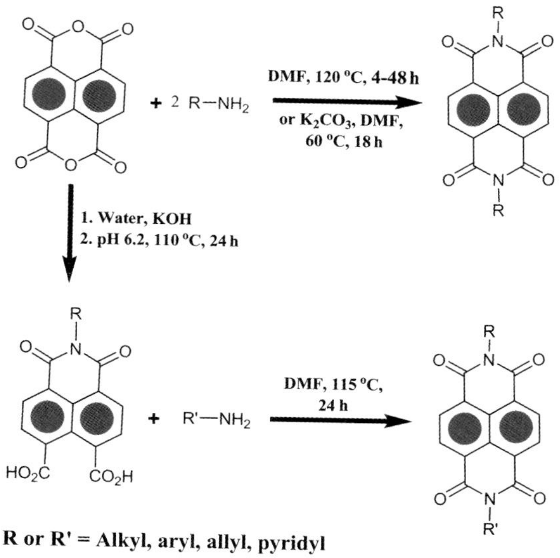

Scheme 11.1 General synthetic approach to symmetric and asymmetric naphthalene diimides.

R or R' = Alkyl, aryl, allyl, pyridyl

4 hours to 48 hours and the reaction can be monitored by thin-layer chromatography. The advantages of this new method (using K_2CO_3) are that low-boiling or heat-sensitive amines can be employed in the preparation of NDIs. Asymmetric NDIs have been synthesized using KOH and H_3PO_4 in water, as shown in Scheme 11.1. A typical reaction strategy uses a suspension of the 1,4,5,8-naphthalene-tetracarboxylic acid dianhydride dissolved in water in the presence of 1 M KOH solution, followed by acidification with 1 M H_3PO_4. Reaction with heating at 110 °C overnight with 1 equivalent of the amine in the same reaction vessel leads to the formation of the monoimides. These diacid monoimides can then be directly converted to asymmetric diimides by heating in DMF with an appropriate amine at temperatures above 110 °C [16a]. Recently the synthesis of symmetric and asymmetric NDIs has been achieved in microwave reactors, where the naphthalenetetracarboxylic dianhydride along with an α-amino acid is heated in a microwave reactor in a pressure-resistant, tightly closed reaction vessel [16b]. The reaction is carried out at 140 ± 5 °C for 5 minutes and yields are quantitative.

A proposed mechanism for the formation of the diimides is shown in Scheme 11.2. The reaction begins with attack of the nucleophilic amine on one of the carbonyls of the dianhydride which, following a series of proton transfers, leads to the amic acids and eventual loss of water to produce the diimide. The role of the potassium carbonate is speculative, but presumably it acts as a dehydrating agent catalyzing the conversion of the amic acid to the diimide and is thought to be the rate-determining step within the reaction [16c].

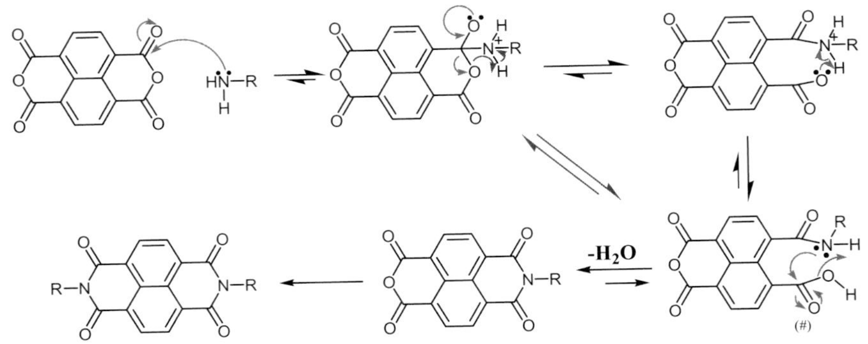

Scheme 11.2 Proposed mechanism of diimide formation.

11.2.1
Synthesis of Core-substituted NDIs

Unsubstituted NDIs are colorless nonfluorescent electron traps. However, recent findings show that a minor change on the structural level suffices to introduce all of the characteristics needed for photosynthetic activity. The disappointing optical properties of early NDIs with arylamino core substituents found by Vollmann et al. may be a reason why core-substituted NDIs have never been much studied [17]. This situation changed dramatically in the year 2002 with the introduction of alkylamino and alkoxy core substitution (Scheme 11.3) [18, 19]. Colored NDIs were obtained [18, 19] by the following synthetic method: Pyrene was converted into the dicholoro dianhydride **4** in four steps modified from previously reported procedures by Vollmann [17]. The conversion of dianhydride **4** into diimide **5** was achieved in 44% yield in acetic acid at 120 °C. Core substitution with *n*-alkylamine at 85 °C gave the core-substituted *N,N*-NDI **6** in 45% yield. The core-substituted *O,O*-NDI **7** was prepared in 51% yield by treating **5** with sodium ethanolate or sodium methanolate at room temperature. Partial nucleophilic aromatic substitution of **5**, first with sodium ethanolate at room temperature occurs in 46% yield; subsequent treatment with *n*-alkylamine at 0 °C gives the asymmetric core-substitution in *N,O*-NDI **8** in 47% yield (Scheme 11.3) [18, 19].

11.2.2
General Chemical and Physical Properties

Unsubstituted NDIs are seen as attractive chromophoric units because of their electronic complementarity to ubiquinones and their ease of formation from commercially available starting materials, forming symmetrical and asymmetric analogues relatively efficiently. Simple aromatic diimides undergo single reversible reduction processes either chemically or electrochemically to yield the corresponding radical anion in high yield ($E^1_{red} = -1.10$ V vs. Fc/Fc$^+$) (Fc: ferrocene) [20a].

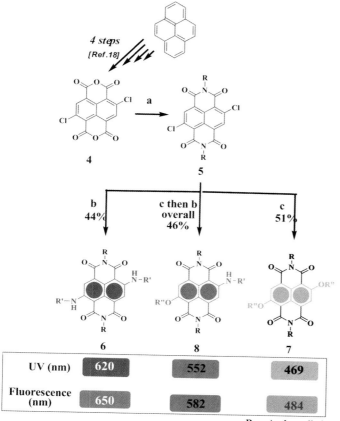

Scheme 11.3 Synthesis of core substituted NDIs.

These reduction potentials are similar to that of trimethylbenzoquinone (E^1_{red} = −1.20 V vs. Fc/Fc$^+$), implying similar energetics between the natural porphyrin-quinone and the synthetic porphyrin-diimide. The absorption and emission properties of NDIs in CH$_2$Cl$_2$ comprise strong, structural absorptions below 400 nm, and a weak, mirror-image emission with a 7 nm Stokes shift. The radical anions of most NDIs are good chromophores that exhibit intense and characteristic visible and near-infrared absorption bands [20a]. NDIs have low fluorescence quantum yields (0.002–0.006) and short fluorescence lifetimes (5–18 ps) most likely due to intersystem crossing processes from excited singlet states. Quenching of fluorescence emissions of aromatic donor molecules—naphthalene, pyrene, and phenanthrene—at rates reaching the diffusion-controlled limit in acetonitrile

$(2–8 \times 10^{10}\, M^{-1}\, s^{-1})$, has demonstrated the electron acceptor capacities of naphthalene diimides. The NDIs studied were found to display absorption spectra consisting of a narrow band at 236 nm, a broad absorption band composed of a shoulder around 340–345 nm, and two distinct maxima (situated at 355–365 nm and 375–385 nm, depending on the solvent used) [20b]. The radical anions of naphthalene diimides, in particular, are characterized by dramatic changes in absorption and electron paramagnetic resonance (EPR) spectra. Strong and highly structured EPR signals of the radical anions of simple NDIs have been obtained. While not useful currently, time-resolved EPR may be useful in the future as a technique for probing electron transfer processes. The radical cations of pyromellitic diimides have a characteristic absorption signal at ~715 nm [21].

In striking contrast to arylamino core substitution, brilliant colors and intense fluorescence were found with alkylamino **6** and alkoxy **7** core substituents. Introduction of the weaker alkoxy π-donors converted the colorless NDI into the yellow-colored and green-fluorescent O,O-NDI **7** (λ_{abs} = 469 nm; λ_{em} = 484 nm). Stronger alkylamino π-donors resulted in a red color and orange fluorescence for N,O-NDI **8** (λ_{abs} = 534 nm; λ_{em} = 564 nm) as well as the blue color and red fluorescence for N,N-NDI **6** (λ_{abs} = 620 nm; λ_{em} = 650 nm), shown in Figure 11.2. The redox potentials decreased accordingly. Photoluminescence showed quantum yields of 22% for green **7**, 64% for orange **8**, and 76% for red **6**, increasing with decreasing solvent polarity, which is the important property for core-substituted NDIs to become electron-rich as compared with unsubstituted NDIs. Importantly with regard to supramolecular architecture (but differing from core-substituted PDIs), core-substituted NDIs are planar, as shown by x-ray structures. This planarity can be expected to maximize co-facial π-stacking and therefore, as demonstrated by the chlorophyll special-pair, maximize the charge separation in a photoactive nanoarchitecture [18]. All NDIs (**6–8**) exhibit very good solubility and do not show any aggregation, even in solvents such as methanol or n-hexane [19].

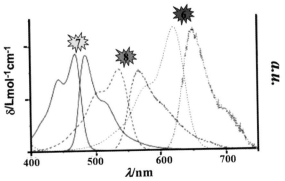

Figure 11.2 Absorption and fluorescence spectra of core-substituted NDIs. Adapted from reference [18].

11.3
Redox and Optical Properties of NDIs

The optical properties of NDIs show that substituents at the naphthalene core have dramatic effects on the electronic properties of these dyes. Further insight into the influence of core substituents on the electronic properties of NDIs was gained by cyclic voltammetry, which was carried out for a selection of NDIs bearing different substitution patterns at the naphthalene core. The results are summarized in Table 11.1. The cyclic voltammograms of compounds **i**, **ii**, and **iii**, bearing *n*-octylamine substituents at the imide positions showed two reversible reduction waves, implying the formation of radical anions and dianions, but no reversible oxidation of these compounds could be observed up to 1.2 V vs. Fc/Fc$^+$ [18, 19]. Within this series, the dichloro-substituted compound **ii** exhibited the most positive reduction potentials of −0.95 and −1.38 V. A gradual shift of the reduction potentials to more negative values was observed for the core-unsubstituted compound **i** (−1.51 and −1.10 V) and the monochloromonoalkylamino derivative **iii** (−1.64 and −1.24 V), revealing a direct relationship between the electron-donating properties of the core substituents and the electrochemical properties of NDIs. Indeed, for the most electron-rich dialkylamino-substituted naphthalene diimide **iv**, significantly lower reduction potentials (−1.80 and −1.40 V) were observed, and the reductions are irreversible. An additional irreversible oxidation wave was observed for **iv** at around +0.6 V vs. Fc/Fc$^+$ [18]. It is noteworthy that at higher scan speeds no quasireversible

Table 11.1 Redox properties of alkylamino- (**i**, **ii**, **iii**, and **iv**) NDIs bearing different patterns of donor and acceptor substituents at the naphthalene core.[a]

	$E_{1/2}$ (V vs. Fc/Fc$^+$)			
	X$^-$/X^{2-}	X/X$^-$	X/X$^+$	X$^+$/X^{2+}
i	−1.51	−1.10	b	b
ii	−1.38	−0.95	b	b
iii	−1.64	−1.24	b	b
iv	−1.80	−1.40	0.60	b

a All measurements were conducted in dichloromethane; scan rate 100 mV s^{-1}, concentration 1.0 mM; supporting electrolyte, tetrabutylammonium hexafluorophosphate (*n*-Bu$_4$PF$_6$, 100 mM).
b Not observed.

Table 11.2 Optical properties of NDIs in dichloromethane.

Color (solid)[a]	λ_{abs} (nm)	ε_{max} (l mol^{-1} cm^{-1})	λ_{em} (nm)	ϕ_{em} (%)
5. Yellowish white	402	13 900	–	<0.1
6. Blue	620	23 300	650	42
7. Yellow	469	18 200	484	22
8. Violet	552	17 700	582	76

a The colors are given for the crystalline materials because the visual color for solutions of some of these dyes (Scheme 11.1) is influenced by their strong photoluminescence.

behavior was observed for **iv**, which suggests that the charged species of this compound is of lower stability than those of other naphthalene diimides studied here. Optical properties of naphthalene diimides (**5–8**) exhibit photoluminescence with quantum yields between 22% and 76% for green–red in dichloromethane (Table 11.2) [18, 19].

11.3.1
NDIs in Host–Guest Chemistry

Proteins adopt helical and sheet-type secondary structures that depend on their primary sequence of amino acids. To understand these protein structures as well as investigate various types of noncovalent interactions, the group of Iverson describe two types of structures: foldamers and aedamers [22a].

11.3.2
NDI-DAN Foldamers

Foldamers and aedamers are molecules designed to utilize noncovalent interactions in order to stabilize well-defined conformations in solution. They were constructed from electron-rich and electron-deficient units in an alternating array via amino acid residues [22b]. In aqueous solutions, aedamers adopt a compact pleated structure in which the aromatic moieties (donor and acceptor) stack in face-to-face geometry. Furthermore, for stacking of aromatic units in aqueous solution, the hydrophobic effect is important [22c]. Complexation of 1,5-dialkoxynaphthalene (DAN) with NDI was found to be favored over DAN:DAN and NDI:NDI self-associations in aqueous solutions due to the complementary electronic differences of the naphthyl rings that provide a driving force favoring aromatic stacking [22d]. π-Electron-deficient NDIs are complementary to π-electron-rich benzenes and naphthalenes, forming co-crystals based on stabilizing factors such as π–π stacking and partial charge transfer interaction [23]. This motif π–π stacking and partial charge transfer interaction dictated the secondary structure of synthetic oligomers in solution (Figure 11.3) [22d]. The system consisted of electron acceptor (NDI) and electron donor (DAN) units tethered by an amino acid linker that

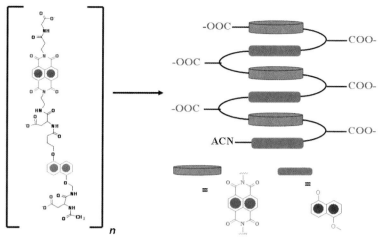

Figure 11.3 Structure of aedemers (n = 1, 2, 3) and the general secondary structure formed in phosphate buffer solution (20 mM).

allowed overlap of aromatic units. The aedemers were prepared by solid-phase synthesis and the co-crystals of the two monomers showed a pleated structure of folded aedamers, resulting in a stacked aromatic core with derivatizable, flexible linkers on the periphery. Upon heating the amphiphilic aedamer, viscosity and spectroscopic changes were observed due to the kinetically irreversible conversion from a folded to a tangled state.

11.3.3
Ion Channels

In the last few years, several groups have been involved in the synthesis of aromatic donor–acceptor complexes of electron-rich DAN and electron-deficient NDIs for the studies in foldamers, catenanes, and rotaxanes and also in ligand-gated ion channels. DAN:NDI complexation was found to be favored over DAN:DAN and NDI:NDI self-associations in aqueous solutions due to complementary electronic differences of the naphthyl rings in host–guest chemistry [11, 22–26, 30]. To stop a process is always easier than to start one. This common wisdom was demonstrated by our group in the field of synthetic ion channels and pores where the inability to create such ion channels or pores that open (**11**) rather than close (**9**) in response to chemical stimulation remains a major challenge. This situation has also been observed in the case with rigid-rod π-stacking barrel-staves architecture. Conversely, with the intercalation of aromatic ligands into a rigid-rod π-helix, the opposite situation is observed (Figure 11.4) [27]. Ligand gating by aromatic electron donor–acceptor interactions was designed on the basis of a hoop–stave mismatch. The architecture of a barrel-stave ion channel requires approximate hydrophobic

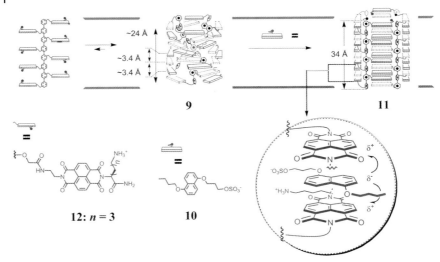

Figure 11.4 A ligand-gated ion channel based on NDI:DAN interactions.

matching with regard to the surrounding membrane (2.5–3.5 nm) on the one hand, and on the other precise matching of the repeat distances in the stave with the repeat distances between the "hoops" that hold the staves together. Self-assembly with the hoop–stave mismatch produces helical rather than barrel-stave supramolecules. π-Stacks with shorter repeat distances therefore result in a rigid-rod π-helix **9**. The intercalation of ligands **10** into the closed π-helix **9**, however, reduces the hoop–stave mismatch and causes an untwisting helix–barrel transition to open up into a barrel-stave ion channel **11** (Figure 11.4). The validity of this adventurous concept was probed with electron-poor NDIs to ensure favorable self-assembly of monomeric *p*-octaphenyl rods **12** into the π-helix **11**.

A sulfonate was added at one side of ligand **10** to support the DAN:NDI charge transfer complexes by ion pairing, and an alkyl tail at the other for favorable interactions with the surrounding lipid bilayer membrane. In spherical egg yolk phosphtidylcholine large unilamellar vesicle (EYPC-LUV) bilayers, only the closed channel **9** and/or ligand **10** were inactive. When they were mixed together, highly cooperative (n = 6.5, EC_{50} = 13 μM) intercalation of donors **10** resulted in the expected helix–barrel transition and produced surprisingly homogenous, ohmic, low-conductance (expected, $d \approx 5$ Å; found, d_{Hille} = 3.5 Å; g = 94 ps) anion channels **11** (P_{Cl^-}/P_{K^+} = 1.38). Operational aromatic donor–acceptor interactions **6** were visible to the naked eye as a plum color of the ion channel [27]. Recently our group synthesized the NDI rigid-rod and used these π–rigid-rods for anion-π-slide activity measurements (Figure 11.5), which were carried out in EYPC-LUVs loaded with HPTS pH-sensitive fluorescent dye and were exposed to pH gradients to evaluate the anion slides [28].

We found that the anion transport by neutral NDI rigid-rod discriminates a halide-selective sequence ($Cl^- > F^- > Br^- > I^-$) and the NDI rigid-rod with one

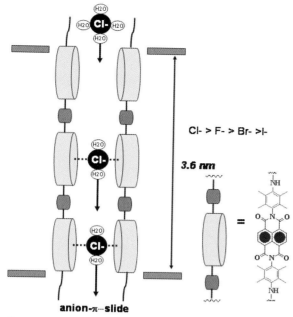

Figure 11.5 Rigid NDI rods as transmembrane anion-π-slides.

ammonium cationic terminus with reduced selectivity. The selectivity for the cation-terminal NDI-rigid-rod shifted from halide I⁻ to a weaker halide Cl⁻ with selectivity in the sequence Cl⁻ > Br⁻ > F⁻ > I⁻. In conclusion, the results suggested that increasing the proximity between transmembrane NDI slides increases selectivity at reduced activity [28].

11.3.4
NDIs in Material Chemistry

Lindner and co-workers describe the influence of π–π aromatic interactions on the chromatographic behavior of hybrid reversed-phase materials containing electron donor–acceptor systems (Figure 11.6). The chromatographic behavior of NDIs was elucidated in terms of hydrophobic, silanophilic, and π–π interactions, employing common column-chromatographic test methods and also mixtures of geometrical and functional aromatic isomers [29]. It was found that for large polyaromatic compounds (for example, pyrene, naphthacene, chrysene, and so on) with extended π-systems, π–π interactions are more dominant, however, for small aromatic compounds (benzene, naphthalene, biphenyl, phenanthrene, anthracene, and so on), hydrophobic properties are more effective. It is notable that the retention behavior depends not only on π–π interactions but also on hydrogen-bonding and carbonyl–π-interactions [29].

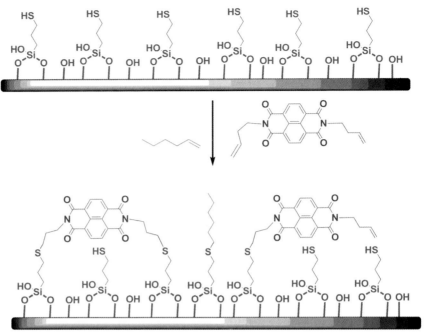

Figure 11.6 Chromatographic behavior of NDIs.

11.4
Catenanes and Rotaxanes

One of the key goals of supramolecular chemistry is to assemble structural building blocks into arrays with new properties that function only in supramolecular architectures. Catenanes and rotaxanes have for some time been the subject of intense study because of fascinating architectures and they have also useful properties as molecular-level switches and sensors [30a–c]. These new properties may improve our understanding of noncovalent interactions, or they may provide useful functionality for practical applications (catalysis, sensors, and molecular machines or computers). However, to achieve this aim we need powerful methods for the synthesis of supramolecular systems; and while there has been dramatic progress in the past 35 years, we still have much to learn. Some groups have pioneered a donor–acceptor recognition system based on neutral π-electron-accepting sites, as well as neutral π-electron-donating sites, for the efficient assembly of neutral catenanes and rotaxanes using template-directed synthesis [11a, 30a]. A typical example of noncovalent interaction [30a] used a simple high-dilution approach to synthesize the first macrocycle containing naphthalene diimides, albeit in a modest 10% yield, by condensation of 1,8-diaminooctane and 1,4,5,8-naphthalenetetracarboxylic dianhydride (Scheme 11.4). Interestingly, this π-electron-deficient macrocycle was intercalated by a nitrobenzene solvent molecule

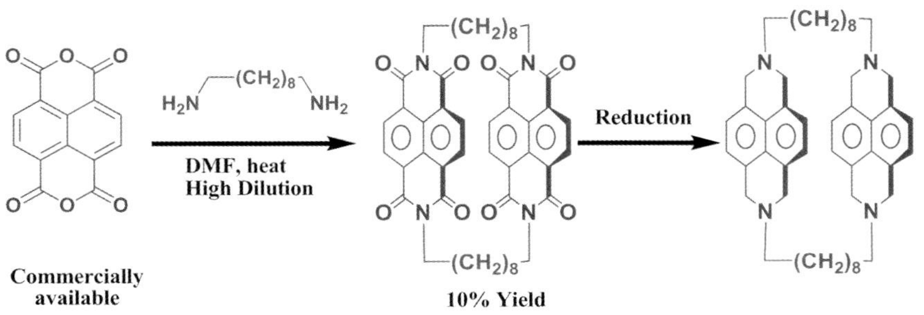

Scheme 11.4 Preparation and reduction of NDI macrocycle.

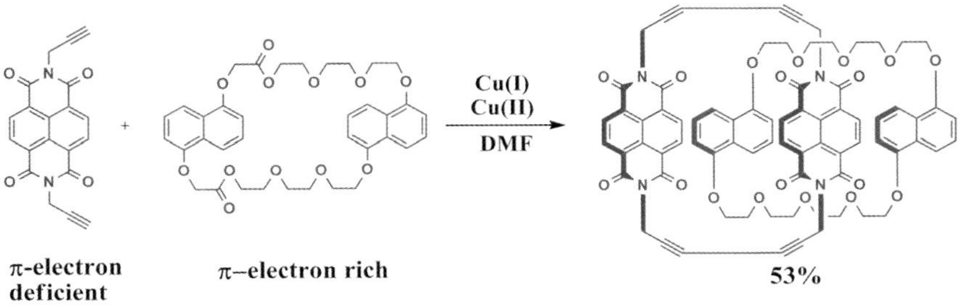

Scheme 11.5 Synthesis of neutral [2]catenane.

within the cavity of the macrocycle. Reduction of the macrocycle using $AlCl_3$/$LiAlH_4$ resulted in the corresponding tetraamine, and its subsequent protonation (using HCl) gave the corresponding water-soluble tetraammonium salt. The above result provided the impetus for the preparation of supramolecular macrocycles utilizing NDIs as a core.

However, produced a "neutral" [2]catenane (Scheme 11.5) [11b] bearing a mechanically interlocked bis-NDI macrocycle in 53% yield. This yield is far more attractive for a bis-NDI macrocycle than the 10% yield originally obtained in macrocycles [30a]. This approach involved an oxidative coupling of two acetylenic naphthalene diimides in the presence of crown ether. The efficiency of the catenation can be attributed to the crown ether component acting as a "permanent template" for the forming cyclophane.

11.4.1
NDIs Used as Sensors

The N,N-substituted naphthalenediimide (gelator) molecule **13** (Figure 11.7) allows for sensing of aromatic molecules by a combination of π–π-stacking, H-bonding, and van der Waals forces and through the detection of charge transfer (CT) complexes by the naked eye. A series of naphthalene derivatives (**15a–g** and **16a–c**) was tested, as they are known to form strong donor–acceptor complexes with NDIs. In

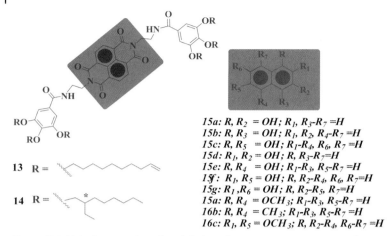

Figure 11.7 Naked eye sensing of naphthalene derivatives using NDI as a gelator.

the case of the protected methoxy substituted naphthalene **16a–c** and even pyrene (Figure 11.7), no color change was observed even after keeping the sample for a few weeks, which indicates that no donor–acceptor complex is present in the gel matrix. However, with the application of seven structural isomers of dihydroxynaphthalene **(15a–g)** in the gel matrix, all of the isomers form donor–acceptor complexes which can be detected with the naked eye. Importantly, even minute quantities (0.2 equiv in 5 mM of gelator) of the dihydroxynaphthalenes **(15a–g)** can be sensed with the naked eye with the NDI gelator **13** and **14** (Figure 11.7). Within this system, it is possible to distinguish clearly several isomers of dihydroxynaphthalenes with the naked eye. The color change observed with the various dihydroxynaphthalene isomers and the inability of other hydrophilic and insoluble naphthalene derivatives to complex, explains that not only the solvophobic effect, but also hydrogen bonding, and donor–acceptor interactions facilitate the binding of the dihydroxynaphthalene in the sensing of the organogel matrix [8].

NDIs bearing ferrocenyl moieties (Scheme 11.6) at the termini of imide substituents were applied to DNA sensing on a gold surface. They were shown to bind intact calf thymus DNA more strongly than the denatured DNA, and the complex also dissociates more slowly than that with denatured DNA [31]. In a typical sensing procedure, the DNA-modified electrode was soaked in aqueous solution of **17** for 5 minutes and washed with deionized water and then subjected to electrochemical study. Ferrocenylnaphthalene diimide **17** underwent hypochromic and bathochromic shifts of the typical absorption band of the NDI chromophore, at 383 nm upon binding to DNA. This was indicative of DNA intercalation. Applying the same procedure, ferrocenylnaphthalene diimide **17** binds to double-stranded DNA, with the reversibility of the binding demonstrated in the presence of SDS (1% sodium dodecyl sulfate) with a large hypochromic shift (thus returning to that of free **17**) [31].

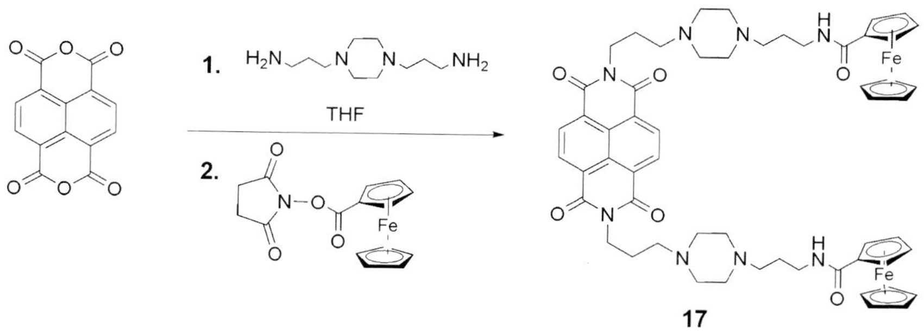

Scheme 11.6 Synthesis of ferrocenylnaphthalene diimide **17**.

Charge separation in DNA is of current interest because of its involvement in oxidative DNA damage and electronic molecular devices. NDI has been used in the formation of the charge-separated state between NDI and (pentylenetetrazole) (PTZ) in DNA by utilizing the consecutive adenine (A) sequence as a bridge [32]. It was found that the yields of the charge separation and recombination processes between NDI and PTZ were dependent upon the number of A bases between the two chromophores, with the charge recombination strongly dependent upon distance. Charge separation over 300 µs was observed when NDI was separated from PTZ by eight A bases [32]. Our group pioneered these electron-deficient NDIs, which have been utilized as π-clamps for molecular recognition within synthetic multifunctional pores by aromatic electron donor–acceptor interactions [33].

11.4.2
Nanotubes

Three different synthetic routes were employed for cyclic peptides bearing NDI side-chains for formation of nanotubes [16, 34]. It has been shown that these nanotubes can be useful as model systems for the study of NDI–NDI charge transfer by probing hydrogen bond-directed intermolecular NDI–NDI interactions (3.6 Å) in solution. Tight packing is required for efficient charge delocalization, as characterized by fluorescence measurements in the self-assembled cyclic D,L-α-peptide nanotubes (Figure 11.8b) [16]. Nanotube formation has been supported by the directed backbone hydrogen-bonding interactions and self-assembling cyclic D,L-α-peptide, and characterized by atomic force microscopy (AFM) on a mica surface. However self-assembled peptide nanotubes can also be adsorbed onto other surfaces, for example, graphite or silicon oxide. The observation of adsorbed peptide nanotubes on distinct surfaces strongly suggests that nanotubes are formed in solution in the presence of sodium dithionite (Figure 11.8a) and then physisorbed onto the surfaces; in the absence of the reducing agent (sodium dithionate) only amorphous structures were observed.

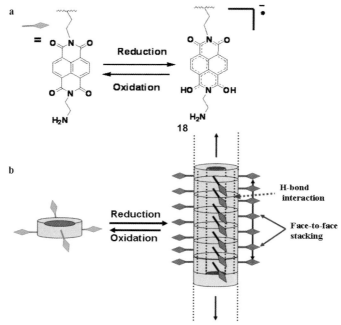

Figure 11.8 Self-assembled cyclic-D,L-α-peptide nanotubes containing NDI as the core. (a) Reversible reduction of the NDIs with sodium dithionate. (b) Self-assembly of the peptide NDI nanotube after reduction reaction via chemical or electrochemical methods.

The self-assembling cyclic D,L-α-peptide nanotubes described demonstrate high stability on surfaces even after two months' exposure to ambient temperature. NDI peptide nanotubes **18** may provide a facile method for the preparation of a new class of synthetic biomaterials [16b, 34a]. Recently Sanders and co-workers demonstrated the formation of amino acid-derived NDI hydrogen-bonded supramolecular organic M-helical nanotubes in nonpolar solvents and also in the solid state [34b]. The hydrogen-bonded supramolecular nature of the helical nanotubes was confirmed by the circular dichroism (CD) spectrum in chloroform; with the addition of methanol, destruction of the supramolecular nanotubes was observed, due to the capabilities of such an aprotic solvent to compete for hydrogen-bond interactions [34b].

Because NDIs are a popular choice of acceptor in electron transfer studies, the NDI redox and photophysical properties are well documented and, more importantly, absorbance and fluorescence maxima are known for both the neutral and radical anion species. This has led the group of Miller [25] to exploit their properties in forming conducting wires based on the doping of water-soluble NDIs by peripheral modification on the polyamidoamine (PAMAM) dendrimers **19** (Figure 11.9).

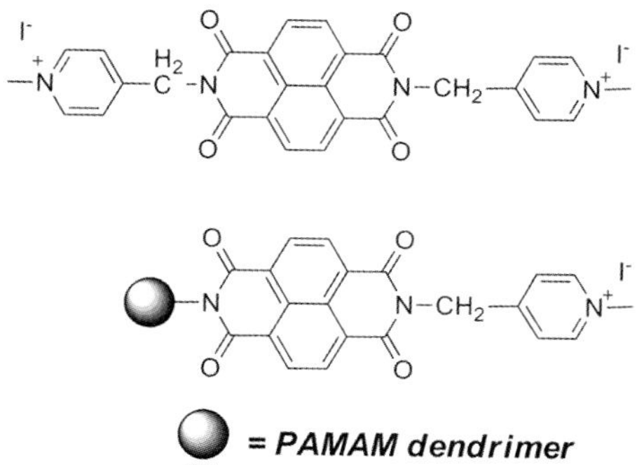

Figure 11.9 Water-soluble NDIs modified on poly(aminodoamine) (PAMAM) dendrimer **19**.

It has been observed that the conductivity was actually higher than the conductivity reported for pure diimide anion radical salts, because dendrimeric structures may produce π-stacking formation in three dimensions, a structural feature differing from the usual one-dimensional structures of small molecular stacks.

11.5
NDIs in Supramolecular Chemistry

11.5.1
Energy and Electron Transfer

NDIs have been used to mimic long-lived charge separation [2–6]. Porphyrin–aromatic diimide systems have been used as models of photosynthetic processes, with a view to rationalizing some design features for fabricating energy transduction systems in which efficiency, longevity, and pragmatism are equal players. This chapter offers a unique perspective by comparing charge-separation dynamics in covalent and supramolecular constructs in which the same chromophores are used.

11.5.2
Covalent Models

Intramolecular electron transfer (ET) is the subject of many studies that aimed to elucidate the role of the various parameters governing the rate of ET [35]. Work with second-generation porphyrin–diimide systems concentrated on manipulating the dynamics of ET toward attaining long-lived (μs) charge–separated states. The first example of an NDI-derivatized linear π-conjugated system (**20**) was reported in 1993, having fixed-distance triads consisting of zinc(II)porphyrin (ZnP) and free base porphyrin (H$_2$P) moieties which are bridged by aromatic spacers such as 1,4-phenylene or 4,4′-biphenylene. The center-to-center distance between the ZnP and

H$_2$P was fixed to 13 and at the other end NDI was fixed to a distance of 17.2 Å from H$_2$P as an electron acceptor (Figure 11.10a) [36]. The formation of photoinduced long-lived charge-separated states upon photoexcitation of these triads is in the order of 0.14–80 μs in THF by a relay approach, exciting the primary electron donor ZnP at 532 nm. The steady-state fluorescence spectra indicated the occurrence of intramolecular singlet–singlet energy transfer (EnT) from the ZnP to the H$_2$P, and subsequent charge separation between H$_2$P and NDI (6.7 × 10^9 s^{-1} in THF) resulting in the formation of ZnP–(H$_2$P)–(NDI)$^-$, in which a slightly exothermic hole transfer from the H$_2$P to the ZnP moiety, in competition with an energy wasteful charge recombination to the ground state, finally leads to the generation of (ZnP)$^+$–H$_2$P–(NDI)$^-$. This process occurs within the lifetime of several hundred nanoseconds to microseconds [36]. The lifetimes and efficiency of the charge-separated states (ϕ_{1a} = 0.79) across **20** exhibited distance-dependent behavior, as defined by the rigid bridges **20a–d**. On the basis of the present results, further work is needed toward achieving longer-lived ion pair states via more efficient sequential electron transfer and recombination.

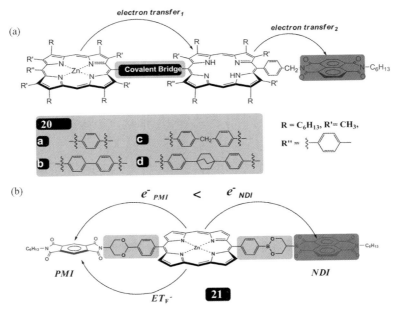

Figure 11.10 (a) Covalently linked triads capable of microsecond lifetimes for the charge-separated state through a relay approach. (b) The triad **21**, bearing two acceptors, was used to demonstrate ET switching between NDI and PMI groups. The system exploits the change in electronic coupling between the porphyrin and NDI group through the fluoride-responsive boronate ester.

As an interesting development, the triad **21** (Figure 11.10b) bearing a ZnP donor and PMI and NDI groups antipodally placed on the porphyrin periphery has been examined as a means of controlling the direction of intramolecular ET through the acid–base properties of the boronate bridge between the NDI and PMI acceptors [37]. Photoexcitation preferentially led to a charge-separated state involving the NDI radical anion as a result of its better accepting properties and greater ET efficiency. Upon the addition of fluoride, which complexes to the boron center, the electronic coupling in this pathway is switched OFF causing ET to occur completely in the direction of the PMI [37]. Such systems, in which a change in the electronic properties of the bridge is possible, offer specific directionality of ET within the photosynthetic reaction center.

To determine the driving force of electron-transfer processes in the presence of metal ion salts, zinc porphyrin–naphthalene diimide dyads **22** and zinc porphyrin–pyromellitic diimide-naphthalenediimide triad **23** shown in Figure 11.11 have been employed by the group of Fukuzumi [37b]. The two salts chosen, Sc(OTf)$_3$ or Lu(OTf)$_3$, are able to bind with the radical anion of the NDI and hence modulate the rate of charge recombination relative to charge separation. As a result, the lifetimes of the charge-separated states in the presence of millimolar concentrations of Sc^{3+} (14 μs for **22**, 8.3 μs for **23**) are far longer than those in the absence of metal ions (1.3 μs for **22**, 0.33 μs for **23**) [37b]. In contrast, the rate constants of the charge-separation step determined by the fluorescence lifetime measurements are the same, irrespective of the presence or absence of metal ions. These results indicate that photoinduced electron transfer precedes complexation. Hence, the driving force of the charge-separation process is the same as that in the absence of metal ions, whereas the driving force of the charge recombination

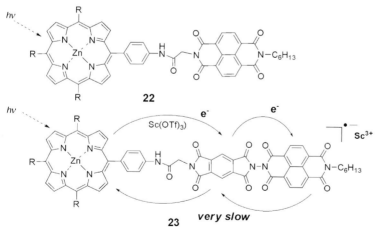

Figure 11.11 Photoinduced electron transfer from zinc porphyrin to diimide in the presence of Sc(OTf)$_3$, affording the diimide anion/Sc^{3+} complex.

process decreases with increasing metal ion concentration. This study provides a new tool for controlling back electron-transfer processes of the charge-separated states by complexation with metal ion.

More recently, the group of Johansson and Hammarström has synthesized the first synthetically linked electron donor–sensitizer–acceptor triad in which a manganese complex plays the role of the donor. EPR spectroscopy was used to directly demonstrate the light-induced formation of both products: the oxidized manganese dimer complex ($Mn_2^{II,III}$) and the reduced naphthalenediimide (NDI). They found that the charge separation state was very long-lived, with an amplitude-weighted average lifetime of ~600 ms in room temperature solution, which is at least two orders of magnitude longer than those previously observed [37c].

11.5.3
Noncovalent Models

In recent years there has been a considerable upsurge in interest in the study of electron transfer within hydrogen-bonded donor–acceptor pairs in relation to its importance in biological energy conversion systems [35]. The construction of artificial photosystems for solar energy conversion using noncovalent bonding parallels that based on covalent bonds [35]. The chemical structure of the imide group and its structural and electronic relationship to the nucleic base thymine make it particularly amenable to three-point hydrogen bonding with a complementary acceptor. Each of the porphyrin–aromatic diimide examples herein utilizes this approach as a means of maximizing the association of donor and acceptor components. An advantage of using noncovalently assembled structures is the simple and fast interchange of components within the systems and the fact that it does not required lengthy synthetic routes as compared with covalent systems [35–37]. Further, the advantage of such noncovalent systems for the conversion of light energy to chemical energy is the contribution to the photoinduced charge separation over long distances, which could only be provided by movements in the weakly bound noncovalent array. A very demanding design and the synthesis of covalently assembled diads, triads, tetrads, or even pentads bearing the donor and acceptor (NDI) at the extremities of long arrays is often necessary to optimize these systems [35–37].

In this section we will look at noncovalent approaches that have been investigated as a theme for preparation of dyads for pragmatic applications. There has been a considerable increase in the use of noncovalent systems using hydrogen-bond or axial coordination to study ET between donors and acceptors. A second-generation supramolecular ensemble has been prepared, whereby the imide acceptor motif is directly attached to the *meso* position of a ZnP **24** and both reduces the conformational freedom of the ensemble and provides a more suitable "through-bond" pathway for ET (Figure 11.12) [38]. Steady-state fluorescence studies in C_6H_6, using a substantial excess of the NDI, showed significant fluorescence quenching (70%) compared to the porphyrin alone. Time-resolved picosecond transient absorption studies (λ_{ex} 532 nm, benzene) provided some evidence

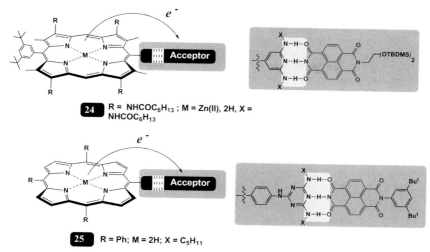

Figure 11.12 Novel noncovalent assemblies, employing hydrogen bonding to mediate ET processes.

for the formation of an NDI radical anion (k_{CS} = 4.1 × 10^{10} s^{-1}) and decay over 270 ps. A deuterium effect (k_H/k_D) of ~1.5 for the rates of charge separation (CS) and charge recombination (CR) also provides evidence for complexation. The synthetic utility of triazines and the inherent three-point hydrogen-bonding motif of 2,4-diaminotriazines were exploited to prepare the complex **25** (Figure 11.12) [39a]. Quenching of the H$_2$P fluorescence relative to an N,N'-dipentylnaphthalene diimide by addition of the two components that make up **25** indicated direct excitation of the supramolecular dyad over the components in solution. Time-resolved studies indicated a biexponential decay in which the shorter component (0.27 ns) could be attributed to the complex and the longer component to the uncomplexed porphyrin unit. The rate of ET (k_{CS}) for the complex was calculated to be 3.6 × 10^9 s^{-1} over an estimated center-to-center interchromophore separation of 18.4 Å. Nanosecond resolution transient absorption spectroscopic experiments indicated that charge recombination occurred below the resolution of the instrument (<10 ns) [39a].

Another method showing success in supramolecular self assembly is the formation of stable coordination complexes of ZnP by axial ligation using pyridine ligands (K_a ≈ 10^3–10^4 M^{-1}), providing a synthetically easy method of assembling simple dyads in which the NDI acceptor occupies a position distal (time averaged) to the plane of the porphyrin donor. The use of ditopic bis-zinc(II) porphyrin tweezers to increase the binding strength of the NDI acceptor and subsequent charge transfer was studied using both flexible and rigid bridging units **26** and **27** (Figure 11.13) [39b]. The supermolecule **26**, bearing an NDI acceptor, is formed with an association constant (K_a) of ~7 × 10^7 M^{-1} [39b]. The rigid nature of the norbornylogous bridging unit between porphyrin units implies a center-to-center

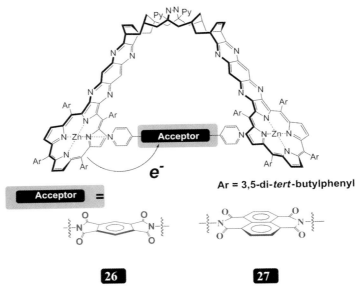

Figure 11.13 Use of two coordinative bonds in these tweezerlike compounds holds the acceptor units bearing pyridine linkers in well-defined orientations.

distance of 22 Å. The higher association constant in the case of **26** compared to that found for **27** can be attributed to both the rigidity of the bisporphyrin host and optimized distances between the porphyrins for the dipyridyl–diimide acceptors.

Steady-state fluorescence spectra recorded after the addition of the NDI or PMI acceptor to the bisporphyrin tweezer (λ_{max} = 660 nm), demonstrated substantial quenching (75%) with increasing quantities of the NDI or PMI acceptors. Time-resolved emission spectra recorded in toluene for the complex **26** were biexponential; containing a dominant short-lived CS components (80 ps, ~95%) attributed to photoinduced ET from donor porphyrin to NDI, and a minor long-lived component (1 ns, 5%). The lifetime of the dominant short-lived CS state is increased two- to threefold relative to covalently linked systems under similar conditions of solvent, donor–acceptor distance and thermodynamics [37]. Charge recombination rates from 1.4 to 3.8 × $10^9 s^{-1}$ were observed, depending on whether the NDI or PMI acceptor was bound within the cavity.

The more recent example was the synthesis of axial coordination of Zn(II)P and an N,N'-diethylpyridyl-terminated NDI (Figure 11.12) in order explore electron transfer in a complex. This simple supramolecular dyad forms in solution through axial coordination of Zn(II)TPP and a N,N'-diethylpyridyl-terminated NDI (Figure 11.14) [40]. Photoexcitation in nonpolar solvents at 604 nm, an absorption unique to this complex, results in charge separation with lifetimes of 1 to 10 µs depending

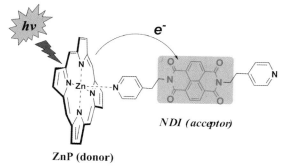

Figure 11.14 Simple dyad ensembles utilizing Zn(II)···N coordination bonds to facilitate ET between ZnP and NDI. In the above, this occurs with microsecond lifetimes.

on the solvent used. In this case, both CS and CR were much faster ($\tau_{CS} < 0.01\,\mu s$) and comparable to that observed in **26**. The electron recombination from the triplet charge-transfer state (CT^T) is formally forbidden as it involves an inversion of electron spin and a longer-lived CT state would result [40].

11.6
Applications of Core-Substituted NDIs

Core-substituents have been shown to greatly influence the redox properties of NDIs, thus offering the possibility of a huge range of tunable fluorophores. In contrast, substituents at the imide position do not have any significant effect on the chromophore properties [19].

An improved synthesis of 2,6-dichloronaphthalene diimides (with aryl/alkyl amines) has been developed, whereby their optical and redox properties provide a series of intensely colored red and blue NDI chromophores. As the dichlorosubstituted NDI **4** possesses a low reduction potential (Table 11.1), it is a promising candidate for *n*-type organic semi-conducting materials, which are still very rare. On the other hand, when a dichlorochromophoric conjugate, made up of a red amino-core-substituted dye and a blue carboxy substituted dye, is covalently assembled, FRET (fluorescence resonance energy transfer) studies reveal that the ratio of the integrated emission bands at <600 nm (red dye) and >600 nm (blue dye) lead to an energy transfer efficiency of about 96% and a fluorescence quantum yield of 30% (compared with direct excitation of the blue dye itself at 622 nm) [18]. The self-organization, via hydrogen-bonding, of an unsymmetrically substituted 2,6-diamino-functionalized NDI has been well studied by NMR, UV-vis, and fluorescence titration and at the solution–graphite interface, both in the presence and in the absence of complementary melamines (Figure 11.15) [41]. Self-assembly of core-substituted NDIs with ZnChl (zinc chlorophyll) chromophores

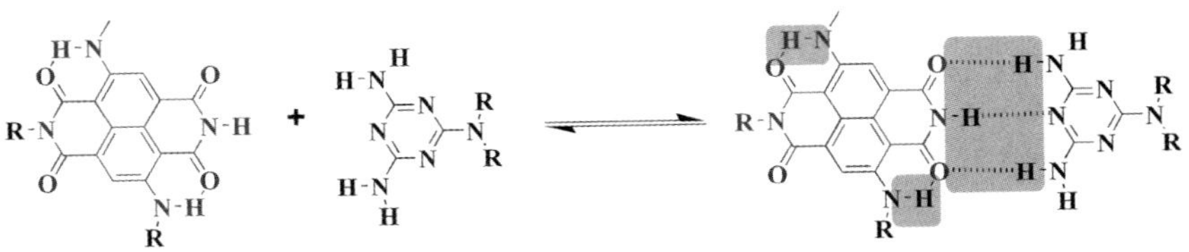

Figure 11.15 Hydrogen-bonded complex of 1:1 stoichiometry between NDI and melamine, and intramolecular hydrogen bonds in NDI itself.

leads to a well-defined rod aggregate with light-harvesting functionality. These systems thereby operate as bio-inspired light-harvesting mimics, with promise to bridge the "green gap" [42].

Knowing all these properties of core-substituted NDIs, we have recently explored the synthesis and use of N,N-core-substituted NDIs for artificial photosynthesis [43]. An example of such a system in the blue-colored, red fluorescent rigid-rod π-M-helix **28**, the production of which aims to tackle this challenge in a lipid bilayer membrane. Photosystem **28** was prepared by self-assembly of four *p*-octiphenyls **31**, each bearing eight amino-N,N-core-substituted NDIs along the rigid-rod scaffold (Figure 11.16) [43]. Photosystem **28** was characterized in vesicles equipped with the external electron donors ethylenediaminetetraacetate (EDTA), internal electron acceptors 1,4-naphthoquinone-2-sulfonate **32** (Q), and 8-hydroxypyrene-1,3,6-trisulfonate (HPTS) as an internal fluorescent pH meter (Figure 11.16) [43]. The addition of the aromatic ligand **33** caused the immediate collapse of the proton gradient, which was created by excitation of **28**. This finding was consistent with smart supramolecular architecture, that is, the programmed transition from active electron transport with photosystem **28** to passive anion transport with ion channel **29** in response to chemical stimulation with ligand **33** (Figure 11.16).

Irradiation of externally added NDI rods **31** at 635 nm was detected as an increase in intervesicular pH in response to a reduction of internal quinone (Figure 11.17). The change in HPTS emission, which reflects internal proton consumption, is analyzed with Hill's equation ($n = 3.9 \pm 1.2$), which indicates that the active helix photosystem **28** is quadruple (four rods form the helix) with an effective concentration of 1.3 μM. However, the shortened length of the rigid-rod *p*-barrel scaffold to the NDI dimer **30** is inactive (Figure 11.17) [43].

We have found that transmembrane charge separation **28**^{CS} in response to the irradiation of helix **28** is translated into external EDTA oxidation [$E_{1/2}$ (NHE) ≈ +430 mV] and internal quinone reduction [$E_{1/2}$ (NHE) ≈ −60 mV] for the fluorometric detection of photoactivity as intravesicular deacidification with light was detected utilizing the HPTS dye. The activity of photosystem **28** was perfectly reflected in nearly quantitative ultrafast formation (<2 ps, >97%) and a long lifetime of the charge-separated state **28**^{CS} (61 ps). The optical signature of this crucial **28**^{CS} was identified with the appearance of a new band at 480–540 nm in the transient

11.6 Applications of Core-Substituted NDIs

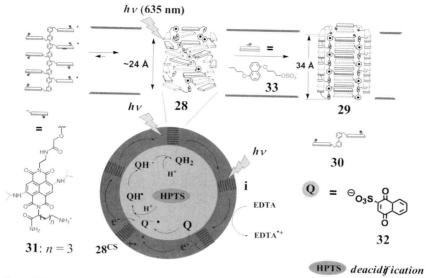

Figure 11.16 Rigid-rod π-M-helix **28** as a supramolecular photosystem that can open up into an ion channel **29** after intercalation with the aromatic ligand **33**. Transmembrane photoinduced electron transfer from EDTA donors to quinone acceptors Q is measured as formal proton pumping with light across lipid bilayers. HPTS is used to measure intravesicular deacidification with light.

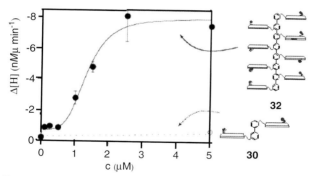

Figure 11.17 Intervesicular proton consumption during irradiation at 635 nm as a function of the concentration of NDI rigid rod **32** and NDI dimer **30**, detected with change in pH of HPTS. Figure is adapted from reference 9.

absorption spectrum of a monomeric N,N-NDI in the presence of an electron donor (DMA, N,N-dimethylaniline). Stabilization of the radical cation N,N-NDI$^{+\bullet}$ with acceptor TCNE (tetracyanoethylene) was not detectable. The transient N,N-NDI$^{-\bullet}$ band of photosystem **28** was, therefore, used to characterize the charge-separated state **28**^{CS}. The disappearance of activity upon shortening of the rigid-rod scaffold in N,N-NDI biphenyl **30** proved the importance of the π-helix for the photosynthetic reaction center [43].

11.7
Prospects and Conclusion

The formation of functional supramolecular materials became possible with the introduction of aromatic electron donor–acceptor interactions between DAN donors and NDI acceptors; these allowed detailed study using covalent and non-covalent systems of host–guest chemistry, including catanenes, rotaxanes, and ion channel ligand gating. NDIs have been used to prepare thin films with zirconium phosphonate, peptide nanotubes, anion transport systems, and construction for use in sensing. The core-substituted NDIs are conducting and colorful, with charge mobilities that depend on the precise nature of the supramolecular organization. The usefulness of these characteristics in creating function in lipid bilayer membranes was confirmed with the synthesis of smart photosystems, and also the potential to fill "green gaps" in light-harvesting systems. Synthetic access to functional π-stacking architecture in lipid bilayer membranes may open several doors. Research on functional NDI architecture will in any case contribute to the major effort in basic research towards smart optoelectronic nanomaterials that, according to reports [44], are needed today to meet tomorrow's energy demands in a sustainable way. From a fundamental viewpoint, well-defined molecular non-covalent models remain ideal systems for the investigation of the relationships between structural parameters, such as the ratio between the donor and the acceptor, their distance and relative orientation, and the nature of the link between the donor and the acceptor (photoinduced energy or/and electron transfer, and long-lived charge separation) [45].

11.8
Acknowledgment

I am deeply indebted to Prof. Steven Langford for his help, stimulating suggestions and encouragement. I would thank the School of Chemistry, Monash University, Clayton for their support and facilities. I am also grateful to Ms. Katrina Lee and Mr. Ish Kumar for comments and corrections. Last but not least, I would like to give my thanks to my wife Madhuri Jawalge, whose patience and love enabled me to achieve this milestone, and I thank my brother Sidhanath Bhosale and my family members for their support.

References

1. 1a. Lokey, R.S., Kwok, Y., Guelev, V., Pursell, C.J., Hurley, L.H. and Iverson, B.L. (1997) *Journal of the American Chemical Society*, **119**, 7202. b. Takenaka, S. and Takagi, M. (1999) *Bulletin of the Chemical Society of Japan*, **72**, 327.
2. 2a. Gosztola, D., Niemczyk, M.P., Svec, W.A., Lukas, S. and Wasielewski, M.R. (2000) *The Journal of Physical Chemistry. A*, **104**, 6545. b. Greenfield, S.R., Svec, W.A., Gosztola, D. and Wasielewski, M.R. (1996) *Journal of the American Chemical Society*, **118**, 6767.
3. Angadi, M.A., Gosztola, D. and Wasielewski, M.R. (1998) *Journal of Applied Physiology*, **83**, 6187.
4. Levanon, H., Galili, T., Regev, A., Wiederrecht, G.P., Svec, W.A. and Wasielewski, M.R. (1998) *Journal of the American Chemical Society*, **120**, 6366 [and see references therein].
5. Debreczeny, M.P., Svec, W.A., Marsh, E.M. and Wasielewski, M.R. (1996) *Journal of the American Chemical Society*, **118**, 8174.
6. 6a. Lee, S.K., Herrmann, Y., Zu, A., Geerts, A., Mullen, K. and Bard, A.J. (1999) *Journal of the American Chemical Society*, **121**, 3513. b. Angadi, M.A., Gosztola, D. and Wasielewski, M.R. (1999) *Materials Science and Engineering. B*, **B63**, 191.
7. 7a. Wiederrecht, G.P. and Wasielewski, M.R. (1998) *Journal of the American Chemical Society*, **120**, 3231. b. Wiederrecht, G.P., Yoon, B.A. and Wasielewski, M.R. (1995) *Science (Washington, DC)* **270**, 1794.
8. Mukhopadhyay, P., Iwashita, Y., Shirakawa, M., Kawano, S., Fujita, N. and Shinkai, S. (2006) *Angewandte Chemie (International Ed. in English)*, **45**, 1592 [for sensing see references therein].
9. Bhosale, S., Sisson, A.L., Sakai, N. and Matile, S. (2006) *Organic and Biomolecular Chemistry*, **4**, 3031 [for π–π-stacking see references therein].
10. Katz, H.E., Lovinger, A.J., Johnson, J., Kloc, C., Siegrist, T., Li, W., Lin, Y.Y. and Dodabalapur, A. (2000) *Nature*, **404**, 478 [see references therein].
11. 11a. Fallon, G.D., Lee, M.A.-P., Langford, S.J. and Nichols, P.J. (2004) *Organic Letters*, **6**, 655. b. Hamilton, D.G., Davies, J.E., Prodi, L. and Sanders, J.K.M. (1998) *Chemistry–A European Journal*, **4**, 608.
12. Miller, L.L., Duan, R.G., Hong, Y. and Tabakovic, I. (1995) *Chemistry of Materials: A Publication of the American Chemical Society*, **7**, 1552.
13. Stewart, W.W. (1981) *Nature*, **292**, 17.
14. Vicic, D.A., Odom, D.T., Nunez, M.E., Gianolio, D.A., McLaughlin, L.W. and Barton, J.K. (2000) *Journal of the American Chemical Society*, **122**, 8603.
15. Cammarata, V., Atanasoska, L., Miller, L.L., Kolaskie, C.J. and Stallman, B.J. (1992) *Langmuir: The ACS Journal of Surfaces and Colloids*, **8**, 876.
16. 16a. Horne, W.S., Ashkenasy, N. and Ghadiri, M.R. (2005) *Chemistry–A European Journal*, **11**, 1137. b. Pengo, P., Panto, G.D., Otto, S. and Sanders, J.K.M. (2006) *The Journal of Organic Chemistry*, **71**, 7063. c. Lee, M.A.-P. (2004). "Molecular Assemblies and Supramolecular Arrays Based on Napthalene Diimides and Sn (IV) porphyrin phenolates," PhD thesis, Monash University, Australia.
17. Vollmann, H., Becker, H., Corell, M. and Streeck, H. (1937) *Justus Liebigs Annalen Der Chemie*, **531**, 1.
18. Würthner, F., Ahmed, S., Thalacker, C. and Debacrdemacker, T. (2002) *Chemistry–A European Journal*, **8**, 4742.
19. Thalacker, C., Roger, C. and Würthner, F. (2006) *The Journal of Organic Chemistry*, **71**, 8098.
20. 20a. Andric, G., Boas, J.F., Bond, A.M., Fallon, G.D., Ghiggino, K.P., Hogan, C.F., Hutchison, J.A., Lee, M. A.-P., Langford, S. J., Pilbrow, J.R., Troup, G.J. and Woodward, C.P. (2004) *Australian Journal of Chemistry*, **57**, 1011 [for detailed photophysical study see references therein]. b. Aveline, B.M., Matsugo, S. and Redmond, R.W. (1997) *Journal of the American Chemical Society*, **119**, 11785.
21. Osuka, A., Zhang, R-P., Maruyama, K., Ohno, T. and Nozaki, K. (1993) *Bulletin of the Chemical Society of Japan*, **66**, 3773.

22 22a. Cubberley, M.S. and Iverson, B.L. (2001) *Journal of the American Chemical Society*, **123**, 7560. b. Gabriel, G.J. and Iverson, B.L. (2002) *Journal of the American Chemical Society*, **124**, 15174. c. Gabriel, G.J., Sorey, S. and Iverson, B.L. (2005) *Journal of the American Chemical Society*, **127**, 2637. d. Zych, A.J. and Iverson, B.L. (2000) *Journal of the American Chemical Society*, **122**, 8898.

23 23a. Hamilton, D.G., Lynch, D.E., Byriel, K.A., Kennard, C.H.L. and Sanders, J.K.M. (1998) *Australian Journal of Chemistry*, **51**, 441. b. Hansen, J.G., Feeder, N., Hamilton, D.G., Gunter, M.J., Becher, J. and Sanders, J.K.M. (2000) *Organic Letters*, **2**, 449.

24 24a. Hunter, C.A. (1994) *Chemical Society Reviews*, **23**, 101. b. Hunter, C.A. (1993) *Angewandte Chemie (International Ed. in English)*, **32**, 1584. c. Hunter, C.A. and Sanders, J.K.M. (1990) *Journal of the American Chemical Society*, **112**, 5525.

25 25a. Duan, R.G., Miller, L.L. and Tomalia, D.A. (1995) *Journal of the American Chemical Society*, **117**, 10783. b. Miller, L.L. and Mann, K.R. (1996) *Accounts of Chemical Research*, **29**, 417.

26 Reczek, J.J. and Iverson, B.L. (2006) *Macromolecules*, **39**, 5601.

27 27a. Talukdar, P., Bollot, G., Mareda, J., Sakai, N. and Matile, S. (2005) *Journal of the American Chemical Society*, **127**, 6528. b. Talukdar, P., Bollot, G., Mareda, J., Sakai, N. and Matile, S. (2005) *Chemistry – A European Journal*, **11**, 6525.

28 Gorteau, V., Bollot, G., Mareda, J., Perez-Velasco, A. and Matile, S. (2006) *Journal of the American Chemical Society*, **128**, 14788.

29 Horak, J., Maier, N.M. and Lindner, W. (2004) *Journal of Chromatography. A*, **1045**, 43.

30 30a. Blacker, A.J., Jazwinski, J., Lehn, J.-M., Cesario, M., Guilhem, J. and Pascard, C. (1987) *Tetrahedron Letters*, **28**, 6057. b. Kaiser, G., Jarrosson, T., Otto, S., Ng, Y.-F., Bond, A.D. and Sanders, J.K.M. (2004) *Angewandte Chemie (International Ed. in English)*, **43**, 1959. c. Iijima, T., Vignon, S.A., Tseng, H.R., Jarrosson, T., Sanders, J.K.M., Marchioni, F., Venturi, M., Apostoli, E., Balzani, V. and Stoddart, J.F. (2005) *Chemistry – A European Journal*, **10**, 6375.

31 Takenaka, S., Yamashita, K., Takagi, M., Uto, Y. and Kondo, H. (2000) *Analytical Chemistry*, **72**, 1334.

32 Takada, T., Kawai, K., Cai, X., Sugimoto, A., Fujitsuka, M. and Majima, T. (2004) *Journal of the American Chemical Society*, **126**, 1125.

33 Tanaka, H., Livinchuk, S., Tran, D.-H., Bollot, G., Mareda, J., Sakai, N. and Matile, S. (2006) *Journal of the American Chemical Society*, **128**, 16000.

34 34a. Ashkenasy, N., Horne, W.S. and Ghadiri, M.R. (2006) *Small*, **2**, 99. b. Pantos, G.D., Pengo, P. and Sanders, J.K.M. (2007) *Angewandte Chemie (International Ed. in English)*, **46**, 195.

35 35a. Hayashi, T., Miyahara, T., Hashizume, N. and Ogoshi, H. (1993) *Journal of the American Chemical Society*, **115**, 4013. b. Wasielewski, M.R. (1992) *Chemical Reviews*, **92**, 435 [and see references therein].

36 Osuka, A., Zhang, R.P., Maruyama, K., Ohno, T. and Nozaki, K. (1993) *Bulletin of the Chemical Society of Japan*, **66**, 3773.

37 37a. Shiratori, H., Ohno, T., Nozaki, K., Yamazaki, I., Nishimura, Y. and Osuka, A. (2000) *The Journal of Organic Chemistry*, **65**, 8747. b. Okamoto, K., Mori, Y., Yamada, H., Imahori, H. and Fukuzumi, S. (2004) *Chemistry – A European Journal*, **10**, 474. c. Borgström, M., Shaikh, N., Johansson, O., Anderlund, M.F., Styring, S., Åkermark, B., Magnuson, A. and Hammarström, L. (2005) *Journal of the American Chemical Society*, **127**, 17504.

38 Osuka, A., Yoneshima, R., Shiratori, H., Okada, T., Taniguchi, S. and Mataga, N. (1998) *Chemical Communications*, **15**, 1567.

39 39a. Ghiggino, K.P., Hutchison, J.A., Langford, S.J., Latter, M.J. and Takezaki, M. (2006) *Journal of Physical Organic Chemistry*, **19**, 491. b. Flamigni, L., Johnston, M.R. and Giribabu, L. (2002) *Chemistry – A European Journal*, **8**, 3938.

40 Ghiggino, K.P., Hutchison, J.A., Langford, S.J., Latter, M.J., Lee, M.A.-P. and Takezaki, M. (2006) *Australian Journal of Chemistry*, **59**, 179 [and see references therein].

41 Thalacker, C., Miura, A., De Feyter, S., De Schryver, F.D. and Würthner, F. (2005) *Organic and Biomolecular Chemistry*, **3**, 414.

42 Roger, C., Muller, M.G., Lysetska, M., Milsolavina, Y., Holzwarth, A.R. and Würthner, F. (2006) *Journal of the American Chemical Society*, **128**, 6542.

43 Bhosale, S., Sission, A.L., Talukdar, P., Furstenberg, A., Banerji, N., Vauthey, E., Bollot, G., Mareda, J., Roger, C., Würthner, F., Sakai, N. and Matile, S. (2006) *Science*, **313**, 84.

44 Hess, G. (2005) Basic research needs for solar energy utilization, US Department of Energy, *Chemical and Engineering News*, **83**, 12.

45 Bhosale, S.V., Jani, C. and Langford, S. (2008) *Chemical Society Reviews*, DOI: 10.1039/b615857a.

12
Coordination Chemistry of Phosphole Ligands Substituted with Pyridyl Moieties: From Catalysis to Nonlinear Optics and Supramolecular Assemblies

Christophe Lescop and Muriel Hissler

12.1
Introduction

Linear π-conjugated oligomers and polymers based on a planar backbone of sp^2-bonded carbon atoms have attracted increasing interest in recent years owing to their potential application in electronic devices (e.g., light-emitting diodes, thin-film transistors, photovoltaic cells) [1]. The optical and electronic properties of these macromolecules depend significantly on their HOMO (highest occupied molecular orbital)–LUMO (lowest occupied molecular orbital) gap and the electron density of the carbon backbone. Several strategies have been developed to vary these parameters with the aim of preparing novel conjugated frameworks with enhanced performance. A successful strategy to obtain such derivatives involves the incorporation of aromatic heterocycles in the backbone of π-conjugated systems [1d, 2]. The possibility of incorporating heterocyclic building blocks with different aromatic character and electronic nature allows for engineering at the molecular level. Indeed, low aromaticity favors electron delocalization along the main chain, while maximum conjugation is obtained in copolymers with alternating electron-rich and electron-deficient subunits due to intramolecular charge transfer (ICT) [2a–2d]. Heterocyclopentadienes have been used extensively for such purposes, since their electronic properties depend on the nature of the heteroatoms. For example, metalloles of group 14 elements (siloles, germoles, stannoles) exhibit a high electron affinity [2i–2l], whereas those of groups 15 (pyrrole) and 16 (furan, thiophene) have electron-rich aromatic π-systems [2f].

Although the chemistry of phospholes is well developed [3], these phosphacyclopentadienes have received little attention as building blocks for the design of π-conjugated systems [4]. The scarcity of such materials is surprising, since phospholes have certain properties that make them attractive synthons for the construction of π-conjugated systems [5a, 5b]. In this heterocycle, the phosphorus atom does not readily form sp hybrids and mainly employs p electrons for bonding. This leads to a pyramidal geometry of the tricoordinate phosphorus atom and a pronounced s character of the lone pair [3, 5]. These geometric and electronic features

Tomorrow's Chemistry Today. Concepts in Nanoscience, Organic Materials and Environmental Chemistry.
Edited by Bruno Pignataro
Copyright © 2008 WILEY-VCH Verlag GmbH & Co. KGaA, Weinheim
ISBN: 978-3-527-31918-3

prevent efficient interaction of the lone pair on the phosphorus atom with the endocyclic diene system. Hence, phospholes possess (i) a weak aromatic character, which should favor delocalization of the π-system, and (ii) a reactive heteroatom, which potentially offers the possibility of tuning the HOMO and the LUMO levels by chemical modification. Moreover, the presence of this reactive P-atom allows the use of the tool of coordination chemistry for the three-dimensional organization of π-conjugated chromophores. This organization in order to control and understand the electronic interactions between individual chromophoric subunits is of fundamental importance for the engineering of organic materials relevant to electronic or optoelectronic applications. New supramolecular assemblies have thus been characterized whose physical properties depend on the intrinsic properties of the phosphole-based ligand as well as on their controlled specific organization. Here we give an account on our recent work devoted to the synthesis, the organization and the applications of co-oligomers alternating phosphole and thiophene and/or pyridine rings.

12.2
π-Conjugated Derivatives Incorporating Phosphole Ring

12.2.1
Synthesis and Physical Properties

2,5-Di(heteroaryl)phospholes **3** (Scheme 12.1) were prepared via the "Fagan–Nugent method," a general and efficient organometallic route to phosphole moieties [6]. The intramolecular oxidative coupling of functionalized 1,7-diynes **1**, possessing a $(CH_2)_4$ spacer in order to obtain the desired 2,5-substitution pattern, with "zirconocene" provides the corresponding zirconacyclopentadienes **2** [7]. These organometallic intermediates react with dihalogenophosphines to give the corresponding phospholes **3** in medium to good yields. The stability of these phospholes **3** is intimately related to the nature of the P-substituent. 1-Phenylphospholes **3a–g** (Scheme 12.1) can be isolated as air-stable solids following flash column chromatography on basic alumina. In contrast, 1-cyclohexyl phosphole **3b′**, 1-alkylphosphole **3b″**, and 1-aminophosphole **3b‴** are extremely air- and moisture-sensitive compounds and they can only be handled under inert atmosphere [7d].

It is noteworthy that this synthetic route allows the preparation of symmetric (Ar = Ar′) and unsymmetric (Ar ≠ Ar′) 2,5-disubstituted 1-phenylphospholes. This dissymmetry can be easily introduced at the level of the functionalized 1,7-diynes **1** through successive Sonogashira coupling reactions [8]. A large variety of functional groups can be introduced, leading to numerous 1-phenylphosphole derivatives **3a–f** with markedly different physical properties [8]. Moreover, oligo(heterocycle-phosphole) derivative **3g** with precise length has also been obtained by this synthetic procedure [9].

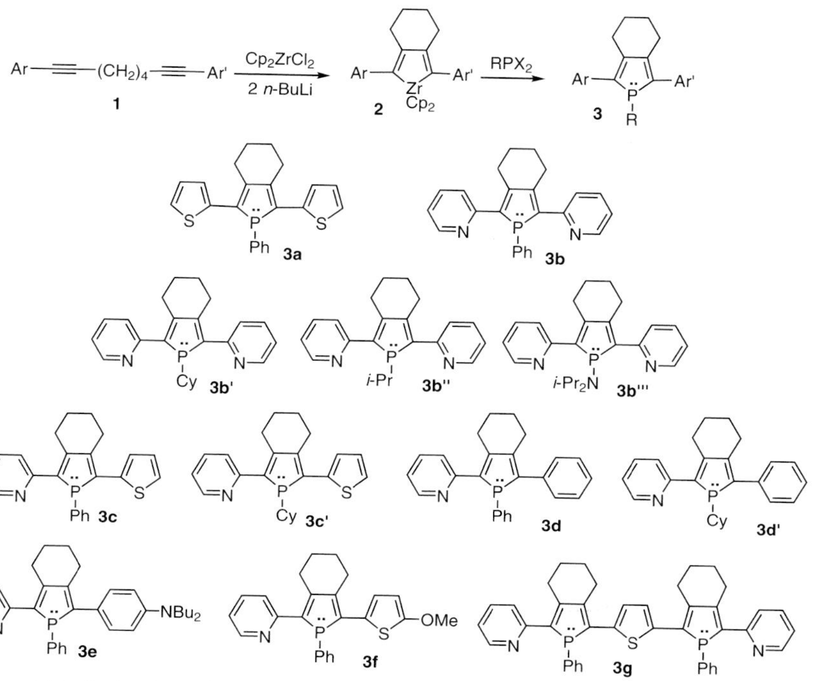

12.2 π-Conjugated Derivatives Incorporating Phosphole Ring | 297

Scheme 12.1 Synthesis of phosphole derivatives **3**.

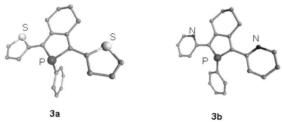

3a   **3b**

Figure 12.1 Perspective views of the crystallographic structures of the phosphole-based ligands **3a** and **3b**. Hydrogen atoms have been omitted for clarity.

The σ^3,λ^3-phospholes **3a** [7b, 7d] and **3b** [7c, 7d] (Scheme 12.1) bearing electron-rich and electron-deficient substituents, respectively, were characterized by x-ray diffraction studies (Figure 12.1). In spite of the different electronic natures of the two 2,5-substituents, compounds **3a** and **3b** share some important structural features in the solid state. Their molecular crystal structures reveal that the three heterocycles are almost coplanar while the phosphorus atom's environment is strongly pyramidalized [7b–7d].

Moreover, the carbon–carbon lengths between the rings are in the range expected for Csp^2–Csp^2 bonds. These solid state data indicate that an efficient delocalization of the π-system takes place over the three heterocycles in oligomers **3a,b**, as supported by theoretical studies [7d].

The presence of an extended π-conjugated system in phospholes **3a,b** was also confirmed by the observation of a broad absorption in the visible region of their UV-vis spectrum. The energy of these absorptions, attributed to π–π* transitions, depends dramatically on the nature of the 2,5-substituents of the phosphole ring [7d]. The values of λ_{max} and the optical end absorption λ_{onset} (the solution optical "HOMO–LUMO" gap) recorded in solution [7d], become red-shifted on replacing the 2-pyridyl groups (λ_{max} = 390 nm) by 2-thienyl rings (λ_{max} = 412 nm). Theoretical studies showed that this bathochromic shift is due to a better interaction between the HOMO of phosphole with the HOMO of thiophene, compared to that of pyridine [7d, 10]. It is noteworthy that the λ_{max} value recorded for **3a** (412 nm) [7d] is considerably more red-shifted than that of the related *tert*-thiophene (355 nm) [11]. This observation is in agreement with theoretical studies predicting that heterocyclopentadienes with a low aromatic character are optimal building blocks for the synthesis of extended π-conjugated systems with low HOMO–LUMO gap [1d, 7a].

Phospholes **3a,b** are also fluorophores and their λ_{em} values depend on the nature of the 2,5-substituents. A blue emission is observed for 1-phenyl-2,5-di(2-pyridyl)phosphole **3b** (λ_{em} = 463 nm) whereas the emission of 1-phenyl-2,5-di(2-thienyl)phosphole **3a** is red-shifted ($\Delta\lambda_{em}$ = 35 nm). Cyclic voltammetry (CV) performed at 200 mV s^{-1} revealed that the redox processes observed for all the σ^3-phospholes **3** are irreversible and that their redox properties are related to the electronic properties of the phosphole substituents [7d]. For example, derivative **3a** featuring electron-rich thienyl substituents is more easily oxidized than compound **3b** which possesses electron-deficient pyridyl substituents [7d].

12.2.2
Fine Tuning of the Physical Properties via Chemical Modifications of the Phosphole Ring

One of the appealing properties of phosphole rings is the versatile reactivity of the endocyclic heteroatom that allows creation of structural diversity in these types of systems. This feature offers a direct access to a broad range of new π-conjugated systems from single P-containing chromophores, without the need for additional multistep syntheses, as illustrated with dithienyl-phospholes **3a** (Scheme 12.2).

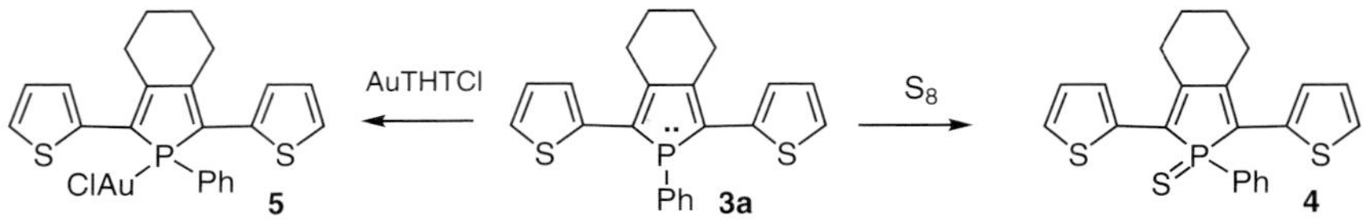

Scheme 12.2 Chemical modifications on the nucleophilic P center.

The chemical modifications of the nucleophilic P-center have a profound impact on the optical and electrochemical properties of the phosphole oligomers as a whole. For example, upon modification of the σ^3-phosphole **3a** toward the neutral σ^4-derivatives **4** and **5** (Scheme 12.2), a red shift in their emission spectra recorded in solution, together with an increase and decrease of their oxidation and reduction potentials, respectively, is observed. It is also noteworthy that the quantum yield of the gold complex **5** ($\Phi = 14\%$) is much higher than those of the corresponding phosphole **3a** ($\Phi = 5\%$) or its thioxo-derivative **4** ($\Phi = 4.6\%$), illustrating that the nature of the P-modification has a considerable impact on the optical properties of these fluorophores [7d, 12]. Comparison of the solution and thin-film optical properties of the thioxophosphole **4** shows that this compound does not form aggregates in the solid state. This property, which is supported by an x-ray diffraction study, results in an enhancement of the fluorescence quantum yields in the solid state. In contrast, the solution and thin-film emission spectra of the gold complex **5** are different. Two broad emission bands are observed for the thin films, one at a wavelength similar to that of the solution spectrum (550 nm), and a second which is considerably red-shifted (690 nm). The excitation spectra of **5** in thin film, recorded for emission at 550 nm or 700 nm, are similar to its absorption spectrum, clearly indicating that both emissions are due to the (phosphole)gold complex **5**. The low-energy luminescence band obtained in thin film with phosphole-Au(I) complex **5** very probably arises from the formation of aggregates.

Single-layer and multilayer organic light-emitting diodes (OLEDs) using these P-derivatives as the emissive layer have been fabricated. The emission color of these devices and their performances vary with the nature of the P material. Upon sublimation, phosphole **3a** decomposed while the more thermally stable thioxo-derivative **4** formed homogeneous thin films on an indium–tin oxide (ITO) semi-transparent anode, allowing a simple layer OLED to be prepared [12]. This device exhibits yellow emission for a relatively low turn-on voltage of 2 V. The comparatively low maximum brightness (3613 cd m^{-2}) and electroluminescence (EL) quantum yield (0.16%) can be increased by nearly one order of magnitude using a more advanced device, in which the organic layer consisting of **4** was sandwiched between hole- and electron-transporting layers (α-NPD and Alq$_3$, respectively). Upon doping the "phosphole" layer with a red-emitting dopant (DCJTB), the EL efficiency is further enhanced to 1.83% with a maximum brightness of ~37 000 cd m^{-2} [12]. The ligand behavior of **3a** can be exploited in an innovative approach whereby metal complexes are investigated as materials for OLEDs. The Au(I) complex **5** (Scheme 12.2) is thermally stable enough to give homogeneous thin films upon sublimation in high vacuum. The corresponding single-layer device exhibited an EL emission covering the 480–800 nm domain [12], a result which is encouraging for the development of white-emitting OLEDs based on phosphole–Au complexes. Note that the low-energy emissions are very probably due to the formation of aggregates in the solid state [12]. These results constitute the first application in optoelectronics of π-organophosphorus materials.

In conclusion, co-oligomers with a phosphole core exhibit very interesting properties. They are air-stable providing that the P-substituent is a phenyl group; they exhibit low optical gap; and their electronic properties can be tuned by varying the nature of the 2,5-substituents (2-thienyl or 2-pyridyl) or chemical modification of the phosphole ring. Exploitation of this method of tailoring π-conjugated systems, which is not possible with pyrrole and thiophene units, has led to the optimization of the properties of thiophene–phosphole co-oligomers for their use as materials suitable for OLED fabrication [12].

12.3
Coordination Chemistry of 2-(2-Pyridyl)phosphole Derivatives: Applications in Catalysis and as Nonlinear Optical Molecular Materials

Compared with organic molecules, metal complexes offer a larger variety of molecular structures and a diversity of electronic properties [13]. Since it is well known that phosphole and pyridines behave as classical two-electron donors toward transition metals [13, 14], we have explored the coordination behavior of 2-pyridyl-substituted phospholes with the aim of obtaining complexes exhibiting interesting physical properties. Moreover, original structures can be expected since these heteroditopic P,N-ligands possess two coordination centers with different stereoelectronic properties. The performances of such complexes have been evaluated as catalysts as well as new materials for optoelectronics.

12.3.1
Syntheses and Catalytic Tests

The coordination chemistry of 2-(2-pyridyl)phosphole ligands toward Pd(II) centers was investigated with the aim of gaining insights into their coordination behavior. Thus, the reaction of **3b–d** with [(cod)PdMeCl] (cod = 1,5-cyclo-1,5-octadiene) afforded the corresponding air-stable complexes **6b–d** as single diastereoisomers in excellent yields (Scheme 12.3) [15].

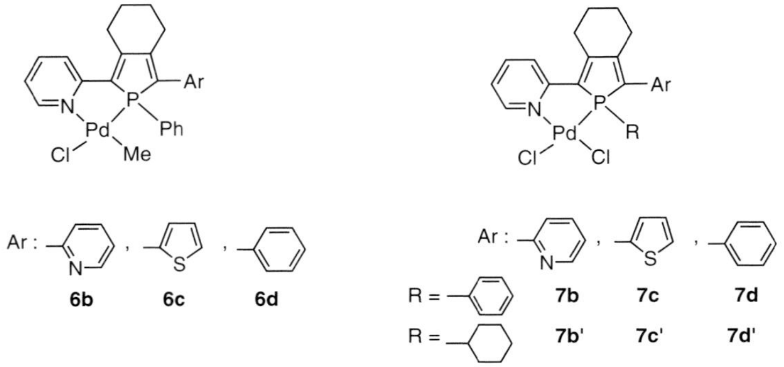

Scheme 12.3 Palladium complexes.

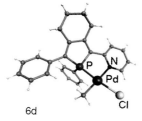

Figure 12.2 View of the crystallographic structure of derivative **6d**.

X-ray diffraction studies performed on complex **6d** (Figure 12.2) revealed a *cis*-arrangement of the phosphorus atom and the methyl group, as a consequence of the classical *trans-effect* occurring in the coordination sphere of square planar d^8 metal ions [16].

Very few P,N-ligands incorporating phosphole rings have been used in homogeneous catalysis, while bidentate P,N-ligands combining σ^3,λ^3-phosphorus and sp^2-hybridized nitrogen donor centers have been particularly studied [17]. Owing to the different stereoelectronic properties of the two coordination sites, these can indeed act as hemilabile ligands and induce selective processes, allowing control over the reactivity of the metal center. Mixed P,N-chelates are also efficient ligands for important catalytic reactions, including reduction [18], allylic alkylation [19], and the Heck reaction [19c]. In several cases, owing to their specific properties, catalytic activities or selectivities higher than those observed with diphosphines or dinitrogen donor ligands have been achieved. It is noteworthy that the cationic version of complex **6d** turned out to be among the most efficient mixed P,N-ligand catalyst for carbon monoxide–olefin copolymerization [15].

12.3.2
Isomerization of Coordinated Phosphole Ring into 2-Phospholene Ring

As described in the case of complexes **6b–d**, the reaction of **3b–d** and of their P-cyclohexyl-substituted analogues **3b′–d′** with [(CH$_3$CN)$_2$PdCl$_2$] allows the characterization of the neutral complexes **7b–d**, **7b′–d′** (Scheme 12.3) as single diastereoisomers in excellent yields.

It is interesting that x-ray diffraction studies of complex **7c′** revealed a significant degree of ring strain imposed by the formation of the five-membered metallacycle (Figure 12.3) [8c, 8d] as shown by the nonplanarity of the sp^2 carbon C(1) (see Figure 12.3).

This ring strain permits a base-catalyzed stereospecific [1,3]-H migration to afford complexes **8b–d**, **8b′–d′** containing 2-phospholene ligands (Scheme 12.4) [8c, 8d]. In this family of complexes, ligands featuring a pendant pyridyl group spontaneously evolve to the corresponding 2-pyridyl-2-phospholene derivatives (Scheme 12.4) in the coordination sphere of Pd(II). The isomerization of ligands lacking pendant pyridyl groups was accomplished by adding bases such as

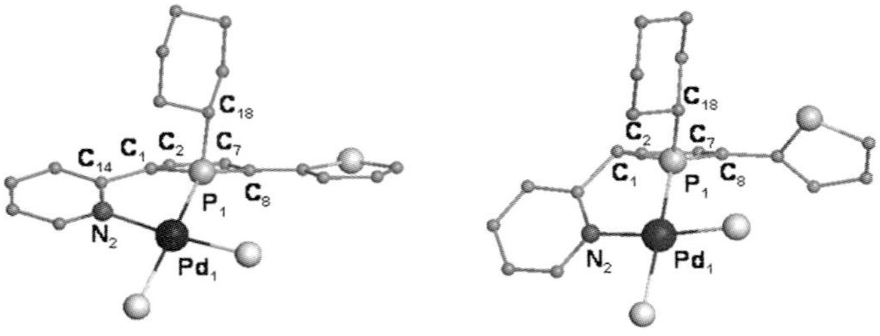

Figure 12.3 Simplified views of the solid-state structures of phosphole (**7c′**) and phospholene (**8c′**) PdCl$_2$ complexes; The cyclohexyl ring fused to the phosphole ring in the case of **7c′** and to the 2-phospholene ring in the case of **8c′** have been omitted as well as the hydrogen atoms.

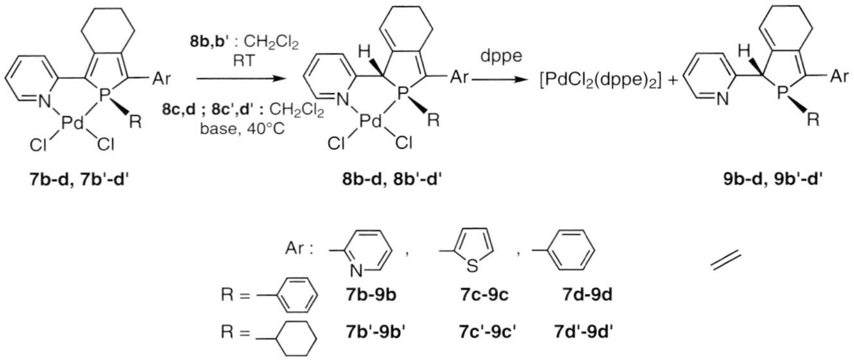

Scheme 12.4 Synthesis of the 2-phospholene derivatives.

pyridine or diethylamine to the reaction media. To the best of our knowledge, complexes **8b–d, 8b′–d′** are the first P,N-chelates that incorporate a 2-phospholene moiety. Ligands based on 2-phospholenes have been indeed poorly investigated owing to the somewhat underdeveloped chemistry of this P-heterocycle [20, 21]. Furthermore, their syntheses via isomerization of the corresponding 2-(2-pyridyl)phospholes are a very attractive route for several reasons. Firstly, the phosphole precursors are readily available and their substitution pattern can easily be varied [7d, 15]. Secondly, the isomerization is not sensitive to the nature of the P-substituent, allowing the stereoelectronic properties of the P-donor to be tuned. Lastly, the [1,3]-H shift leading to **8b–d, 8b′–d′** creates a new stereogenic center (the C(1) carbon atom, Figure 12.3), and the fact that only one diastereoisomer out of the two possible is detected by NMR spectroscopy shows that this process is stereoselective. X-ray diffraction studies of complex **8c′** showed that the H-atom

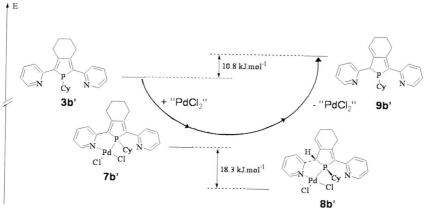

Figure 12.4 DFT-calculated relative energies of isomers **3b'** and **9b'** and of the corresponding Pd(II) complexes **7b'** and **8b'**.

linked to the C(1)-center and the P-substituent adopt a mutually *cis*-configuration toward the 2-phospholene cycle (Figure 12.3).

This transformation did not occur with free 2-(2-pyridyl)phospholes, even at high temperatures in the presence of pyridine [8d]. Density functional theory (DFT) calculations performed on the basis of the P-cyclohexyl derivative **3b'** analogue of the ligand **3b** confirmed that coordination to a Pd(II) center is required to make this isomerization thermodynamically feasible (Figure 12.4) [8c, d]. This isomerization has also been demonstrated with Pt(II) precursors [8d].

Finally, the free 2-pyridyl-2-phospholenes **9b–d**, **9b'–d'** were obtained by reacting the complexes **8b–d**, **8b'–d'** with one equivalent of 1,2-bis(diphenylphosphino)ethane (dppe, Scheme 12.4). No inversion of the P-atom of the phospholene ring was observed up to 90 °C, indicating that this novel family of P,N-chelates is a promising class of ligand for homogeneous catalysis. Thus, 2-(2-pyridyl)-2-phospholenes are efficient ligands for the palladium-catalyzed telomerization of isoprene with diethylamine [22]. It is noteworthy that this base-promoted isomerization of coordinated phospholes into 2-phospholenes can be a serious drawback since the extended π-conjugated system of these phosphole-based ligands collapses in the corresponding 2-phospholene derivatives in such media.

12.3.3
Square-Planar Complexes Exhibiting Nonlinear Optical Activity

The pyridylphosphole ligands possess two coordination centers with different stereoelectronic properties which, in accordance with the Pearson antisymbiotic effect [16c], allows the control the orientation of a second chelating ligand in the

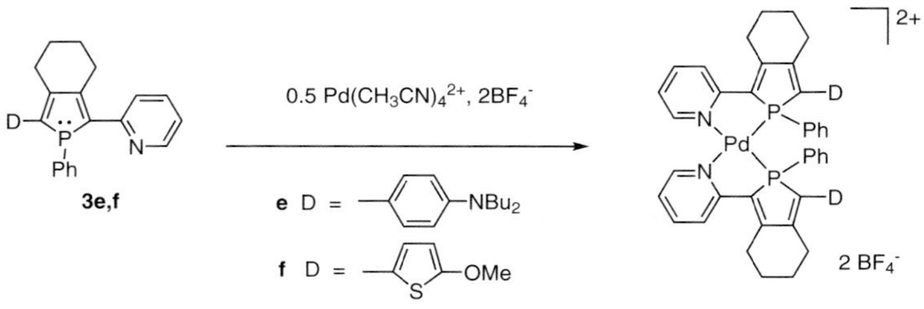

Scheme 12.5 Synthesis of complexes exhibiting high NLO activities.

coordination sphere of a square planar d^8-metal center. This property has been exploited in order to control the in-plane parallel arrangement of 1D-dipolar chromophores. Phospholes **3e** and **3f** (Scheme 12.1), have a typical "D–(π-bridge)–A" dipolar topology incorporating a 1,4-P,N-chelate. They exhibit moderate nonlinear optical activities ($\beta_{1.9\mu m} \not\subset 30 \times 10^{-30}$ e.s.u.) compared with classical chromophores such as Dispersed Red 1 (DR1) ($\beta \not\subset 50 \times 10^{-30}$ e.s.u., where 2.694×10^{-30} e.s.u = $10^{-50} Cm^3 V^{-2}$) [8a]. These low values are consistent with the weak acceptor character of the pyridine group. However, their potential in nonlinear optics (NLO) is considerably increased by their coordination behavior toward Pd(II) centers. The reaction of two equivalents of 2-(2-pyridyl)phospholes **3e,f** with [Pd(CH$_3$CN)$_4$][BF$_4$]$_2$ afforded complexes **10e,f** as single diastereoisomers (Scheme 12.5) [8a], with the P and N donor in a mutual *trans*-arrangement. Clearly, the *trans*-influence can overcome the natural tendency of 1D-dipolar chromophores to adopt an antiparallel alignment as the square-planar metal center acts as a template imposing a noncentrosymmetric assembly of two identical 1D-chromophores **3e,f**. Furthermore, the metal plays a puzzling role since a considerable enhancement of the NLO-activities is observed upon complexation. For example, complex **10e** exhibits fairly high nonlinear optical activities with β value of 180×10^{-30} e.s.u; a value which is much higher than the sum over the contribution of two sub-chromophores **3e**. In a first approach, this effect could be related to an increase of the acceptor character of the pyridine groups [23]. However, it is very likely that the origin of this large β enhancement is due to metal-to-ligand charge transfers (MLCT) or ligand-to-metal-to-ligand charge transfers (LMLCT), evidenced by UV-vis spectroscopy [8a], that contribute coherently to the second-harmonic generation. In conclusion, the coordination chemistry offers a simple synthetic methodology for controlling the in-plane parallel arrangement of P,N-dipoles in a molecular assembly.

12.3.4
Ruthenium Complexes

Great attention has been paid to the synthesis and photophysical properties of dinuclear ruthenium complexes containing bridging π-conjugated ligands [24].

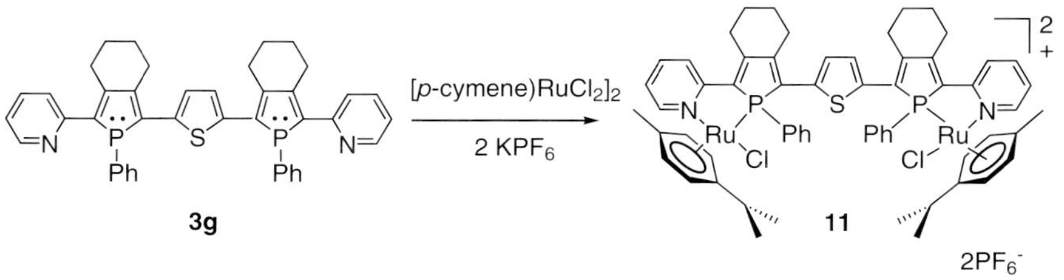

Scheme 12.6 Synthesis of ruthenium complex **11**.

The design of the bridging ligands allows tuning of the electronic properties of these dyads and has led to important progress in the field of photoinduced electron or energy transfer processes. These properties prompted us to investigate a new family of derivatives possessing two terminal (2-pyridyl)phosphole moieties joined by a 2,5-thienyl spacer as potential bridging ligands to build bimetallic ruthenium complexes.

The target complex **11** was obtained by addition of [(p-cymene)RuCl$_2$]$_2$ and two equivalents of KPF$_6$ to the ligand **3g** (Scheme 12.6) [9]. This coordination is diastereoselective, each diastereoisomer of **3g** giving one diastereoisomer of **11**. The UV-vis spectrum of **3g** exhibits an absorption maximum in the visible region at 482/nm attributed to π–π^* transitions of the π-conjugated system [9]. This value is considerably red-shifted ($\Delta\lambda_{max}$ = 84 nm) compared to that of **3c** (Scheme 12.1). The coordination of the 2-pyridylphosphole moieties of **3g** has almost no influence on the λ_{max} observed for complex **11**. This is probably due to the symmetrical structure of the complex **11**, which prevents efficient interligand charge transfer. It is noteworthy that this broad ligand-centered transition overlaps the possible MLCT transition arising from Ru(II) to π^*-orbital of the coordinated 2-pyridylphosphole moiety.

12.4
Coordination Chemistry of 2,5-(2-Pyridyl)phosphole Derivatives: Complexes Bearing Bridging Phosphane Ligands and Coordination-driven Supramolecular Organization of π-Conjugated Chromophores

12.4.1
Bimetallic Coordination Complexes Bearing a Bridging Phosphane Ligand

In comparison to the other 2-(2-pyridyl)-phosphole P,N-ligands **3c–f**, 2,5-bis(2-pyridyl)-phosphole ligand **3b** revealed a very original coordination chemistry toward Pd(I), Pt(I), and Cu(I) metal centers. A family of coordination complexes exhibiting a bridging phosphane coordination mode was thus evidenced, a very rare coordination mode for such a commonly used family of ligands.

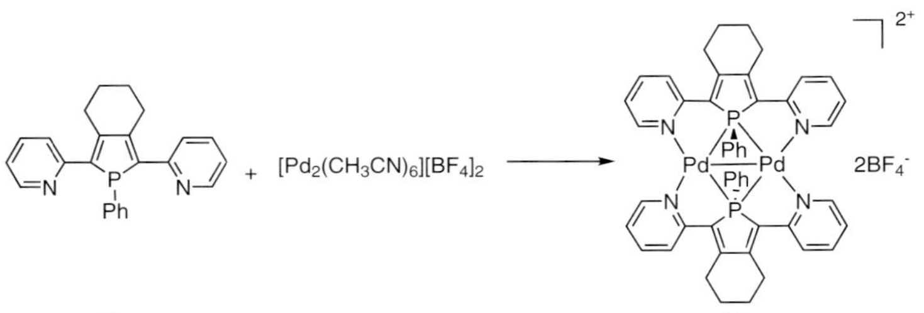

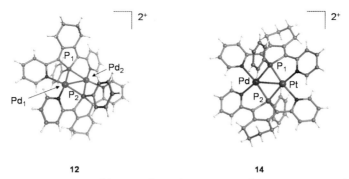

Scheme 12.7 Synthesis of the Pd(I) dimer **12**.

Figure 12.5 Views of the crystallographic structures of the dicationic complexes **12** and **14**.

12.4.1.1 Pd(I) and Pt(I) Bimetallic Complexes

The reaction of derivative **3b** with [Pd$_2$(CH$_3$CN)$_6$][BF$_4$]$_2$ afforded complex **12** in a 85% yield (Scheme 12.7) [25]. The x-ray crystal structure of the complex **12** (Figure 12.5) shows that the Pd(I) dication contains two square-planar metal centers capped by two 2,5-bis(2-pyridyl)phospholes **3b** acting as 6-electron μ-1kN:1,2kP:2kN donors.

Derivative **3b** bridges thus the two Pd(I) via a symmetrically bridging phosphane coordination mode [Δ(Pd–P) = 0.01 Å], a very rare coordination mode for a phosphane ligand that was only observed previously in the case of a family of Rh(I) complexes [26]. Such a discovery that tertiary phosphanes R$_3$P can act as bridging ligands is a breakthrough in coordination chemistry [27], since binucleating ligands potentially allow the synthesis of dinuclear and polynuclear complexes that are of great interest in manifold fields such as catalysis and bio-inorganic and materials sciences [28].

Theoretical calculations have identified that Pd–Pd and Pd–P bonds in **12** are highly delocalized with strong σ interactions in a system of four centers and six electrons [25a]. Owing to the original coordination mode of the ligand **3b**, the π-conjugated systems of the two equivalents of **3b** are orientated in a strictly parallel arrangement.

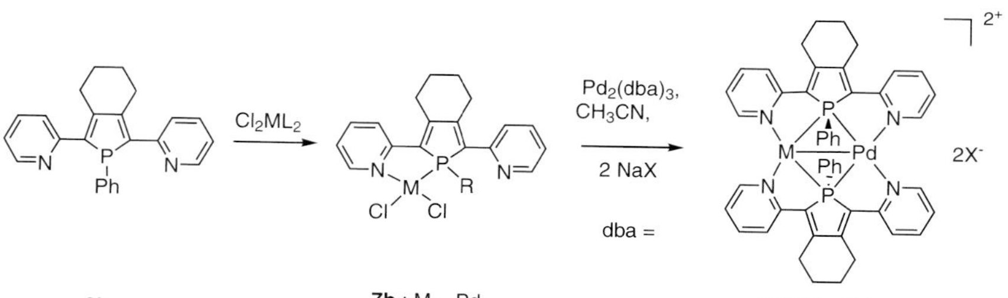

12.4 Coordination Chemistry of 2,5-(2-Pyridyl)phosphole Derivatives | 307

Scheme 12.8 Stepwise synthesis of the complexes **12** and **14**.

Complex **12** can be also synthesized via a stepwise method (Scheme 12.8) that evidences a reversible oxidative addition of the phosphole P–Ph bond on Pd(0) centers [25b].

With this method, a sequential introduction of the metal centers was undertaken in order to synthesize a heteronuclear Pd-Pt analogue **14** of complex **12** [29]. X-ray diffraction studies of this complex (Figure 12.5) revealed almost identical geometric parameters for **12** and **14**. Nevertheless, the geometry of the M_2P_2 core differs markedly in these two solid-state structures, as in the complex **14** each μ-P atom asymmetrically binds the two metal centers [Δ(μP–M): 0.083(3) Å]. Thus, in spite of the tridentate coordination mode of the N,P,N-ligand **3b**, the P atom can adopt a nonsymmetrical coordination mode. In fact, the μ-P centers of the heterobimetallic complex **14** can adopt a geometry that is intermediate between a symmetrically bridging and a semibridging coordination mode; such a result suggests that there is no substantial discontinuity between these two coordination modes for μ-PR_3 ligands. DFT calculations were performed [29] and led to the conclusion that in such complexes there is little energy cost for a bridging P center to move from a symmetrical to a substantially asymmetrical bridging position. Tiny changes such as in the metal's electronic requirements, the environment about the metal, and/or the crystal packing can be invoked as the cause for a switch from a symmetrical to an asymmetrical bridging coordination mode in these complexes.

12.4.1.2 Cu(I) Bimetallic Complexes

In an attempt to generalize this original bridging phosphane coordination mode to other metal ions, we have investigated the coordination chemistry of ligand **3b** with Cu(I) salts [29]. Thus, reaction of di(2-pyridyl)phosphole **3b** with $Cu(CH_3CN)_4PF_6$ in a 1:2 ratio afforded quantitatively the dimetallated complex **15** (Scheme 12.9).

X-ray diffraction study revealed that compound **15** is a $[Cu_2(\mathbf{3b})(CH_3CN)_4]^{2+} \cdot 2PF_6^-$ complex (Figure 12.6) in which two Cu(I) atoms are capped by a 2,5-bis(2-pyridyl)phosphole ligand **3b** acting again as a 6-electron μ-1κN:1,2κP:2κN donor.

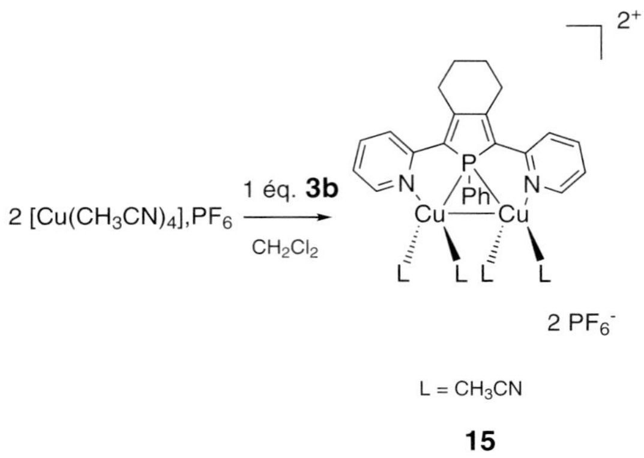

Scheme 12.9 Synthesis of the Cu(I) bimetallic complex **15**.

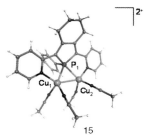

15

Figure 12.6 View of the crystallographic structure of the dicationic complex **15**.

In this complex, the P atom adopts an almost symmetrical coordination mode [$\Delta(\mu P$–Cu): 0.010(2) Å] and short intermetallic distance [2.568(10) Å] is observed, indicating metallophilic interactions between the two metal centers. Derivative **15** is the first complex in which: (i) two d^{10}-metal centers are bridged by a phosphane donor, and (ii) two metals are held together by one bridging phosphane and no other supporting ligands. These results suggest that bridging phosphane ligands are able to stabilize a large variety of bimetallic complexes and confirm that "there is no inherent thermodynamic instability associated with this bonding situation" [27a].

The complex **15** can be reacted either with one equivalent of phosphole **3b** [29] or with one equivalent of diphenylphosphinomethane ligand (dppm) [30] giving rise in very good yields to the air-stable complexes **16a** and **16b** (Scheme 12.10). Furthermore, **16a** can be reacted with a third equivalent of phosphole **3b** to afford the air-stable bimetallic complex **17** (81% yield) [29].

X-ray diffraction studies (Figure 12.7) revealed that these dicationic Cu(I)-dimers still feature one 2,5-bis(2-pyridyl)phosphole ligand **3b** acting as a µ-1kN:1,2kP:2kN

12.4 Coordination Chemistry of 2,5-(2-Pyridyl)phosphole Derivatives | 309

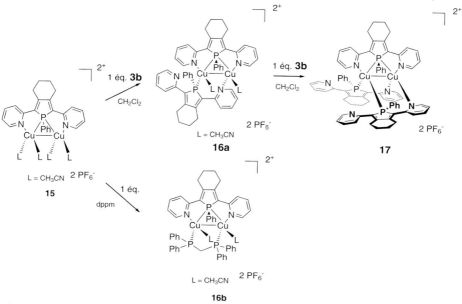

Scheme 12.10 Synthesis of the complexes **16a,b** and **17**.

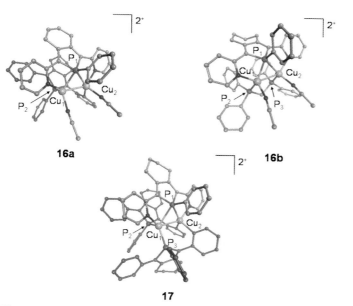

Figure 12.7 Views of the crystallographic structures of the dicationic complexes **16a**, **16b**, **17**. Hydrogen atoms have been omitted for clarity.

donor. The fact that the μ-bridging coordination mode of the P center is retained in this series highlights the robustness of ligand-bearing bridging phosphanes as binucleating ligands. The tetrahedral geometry of the Cu(I) atoms in **15** prevents the formation of a doubly bridged structure of the type observed in the Pd(I) and Pd(I)-Pt(I) dimers **12** and **14** (Figure 12.5), which requires square-planar metal centers.

This series of complexes **15–17** gives the unique opportunity to study the evolution of the bridging coordination mode of a phosphane upon decreasing the local symmetry of the bimetallic core. It is striking to observe that the difference between the two μP–Cu distances increases continuously in going from complexes **15** to **16a** and **17** [Δ(μP–Cu): **15**, 0.010(2) Å; **16a**, 0.094(2) Å; **17**, 0.295(2) Å] while the addition of the symmetrical dppm ligand to afford complex **16b** does not induce major change [Δ(μP–Cu): **16b**, 0.023(2) Å]. As observed in the case of the Pd(I)-Pt(I) heterometallic dimer **14**, the P atom of the N,P,N-pincer **3b** can adopt a nonsymmetrical bridging geometry. Moreover, these structural data confirm experimentally that there is no substantial discontinuity between symmetrical bridging and semibridging coordination modes for μ-PR$_3$ ligands. This situation is reminiscent of that of CO, the prototypical bridging ligand, and is a clue to understanding many of their key properties (fluxional behavior, bonding modes, etc.) [27a, 28d, 31]. Finally, these results show that chelates featuring bridging phosphanes are versatile binucleating ligands and that phosphanes have to be considered as potentially classical bridging ligands.

12.4.2
Supramolecular Organization of π-Conjugated Chromophores via Coordination Chemistry: Synthesis of Analogues of [2.2]-Paracyclophanes

Owing to their specific topological properties, the bimetallic Cu(I) complexes **16a** and **16b** have been efficiently used as molecular clips [32] in order to gather various π-conjugated systems into rectangular supramolecular assemblies. In such derivatives, the incorporated π-conjugated systems are forced to interact together, affording potential interest for materials in the field of molecular devices for optoelectronics. Indeed, the synthesis of π-stacked molecular assemblies is of great importance in understanding the electronic interactions between individual chromophores. One fruitful approach for probing co-facial π–π interactions involves the assembly of chromophore pairs into well-defined [2,2]paracyclophanes [33]. These molecules provide incisive insight into bulk properties of conjugated systems and are suitable π-dimer models. However, straightforward routes to these assemblies as well as for tailoring their structure remain a challenge to chemical synthesis. Hence, we have investigated a novel synthetic route to analogues of [2,2]paracyclophanes using supramolecular coordination-driven chemistry. Following the concepts of the "directional-bonding approach" [34], the construction of metalloparacyclophanes with π-stacked walls requires a bimetallic clip possessing two *cis* coordinatively labile sites that are closely aligned (Scheme 12.11).

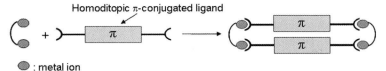

Scheme 12.11 Principle of synthesis of supramolecular rectangles.

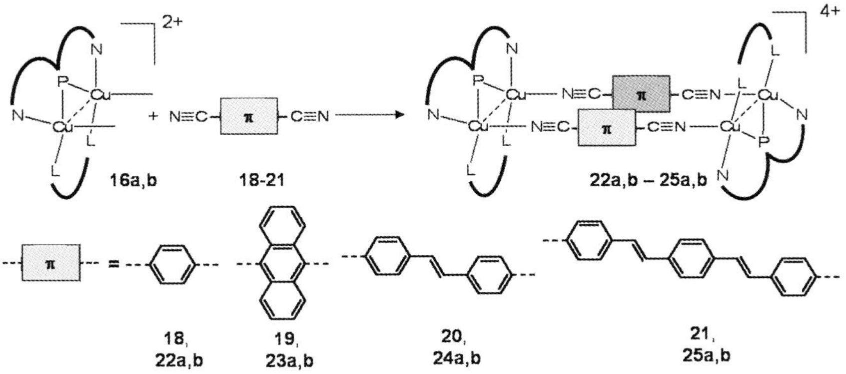

Scheme 12.12 Synthesis of the supramolecular assemblies **22a–25b**.

The two complexes **16a** [29] and **16b** [30] possess two acetonitrile ligands that have a *cisoid* arrangement. Furthermore, the bridging coordination mode of the phosphole ligand **3b** imposes a short intermetallic distance [**16a**, 2.555(1) Å; **16b**, 2.667(1) Å] resulting in a close proximity of the two kinetically labile acetonitrile ligands (N···N distances, ~3.2 Å). These geometrical characteristics are prerequisite for obtaining the organization of homoditopic ligands coordinated on **16a,16b** in new supramolecular assemblies in which electronic interactions could be forced between the cores of these homoditopic ligands (see Scheme 12.11).

Thus, bimetallic derivatives **16a,b** were reacted with linear homoditopic ligands **18–21** incorporating π-conjugated systems affording the new supramolecular rectangles **22a–25b** (Scheme 12.12) [30]. X-ray diffraction studies of these supramolecular assemblies (Figures 12.8, 12.9, and 12.10) showed that the metric data of the dimetallic clips **16a,b** do not change significantly upon their incorporation into the self-assembled structures, demonstrating the conformational rigidity of the Cu(I)-based subunits. In all cases, the four Cu-atoms lie in the same plane, defining a rectangle, and the aromatic moieties of the chromophores are parallel as result of hindered rotation. Moreover, owing to the short Cu(I)–Cu(I) intermetallic distance imposed by the bridging phosphane coordination mode, these aromatic moieties participate in face-to-face π-interactions (phenyl centroid–centroid distances: 3.4–3.5 Å) with small lateral offsets. The dimensions of these

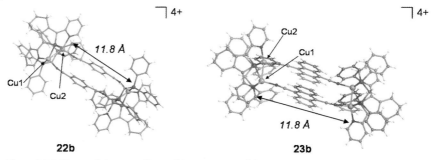

Figure 12.8 Views of the crystallographic structures of the tetracationic supramolecular rectangles **22b** and **23b**.

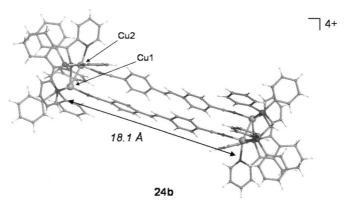

Figure 12.9 Views of the crystallographic structures of the tetracationic supramolecular rectangle **24b**.

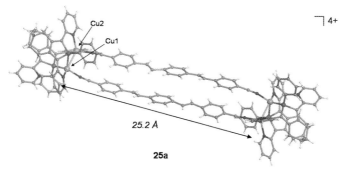

Figure 12.10 Views of the crystallographic structures of the tetracationic supramolecular rectangle **25a**.

nanosized rectangles are fixed by the size of the ditopic ligands and reach 18.1 Å for **24a,b** and 25.2 Å for **25a** (Cu–Cu distances) with an overall length for **26a** of about 39.0 Å [30].

These results show that, in spite of the repulsive interactions between the closed-shell π-clouds of the homoditopic ligands, the molecular clips **16a,b** can force face-to-face π-stacking of aromatic derivatives upon coordination into well-defined supramolecular metalloparacyclophanes. This proves that, owing to their rigidity and unique topology, complexes **16a,b** are unique building blocks for the synthesis of π-stacked molecular assemblies having a [2.2]paracyclophane-like topology. This result opens appealing perspective for the design of multifunctional molecular material via this supramolecular assembly approach.

The stacking pattern of the metalloparacyclophanes **24a,b** and **25a** is remarkable. These supramolecular rectangles make columns with short intermolecular distances (~3.6 Å). In these columns, the cationic parts of the rectangles have a parallel-displaced arrangement along the a-axes. Infinite columnar stacks resulting from *intra*- and *inter*-molecular π–π interactions of (*para*-phenylenevinylene)-based chromophores **20**, **21** are thus formed at the macroscopic scale with these novel metalloparacyclophanes (Figures 12.11 and 12.12) [30]. Organization of the π-conjugated chromophores is then observed at two hierarchical levels, the first within the metalloparacyclophanes and the second in the infinite one-dimensional stack of these metalloparacyclophanes.

Complexes **16a,b** are thus versatile molecular clips to control the organization of π-conjugated systems in supramolecular assemblies having a [2.2]paracyclophane topology. The elucidation of the electronic properties of these novel π-stacked molecular assemblies and the use of other ditopic conjugated systems to construct novel metallo[2.2]paracyclophane are under active investigations.

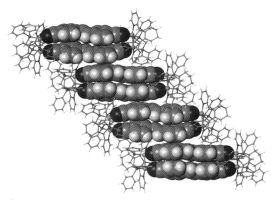

Figure 12.11 View of the packing of **24b** along the a-axes (H atoms, counteranions and solvent molecules have been omitted for clarity).

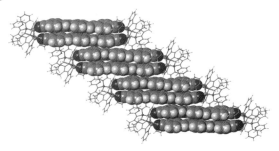

Figure 12.12 View of the packing of **25a** along the a-axes (H atoms, counteranions and solvent molecules have been omitted for clarity).

12.5
Conclusions

Our studies have demonstrated that oligomers based on phosphole rings are valuable building blocks for the preparation of π-conjugated systems. The optical and electrochemical properties of 2,5-diheteroarylphospholes can easily be tuned over a wide range by tailoring the 2,5-substituents or by performing simple chemical modifications involving the nucleophilic P-atom. These small oligomers allowed structure–property relationships to be established on the basis of experimental and theoretical data. In this field, the good stability of the phosphole ring, along with its specific electronic properties opens new perspectives in the chemistry of π-conjugated systems incorporating heavy heteroatoms.

In addition, the association of the phosphole rings with the pyridine rings has allowed the synthesis of new P,N- and N,P,N-ligands. We have demonstrated the original coordination chemistry of these derivatives and characterized new coordination complexes exhibiting appealing potential for catalysis as well as for the development of molecular materials for optoelectronics. Moreover, the discovery of the bridging phosphane coordination mode opens new perspectives as it gives the access to unprecedented dinuclear complexes that are of great interest in manifold fields such as catalysis and bio-inorganic and materials sciences.

Finally, in order to achieve operative molecular devices with specific properties, an important step lies in the control of the organization of the chromophores in the solid state. We have shown that simple reactions of organic and coordination chemistry allow the access to a variety of molecular materials where specific local and/or global organization of π-conjugated systems are induced. Coordination of P,N-chelates integrated in phosphole based π-conjugated systems afford monometallic complexes where NLO activities are generated. Moreover, starting from original coordination complexes bearing bridging phosphole ligands, a variety of organic chromophores can be organized in nanometric supramolecular assemblies where π-conjugated systems are in interaction at the level of the bulk crystalline solid.

12.6
Acknowledgments

We thank all researchers, including collaborators, colleagues, and co-workers, whose work has been presented in this chapter. We thank especially Prof. R. Réau for his valuable inputs and critical discussions giving rise to new ideas. Moreover, we thank him for giving us the opportunity to develop our scientific projects and for showing us that phosphorus chemistry can be a very powerful tool for the development of new structures. We thank the Ministère de l'Education Nationale, de la Recherche et de la Technologie, the Centre National de la Recherche Scientifique and the conseil régional de Bretagne.

References

1 a. Kraft, A., Grimsdale, A.C. and Holmes, A.B. (1998) *Angewandte Chemie (International Ed. in English)*, **37**, 403–28. b. Müllen, K. and Wegner, G. (1998) *Electronic Materials: The Oligomer Approach*, Wiley-VCH Verlag GmbH, Weinheim. c. Skotheim, T.A., Elsenbaumer, R.L. and Reynolds, J.R. (eds) (1998) *Handbook of Conducting Polymers Reynolds*, Marcel Dekker, New York. d. Roncali, J. (1997) *Chemical Reviews*, **97**, 173–205. e. Garnier, F. (1998) *Accounts of Chemical Research*, **32**, 209–15. f. Tour, J.M. (2000) *Accounts of Chemical Research*, **33**, 791–804. g. Mitschke, U. and Bäuerle, P. (2000) *Journal of Materials Chemistry*, **10**, 1471–507. h. Segura, J.L. and Martin, N. (2000) *Journal of Materials Chemistry*, **10**, 2403–35. i. Martin, R.E. and Diederich, F. (1999) *Angewandte Chemie (International Ed. in English)*, **38**, 1350–77. j. Fichou, F. (2000) *Journal of Materials Chemistry*, **10**, 571–88. k. Nguyen, T-Q., Wu, J., Doan, V., Schwartz, B.J. and Tolbert, S.H. (2000) *Science*, **288**, 652–6. l. Baldo, M.A., Thompson, M.E. and Forrest, S.R. (2000) *Nature*, **403**, 750–2. m. Shi, Y., Zhang, C., Zhang, H., Bechtel, J.H., Dalton, L.R., Robinson, B.H. and Steier, W.H. (2000) *Science*, **288**, 119–22.

2 a. Irvin, D.J., Dubois, C.J. and Reynolds, J.R. (1999) *Chemical Communications*, 2121–2. b. Zhou, Z., Maruyama, T., Kanbara, T., Ikeda, T., Ichimura, K., Yamamoto, T. and Tokuda, K. (1991) *Journal of the Chemical Society. Chemical Communications*, 1210–2. c. Yamamoto, T., Zhou, Z.-H., Kanbara, T., Shimura, M., Kizu, K., Maruyama, T., Nakamura, Y., Fukuda, T., Lee, B.L., Ooba, N., Tomura, S., Kurihara, T., Kaino, T., Kubota, K. and Sasaki, S. (1996) *Journal of the American Chemical Society*, **118**, 10389–99. d. Zhang, Q.T. and Tour, J.T. (1998) *Journal of the American Chemical Society*, **120**, 5355–62. e. Lee, C.-F., Yang, L.-M., Hwu, T.-Y., Feng, A.-S., Tseng, J.-C. and Luh, T.-Y. (2000) *Journal of the American Chemical Society*, **122**, 4992–3. f. Albert, I., Marks, T. and Ratner, M. (1997) *Journal of the American Chemical Society*, **119**, 6575–82. g. Jiang, B. and Tilley, T.D. (1999) *Journal of the American Chemical Society*, **121**, 9744–5. h. Suh, M.C., Jiang, B. and Tilley, T.D. (2000) *Angewandte Chemie (International Ed. in English)*, **39**, 2870–3. i. Yamagushi, S., Itami, Y. and Tamao, K. (1998) *Organometallics*, **17**, 4910–16. j. Yamagushi, S., Endo, T., Uchida, M., Izumizawa, T., Furukawa, K. and Tamao, K. (2000) *Chemistry–A European Journal*, **6**, 1683–92. k. Yamagushi, S., Goto, T.

and Tamao, K. (2000) *Angewandte Chemie (International Ed. in English)*, **39**, 1695–7. l. Tamao, K., Yamagushi, S., Shiozaki, M., Nakagawa, Y. and Ito, Y. (1992) *Journal of the American Chemical Society*, **114**, 5867–9. m. Yamagushi, S. and Tamao, K. (1998) *Journal of the Chemical Society. Dalton Transactions*, 3693–702. n. Corriu, R.J.P., Douglas, W.E. and Yang, Z.-X (1993) *Journal of Organometallic Chemistry*, **456**, 35–9. o. Ferraris, J.P., Andrus, R.G. and Hrncir, D.C. (1989) *Journal of the Chemical Society. Chemical Communications*, 1318–20. p. Hucke, A. and Cava, M.P. (1998) *The Journal of Organic Chemistry*, **63**, 7413–17. q. Lucht, B.L., Buretea, M.A. and Tilley, T.D. (2000) *Organometallics*, **19**, 3469–75.

3 a. Mathey, F. (1988) *Chemical Reviews*, **88**, 429–53. b. Quinin, L.D. (1996) *Comprehensive Heterocyclic Chemistry* (ed. A.R. Katritzky), Pergamon, Oxford, pp. 757–856.

4 a. Deschamps, E., Ricard, L. and Mathey, F. (1994) *Angewandte Chemie (International Ed. in English)*, **33**, 1158–61. b. Bevière, M.-O., Mercier, F., Ricard, L. and Mathey, F. (1990) *Angewandte Chemie (International Ed. in English)*, **29**, 655–7. c. Bévierre, M.-O., Mercier, F., Mathey, F., Jutand, A. and Amatore, C. (1991) *New Journal of Chemistry* **15**, 545–50. d. Mao, S.S.H. and Tilley, T.D. (1997) *Macromolecules*, **30**, 5566–556.

5 a. Salzner, U., Lagowski, J.B., Pickup, P.G. and Poirier, R.A. (1998) *Synthetic Metals*, **96**, 177–89. b. Delaere, D., Dransfeld, A., Nguyen, M.N. and Vanquickenborne, L.G. (2000) *The Journal of Organic Chemistry*, **65**, 2631–6. c. Nyulazi, L., Vespremi, T., Réffy, J., Burhardt, B. and Regitz, M. (1992) *Journal of the American Chemical Society*, **114**, 9080–4. d. Schleyer, P.vonR., Maerker, C., Dransfeld, A., Jiao, H. and Von Eikema Hommes, N.J.R. (1996) *Journal of the American Chemical Society*, **118**, 6317–18. e. Schleyer, P.vonR., Freeman, P.K., Jiao, H. and Goldfuss, B. (1995) *Angewandte Chemie (International Ed. in English)*, **34**, 337–40. f. Chesnut, B.D. and Quin, L.D. (1994) *Journal of the American Chemical Society*, **116**, 9638–43. g. Dransfeld, A., Nyulaszi, L. and Schleyer, P.vonR. (1998) *Inorganic Chemistry*, **37**, 4413–20. h. Mattmann, E., Simonutti, D., Ricard, L., Mercier, F. and Mathey, F. (2001) *The Journal of Organic Chemistry*, **66**, 755–8. i. Keglevich, G., Quin, L.D., Böcskei, Z., Keserü, G.M., Kalgutkar, R. and Lahti, P.M. (1997) *Journal of Organometallic Chemistry*, **532**, 109–16. j. Keglevich, G., Chuluunbaatar, T., Dajka, B., Dobó, A., Szöllösy, A. and Töke, L. (2000) *Journal of the Chemical Society. Perkin Transactions*, **1**, 2895–7. k. Nyulászi, L. (2001) *Chemical Reviews*, **101**, 1229–46.

6 a. Fagan, P.J. and Nugent, W.A. (1988) *Journal of the American Chemical Society*, **110**, 2310–12. b. Fagan, P.J., Nugent, W.A. and Calabrese, J.C. (1994) *Journal of the American Chemical Society*, **116**, 1880–9.

7 a. Hissler, M., Dyer, P. and Réau, R. (2003) *Coordination Chemistry Reviews*, **244**, 1–44. b. Hay, C., Fischmeister, C., Hissler, M., Toupet, L. and Réau, R. (2000) *Angewandte Chemie (International Ed. in English)*, **10**, 1812–15. c. Le Vilain, D., Hay, C., Deborde, V., Toupet, L. and Réau, R. (1999) *Chemical Communications*, 345–6. d. Hay, C., Hissler, M., Fischmeister, C., Rault-Berthelot, J., Toupet, L., Nyulazi, L. and Réau, R. (2001) *Chemistry–A European Journal*, **7**, 4222–36. e. Hay, C., Fave, C., Hissler, M., Rault-Berthelot, J. and Réau, R. (2003) *Organic Letters*, **5**, 3467–70.

8 a. Fave, C., Hissler, M., Sénéchal, K., Ledoux, I., Zyss, J. and Réau, R. (2002) *Chemical Communications*, 1674–5. b. Le Sauthier, M., Guennic, B., Deborde, V., Toupet, L., Halet, J.-F. and Réau, R. (2001) *Angewandte Chemie (International Ed. in English)*, **40**, 228–32. c. Leca, F., Sauthier, M., le Guennic, B., Lescop, C., Toupet, L., Halet, J.-F. and Réau, R. (2003) *Chemical Communications*, 1774–5. d. Leca, F., Lescop, C., Toupet, L. and Réau, R. (2004) *Organometallics*, **23**, 6191–201.

9 Hay, C., Sauthier, M., Deborde, V., Hissler, M., Toupet, L. and Réau, R. (2002) *Journal of Organometallic Chemistry*, **643**, 494–7.

10 Delaere, D., Nguyen, M.N. and Vanquickenborne, L.G. (2003) *The Journal of Physical Chemistry. A*, **107**, 838–46.
11 Yamagushi, S., Itami, Y. and Tamao, K. (1998) *Organometallics*, **17**, 4910–16.
12 a. Fave, C., Cho, T.-Y., Hissler, M., Chen, C.-W., Luh, T.-Y., Wu, C.-C. and Réau, R. (2003) *Journal of the American Chemical Society*, **125**, 9254–5. b. Su, H.-C., Fadhel, O., Yang, C.-J., Cho, T.-Y., Fave, C., Hissler, M., Wu, C.-C. and Réau, R. (2006) *Journal of the American Chemical Society*, **128**, 983–95.
13 a. Marder, S.R. (1992) *Inorganic Materials*, John Wiley & Sons, Inc, New-York. b. Gerloch, M. and Constable, E.C. (1994) *Transition Metal Chemistry*, VCH, New York.
14 Dillon, K.B., Mathey, F. and Nixon, J.F. (1998) *Phosphorus: The Carbon Copy*, John Wiley & Sons, Ltd, Chichester.
15 Sauthier, M., Leca, F., Toupet, L. and Réau, R. (2002) *Organometallics*, **21**, 1591–602.
16 a. Harvey, J.N., Heslop, K.M., Orpen, A.G. and Pringle, P.G. (2003) *Chemical Communications*, 278–9. b. Buey, J., Coco, S., Diez, L., Espinet, P., Martin-Alvarez, J.M., Miguel, J.A., Garcia-Granda, S., Tesouro, A., Ledoux, I. and Zyss, J. (1998) *Organometallics*, **17**, 1750–5. c. Pearson, R.G. (1973) *Inorganic Chemistry*, **2**, 712–13.
17 a. Braunstein, P. and Naud, F. (2001) *Angewandte Chemie (International Ed. in English)*, **40**, 680–99. b. Helmchem, G. and Pfaltz, A. (2000) *Accounts of Chemical Research*, **33**, 336–45. c. Drury, W.J., Zimmermann, N., III, Keenan, M., Hayashi, M., Kaiser, S., Goddard, R. and Pfaltz, A. (2004) *Angewandte Chemie (International Ed. in English)*, **43**, 70–4. d. Chelucci, G., Orru, G. and Pinna, G.A. (2003) *Tetrahedron*, **59**, 9471–515. e. Espinet, P. and Soulantica, K. (1999) *Coordination Chemistry Reviews*, **193–195**, 499–556. f. Braunstein, P., Knorr, M. and Stern, C. (1998) *Coordination Chemistry Reviews*, **178–180**, 903–65. g. Bianchini, C. and Meli, A. (2002) *Coordination Chemistry Reviews*, **225**, 35–66. h. Guiry, P.J. and Saunders, C.P. (2004) *Advanced Synthesis and Catalysis*, **346**, 497–537.
18 a. Bunlaksananusorn, T., Polborn, K. and Knochel, P. (2003) *Angewandte Chemie (International Ed. in English)*, **42**, 3941–3. b. Tang, W., Wang, W. and Zhang, X. (2003) *Angewandte Chemie (International Ed. in English)*, **42**, 943–6. c. Thoumazet, C., Melaimi, M., Ricard, L., Mathey, F. and Le Floch, P. (2003) *Organometallics*, **22**, 1580–1.
19 a. Shintani, R., Lo, M.M-C. and Fu, G-C. (2000) *Organic Letters*, **2** (*23*), 3695–7. b. Delapierre, G., Brunel, J.M., Constantieux, T., Labande, A., Lubatti, F. and Buono, G. (2001) *Tetrahedron, Asymmetry*, **12**, 1345–52. c. Gilbertson, S.R., Genov, D.G. and Rheingold, A.L. (2000) *Organic Letters*, **2** (*18*), 2885–8. d. Zablocka, M., Koprowski, M., Donnadieu, B., Majoral, J.P., Achard, M., Buono, G. and G. (2003) *Tetrahedron Letters*, **44**, 2413–15.
20 Mathey, F. (2001) *Phosphorus–Carbon Heterocyclic Chemistry: The Rise of a New Domain*, Elsevier Science Ltd, Oxford.
21 a. Deschamps, B., Ricard, L. and Mathey, F. (2003) *Organometallics*, **22**, 1356–7. b. Tran Hyu, N.H. and Mathey, F. (1994) *Organometallics*, **13**, 925–8. c. Redwine, K.D. and Nelson, J.H. (2000) *Organometallics*, **19**, 3054–61. d. Klärner, F.-G., Oebels, D. and Sheldrick, W.S. (1993) *Chemische Berichte*, **126**, 473–84. e. Wilson, W.L., Rahn, J.A., Alcock, N.W., Fischer, J., Frederick, J.H. and Nelson, J.H. (1994) *Inorganic Chemistry*, **33**, 109–17. f. Wilson, W.L., Fischer, J., Wasylishen, R.E., Eichele, K., Catalano, V.J., Frederick, J.H. and Nelson, J.H. (1996) *Inorganic Chemistry*, **35**, 1486–96.
22 Leca, F. and Réau, R. (2006) *Journal of Catalysis*, **238**, 425–9.
23 a. Coe, B.J., Harris, J.A., Clays, K., Perssoons, A., Wostyn, K. and Brunschwig, B.S. (2001) *Chemical Communications*, 1548–9. b. Maury, O., Guégan, J.-P., Renouard, T., Hilton, A., Dupau, P., Sandon, N., Toupet, L. and Le Bozec, H. (2001) *New Journal of Chemistry*, **25**, 1553–66. c. Renouard, T. and Le Bozec, H. (2000) *European Journal of Inorganic Chemistry*, **1**, 229–39. d. Di Bella, S. (2001) *Chemical Society Reviews*, **30**,

355–66. e. Lacroix, P.G. (2001) *European Journal of Inorganic Chemistry*, 339–48.

24 a. Balzani, V., Juris, A., Venturi, M., Campagna, S. and Serroni, S. (1996) *Chemical Reviews*, **96**, 759–834. b. Ziessel, R., Hissler, M., El-ghayoury, A. and Harriman, A. (1998) *Coordination Chemistry Reviews*, **178–80**, 1251–98. c. Lehn, J.M. (1995) *Supramolecular Chemistry*, VCH, Weinheim. d. Barigelletti, F., Flamigni, L., Balzani, V., Colin, J.-P., Sauvage, J.-P., Sour, A., Constable, E.C. and Cargill-Thompson, A.M.W. (1994) *Journal of the American Chemical Society*, **104**, 7692–9. e. Long, N.J. and Williams, C.K. (2003) *Angewandte Chemie (International Ed. in English)*, **42**, 2586–617.

25 a. Sauthier, M., Le Guennic, B., Deborde, V., Toupet, L., Halet, J.-F. and Réau, R. (2001) *Angewandte Chemie (International Ed. in English)*, **40**, 228–31. b. Leca, F., Sauthier, M., Deborde, V., Toupet, L. and Réau, R. (2003) *Chemistry–A European Journal*, **9**, 3785–95.

26 a. Pechmann, T., Brandt, C.D. and Werner, H. (2000) *Angewandte Chemie (International Ed. in English)*, **39**, 3909–11. b. Pechmann, T., Brandt, C.D., Röger, C. and Werner, H. (2002) *Angewandte Chemie (International Ed. in English)*, **41**, 2301–3. c. Pechmann, T., Brandt, C.D. and Werner, H. (2003) *Chemical Communications*, 1136–7. d. Pechmann, T., Brandt, C.D. and Werner, H. (2004) *Chemistry–A European Journal*, **10**, 728–36. e. Pechmann, T., Brandt, C.D. and Werner, H. (2004) *Dalton Transactions (Cambridge, England: 2003)*, 959–66.

27 a. Braunstein, P. and Boag, N.M. (2001) *Angewandte Chemie (International Ed. in English)*, **40**, 2427–33. b. Werner, H. (2004) *Angewandte Chemie (International Ed. in English)*, **43**, 938–54.

28 a. Gavrilova, A.L. and Bornich, B. (2004) *Chemical Reviews*, **104**, 349–84. b. Braunstein, P. and Rose, J. (1998) *Catalysis by Di- and Polynuclear Metal Cluster Complexes* (eds R.D. Adams and F.A. Cotton), VCH, New York. c. Whealey, N. and Kalck, P. (1999) *Chemical Reviews*, **99**, 3379–420. d. Braunstein, P., Oro, L.A. and Raithby, P.R. (eds) (1999) *Metal Clusters in Chemistry*, Wiley-VCH Verlag GmbH, New York. e. Carson, E.C. and Lippard, S.J. (2004) *Journal of the American Chemical Society*, **126**, 3412–13. f. Ochiai, M., Lin, Y.-S., Yamada, J., Misawa, H., Arai, S. and Matsumoto, K. (2004) *Journal of the American Chemical Society*, **126**, 2536–45. g. Goto, E., Begum, R.A., Zhan, S., Tanase, T., Tanigaki, K. and Sakai, K. (2004) *Angewandte Chemie (International Ed. in English)*, **43**, 5029–32.

29 Leca, F., Lescop, C., Rodriguez, E., Costuas, K., Halet, J.-F. and Réau, R. (2005) *Angewandte Chemie (International Ed. in English)*, **44**, 4362–5.

30 Nohra, B., Graule, S., Lescop, C. and Réau, R. (2006) *Journal of the American Chemical Society*, **128** (*11*), 3520–1.

31 a. Macchi, P., Garlaschelli, L. and Sironi, A. (2002) *Journal of the American Chemical Society*, **124**, 14173–84. b. Orpen, A.G. (1993) *Chemical Society Reviews*, **22**, 191–7. c. Xie, Y., Schaefer, H.F. and King, R.B., III (2000) *Journal of the American Chemical Society*, **122**, 8746–61. d. Ignatey, I.S., Schaefer, H.F., King, R.B., III and Brown, S.T. (2000) *Journal of the American Chemical Society*, **122**, 1989–94.

32 Leininger, S., Olenyuk, B. and Stang, P.J. (2000) *Chemical Reviews*, **100**, 853–908.

33 a. Gleiter, R. and Hopf, H. (eds) (2004) *Modern Cyclophane Chemistry*, Wiley-VCH Verlag GmbH, Weinheim. b. Hong, J.W., Woo, H.Y., Liu, B. and Bazan, G.C. (2005) *Journal of the American Chemical Society*, **127**, 7435–43. c. Woo, H.Y., Korystov, D., Mikhailovsky, A., Nguyen, T.-Q. and Bazan, G.C. (2005) *Journal of the American Chemical Society*, **127**, 13794–5. d. Sakai, T., Satou, T., Kaikawa, T., Takimiya, K., Otsubo, T. and Aso, Y. (2005) *Journal of the American Chemical Society*, **127**, 8082–9.

34 a. Seidel, R.S. and Stang, P.J. (2002) *Accounts of Chemical Research*, **35**, 972–83. b. Holliday, G.J. and Mirkin, C.A.

(2001) *Angewandte Chemie (International Ed. in English)*, **40**, 2022–43. c. Fujita, M., Tominaga, M., Aoai, A. and Therrien, B. (2005) *Accounts of Chemical Research*, **38**, 369–78. d. Gianneschi, N.C., Masar, M.S. and Mirkin, C.A., III (2005) *Accounts of Chemical Research*, **38**, 825–37. e. Cotton, F.A., Lin, C. and Murillo, C.A. (2001) *Accounts of Chemical Research*, **34**, 759–71. f. Puddephatt, R.J. (2001) *Coordination Chemistry Reviews*, **216–217**, 313–32.

13
Selective Hydrogen Transfer Reactions over Supported Copper Catalysts Leading to Simple, Safe, and Clean Protocols for Organic Synthesis

Federica Zaccheria and Nicoletta Ravasio

Reduction of carbonyl compounds and oxidation of alcohols have always been key transformations of organic chemistry. In both cases, the use of stoichiometric toxic reagents is still widespread and new methods and catalysts offering greater activity, selectivity, and safeness are constantly being sought.

Thus, in the fine chemicals industry, reduction of ketones and aldehydes relies mainly on the use of complex metal hydrides that require time-consuming work-up of reaction mixtures and produce significant amounts of inorganic and organic wastes. Similarly, the oxidation of alcohols into carbonyls is traditionally performed with stoichiometric inorganic oxidants, notably Cr(VI) reagents or a catalyst in combination with a stoichiometric oxidant [1].

Of course, metal-based heterogeneous catalysts for hydrogenation with molecular H_2 or oxidation with O_2 offer several advantages from the practical and environmental point of view, as they minimize wastes and reduce the work-up procedures.

On the other hand, hydrogenation and oxidation are particularly challenging when other functional groups are present in the substrate, and in some cases the catalysts employed, especially the noble metal-based ones, lack in selectivity. In this context, hydrogen transfer reactions with a hydrogen donor or acceptor other than H_2 or O_2 enable one to obtain highly selective transformations and to work under very mild experimental conditions. Moreover, they are industrially attractive in terms of safety, engineering, and economic considerations, allowing one to overcome problems and costs related with the use of hydrogen or molecular oxygen [2, 3].

Among the hydrogen transfer reactions, the Meerwein–Ponndorf–Verley reduction and its counterpart, the Oppenauer oxidation, are undoubtedly the most popular. These are well-established selective and mild redox reactions and they have been studied extensively [4, 5]. Nevertheless, traditional Meerwein–Ponndorf–Verley–Oppenauer (MPVO) reactions have some drawbacks, as they usually suffer from poor reactivity of the traditional $Al(OiPr)_3/iPrOH$ system, for which continuous removal of the produced acetone is necessary in order to shift the equilibrium between reduction of the ketone and oxidation of the donor alcohol.

Tomorrow's Chemistry Today. Concepts in Nanoscience, Organic Materials and Environmental Chemistry.
Edited by Bruno Pignataro
Copyright © 2008 WILEY-VCH Verlag GmbH & Co. KGaA, Weinheim
ISBN: 978-3-527-31918-3

Moreover, MPVO reactions are traditionally performed with stoichiometric amounts of Al(III) alkoxides. Some improvements came from the use of dinuclear Al(III) complexes that can be used in catalytic amount [6, 7]. This is why there has been an ever-increasing interest in catalytic MPVO reactions promoted by lanthanides and transition-metal systems [8]. In these cases, it is believed that reaction proceeds via formation of a metal hydride, in contrast with the mechanism accepted for traditional aluminum alkoxide systems, which involves direct hydrogen transfer by means of a cyclic intermediate [9]. As well as La, Sm, Rh and Ir complexes, Ru complexes have been found to be excellent hydrogen transfer catalysts. The high flexibility of these systems makes them very useful not only for MPVO-type reactions, but also for isomerization processes [10].

Significant improvements have also been introduced with the use of heterogeneous catalysts that are less water-sensitive than homogeneous Lewis acids and more convenient because of easier reaction mixture work-up. An important class of MPVO solid catalysts consists of zeolite beta and its metal-containing derivatives, especially Sn-, Zr- and Ti-beta. Several examples are known and the reduction or oxidation can be performed either in the gas phase [11, 12] or in solution [13, 14]. A very recent paper also reports the use of a bifunctional Zr-beta-supported Rh catalyst able to promote both arene and carbonyl reduction [15].

Amorphous solids such as MgO and Mg-Al mixed oxides gave some interesting results in gas-phase hydrogen transfer reduction, though all these catalytic systems require activation temperatures of at least 500 °C [16–18]. In contrast, very few cases have been reported of hydrogen transfer reactions mediated by a supported metal catalyst, the most efficient being $Ru(OH)_x/Al_2O_3$ [19].

We recently reported that a heterogeneous copper catalyst prepared with a nonconventional chemisorption–hydrolysis technique is able to promote a hydrogen transfer reduction using a donor alcohol. In this case, the role of copper is cricial, both for activity and selectivity [20].

The most active among the copper catalysts tested in the reduction of 4-*tert*-butylcyclohexanone with 2-propanol as H_2 donor (Scheme 13.1), is an alumina-supported one and, though alumina is known to promote hydrogen transfer itself [21, 22], in our conditions the metal catalyst shows a remarkable activity with respect to the bare support. In addition, the stereoisomeric distribution of products is considerably different (see Figure 13.1). Actually, pure Al_2O_3 pretreated as Cu/Al_2O_3 at 270 °C, slowly reduces the ketone giving a 89/11 eq/ax ratio, in agreement with the activity and the 91/9 eq/ax ratio obtained with γ-Al_2O_3 reported by Creighton *et al.* [23].

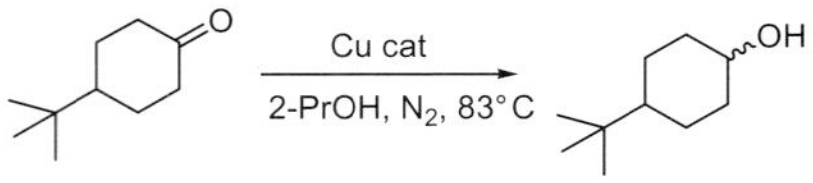

Scheme 13.1 4-*tert*-butylcyclohexanone reduction under hydrogen transfer conditions with heterogeneous copper catalysts.

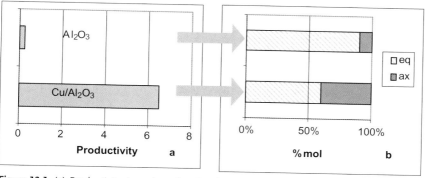

Figure 13.1 (a) Productivity (mmol product/(g_{cat} h)) in the hydrogen transfer reaction obtained with Cu/Al$_2$O$_3$ and Al$_2$O$_3$. (b) Stereoselectivity obtained with Cu/Al$_2$O$_3$ and Al$_2$O$_3$ respectively.

Conversely, the stereoisomeric distribution observed with the copper catalyst (59/41 eq/ax) is consistent with that obtained when the reaction is carried out with molecular hydrogen (58/42 eq/ax) [24]. Similar behavior, though with lower activity, was observed when carrying out the reaction with Cu/MgO and MgO [20].

A deep study of the 4-*tert*-butylcyclohexanone reduction aimed at understanding the effect of the donor alcohol structure revealed the existence of a two-step mechanism based on donor alcohol dehydrogenation and ketone hydrogenation. In particular, when the reaction was carried out in the presence of Cu/SiO$_2$, in order to exclude a contribution from the support, all the alcohols used as donors were capable of transferring H$_2$, and in the case of (iPr)$_2$CHOH and 3-octanol, not only was the formation of the corresponding ketone observed but it continued after complete conversion of the substrate.

Such high activity in hydrogen transfer reactions smoothes the way for a series of selective transformations of synthetic interest. Thus, a versatile and efficient catalyst could potentially promote hydrogenation, dehydrogenation, and a combination of the two reactions in order to set up isomerization reactions.

Here we summarize some of our results obtained by exploiting the hydrogen transfer ability of heterogeneous copper catalysts and therefore their activity in the reduction of polyunsaturated compounds, racemization and dehydrogenation of unactivated secondary alcohols, and isomerization of allylic alcohols.

13.1
Chemoselective Reduction of Polyunsaturated Compounds via Hydrogen Transfer

As already mentioned, Meerwein–Ponndorf–Verley reductions are intrinsically selective, since only carbonyl group can coordinate with the Lewis acid reaction center, while C═C double bonds remain unactivated. In contrast, the selectivity of the copper-catalyzed hydrogen transfer can be tuned depending on the substrate, by means of both the two-step dehydrogenation–hydrogenation mechanism described above and the peculiar selectivity of copper toward hydrogenation of conjugated double bonds with respect to isolated ones [25, 26]. Thus, while unsaturated, unconjugated ketones can be reduced to the corresponding unsaturated alcohols, α,β-unsaturated ones are selectively reduced to the saturated ketones.

Some selected results are collected in Table 13.1. Entries 1 and 2 refer to the selective reduction of carbonyl groups in the presence of unconjugated C=C double bonds. Both the ketones can be converted into the corresponding alcohols with excellent yields, leaving the other functionalities unaffected.

It is worth noting that comparable results for geranylacetone (entry 1) are reported only through reduction with stoichiometric amounts of polymethylhydroxysiloxane (PMHS) in the presence of catalytic amount of active zinc compound [31, 32].

Table 13.1 Hydrogen transfer reactions carried out in 2-propyl alcohol as solvent and inert atmosphere.

Entry	Substrate	Product	Catalyst	Yield (%)	Reference
1			Cu/SiO$_2$	95	[27]
2			Cu/SiO$_2$	94	[27]
3			Cu/SiO$_2$	97	[26]
4			Cu/SiO$_2$	98	[28]
5			Cu/Al$_2$O$_3$	98	[29]
6			Cu/Al$_2$O$_3$	98	[30]

In the same way, 6-methyl-5-hepten-2-one (entry 2) can conveniently be converted into 6-methyl-5-hepten-2-ol (sulcatol), which in its (R) configuration is an insect pheromone.

The use of 2-propanol for these substrates, besides its role as hydrogen source, allows suppression of cyclization reactions promoted by the acidity of the catalyst. Nevertheless, when cyclization leads to valuable products, as in the case of 3,3,5-trimethylpyran derived from sulcatol, it is possible to steer the reaction in this direction by using molecular hydrogen and a hydrocarbon solvent [27].

On the other hand, reduction of α,β-unsaturated compounds under the same conditions leads to the corresponding saturated ketones. This result can be reached in two cases of industrial interest such as the reduction of β-ionone into dihydro-β-ionone (entry 3), a valuable intermediate for the flavors and fragrances industry, and the reduction of 4-(6-methoxy-2-naphthyl)-3-buten-2-one into nabumetone (entry 4), a nonsteroidal anti-inflammatory drug [33]. The selective hydrogenation of β-ionone can be performed in 89% yield using Raney nickel alloy treated with sodium hydroxide [34], with a Ru-C catalyst [35] or with Ph_3SnH [36]. On the other hand Pd/C [37, 38] and $Rh(TOA)/Al_2O_3$ systems [39] are reported to be selective for the preparation of nabumetone.

The very high activity of Cu/Al_2O_3 in hydrogen transfer from 2-propanol can be exploited to obtain carbonyl reduction under very mild conditions where traditional catalytic hydrogenation fails or requires additives in order to suppress the formation of undesired products. 1-(p-Isobutyl)phenylethanol (entry 5) can be obtained easily from p-isobutylacetophenone with excellent selectivity without any basic additive. It is worth noting that this product is the intermediate in the Hoechst–Celanese process for the synthesis of ibuprofen [40] and the hydrogenation of the parent acetophenone is not trivial, as it requires the use of amines or NaOH [41] or the design of bimetallic catalysts [42] in order to suppress hydrogenolysis of the product.

In the same way it is possible to selectively reduce citronellal into citronellol with a yield up to 98%, thus avoiding both the hydrogenation of C=C double bonds and the formation of isopulegol by means of the acid-catalyzed *ene* reaction. A 91% yield of citronellol is reported using a Ru/Fe/C catalyst working in methanol–trimethylamine solution [43].

13.2
Alcohol Dehydrogenation

The strong dehydrogenation of the donor alcohol observed in hydrogen transfer reductions and the need for heterogeneous and simple systems for the oxidation of hydroxyl groups under liquid-phase conditions prompted us to investigate the activity of copper catalysts in alcohol dehydrogenation. As already mentioned, the development of catalytic methods for alcohol oxidation with oxygen or air has been one of the most avidly pursued targets in the last few years, owing to the urgency of substituting stoichiometric reagents. Very active homogeneous copper catalysts have been set up [44, 45], whereas the heterogeneous ones mainly rely on the use of noble metals [46–49].

The use of anaerobic conditions would overcome safety concerns relating to oxygen. Homogeneous Ru and Ir complexes have recently been reported to be efficient systems for acceptor-free dehydrogenation of alcohols [50–52], while very few heterogeneous catalysts are reported to be active in transfer dehydrogenation, the most recent being Pd [53–55] and Ru [56] systems, active only with benzylic or allylic alcohols.

Results obtained using copper catalysts in 3-octanol dehydrogenation showed good activity, although an equilibrium situation was reached between dehydrogenation of the alcohol and hydrogenation of the product ketone (Table 13.2, entry 1) [57]. However, when hydrogen was vented off from the reactor, it was apparent that the dehydrogenation reaction could go to completion (Table 13.2 entries 2 and 4).

To improve the synthetic potential of this reaction it seemed useful to adopt transfer dehydrogenation conditions by exploiting an already-mentioned peculiarity of these catalytic systems – that is, their specificity toward hydrogenation of a conjugated system rather than an isolated one. Thus, by adding styrene in equimolar ratio with respect to the substrate as a hydrogen acceptor, it is possible to completely oxidize alcohol with excellent activity and selectivity, particularly by using Cu/Al_2O_3 (Table 13.2, entries 3 and 5). The hydrogen uptake by styrene is irreversible, thus leading to complete conversion of the substrate in short reaction times (Scheme 13.2).

The by-product ethylbenzene can be easily removed from the reaction mixture together with the solvent. This reaction protocol is general for secondary alcohols

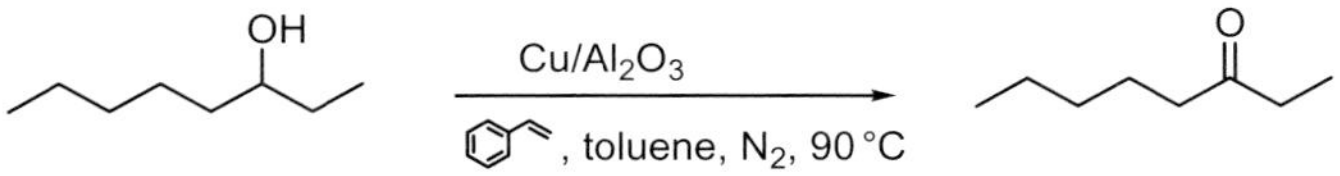

Scheme 13.2 Complete oxidation of 3-octanol over Cu/Al_2O_3.

Table 13.2 Oxidation of 3-octanol with different 8% Cu catalysts.

Entry	Catalyst	Time (h)	Conversion (%)	Selectivity (%)
1	Cu/SiO_2[a]	48	60	100
2	Cu/SiO_2[b]	20	100	100
3	Cu/SiO_2[c]	3	100	100
4	Cu/Al_2O_3[b]	12	84	100
5	Cu/Al_2O_3[c]	1.5	100	100
6	Al_2O_3	24	0	–

a N_2 (1 atm), 90 °C.
b Reaction carried out by venting reactor at fixed times.
c Reaction carried out by adding styrene as hydrogen acceptor.

and particularly efficient for unactivated ones. Selected results obtained with this procedure are reported in Table 13.3.

Three main features are apparent from the results reported: (i) the incomplete conversion of benzylic alcohols (entry 4); (ii) the lack of activity toward primary alcohols (entries 2 and 5); and (iii) the high activity toward unactivated secondary alcohols [58].

This trend is markedly different from that observed over almost all the catalytic oxidation systems reported so far, both based on a metal and on a radical precursor, which always preferentially convert benzylic alcohols with respect to the others, suggesting that a different mechanism is operating in the present case.

Nevertheless, dehydrogenation of benzyl alcohols can help to elucidate the reaction mechanism. Competitive oxidation of differently substituted benzyl alcohols produced a linear plot of $\log(k_x/k_H)$ versus the Brown–Okamoto σ^+, with a slope corresponding to a Hammett ρ^+ value of −0.75 (Figure 13.2). The moderate negative value of ρ^+ can be interpreted in terms of a positively charged transition state. The existence of an incipient carbenium ion intermediate in the reaction pathway can account for the experimental evidence and in particular for the faster oxidation of 3-octanol respect to 2-octanol (entry 5 in Table 13.2 vs. entry 1 in Table 13.3),

Table 13.3 Transfer dehydrogenation of different substrates using Cu/Al_2O_3.[a]

Entry	Substrate	Time (h)	Conversion (%)	Selectivity (%)
1	2-Octanol	4	100	100
2	1-Octanol	24	4	100
3	2,4-Dimethylpentan-3-ol	6	100	100
4	Benzyl alcohol	20	51.5	100
5	2-Phenylethanol	24	1	100
6	Cyclohexanol	3	99	98
7	2-Methylcyclohexanol	3.5	99	100
8	3-Methylcyclohexanol	3	100	100
9	4-Methylcyclohexanol	1.5	100	100
10	Carveol	2.5	100	88
		1.5[b]	100	95
11	Perillyl alcohol	6	95	91
12	Cyclooctanol	0.5	100	100
13	Cyclododecanol	2	97	100
14	Adamantanol	1.5	100	100
15	5α,3β-Androsterone	2.5	100	96
16	(−)-Menthol	48	50	100
17	Neomenthol	6	97	100

a Styrene/substrate = mol/mol, N_2 (1 atm), 90 °C.
b Reaction carried out using 2 equivalents of hydrogen acceptor.

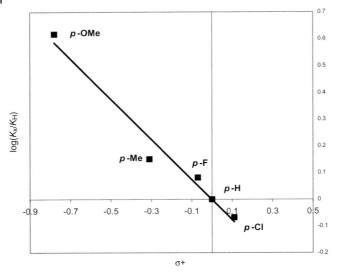

Figure 13.2 Hammett plot for competitive dehydrogenation of benzyl alcohol and *p*-substituted benzyl alcohols. Reaction conditions: benzyl alcohol (0.5 mmol), *p*-substituted benzyl alcohol (0.5 mmol), styrene (1 mmol), 8% Cu/Al$_2$O$_3$ (100 mg), toluene (8 ml), 363 K, N$_2$ atmosphere.

for the lack of activity of primary alcohols (Table 13.3, entries 2 and 5) but not of allylic ones, and for the faster oxidation of allylic alcohols respect to the homologous saturated ones [58].

On this hypothesis, benzylic alcohols should exhibit the fastest reaction rates, owing to the higher stability of their corresponding carbocations. It is therefore apparent that the incomplete conversion observed under our experimental conditions is due not to an intrinsically poor reactivity of the substrates but to the inadequacy of styrene as hydrogen acceptor, unable to prevail over the aromatic aldehyde formed, highly activated toward hydrogenation with Cu/Al$_2$O$_3$.

Although a suitable acceptor for the transfer dehydrogenation of benzylic alcohols has not yet been found, under the present conditions the low conversion of benzylic alcohols is only an apparent drawback. Indeed, it has a positive side as it allows us to fine-tune the system's selectivity. This makes the catalytic system unique among all the others known, operating both under aerobic and anaerobic conditions, that preferentially oxidize benzylic alcohols with respect to nonactivated secondary ones.

Moreover, it is possible to selectively oxidize secondary alcohols in the presence of primary alcohols without using any protecting group, as shown in the case of competitive oxidation of cyclooctanol and 1-octanol (Figure 13.3).

It is also worth noting that the use of transfer dehydrogenation allows operation under anaerobic atmosphere, thus avoiding the formation of overoxidation prod-

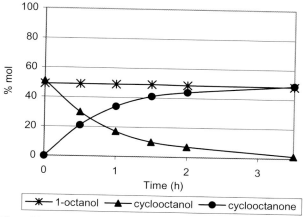

Figure 13.3 Competitive dehydrogenation of cyclooctanol and 1-octanol. Reaction conditions: 1-octanol (0.5 mmol), cyclooctanol (0.5 mmol), styrene (1 mmol), 8% Cu/Al$_2$O$_3$ (100 mg), toluene (8 ml), 363 K, N$_2$ atmosphere.

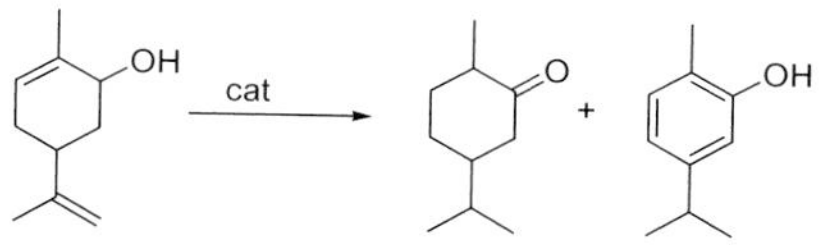

Scheme 13.3 Carveol dismutation over conventional nickel and copper catalysts.

ucts, often observed when using oxygen [59, 60] and the possible concerns linked to the use of O$_2$ in the presence of flammable solvents.

The activity observed with cyclohexanol – which is usually quite refractory to oxidation under mild conditions – is striking. In fact, high-loading copper catalysts are widely employed in cyclohexanol dehydrogenation [61], but working at high reaction temperatures under vapor-phase conditions.

Some other interesting applications are worth underlining. Thus, a carveol mixture was oxidized with 88% selectivity under standard conditions and with 95% selectivity using only two equivalents of hydrogen acceptor, giving carvone (**12** in Scheme 13.7), one of the more sought-after compounds in the flavors and fragrances industry. Nickel and conventional copper-based systems convert carveol **1** into tetrahydrocarvone **2** and carvacrol **3** (Scheme 13.3); therefore, methods proposed for this transformation on the industrial scale rely on the use of Oppenauer oxidation [62] or homogeneous dehydrogenation systems [63, 64]. In the open literature one efficient catalyst for the mixture of stereoisomers is reported, operating in aerobic conditions [65].

Perillyl aldehyde (entry 11 in Table 13.3) is typically obtained from the corresponding alcohol via Oppenauer-type oxidation by using alkylboron compounds and six equivalents of pivalaldehyde [66] or in the presence of Al(OiPr)₃ and more than the stoichiometric amounts of nitrobenzaldehyde [62], whereas under our conditions it can be obtained in 86% yield without formation of any waste but some ethylbenzene.

Steric effects also play a significant role, as is apparent in the monosubstituted cyclohexanols series (entries 7–9 in Table 13.3) and from comparison between linear and branched substrates (entry 1 vs. entry 3). This effect is so strong that in disubstituted cyclohexanols the reaction takes place at reasonable rates only when the —OH group is in axial conformation, as shown by the comparison of (−)-menthol (entry 16) and neomenthol (entry 17). This particular behavior could once more be exploited for synthetic purposes, in order to set up a kinetic resolution process. This has indeed been achieved in the one-pot hydrogenation of pennyroyal oil into menthol [67].

Pennyroyal oil is a very low-value mint oil, mainly consisting of pulegone **4**, a toxic component, usually refractory to hydrogenation. With our copper on alumina it is possible to hydrogenate pulegone directly into menthols **5** under very mild experimental conditions (Figure 13.4). In the menthol mixture produced the isomer with the worst organoleptic tones, the diaxial neoisomenthol initially formed in quite high content, preferentially dehydrogenates, thus giving menthol and neomenthol. This is an example of a multifunctional process combining both hydrogenation and selective dehydrogenation activity in order to set up a dynamic kinetic resolution.

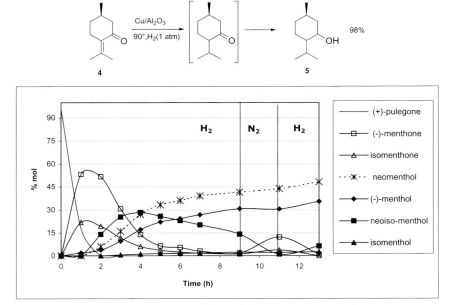

Figure 13.4 Product distribution during pulegone hydrogenation over Cu/Al₂O₃.

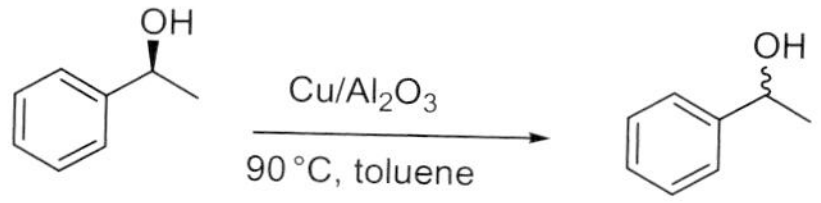

Scheme 13.4 1-(R)-phenylethanol racemization over copper catalyst.

13.3
Racemization of Chiral Secondary Alcohols

The remarkable activity of copper catalysts in carbonyl hydrogenation and alcohol dehydrogenation prompts their use also for the racemization of chiral secondary alcohols. Actually, since the first report on chemoenzymatic dynamic kinetic resolution [68], racemization of alcohols via the corresponding ketone has attracted considerably attention, owing to its role as backbone in this resolution [69, 70].

Resolution of cheap racemic mixtures with enzymes is a common route to enantiomerically pure chemicals on an industrial scale. However, the yield with a classical resolution is limited to 50%. An *in situ* racemization of the undesired enantiomer, combined with the enzymatic kinetic resolution, gives rise to a *dynamic kinetic resolution* (DKR) that should in principle lead to a 100% yield in the desired isomer. In spite of several Ru and Pd homogeneous systems successfully combined with enzymes and successfully applied on industrial scale in DKR [71, 72], few metal-based heterogeneous catalysts active for alcohol racemization have been reported [19, 73, 74].

Zeolite beta has also been used for the racemization of secondary phenylic alcohols in a dynamic kinetic resolution; however, in this case water elimination/addition via a carbenium ion is involved rather than a redox mechanism [75, 76].

Here we make a preliminary report that Cu/Al_2O_3, already shown to be active in hydrogenation and alcohol dehydrogenation, is able to promote alcohol racemization with good activity under reducing conditions. This was successfully obtained for 1-phenylethanol (Scheme 13.4 and Table 13.4) and 2-octanol using Cu/Al_2O_3, under hydrogen atmosphere in order to improve selectivity.

To better appreciate the flexibility of these catalytic systems it is worth mentioning that under transfer dehydrogenation conditions, namely by adding styrene to the reaction mixture, oxidation of (R)-1-phenylethanol proceeds without racemization up to 80% conversion [58]. This proves that it is possible to maximize and exploit the selectivity of the catalyst by tuning reaction conditions and by choosing the proper donor–acceptor couple.

13.4
Isomerization of Allylic Alcohols

Another interesting implication of hydrogen transfer reactions is the isomerization of allylic alcohols to the corresponding saturated ketones, due to its potential for use in organic synthesis. Actually, it forms an elegant shortcut to carbonyl

Table 13.4 Racemization of 1-(R)-phenylethanol over Cu/Al$_2$O$_3$.

Entry	Conditions	Time (h)	Selectivity (%)	ee [%][a]
1	N$_2$	3	78[b]	8
2	H$_2$	3	98	20

a Enantiomeric excess.
b Main by-product is acetophenone.

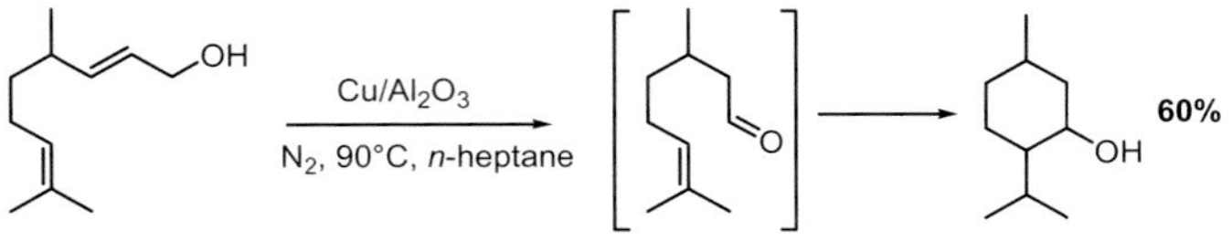

Scheme 13.5 One-pot one-step isomerisation of geraniol into menthols over Cu/Al$_2$O$_3$.

compounds in a completely atom-economical process that offers several useful applications in natural products synthesis and in bulk chemical processes [77], thus conveniently replacing the conventional two-step dehydrogenation and hydrogenation sequence.

As already noted for the other transformations, many homogeneous catalytic systems have been developed, particularly Ru complexes [78–80], but Ru(OH)$_x$/Al$_2$O$_3$ reported by Mizuno [19] is the unique heterogeneous example.

The isomerization activity of Cu/Al$_2$O$_3$ was observed during a detailed study on geraniol transformations [30]. Although isomerization of geraniol to γ-geraniol under catalytic hydrogenation conditions in the presence of a ruthenium chiral complex has been reported [81], achiral ruthenium phosphine complexes failed to isomerize geraniol to citronellal [82].

In the particular case of this substrate the achievement of selective transformations is particularly challenging, owing to its high reactivity toward hydrogenation, dehydrogenation, and cyclization. However, over Cu/Al$_2$O$_3$ the isomerization of geraniol **6** into citronellal **7**, followed by the fast *ene* reaction was observed, thus giving isopulegol mixture **8** in 60% yield (the main by-product being citral and citronellol). In practice, the combination of the hydrogen transfer ability and the acidic properties of the catalyst allows a one-pot one-step transformation (Scheme 13.5).

Highly selective and fast isomerization was also observed for 1-octen-3-ol **9** into 3-octanone **10** under very mild reaction conditions, meaning room temperature and nitrogen atmosphere (Scheme 13.6).

Also dihydrocarvone **11** can be obtained in moderate yield from carveol **1** using Cu/Al$_2$O$_3$. The comparison with transfer dehydrogenation reaction carried out over the same substrate (Table 13.3 entry 10), shows once more the pivotal role and unique properties of the hydrogen acceptor (Scheme 13.7).

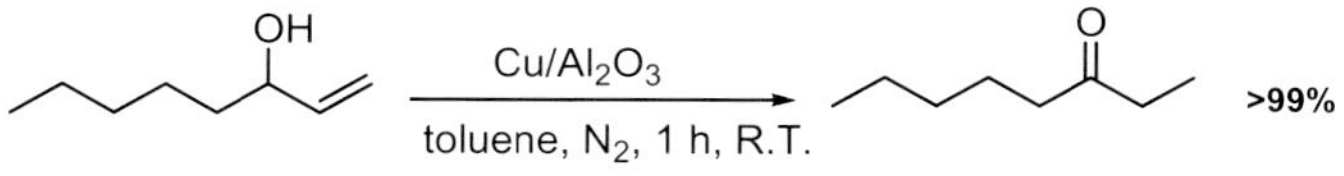

Scheme 13.6 Isomerisation of 1-octen-3-ol into 3-octanone by using copper supported on alumina.

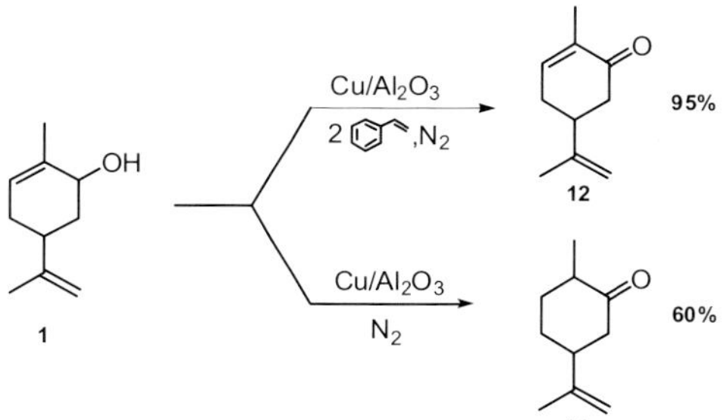

Scheme 13.7 Carveol transformation over Cu/Al$_2$O$_3$ under different conditions.

13.5
Conclusions

Heterogeneous copper catalysts prepared with the chemisorption–hydrolysis technique are effective systems for hydrogen transfer reactions, namely carbonyl reduction, alcohol dehydrogenation and racemization, and allylic alcohol isomerization. Practical concerns argue for the use of these catalysts for synthetic purposes because of their remarkable performance in terms of selectivity and productivity, which are basic features for the application of heterogeneous catalysts to fine chemicals synthesis. Moreover, in all these reactions the use of these materials allows a simple, safe, and clean protocol.

Deep and exhaustive study of the catalyst behavior, especially analyzed by means of reactivity and product selectivity, allows us to finely match the catalytic performances to the synthetic needs.

In particular, by varying the support and the reaction conditions it is possible to set up multifunctional processes as already shown in the case of pennyroyal oil [67] or geraniol [30]. Moreover, by exploiting the ability of the catalysts to modify the adsorbed molecule from the electronic point of view, it is possible to use them in other kinds of reactions with unsuitable substrates. This has indeed been observed in Sonogashira reactions [83].

All the hydrogen transfer reactions shown highlight the potential for application of copper-supported catalysts as hydrogen reservoir systems. This leads to the possibility of applying the concept of catalytic electronic activation introduced by Williams and realizing sequences of domino transformations that accomplish otherwise "impossible" reactions [84, 85].

References

1 Sheldon, R.A., Arends, I.W.C.E., Ten-Brink, G-J. and Dijksman, A. (2002) *Accounts of Chemical Research*, **35**, 774–81.
2 Le Page, M.D., Poon, D. and James, B.R. (2003) *Chemical Industries Series*, in *Catal. Org. React.* Dekker, **89**, pp. 61. (Catal. Org. React.)
3 Anastas, P.T. and Kirchhoff, M.M. (2002) *Accounts of Chemical Research*, **35**, 686–94.
4 Zassinovich, G., Mestroni, G. and Gladiali, S. (1992) *Chemical Reviews*, **92**, 1051–69.
5 de Graauw, C.F., Peters, J.A., van Bekkum, H. and Huskens, J. (1994) *Synthesis*, **10**, 1007–17.
6 Ooi, T., Ichikawa, H. and Maruoka, K. (2001) *Angewandte Chemie (International Ed. in English)*, **40**, 3610–2.
7 Ooi, T., Miura, T., Itagaki, Y., Ichikawa, H. and Maruoka, K. (2002) *Synthesis*, 279–91.
8 Klomp, D., Maschmeyer, T., Hanefeld, U. and Peters, J.A. (2004) *Chemistry – A European Journal*, **10**, 2088–93.
9 Cohen, R., Graves, C.R., Nguyen, S.T., Martin, J.M.L. and Ratner, M.A. (2004) *Journal of the American Chemical Society*, **129**, 14796–803 [and references therein].
10 Bäckvall, J-E. (2002) *Journal of Organometallic Chemistry*, **652**, 105–11.
11 Shabtai, J., Lazar, R. and Biron, E. (1984) *Journal of Molecular Catalysis*, **27**, 35–43.
12 Berkani, M., Lemberton, J.L., Marczewski, M. and Perot, G. (1995) *Catalysis Letters*, **31**, 405–10.
13 Chuah, G.K., Jaenicke, S., Zhu, Y.Z. and Liu, S.H. (2006) *Current Organic Chemistry*, **10**, 1639–54 [and references therein].
14 Boronat, M., Corma, A. and Renz, M. (2006) *The Journal of Physical Chemistry. B*, **110**, 21168–74.
15 Nie, Y., Jaenicke, S., van Bekkum, H. and Chuah, G-K. (2007) *Journal of Catalysis*, **246**, 223–31.
16 Kaspar, J., Trovarelli, A., Zamoner, F., Farnetti, E. and Graziani, M. (1991) *Studies in Surface Science and Catalysis*, **59**, 253.
17 Di Cosimo, J.I., Acosta, A. and Apesteguía, C.R. (2004) *Journal of Molecular Catalysis A: General*, **222**, 87–96.
18 Di Cosimo, J.I., Acosta, A. and Apesteguía, C.R. (2005) *Journal of Molecular Catalysis A: General*, **234**, 111–20.
19 Yamaguchi, K., Koike, T., Kotani, M., Matsushita, M., Shinaci, S. and Mizuno, N. (2005) *Chemistry – A European Journal*, **11**, 6574–82.
20 Zaccheria, F., Fusi, A., Psaro, R. and Ravasio, N. (2005) *Chemical Industries Series*, in *Catal. Org. React.* (ed. J. Sowa), Taylor & Francis, pp. 293.
21 Posner, G.H., Runquist, A.W. and Chapdelaine, M.J. (1977) *The Journal of Organic Chemistry*, **42**, 1202.
22 Gargano, M., D'Orazio, V., Ravasio, N. and Rossi, M. (1990) *Journal of Molecular Catalysis* **58**, L5–L8.
23 Creighton, E.J., Ganeshie, S.D., Downing, R.S. and van Bekkum, H. (1997) *Journal of Molecular Catalysis A: Chemical*, **115**, 457–72.
24 Ravasio, N., Psaro, R. and Zaccheria, F. (2002) *Tetrahedron Letters*, **43**, 3943–5.
25 Ravasio, N., Antenori, M., Gargano, M. and Mastrorilli, P. (1996) *Tetrahedron Letters*, **37**, 3529–33.
26 Ravasio, N., Zaccheria, F., Guidotti, M. and Psaro, R. (2004) *Topics in Catalysis*, **27**, 157–68.
27 Ravasio, N., Psaro, R., Zaccheria, F. and Recchia, S. (2003) *Chemical Industries Series*, Dekker, (Catal. Org. React.) (ed. D. Morrel), pp. 263–71.
28 Ravasio, N., Zaccheria, F., Ercoli, M. and Allegrini, P. (2007) *Catal Today*, **121**, 2–5.
29 Zaccheria, F., Ravasio, N., Psaro, R. and Fusi, A. (2005) *Tetrahedron Letters*, **46**, 3695–7.
30 Zaccheria, F., Ravasio, N., Fusi, A., Rodondi, M. and Psaro, R. (2005) *Advanced Synthesis and Catalysis*, **347**, 1267–72.
31 Mimoun, H. (2001) US Patent 6,245,952 to Firmenich SA.

32 Mimoun, H. (1999) *The Journal of Organic Chemistry*, **64**, 2582–9.
33 Fitch, J.R., Aslam, M., Rios, D.E. and Smith, J.C. (1996) PCT-WO/96-40608 to Hoechst-Celanese.
34 Masuda, H. and Mihara, S. (1984) JP 84-155438 to Ogawa and Co.
35 Chapuis, C. and Jacobi, D. (2001) *Applied Catalysis A: General*, **221**, 93
36 Joung, J.-J., Jung, L.-J. and Cheng, K.-M. (2000) *Tetrahedron Letters*, **41**, 3415–8.
37 Ramachandran, V. and Belmont, S., (1998) PCT-WO/98-49126 to Albemarle Corporation.
38 Cabri, W., Magrone, D., Oldani, E. and Angelini, R. (1999) US Patent 5,955,635 to Secifarma SpA.
39 Piccolo, O. and Verrazzani, A. (2003) PCT-WO/03-037508 to Chemi SpA.
40 Elango, V., Davenport, K.G., Murphy, M.A., Mott, G.N., Zey, E.G., Smith, B.L. and Moss, G.L. (1990) EP 0 400892 A2 to Hoechst-Celanese.
41 Mathew, S.P., Rajasekharam, M.V. and Chaudari, R.V. (1999) *Catal Today*, **49**, 49–56 [and references therein].
42 Casagrande, M., Storaro, L., Talon, A., Lenarda, M., Frattini, M., Rodriguez Castellon, E. and Maireles-Torres, P. (2002) *Journal of Molecular Catalysis A*, **188**, 133–9.
43 Goebbel, H-G., Gerlach, T., Wegner, G., Fuchs, H., Unverricht, S. and Salden, A., (2004) PCT Int. Appl., WO 2004007411 A1
44 Markò, I.E., Gautier, A., Dumeunier, R., Doda, K., Philippart, S., Brown, S.M. and Urch, C.J. (2004) *Angewandte Chemie (International Ed. in English)*, **43**, 1588–91 [and references therein].
45 Gamez, P., Arends, I.W.C.E., Sheldon, R.A. and Reedijk, J. (2004) *Advanced Synthesis and Catalysis*, **346**, 805–11.
46 Hou, Z., Theyssen, N. and Leitner, W. (2007) *Green Chemistry*, **9**, 127–32.
47 Abad, A., Almela, C., Corma, A. and Garcia, H. (2006) *Chemical Communications*, 3178–80.
48 Ebitani, K., Motokura, K., Mizugaki, T. and Kaneda, K. (2005) *Angewandte Chemie (International Ed. in English)*, **44**, 3423–6.
49 Mallat, T. and Baiker, A. (2004) *Chemical Reviews*, **104**, 3037–58 [and references therein].
50 Fujita, K-I., Tanino, N. and Yamaguchi, R. (2007) *Organic Letters*, **9**, 109–11.
51 Kim, W-H., Park, I.S. and Park, J. (2006) *Organic Letters*, **8**, 2543–5.
52 Adair, G.R.A. and Williams, J.M.J. (2005) *Tetrahedron Letters*, **46**, 8233–5.
53 Tanaka, T., Kawabata, H. and Hayashi, M. (2005) *Tetrahedron Letters*, **46**, 4989–91.
54 Keresszegi, C., Mallat, T. and Baiker, A. (2001) *New Journal of Chemistry*, **25**, 1163–7.
55 Hayashi, M., Yamada, K., Nakayama, S., Hayashi, H. and Yamazaki, S. (2000) *Green Chemistry*, **2**, 257–60.
56 Karvembu, R., Prabhakaran, R., Senthilkumar, K., Viswanathamurty, P. and Natarajan, K. (2005) *Reaction Kinetics and Catalysis Letters*, **86**, 211–6.
57 Zaccheria, F., Ravasio, N., Psaro, R. and Fusi, A. (2005) *Chemical Communications*, 253–5.
58 Zaccheria, F., Ravasio, N., Psaro, R. and Fusi, A. (2006) *Chemistry – A European Journal*, **24**, 6426–31.
59 Abad, A., Concepción, P., Corma, A. and García, H. (2005) *Angewandte Chemie (International Ed. in English)*, **44**, 4066–8.
60 Enache, D.I., Knight, D.W. and Hutchings, G.J. (2005) *Catalysis Letters*, **103**, 43–52.
61 Fridman, V.Z., Davydov, A.A. and Titievsky, K. (2004) *Journal of Catalysis*, **222**, 545–57.
62 Tanikawa, S., Matsubayashi, S., Tanikawa, M. and Komatsu, T. (2002) US Pat. Appl. 2002198411 to Nikken Chem. Ltd., Japan.
63 Ooi, T., Otsuka, H., Miura, T., Ichikawa, H. and Maruoka, K. (2002) *Organic Letters*, **4**, 2669–72.
64 Kolomeyer, G. and Oyloe, J.S. (2004) PCT Int. Appl. WO 2004108646 to Millenium Speciality Chem., USA.
65 Ji, H., Mizugaki, T., Ebitani, K. and Kaneda, K. (2002) *Tetrahedron Letters*, **43**, 7179.
66 Ishihara, K., Kurihara, H. and Yamamoto, H. (1997) *The Journal of Organic Chemistry*, **62**, 5664–5.
67 Ravasio, N., Zaccheria, F., Fusi, A. and Psaro, R. (2006) *Applied Catalysis A: General*, **315**, 14–9.

68 Dinh, P.M., Howarth, J.A., Hudnott, A.R., Williams, J.M.J. and Harris, W. (1996) *Tetrahedron Letters*, **37**, 7623–6.
69 Pellissier, H. (2003) *Tetrahedron*, **59**, 8291–327 [and references therein].
70 Huerta, F.F., Minidis, A.B.E. and Bäckvall, J.-E. (2001) *Chemical Society Reviews*, **30**, 321 [and references therein].
71 Rouhi, M.A. (2002) *Chemical and Engineering News*, **80**, 23.
72 Verzijl, G.K., de Vries, J.G. and Broxterman, Q.B. to DSM N.V., The Netherlands. PCT-WO/2001090396 A1 20011129.
73 Wuyts, S., de Vos, D.E., Verpoort, F., Depla, D., de Gryse, R. and Jacobs, P.A. (2003) *Journal of Catalysis*, **219**, 417–24.
74 Kim, W-H., Karvembu, R. and Park, J. (2004) *Bulletin of the Korean Society*, **25**, 931–3.
75 Wuyts, S., de Temmerman, K., de Vos, D. and Jacobs, P.A. (2003) *Chemical Communications*, 1928–9.
76 Wuyts, S., de Temmerman, K., de Vos, D. and Jacobs, P.A. (2005) *Chemistry – A European Journal*, **11**, 386–97.
77 van der Drift, R.C., Bouwman, E. and Drent, E. (2002) *Journal of Organometallic Chemistry*, **650**, 1–24.
78 Crochet, P., Fernández-Zúmel, M.A., Gimeno, J. and Scheele, M. (2006) *Organometallics*, **25**, 4846–9.
79 Martín-Matute, B., Bogár, K., Edin, M., Kaynak, F.B. and Bäckvall, J.-E. (2005) *Chemistry – A European Journal*, **11**, 5832–42.
80 Uma, R., Crévisy, C. and Grée, R. (2003) *Chemical Reviews*, **103**, 27–52.
81 Sun, Y., Le Blond, C., Wang, J., Blackmond, D.G., Laquidara, J. and Sowa, J.R. (1995) *Journal of the American Chemical Society*, **117**, 12647–8.
82 Trost, B.M. and Kulawiec, R.J. (1993) *Journal of the American Chemical Society*, **115**, 2027–36.
83 Biffis, A., Scattolin, E., Zaccheria, F. and Ravasio, N. *Tetrahedron Lett.* (2007) doi:10.1016/j.tetlet.2007.10.005.
84 Black, P.J., Edwards, M.G. and Williams, J.M.J. (2006) *European Journal of Organic Chemistry*, 4367–78.
85 Black, P.J., Edwards, M.G. and Williams, J.M.J. (2005) *Tetrahedron*, **61**, 1363–74.

14
Selective Oxido-Reductive Processes by Nucleophilic Radical Addition under Mild Conditions

Cristian Gambarotti and Carlo Punta

14.1
Introduction

Because the enthalpic effect was considered the sole driving force in free-radical processes, it was long erroneously believed that high reactivity had always to be associated with low selectivity. This is one of the reasons why radical chemistry was thought to be useless for selective synthesis of complex molecules, until Barton, in 1960 [1], showed the synthetic potentiality of free-radical reactions.

In 1968 [2], in a pioneering work, Minisci showed how it was possible to achieve a wide range of selective reactions by the addition of nucleophilic carbon-centered radicals to electron-deficient substrates (such as olefins conjugated with electron-withdrawing groups, protonated heteroaromatic bases, and quinones). In these cases, it was possible to obtain complete conversions with high selectivity owing to the presence of a strong polar effect.

In particular, the alkylation and acylation of protonated heteroarenes under oxidative conditions, commonly known as Minisci reaction, has attracted increasing interest in recent decades because of its synthetic involvement in biochemistry and pharmacology [3]. Beyond the fact that this reaction can be applied to all heteroaromatic bases and almost all carbonyl and alkyl radicals (without electron-withdrawing groups directly bonded to the radical center), the main characteristics of this process are high chemoselectivity and regioselectivity, the substitution usually occurring only in α and γ positions.

This reaction is based on the proposition that the sensitivity to polar effects in free-radical chemistry is the result of polarity and polarizability of both the radical and the substrate. This means that the polarity of the heteroaromatic base plays a key role in the process. Actually, the nucleophilic character of an alkyl radical, for example, is not so marked as to justify the addition to the *N*-heteroaromatic base, and in fact either no substitution occurs or low yields and selectivity are observed.

In contrast, the reaction provides high yields and selectivity when operating in acidic medium. In fact, the protonation of the nitrogen of the heteroaro-

Tomorrow's Chemistry Today. Concepts in Nanoscience, Organic Materials and Environmental Chemistry.
Edited by Bruno Pignataro
Copyright © 2008 WILEY-VCH Verlag GmbH & Co. KGaA, Weinheim
ISBN: 978-3-527-31918-3

matic bases strongly increases their electron-deficient nature, and therefore their reactivity toward nucleophilic species. If we consider the nitrogen of pyridine, for example, as a substituent in the aromatic ring, the high value of 0.93 reported for the Hammett σ constant of the *p*-position shows the electron-deficient character of the heteroarenes [4]. However, for the protonated pyridine, the considerably higher Hammett σ value of 4 was estimated, evidencing the extraordinarily increased reactivity of the substrate toward nucleophiles [5].

This phenomenon cannot be exploited with ionic nucleophilic species, as they would cause the deprotonation of the base as a primary effect. Nucleophilic radicals, on the contrary, are suitable partners under acidic conditions, allowing the development of a wide range of substitutions of great synthetic interest, which reflect the Friedel–Crafts aromatic alkylation and acylation with opposite reactivity and selectivity. High rate constants (10^5–$10^8 \, m^{-1} s^{-1}$) for the addition of alkyl and acyl radicals to the α- and γ-positions of the protonated heterocyclic bases contribute to the synthetic interest [3], the corresponding reactions of the unprotonated bases being much slower ($<10^2 \, m^{-1} s^{-1}$).

More recently, Porta and co-workers [6] applied similar considerations of the polar effects to a new one-pot multicomponent process for the addition of nucleophilic radicals to aldimines, generated *in situ* in the presence of Ti(IV). In analogy with the Minisci reaction, Ti(IV), which acts as a Lewis acid, coordinates the nitrogen of the imine, strongly increasing the electron-deficient character of the carbon in the α-position and thus the reactivity of the imine toward nucleophilic radicals. This reaction, as well as the Minisci one, represents a useful route for the synthesis of a variety of poly-functionalized derivatives of chemical and biochemical relevance.

This chapter will focus on the new developments and our recent contributions to these two important processes.

14.2
Nucleophilic Radical Addition to *N*-heteroaromatic Bases

14.2.1
Acylation of *N*-heteroaromatic Bases

In the classic Friedel–Crafts protocol, electrophilic acyl halogenides react with aromatic rings, in the presence of Lewis acids, leading to the formation of the corresponding acyl derivatives. Electron donor groups on the aromatic ring strongly increase the reaction rate, driving the substitution on the electron-rich position [7]. This method is not so efficient on the electron-poor *N*-heteroaromatic bases and only the less deactivated positions undergo substitution with low reaction rate.

Owing to their nucleophilic character, acyl radicals are easily trapped by protonated *N*-heteroarenes, leading to the formation of the corresponding acyl derivatives, with an inverse selectivity compared with the Friedel–Crafts process (Equation 14.1) [8].

$$\text{(14.1)}$$

The general acylation mechanism involves, as first step, the nucleophilic attachment of the radical generated *in situ* on the protonated N-heteroaromatic base. The intermediate radical-cation loses a proton, affording the neutral radical intermediate (Equation 14.1), which is much less basic ($pK_a \approx 2$) than the corresponding dihydropyridine ($pK_a \approx 7$); moreover, the pyridinyl radical is a highly reducing agent and can be oxidized to the final substituted form by very mild oxidants. Only the α- and γ-positions of the heterocyclic ring are involved; the ratio between α- and γ-substituted products is strongly affected by the structure of the radical, the solvent, and the redox system utilized [3].

Depending on the redox system employed, the intermediate undergoes oxidation to afford the acylated products.

Various systems have been developed in recent decades for the generation of acyl radicals under mild conditions, mostly by Minisci and co-workers. For example, the Fenton-type *tert*-butyl hydroperoxide/Fe(II) system, gives rise *t*-BuO· radical, which is able to abstract the hydrogen from aldehydes, achieving the corresponding acyl radicals (Equations 14.2 and 14.3) [9]. Fe(III), formed in Equation 14.2, acts as oxidant on the intermediate, making the process catalytic in Fe(II) (Equation 14.4).

$$t\text{-BuOOH} + Fe^{2+} \longrightarrow t\text{-BuO·} + OH^- + Fe^{3+} \quad (14.2)$$

$$t\text{-BuO·} + \underset{R}{\overset{O}{\|}}\!\!-\!\!H \longrightarrow \underset{R}{\overset{O}{\|}}\!\!\cdot + t\text{-BuOH} \quad (14.3)$$

$$\text{(14.4)}$$

Another example is represented by the oxidative decarboxylation of α-ketoacids in the presence of the $S_2O_8^{2-}/Ag^+$ redox system, which leads to the formation of acyl radicals by means of the intermediate Ag^{2+} (Equations 14.5 and 14.6) [10]. In this case, the re-aromatization of the ring can occur according to two parallel paths: oxidation by persulfate (Scheme 14.1a) and by Ag(II) (Scheme 14.1b). Thus, this system needs more than the stoichiometric quantity of persulfate, as it both reacts

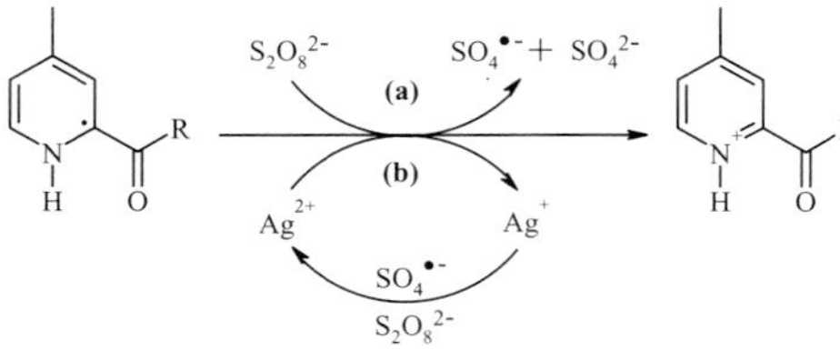

Scheme 14.1 Re-aromatization of the ring mediated by $S_2O_8^{2-}$ (a) and Ag^{2+} (b).

with the intermediate and reoxidizes Ag(I) to Ag(II), which is then utilized in catalytic amount.

$$S_2O_8^{2-} + Ag^+ \longrightarrow 2\,SO_4^{2-} + Ag^{2+} \tag{14.5}$$

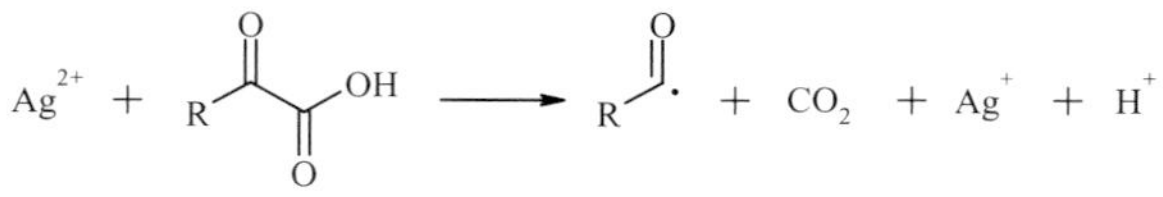

$$Ag^{2+} + R\text{-COOH} \longrightarrow R^\bullet + CO_2 + Ag^+ + H^+ \tag{14.6}$$

The use of amides as source of carbamoyl and α-aminoalkyl radicals has also been widely employed since the 1970s [11]. Fenton-type systems utilizing *tert*-butyl hydroperoxide or hydrogen peroxide in the presence of Fe(II) salts and the $S_2O_8^{2-}/Ag^+$ system have been applied in the functionalization of aromatic bases with amides (Equation 14.7).

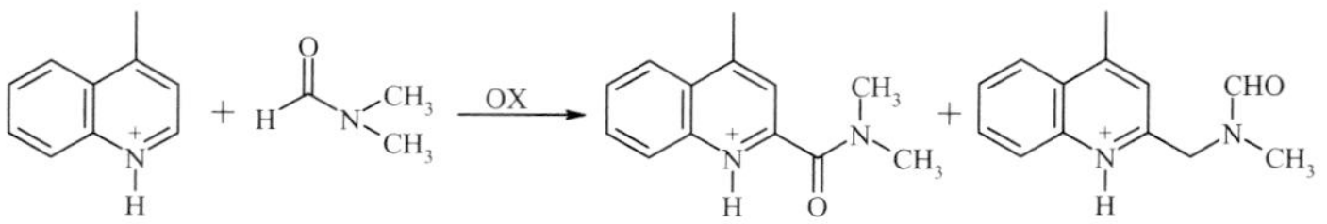

$$\text{(quinoline)} + H\text{-CON(CH}_3)_2 \xrightarrow{OX} \text{(quinoline-CON(CH}_3)_2) + \text{(quinoline-CH}_2\text{N(CH}_3)_2) + \text{(quinoline-CHO-N(CH}_3)_2) \tag{14.7}$$

14.2.2
Acylation of N-heteroaromatic Bases Catalyzed by N-hydroxyphthalimide

Recently, N-hydroxyphthalimide (NHPI) catalysis has been applied to a variety of aerobic oxidations of organic compounds [12]. We have reported how NHPI, in the presence of Co(II) salts, is able to generate the phthalimido N-oxyl (PINO) radical, which rapidly abstracts hydrogen from aromatic and aliphatic aldehydes, in a free-radical chain mechanism under aerobic conditions (Scheme 14.2) [13]. The role of oxygen is to oxidize Co(II) to Co(III), which is also involved in the oxidation of the intermediate (Equation 14.8).

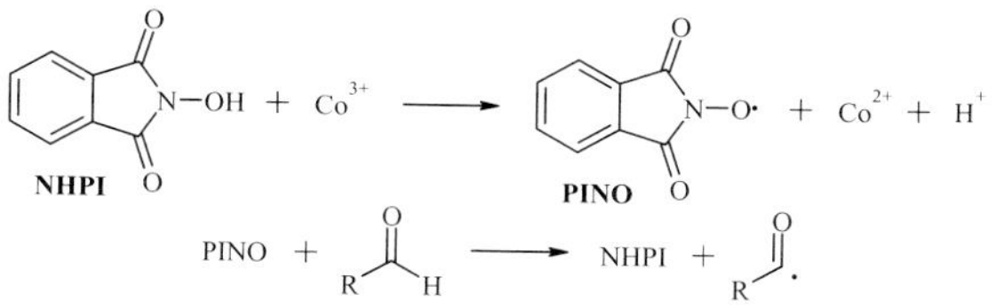

NHPI—OH + Co³⁺ ⟶ PINO—O• + Co²⁺ + H⁺

PINO + RCHO ⟶ NHPI + RC(O)•

Scheme 14.2 Mechanism for the hydrogen abstraction from aldehydes by PINO radical, generated in situ from NHPI in the presence of cobalt salts.

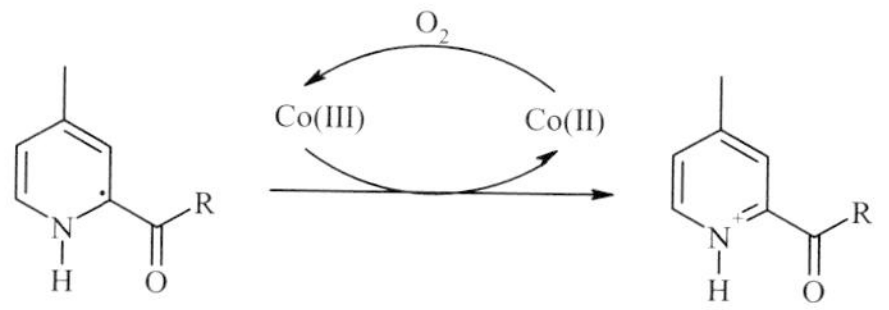

$$\text{(14.8)}$$

The relatively low Bond Dissociation Energy (BDE) value of the C—H bond in the acyl group (94 kcal mol^{-1}) [14] is comparable with the O–H BDE value of NHPI (88.1 kcal/mol), suggesting that also the hydrogen atom abstraction from formamide could be a selective process in the presence of PINO. Actually, when the aerobic oxidation of formamide, catalyzed by NHPI and Co(II) salt, was carried out in the presence of protonated base, the selectivity was complete, but the conversion was low, meaning that the reaction of the carbamoyl radical was faster with oxygen than with protonated N-heteroaromatic base. Thus, cerium(IV) ammonium nitrate (CAN) was used to develop a new carbamoylation system in the presence of NHPI catalysis under anaerobic conditions [15]. Ce(IV) has a double role: first it generates PINO by a fast electron transfer reaction with NHPI; then it is responsible for the oxidation of the intermediate radical species, affording the carbamoylated aromatic base. The global stoichiometry of the process requires two equivalents of Ce(IV) for each equivalent of substrate (Scheme 14.3).

14.2.3
Photoinduced Nucleophilic Radical Substitution in the Presence of TiO$_2$

Heterogeneous photocatalysis on semiconductors has seen strong development in the last twenty years [16]. Recent studies have concentrated mainly on TiO$_2$ because of its stability and environmental tolerability, which make it the ideal material for most applications, despite the fact that it absorbs a small fraction of visible light. In this particular type of heterogeneous photocatalysis, the semiconductors which drive oxidation and reduction reactions are used in dispersed form. The two principal catalytic phases of TiO$_2$ are anatase and rutile [17]. Recent papers have reported enhanced photoactivity in the mixed-phase (anatase and rutile) Degussa P25 TiO$_2$ [18].

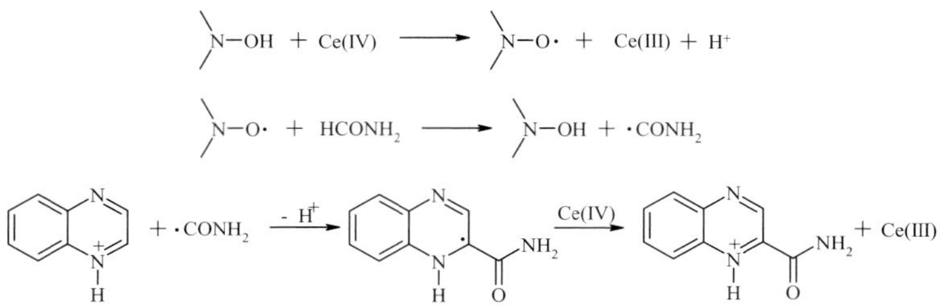

Scheme 14.3 Mechanism for the carbamoylation of N-heteroaromatic bases with NHPI/Ce(IV) system.

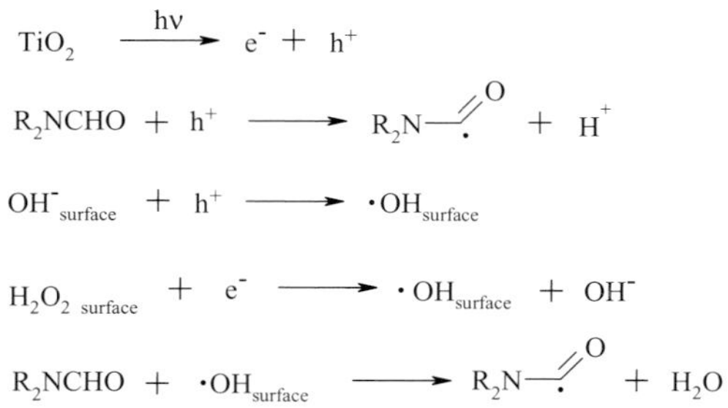

Scheme 14.4 Formation of photo induced carbamoyl radicals in the presence of TiO_2.

In the last few years we have developed new methods for the photochemical carbamoylation of electron-poor N-heteroaromatic bases. The introduction of TiO_2 photocatalysis has supplied new and environmentally friendly synthetic methods that operate under mild conditions (Scheme 14.4) [19, 20].

These processes gave good results when hydrogen peroxide was used as oxidant [19], whereas no reaction occurred when N,N-dialkyl substituted amides, such as dimethylformamide and dimethylacetamide, were employed in the presence of pure oxygen. This could be due to a faster reaction of the α-aminoalkyl radicals with oxygen. For similar reasons, even with formamide, when the reaction was conducted under a flow of oxygen, no products were observed, because of the competition between oxygen and the heteroaromatic base in the reaction with carbamoyl radical. However, when air was employed instead of oxygen, we obtained the desired products of carbamoylation in good yields (Equation 14.9) [20].

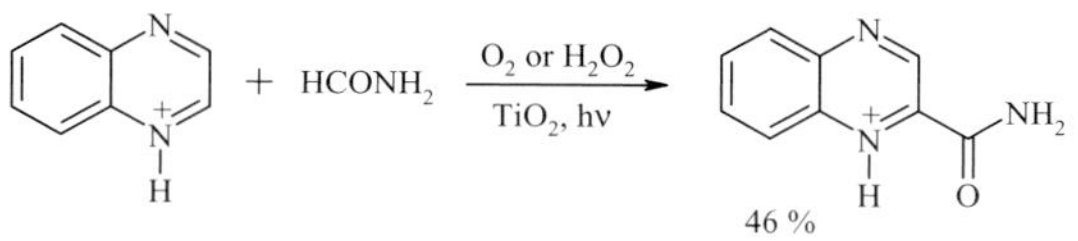

(14.9)

An analogous system has been employed in the presence of ethers, allowing the development of a "green" route for the synthesis of heterocyclic aldehydes (Equation 14.10) [21].

$$\text{(14.10)}$$

More recently we have reported that the TiO_2–sunlight–H_2O_2 system in the presence of aldehydes promotes free-radical acylation by intermediate acyl radical generation (Equation 14.11) [22]. The mechanism of the photoinduced reaction is quite complex as the intermediate could be oxidized on the TiO_2 surface. This step is not as efficient as in the presence of metal salts; in fact, during the period of irradiation, decarbonylation of the acyl radicals occurs, affording the corresponding alkylated products. Decarbonylation of acyl radicals in redox conditions has been largely disclosed [3b]. This behavior could be ascribed to the reversibility of the intermediate radical cation in the reaction conditions, whereas the role of the oxidant H_2O_2 could be only to restores Ti(III) to Ti(IV).

$$\text{(14.11)}$$

14.2.4
Hydroxymethylation of N-heteroaromatic Bases

This group has developed a new selective source of nucleophilic hydroxymethyl (CH_2OH) radical by persulfate oxidation of ethylene glycol catalyzed by Ag(I) salts. This nucleophilic radical is selectively trapped by protonated N-heteroaromatic bases, providing a new general hydroxymethylation protocol (Equation 14.12) [23].

$$\text{(14.12)}$$

This new general synthetic method utilizes the well-known $S_2O_8^{2-}/Ag^+$ redox system. The first step is the formation of the alkoxyl radical by Ag(II) electron transfer oxidation. The fast β-scission gives the nucleophilic CH_2OH radical which adds to the protonated substrate. As already seen, the intermediate is oxidized by $S_2O_8^{2-}/Ag^{2+}$ species, affording the hydroxymethylated products.

14.2.5
Perfluoroalkylation of N-heteroaromatic Bases and Quinones

The introduction of perfluoroalkyl groups in an organic molecule is a challenge in synthesis today. The low toxicity and high biocompatibility of many perfluorinated organic compounds suggested the use of free-radical nucleophilic substitution to synthesize molecules with possible biological activities. Recently, new perfluoroakylation processes of heteroaromatic bases and quinones have been developed (Schemes 14.5 and 14.6) [24, 25].

Perfluoroalkyl radicals were generated according to a procedure developed by this research group, involving iodine abstraction from perfluoroalkyl iodides by phenyl radical (Equation 14.13); this latter derives from thermal decomposition of benzoyl peroxide (Equation 14.14).

$$\text{Ph} \cdot + R_f - I \longrightarrow R_f \cdot + \text{Ph} - I \qquad (14.13)$$

$$(\text{PhCOO})_2 \longrightarrow 2\,\text{Ph} \cdot + CO_2 \qquad (14.14)$$

Owing to the high electronegativity of fluorine atoms, perfluoroalkyl radicals show an electrophilic character; moreover, these radicals are usually much more reactive than the nucleophilic alkyl radicals in the addition to alkenes, aromatic rings, and quinones for enthalpic reasons (Scheme 14.5b).

The perfluoro alkylation of protonated N-heteroaromatic bases (Scheme 14.5a) gives high conversion, because of the high enthalpic effect, which can be ascribed to the stronger C—C bond formed when $R_f \cdot$ rather than R radicals add to the substrate [26]. However, $R_f \cdot$ being an electrophilic radical, low selectivity is observed.

When the reactions are carried out in the presence of electron-rich alkenes (Scheme 14.6), selective introduction of perfluoro-functionalized alkyl groups onto the heteroaromatic bases and quinones take place. This is possible because perfluoroalkyl radicals add more rapidly to alkenes than to the strongly electron-deficient substrates. These radical adducts show a reversed polar character compared to the perfluoroalkyl radicals and thus they react much more rapidly with the electron-deficient substrates, affording products with high selectivity.

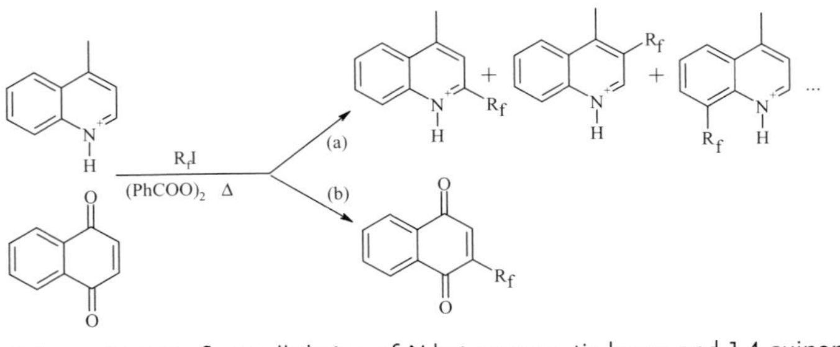

Scheme 14.5 Perfluoroalkylation of N-heteroaromatic bases and 1,4-quinones.

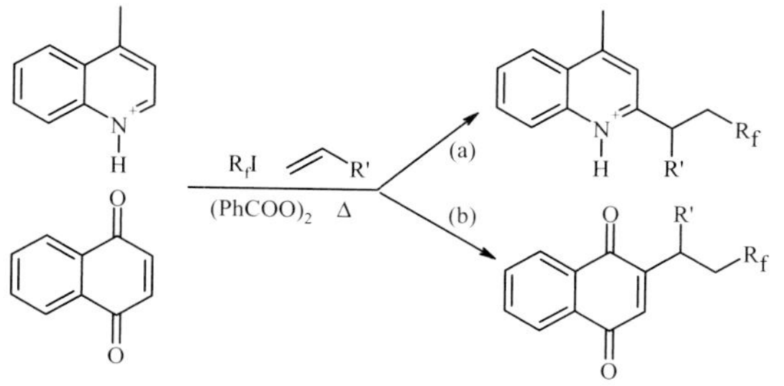

Scheme 14.6 Perfluoroalkylation of N-heteroaromatic bases and 1,4-quinones in the presence of electron rich olefins.

14.3
Nucleophilic Radical Addition to Aldimines

The intermolecular nucleophilic radical addition to C═N double bond of imines has received significant attention only in recent years [27, 28]. Moreover, there are few studies involving the reductive radical addition to aldimines in comparison to those dealing with C═N derivatives containing functional groups (such as oxime ethers, hydrazones, etc.) [27, 28]. The reason for this lack in the literature can be ascribed to three main factors: (a) simple aldimines have a slower radical addition rate because the electrophilicity of the C═N bond is not adequate for a fast addition of nucleophilic radicals, similarly to the behavior of unprotonated heteroaromatic bases; (b) they easily undergo hydrolysis, requiring anhydrous conditions; (c) contrary to what happens with oximes and hydrazones, they do not have the potential for a stabilizing three-electron π-bond in the intermediate aminyl radical.

14.3.1
Nucleophilic Radical Addition Promoted by TiCl$_3$/PhN$_2^+$ Systems

In 1990, Porta and co-workers [6] reported the first nucleophilic radical addition to aldimines promoted by aqueous TiCl$_3$, based on a one-pot tricomponent reaction involving an aromatic amine, a generic aldehyde, and an arene-diazonium salt (Equation 14.15).

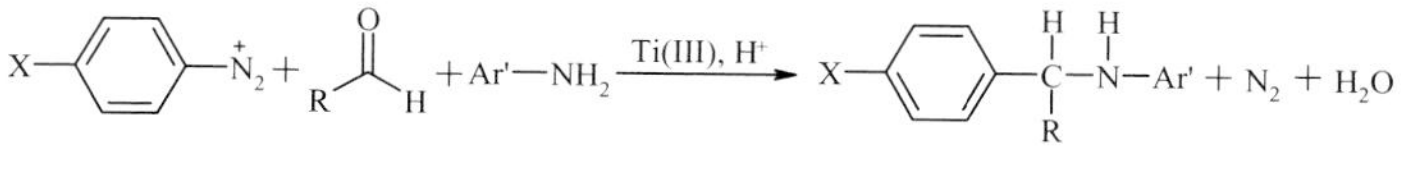

$$\text{X}-\text{C}_6\text{H}_4-\overset{+}{\text{N}}_2 + \underset{\text{R}}{\text{RCHO}} + \text{Ar}'-\text{NH}_2 \xrightarrow{\text{Ti(III), H}^+} \text{X}-\text{C}_6\text{H}_4-\overset{\text{H}}{\underset{\text{R}}{\text{C}}}-\overset{\text{H}}{\text{N}}-\text{Ar}' + \text{N}_2 + \text{H}_2\text{O} \quad (14.15)$$

Titanium plays a key role in this process: owing to its reducing power in the Ti(III) state, it acts as radical initiator generating Ar· by decomposition of arene-diazonium salts (Equation 14.16); owing to its Lewis acid character in the Ti(IV) state, it promotes the formation of the imine [29] from a generic aldehyde and a primary aromatic amine in the aqueous medium (Equation 14.17) and increases the imine reactivity toward nucleophilic radicals (Equation 14.18), even though aryls are very reactive for enthalpic reasons and quite weakly nucleophilic radicals [3]; then Ti(III) terminates the radical process by reducing the final amminium radical intermediate (Equation 14.19).

$$\text{Ar}-\overset{+}{\text{N}}_2 + \text{Ti(III)} \longrightarrow \text{Ar}\cdot + \text{N}_2 + \text{Ti(IV)} \quad (14.16)$$

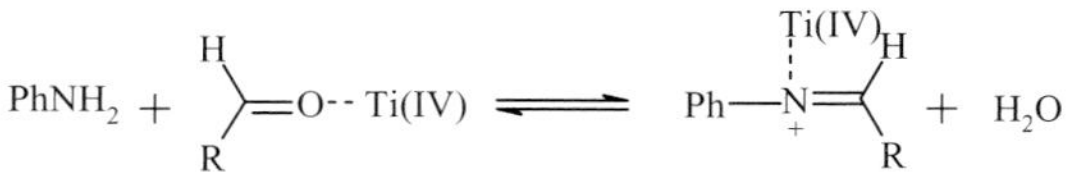

$$\text{PhNH}_2 + \underset{R}{\overset{H}{\diagup}}=\text{O}\cdots\text{Ti(IV)} \rightleftharpoons \text{Ph}-\overset{\overset{\text{Ti(IV)}}{|}}{\underset{+}{N}}=\underset{R}{\overset{H}{\diagup}} + \text{H}_2\text{O} \quad (14.17)$$

$$\text{Ph}-\overset{\overset{\text{Ti(IV)}}{|}}{\underset{+}{N}}=\underset{R}{\overset{H}{\diagup}} + \text{Ar}\cdot \underset{\text{Ti(IV)}}{\overset{H^+}{\rightleftharpoons}} \text{Ph}-\overset{H}{\underset{+\cdot}{N}}\underset{R}{\overset{H}{\diagdown}}\text{Ar} \quad (14.18)$$

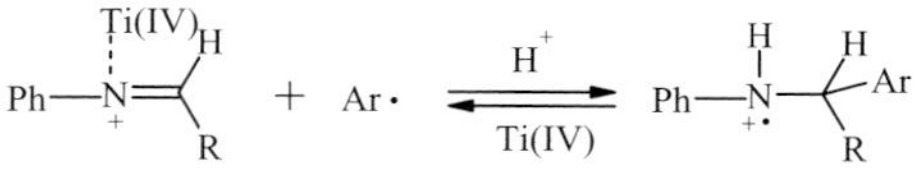

$$\text{Ph}-\overset{H}{\underset{+\cdot}{N}}\underset{R}{\overset{H}{\diagdown}}\text{Ar} \xrightarrow[\text{Ti(IV)}]{\text{Ti(III)}} \text{Ph}-\overset{H}{\underset{R}{N}}\overset{H}{\diagdown}\text{Ar} \quad (14.19)$$

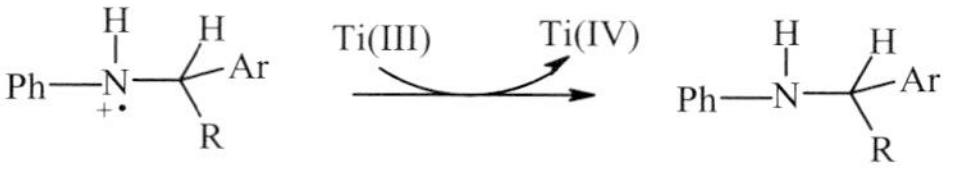

In the last decade the same research group has employed different systems for the generation of a wide range of more nucleophilic radicals, in order to develop new multicomponent synthetic routes of a variety of substrates, according to the general mechanism reported in Scheme 14.7.

By the simple addition of a fourth component to the $\text{TiCl}_3/\text{PhN}_2^+$ system described, it has been possible to enlarge the synthetic potentiality of the process.

In the presence of an alkyl iodide, selective alkyl radical addition to the C-atom of the imine generated *in situ* occurs, overcoming the competitive phenylation reaction (Equation 14.20) [30]. The Ph· radical, generated by decomposition of the diazonium salt, as described before, generates the alkyl radical by selective iodine atom transfer (Equation 14.21).

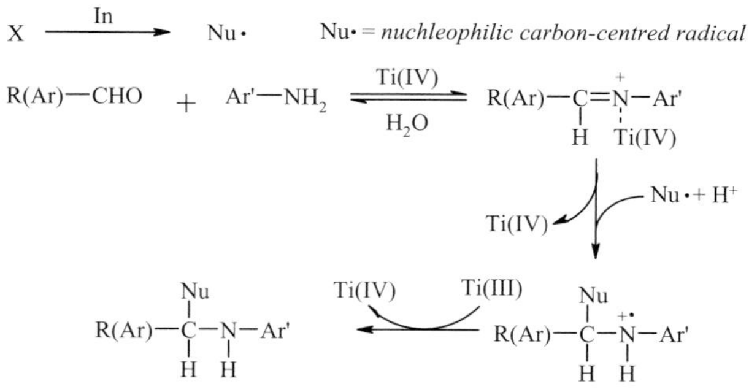

$$X \xrightarrow{\text{In}} \text{Nu}\cdot \qquad \text{Nu}\cdot = \textit{nuchleophilic carbon-centred radical}$$

$$\text{R(Ar)}-\text{CHO} + \text{Ar}'-\text{NH}_2 \underset{\text{H}_2\text{O}}{\overset{\text{Ti(IV)}}{\rightleftharpoons}} \text{R(Ar)}-\underset{H}{\overset{+}{\text{C}}}=\underset{\text{Ti(IV)}}{N}-\text{Ar}'$$

$$\downarrow \text{Ti(IV)} \quad \text{Nu}\cdot + H^+$$

$$\text{R(Ar)}-\underset{H}{\overset{Nu}{\underset{|}{C}}}-\underset{H}{\overset{}{N}}-\text{Ar}' \xleftarrow[\text{Ti(III)}]{\text{Ti(IV)}} \text{R(Ar)}-\underset{H}{\overset{Nu}{\underset{|}{C}}}-\underset{H}{\overset{+\cdot}{N}}-\text{Ar}'$$

Scheme 14.7 General mechanism for the nucleophilic radical addition to imines mediated by Ti (IV).

14.3 Nucleophilic Radical Addition to Aldimines

$$Ph-\overset{+}{N_2} + \underset{Ar}{\overset{O}{\underset{\|}{C}}}_H + Ar'-NH_2 + R-I \xrightarrow{Ti(III), H^+} Ar-\underset{R}{\overset{H}{\underset{|}{C}}}-\overset{H}{\underset{|}{N}}-Ar' + N_2 + H_2O + Ph-I$$

50 - 76 %

(14.20)

$$Ph\cdot + R-I \longrightarrow R\cdot + Ph-I \quad (14.21)$$

When the TiCl$_3$/PhN$_2^+$ system is employed in tetrahydrofuran (THF) solvent, the phenyl radical abstracts an α-H atom from THF (Equation 14.22), leading to the addition of nucleophilic α-alkoxyalkyl radicals to the C-atom of aldimines (Equation 14.23) [31].

$$Ph\cdot + \underset{O}{\text{(THF)}}_H \longrightarrow \underset{O}{\text{(THF)}}\cdot + Ph-H \quad (14.22)$$

$$Ph-\overset{+}{N_2} + \underset{R}{\overset{O}{\underset{\|}{C}}}_H + Ar-NH_2 + THF \xrightarrow{Ti(III), H^+} R-\underset{\text{(THF)}}{\overset{H}{\underset{|}{C}}}-\overset{H}{\underset{|}{N}}-Ar + N_2 + H_2O + Ph-H$$

55 - 85 %

(14.23)

Even imines formed by *in situ* condensation of aromatic amines with either acetaldehyde or formaldehyde afforded the desired products. The successful radical addition to these rather unstable and polymerizable imines in aqueous medium may be ascribed to the lack of steric hindrance at the C-atom.

14.3.2
Nucleophilic Radical Addition Promoted by TiCl$_3$/Pyridine Systems

A few years previously, the same research group reported a one-pot condensation of three components (methylphenylglyoxylate (MPG), aniline, and aromatic aldehyde), promoted by a TiCl$_3$/pyridine system under anhydrous conditions, for the diasteroselective synthesis of β-amino-α-hydroxyesters (Equation 14.24) [32].

$$\text{MeO}\underset{O}{\overset{O}{\underset{\|}{C}}}\text{Ph} + PhNH_2 + \underset{Ar}{\overset{O}{\underset{\|}{C}}}_H \xrightarrow[\text{Pyr, r.t.}]{Ti(III), THF} \text{MeO}\underset{O}{\overset{O}{\underset{\|}{C}}}\underset{\overset{|}{OH}}{\overset{Ph}{\underset{|}{C}}}\underset{Ph}{\overset{NHPh}{\underset{|}{C}}}Ar \quad (14.24)$$

53-67 %

The first step of the reaction involves the dimerization of MPG via single electron transfer process by Ti(III) and the sequential Ti(IV)-catalyzed intramolecular heterolytic cleavage of the dimer, regenerating MPG and the nucleophilic radical (Equations 14.25 and 14.26).

$$2 \text{ MeO} \underset{O}{\overset{O}{\|}} \text{Ph} \xrightarrow[\text{Ti(III)}]{\text{THF, Pyr}} \left[\text{MeO} \underset{\underset{\text{Ph}}{\text{Ph}}}{\overset{\overset{\text{Ti(IV)}}{O\,|\,O}}{\|\,\,\|}} \text{OMe} \right] \xrightarrow{\text{Ti(IV)}} \text{MeO} \underset{O}{\overset{O}{\|}} \text{Ph} + \underset{\text{OH}}{\overset{\text{Ph}\,\,\text{Ph}\,\cdot}{\|}} \text{OMe} \quad (14.25)$$

$$\underset{\text{Ar}}{\overset{\overset{\text{Ti(IV)}}{|}\text{H}}{\text{Ph—N}=}}\,\,+\,\, \underset{\text{OH}}{\overset{\text{Ph}\,\cdot}{\|}}\text{OMe} \xrightleftharpoons{\text{Ti(IV)}} \text{MeO}\underset{\underset{\text{OH}}{\text{Ph}}}{\overset{\overset{\text{NHPh}}{O}}{\|}}\text{Ar} \quad (14.26)$$

This process provides the desired products with yields up to 80% and *syn*-diastereoselectivity. These derivatives are of considerable importance because they are incorporated into many biologically active peptides, such as bestatin, amastatin, norstatin, and taxol [33].

14.3.3
Nucleophilic Radical Addition Promoted by TiCl$_3$/Hydroperoxide Systems

More recently, Porta and co-workers have reported a free-radical Mannich type reaction based on the selective α-CH aminomethylation of ethers by a Ti(III)/*t*-BuOOH system under aqueous acidic conditions (Equation 14.27) [34].

$$\underset{R}{\overset{H}{\underset{|}{N}}}_{R'} + \overset{O}{\underset{H}{\|}}_H + H\overset{}{\frown}_O \xrightarrow[H_2O,\,H^+,\,\text{rt}]{tert\text{-BuOOH/Ti(III)}} \underset{R'}{\overset{R}{\underset{|}{N}}}\frown_O \quad (14.27)$$

The classical Mannich aminomethylation is one of the most important ionic carbon–carbon bond forming reactions in organic chemistry [35]. However, only substituents with electron-withdrawing groups are suitable for the ionic addition. Electron-donating groups directly bonded to the carbon-centered radical favor nucleophilic radical addition to methylene-iminium salts. Thus, the radical-type Mannich reaction provides products which are complementary to those obtained with the classical ionic reaction.

The TiCl$_3$/*t*-BuOOH system is a more practical, efficient, and selective radical precursor of α-alkoxyalkyl radicals from ethers than the previously described phenyl radical (Equations 14.28 and 14.29).

$$tert\text{-BuOOH} + \text{Ti(III)} + \text{H}^+ \longrightarrow tert\text{-BuO}\cdot + \text{H}_2\text{O} + \text{Ti(IV)} \quad (14.28)$$

$$tert\text{-BuO}\cdot + H\frown_O \longrightarrow \cdot\frown_O + tert\text{-BuOH} \quad (14.29)$$

Finally, the development of new methods for the synthesis of natural and non-natural α-amino acids is an area of increasing interest both in synthetic and medicinal chemistry.

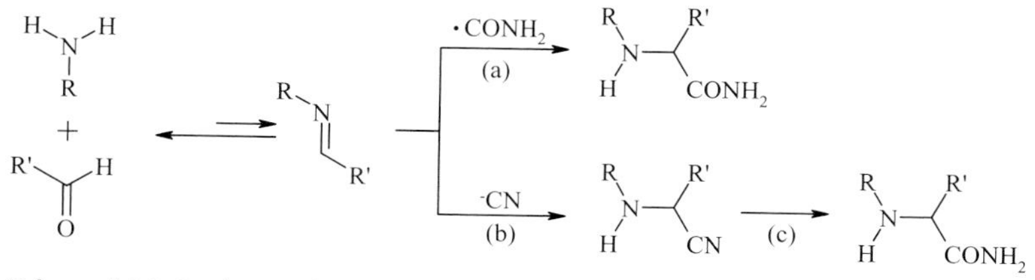

Scheme 14.8 Synthesis of α-aminoamides: radical version (a) versus classical Strecker synthesis (b).

Thus, in collaboration with this research group, we have developed a new free-radical synthesis of α-aminoamides (precursors of amino acids) in high yields, promoted by an aqueous $H_2O_2/TiCl_3/HCONH_2$ system (Equation 14.30) [36]. According to the proposed mechanism, the hydroxyl radical, generated by the Ti(III) one-electron reduction of H_2O_2, abstracts a hydrogen atom from formamide, generating the corresponding carbamoyl radical, which adds to imines generated in situ.

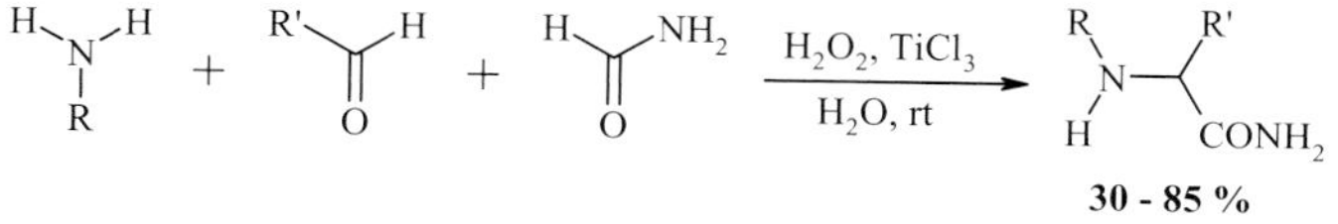

$$\text{(14.30)} \quad 30 - 85 \%$$

This approach, in which a carbamoyl radical acts as a nucleophilic carboxylate synthon (Scheme 14.8a) in place of the nucleophilic ionic cyanide, may be regarded as a more efficient radical version of the Strecker synthesis (Scheme 14.8b).

The successful results and the green credentials of the process (water as co-solvent and only the nontoxic titanium dioxide as final residue) make this reaction an excellent, cheaper, and favorable alternative to the classical Strecker synthesis, which also requires the hydrolysis of cyanide to the amido group (Scheme 14.8c).

The possibility of employing a wide range of nucleophilic radical sources, including alcohols, the choice of ideal protecting groups for amines, and the possibility of extending this process to ketimines by the development of a catalytic system in which $TiCl_4$ is associated with Zn, enables us to anticipate new frontiers for the synthesis of new structural types of α-amino acids and other amino-derivatives of crucial importance for chemistry, medicine, and life.

References

1 Barton, D.H., Beaton, J.M., Geller, L.E. and Pechet, M.M. (1960) *Journal of the American Chemical Society*, **82**, 2640.

2 Minisci, F., Galli, R., Cecere, M., Malatesta, V. and Caronna, T. (1968) *Tetrahedron Letters*, **8**, 5609.

3 a. Minisci, F., Fontana, F. and Vismara, E. (1989) *Heterocycles*, **28**, 489.b. Minisci, F., Fontana, F. and Vismara, E. (1990) *Journal of Heterocyclic Chemistry*, **27**, 79.

4 Jaffé, H.H. (1952) *The Journal of Chemical Physics*, **20**, 1554.

5 Jaffé, H.H. (1955) *Journal of the American Chemical Society*, **77**, 4445.
6 Clerici, A. and Porta, O. (1990) *Tetrahedron Letters*, **14**, 2069
7 Friedel, C. and Crafts, J.M. (1877) *Comptes Rendus*, **84**, 1392, 1450.
8 a. Caronna, T., Gardini, G.P. and Minisci, F. (1969) *Chemical Communications*, **5**, 201. b. Caronna, T. and Minisci, F. (1976) *Reviews on Reactive Species in Chemical Reactions*, **1** (3–4), 263.
9 Caronna, T., Fronza, G., Minisci, F., Porta, O. and Gardini, G.P. (1972) *Journal of the Chemical Society. Perkin Transactions 2*, **10**, 1477.
10 Fontana, F., Minisci, F., Nogueira Barbosa, M.C. and Vismara, E. (1991) *The Journal of Organic Chemistry*, **56**, 2866.
11 Arnone, A., Cecere, M., Galli, R., Minisci, F., Perchinunno, M., Porta, O. and Gardini, G. (1973) *Gazzetta Chimica Italiana*, **103**, 13.
12 Ishii, Y., Sakaguchi, S. and Iwahamab, T. (2001) *Advanced Synthesis and Catalysis*, **343**, 393.
13 Minisci, F., Recupero, F., Cecchetto, A., Punta, C., Gambarotti, C., Fontana, F. and Peduli, G.F. (2003) *Journal of Heterocyclic Chemistry*, **40**, 325–8.
14 Fossey, J., Lefort, D. and Sorba, J. (1995) *Free Radicals in Organic Chemistry*, Masson, Paris, p. 297.
15 Minisci, F., Recupero, F., Punta, C., Gambarotti, C., Antonietti, F., Fontana, F. and Peduli, G.F. (2002) *Chemical Communications*, 2496–7.
16 a. Fox, M.A. and Dulay, M.T. (1993) *Chemical Reviews*, **93**, 341–57.b. Maldotti, A., Molinari, A. and Amadelli, R. (2002) *Chemical Reviews*, **102**, 3811–36.
17 a. Riegel, G. and Bolton, J.R. (1995) *The Journal of Physical Chemistry*, **99**, 4215.b. Yin, H. et al. (2001) *Journal of Materials Chemistry*, **11**, 1694.
18 Hurum, D.C., Agrios, A.G., Gray, K.A., Rajh, T. and Thurnauer, M.C. (2003) *The Journal of Physical Chemistry. B*, **107**, 4545.
19 Caronna, T., Gambarotti, C., Palmisano, L., Punta, C. and Recupero, F. (2003) *Chemical Communications*, 2350.
20 Caronna, T., Gambarotti, C., Mele, A., Pierini, M., Punta, C. and Recupero, F. (2007) *Reviews of Chemical Intermediates*, **33**, 311.
21 Caronna, T., Gambarotti, C., Palmisano, L., Punta, C. and Recupero, F. (2005) *Journal of Photochemistry and Photobiology. A*, **171**, 237.
22 Caronna, T., Gambarotti, C., Palmisano, L., Punta, C. and Recupero, F. (2007) *Journal of Photochemistry and Photobiology. A*, **189**, 322–8.
23 Minisci, F., Porta, O., Recupero, F., Punta, C., Gambarotti, C., Pruna, B., Pierini, M. and Fontana, F. (2004) *Synlett: Accounts and Rapid Communications in Synthetic Organic Chemistry*, **5**, 874.
24 Antonietti, F., Mele, A., Minisci, F., Punta, C., Recupero, F. and Fontana, F. (2004) *Journal of Fluorine Chemistry*, **125**, 205.
25 Antonietti, F., Gambarotti, C., Mele, A., Minisci, F., Paganelli, R., Punta, C. and Recupero, F. (2005) *European Journal of Organic Chemistry*, **20**, 4434.
26 McMillan, D.F. and Golden, D.M. (1982) *Annual Reviews of Physical Chemistry*, **33**, 493.
27 For inter- and intramolecular radical addition to C N bonds see: a. Friestad, G.K. (2005) *European Journal of Organic Chemistry*, **15**, 3157.b. Miyabe, H., Ueda, M. and Naito, T. (2004) *Synlett: Accounts and Rapid Communications in Synthetic Organic Chemistry*, **7**, 1140.c. Ishibashi, H., Sato, T. and Ikeda, M. (2002) *Synthesis*, **6**, 695.d. Friestad, G.K. (2001) *Tetrahedron*, **57**, 5461.e. Naito, T. (1999) *Heterocycles*, **50**, 505–41.
28 a. Miyabe, H., Ueda, M., Nishimura, A. and Naito, T. (2004) *Tetrahedron*, **60**, 4227.b. Friestad, G.K., Shen, Y. and Ruggles, E.L. (2003) *Angewandte Chemie (International Ed. in English)*, **42**, 5061.c. Bertrand, M.P., Coantic, S., Feray, L., Nougier, R. and Perfetti, P. (2000) *Tetrahedron*, **56**, 3951.d. Halland, N. and Jørgensen, K.A.d. (2001) *Journal of the Chemical Society. Perkin Transactions 1*, **11**, 1290.e. Ryu, I., Kuriyama, H., Minakata, S., Komatsu, M., Yoon, J.Y. and Kim, S. (1999) *Journal of the American Chemical Society*, **121**, 12190.

29 a. Weingarten, H., Chupp, J.P. and White, W.A. (1967) *The Journal of Organic Chemistry*, **32**, 3246.b. Desay, M.C. and Thadeio, P.F. (1989) *Tetrahedron Letters*, **30**, 5223.
30 Cannella, R., Clerici, A., Pastori, N., Regolini, E. and Porta, O. (2005) *Organic Letters*, **7**, 645.
31 Clerici, A., Cannella, R., Pastori, N., Panzeri, W. and Porta, O. (2005) *Tetrahedron Letters*, **62**, 5986.
32 Clerici, A., Clerici, L. and Porta, O. (1995) *Tetrahedron Letters*, **33**, 5955.
33 a. Kabayashi, S., Isobe, T. and Ohno, M. (1984) *Tetrahedron Letters*, **25**, 5079.b. Herranz, R., Castro-Pichel, J., Vinuesa, S. and Garcia Lopez, T. (1990) *The Journal of Organic Chemistry*, **55**, 2232.c. Dagger, R.W., Ralbowsky, J.L., Bryant, D., Commander, J., Masset, S.S., Sage, N. and Selvidio, J.R. (1992) *Tetrahedron Letters*, **45**, 6763.d. Jefford, C.W., Wang, J.B. and Hin Lu, Z. (1993) *Tetrahedron Letters*, **47**, 7557.
34 Clerici, A., Cannella, R., Pastori, N., Panzeri, W. and Porta, O. (2006) *Tetrahedron Letters*, **62**, 5986.
35 For reviews on the Mannich reaction see: a. Tramontini, M. (1973) *Synthesis*, **12**, 703.b. Tramontini, M. and Angiolini, L. (1990) *Tetrahedron*, **46**, 1791.c. Azend, M., Westermann, B. and Risch, N. (1988) *Angewandte Chemie (International Ed. in English)*, **37**, 1044.d. Cordova, A. (2004) *Accounts of Chemical Research*, **37**, 102.
36 Cannella, R., Clerici, A., Panzeri, W., Pastori, N., Punta, C. and Porta, O. (2006) *Journal of the American Chemical Society*, **128**, 5358.

Part Three
Health, Food, and Environment

15
Future Perspectives of Medicinal Chemistry in the View of an Inorganic Chemist

Palanisamy Uma Maheswari

15.1
Introduction

Since the discovery of the antitumor activity of cisplatin, *cis*-[Pt(NH$_3$)$_2$Cl$_2$], by Rosenberg in 1965, platinum-based combination chemotherapy is still the major treatment for solid malignancies (especially testicular, ovarian, and small-cell lung cancers) [1–3]. The unique DNA damage delivered by platinum has not, to date, been mimicked by any organic drugs and clearly the metal–biomolecule interaction is critical to the antitumor activity of platinum, as it is, in general, for the activity of any metallodrug. But the serious side-effects, general toxicity, and drug resistance remain as limitations and have provided the motivation for alternative chemotherapeutic pathways for apoptosis of cancerous tissues [4]. For example, *satraplatin* is a platinum(IV) prodrug, that can be orally administered, and is reduced by intracellular biomolecules to yield cytotoxic platinum(II) moieties once inside the cell [5]. Platinum(IV) compounds with functionalized ligands have also been harnessed to defeat glutathione *S*-transferase-mediated drug resistance, to target estrogen receptor-positive breast cancer, and as a prodrug for photochemotherapy (Figure 15.1) [6]. There are also a number of platinum(II) compounds that target aspects of cancer cells to enhance uptake and improve selectivity [7]. But there are compounds which fail to be effective clinically, like the *trans* series of cisplatin analogues (Figure 15.2). Quite surprisingly, of the trinuclear platinum complexes BBR3464 and BBR3499, the latter with *cis* configuration is less active than the former with *trans* configuration (Figure 15.2) [7].

As an alternative to platinum-based drugs, compounds based on ruthenium, in particular, have gained significant prominence. Ruthenium is an attractive alternative to platinum: besides the rich synthetic chemistry, ruthenium has a range of oxidation states (Ru(II), Ru(III), and Ru(IV)) accessible under physiological conditions, which is unique among the platinum-group metals [8]. This feature is significant since the activities of most metal-based anticancer drugs are dependent on their oxidation states. In addition, ruthenium compounds are known to be less toxic than their platinum counterparts [8–10]. This is believed to be due to the ability of ruthenium to mimic iron in the binding to biological molecules, such as

Tomorrow's Chemistry Today. Concepts in Nanoscience, Organic Materials and Environmental Chemistry.
Edited by Bruno Pignataro
Copyright © 2008 WILEY-VCH Verlag GmbH & Co. KGaA, Weinheim
ISBN: 978-3-527-31918-3

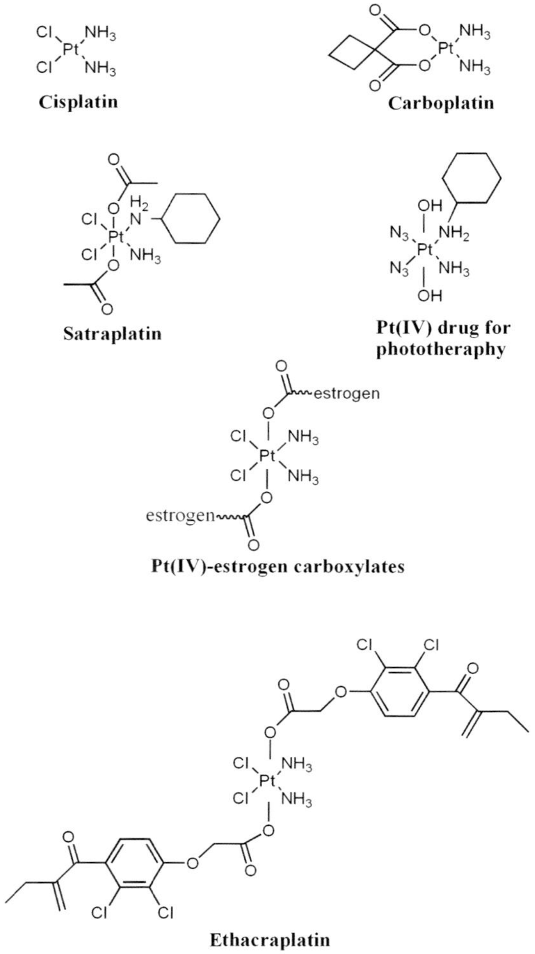

Figure 15.1 Clinical and preclinical platinum-based anticancer drugs.

albumin and transferrin, although platinum drugs can also bind to these proteins. Since rapidly dividing cells, such as cancer cells, have a greater demand for iron, transferrin receptors are overexpressed, thereby allowing ruthenium-based drugs to be more effectively delivered to cancer cells [9, 10]. In addition, the "activation by reduction" mechanism could also account for the lower general toxicity of some ruthenium compounds [9]. Two ruthenium-based anticancer drugs, namely NAMI-A and KP1019 (Figure 15.3), have successfully completed phase 1 clinical trials and are scheduled to enter phase 2 trials in the near future [11–13]. In the past two decades, a new approach to treating cancer, known as targeted therapy,

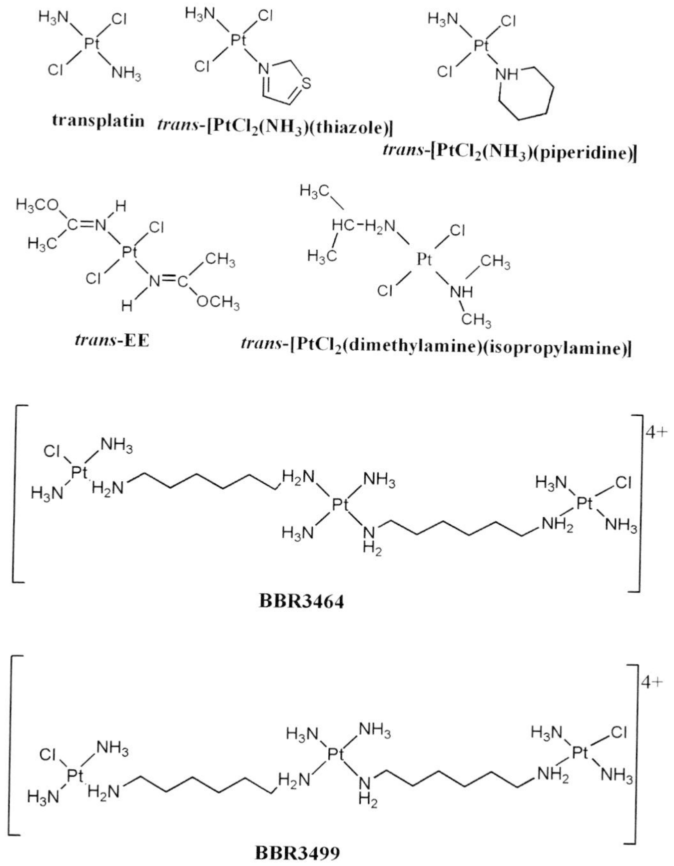

Figure 15.2 *Trans* analogues of cisplatin complexes.

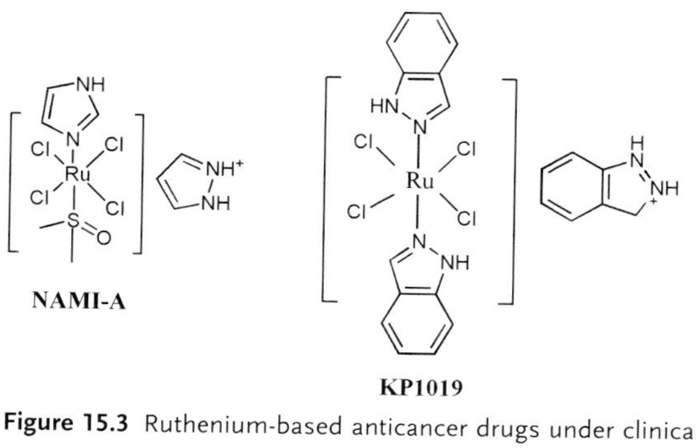

Figure 15.3 Ruthenium-based anticancer drugs under clinical evaluation.

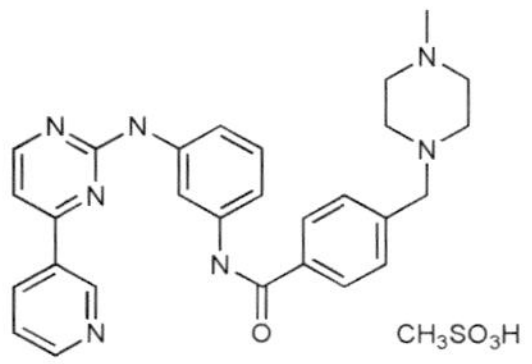

Imatinib mesylate

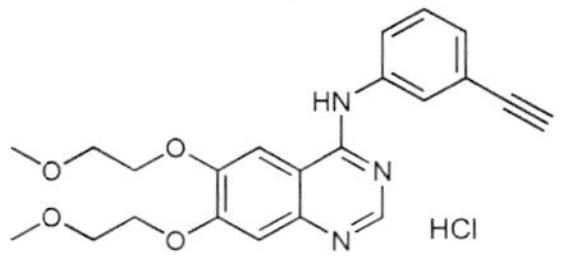

Erlotinib hydrochloride

Figure 15.4 Drugs for targeted therapy in clinical use.

has started to take root [14]. The new strategy involves targeting cellular signaling pathways of cancer cells, yielding highly effective cancer treatments with much less severe side-effects [14, 15]. A few of them, for example, imatinib mesylate and erlotinib hydrochloride (Figure 15.4), demonstrated such potential that their approval processes were fast-tracked.

15.1.1
Conventional versus Targeted Therapy

Conventional chemotherapy, which originated in the 1950s, refers to drugs interfering with replication and mitotic processes of tumor cells [14]. Historically, these were the main "target" of chemotherapy, since cancer biology was not well established at that time and little was known about the causes of cancer and its mechanism, so the general therapeutic strategy was largely premised on the fact that cancer cells replicate their DNA more frequently than normal cells and hence are more susceptible to DNA damage. Examples of cytotoxic substances include alkylating agents, mitosis inhibitors, and topoisomerase inhibitors.

In recent years, with the advent of molecular oncology, the study of cancer at the molecular level has become possible [14]. The discovery of receptors and growth factors such as epidermal growth factor receptor (EGFR), vascular endothelial growth factor (VEGF), cyclin-dependent kinases (CDKs) that are upregulated in cancer cells provides new "targets" for cancer therapy [15, 16]. Whereas the strategy of classical chemotherapy relies on damaging cancerous cells more than normal cells, targeted therapies are far more specific and their toxicity profile is more manageable. Important classes of target therapeutic agents include monoclonal antibodies that interfere with the activities of EGFR, VEGF, or proteasomes, small molecule inhibitors of the tyrosine kinase receptor (which includes EGFR), as well as immunotherapeutic agents that trigger an immune response against

CD20 antigens [15, 17]. The main drawback of classical chemotherapy strategies is that by blocking the metabolism of rapidly dividing malignant cells, they inadvertently inflict damage on healthy cells that divide frequently, such as hair follicles, bone marrow, and cells lining the gastrointestinal tract. Consequently, there are associated side-effects, such as hair loss, anemia, and neutropenia, that are often severe and limit the treatment options [18]. In contrast, targeted therapies focus on specific cellular signaling pathways on which the cancer cells depend for growth, proliferation, metastasis, and angiogenesis, and are therefore much more selective [15]. As it stands, most targeted therapeutic drugs are only effective in fairly specific types of cancer, for example, imatinib mesylate for chronic myelogenous leukemia, erlotinib for advanced non-small-cell lung cancer, and so on. This limits their applicability and most common cancers cannot be treated with "targeted" chemotherapeutic agents, although this is expected to change in the future.

One other strategy is based on the "chemical nucleases" or "artificial nucleases" which target the chromosomal DNA in the cancer cell-lines. These synthetic metal complexes like $[Cu(phen)_2]^+$ which can cleave DNA either hydrolytically or oxidatively in the presence of exogenous agents (H_2O_2, mercaptopropionic acid, dithiothreitol or light) serve as biological tools for molecular biologists in foot-printing and so on. The "self-activated" systems like Fe-bleomycin are applicable as anticancer agents with broad spectrum of activity. More research on these kinds of self-activating systems might lead to another strategy of classical type to target the DNA in cancer cell-lines. In this chapter, the ruthenium drugs and new strategies will be reviewed broadly and the application of self-activated metallonucleases as anticancer drugs will be considered.

15.2
Ruthenium Anticancer Drugs

15.2.1
Ru–Polypyridyl Complexes

Polypyridyl–Ru systems have been exploited extensively as molecular DNA probes, given their photoluminescence properties and the ability of polypyridyl ligands to intercalate DNA. The large, rigid, multidentate polypyridyl ligands confer shape and chirality to the ruthenium complexes that could be exploited to achieve customized DNA-binding properties. This has provided the motivation to develop polypyridyl–Ru complexes as DNA-targeting anticancer agents, and a large number of these complexes have been screened for anticancer activity. Typical polypyridyl ring ligands include 2,2′-bipyridine (bpy), 1,10-phenanthroline (phen) and 2,2′:6′,2″-terpyridine (terpy) as they are commercially available and readily form stable complexes with ruthenium (Figure 15.5). Some of the earliest polypyridyl–Ru complexes studied for potential anticancer properties include cis-$[Ru(bpy)_2Cl_2]$ (both Δ and Λ enantiomers) and mer-$[Ru(terpy)Cl_3]$. In vitro, mer-$[Ru(terpy)Cl_3]$ was

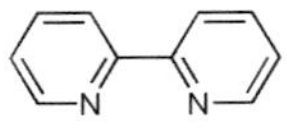

2,2'-bipyridine

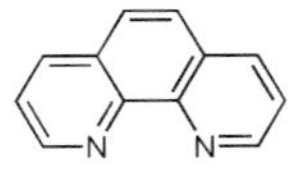

1,10-phenanthroline

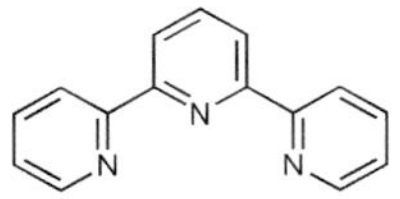

2,2':6',2''-terpyridine

Figure 15.5 Polypyridyl ligands used in ruthenium anticancer complexes.

significantly more cytotoxic (L1210, HeLa) than both enantiomers of cis-[Ru(bpy)$_2$Cl$_2$ [19]. This trend correlates wit the ability of mer-[Ru(terpy)Cl$_3$] to form DNA-interstrand crosslinks, whereas the inactive cis-[Ru(bpy)$_2$Cl$_2$] appears to exhibit no such interactions [19, 20]. There have also been numerous examples of polypyridyl–Ru complexes comprising one or more [Ru(bpy)$_2$L]$^{2+}$ or [Ru(phen)$_2$L]$^{2+}$ units [where L = derivatized quinolines, 2,6-(2'-benzimidazolyl)pyridine/chalcone, aryldiazo-β-diketonate, 4-substituted thiosemicarbazides, 4-substituted thiopicolinanalides, 2-phenylazoimidazole, etc.] in an attempt to improve the DNA-intercalating ability of the complex [21].

A series of DNA-binding ruthenium(II) complexes with tetradentate cyclam rings and varying DNA-intercalating quinonediimide ligands were studied in vitro (KB-3-1, KB-V1). It was noted that cytotoxicity was linked to the ability of the quinonediimide ligands to intercalate, although the IC$_{50}$ values were considered too high to be of interest [22]. Reedijk et al. reported an example of a NO-containing polypyridyl–Ru complex that readily liberates NO upon irradiation under a mercury lamp and that could potentially be applied in phototherapy [23]. The complex cis-(Cl,Cl)-[Ru^{II}(terpy)(NO)Cl$_2$]Cl exhibited good cytotoxicity toward A2780 human ovarian carcinoma cell lines, significantly higher than that of mer-[Ru-(terpy)Cl$_3$], cisplatin, or carboplatin, although their activities under irradiation were not reported. Harding et al. reported the synthesis of ruthenium analogues of streptonigrin, a DNA-targeting antitumor antibiotic [24]. On the basis of structural activity studies, the investigators identified the key structures responsible for their activity and synthesized quinolinide and bipyridine ligands to mimic the active sites. Subsequent dicarbonyl(dichloro)ruthenium(II) complexes [Ru(CO)$_2$Cl$_2$L] (where L = quinolinide and bipyridine mimics of streptonigrin) were prepared, although it appears that no further in vitro studies were undertaken. More recently, a heteronuclear (Pt, Ru) complex, comprising a Ru(terpy) moiety, with a highly flexible bridging chain was developed (Figure 15.6) [25]. The Pt end was designed to bind directly to DNA, leaving the Ru(terpy) end to intercalate, thereby providing additional anchor support. Ultimately, this type of hybrid complex could provide the basis for customized DNA-targeting agents that could form long-range DNA adducts.

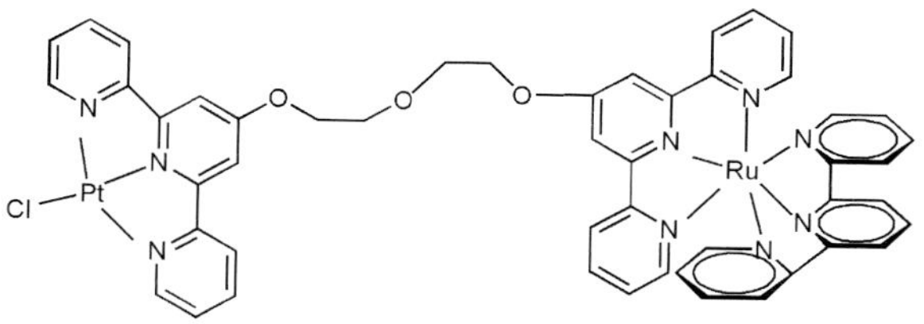

Figure 15.6 A heteronuclear (Pt, Ru) complex that potentially coordinates and intercalates DNA.

15.2.2
Ru–Polyaminocarboxylate Complexes

The use of polyaminocarboxylate (pac) ligands in metallopharmaceutical applications is of growing interest not only because of their ability to bind strongly with metal centers but also because their amino and carboxylate binding entities are parallel to those in biological systems [26]. One of the earlier pac–Ru complexes evaluated for anticancer activity is the highly water-soluble dichloro(1,2-propylenediaminetetraacetate) ruthenium(III), Ru(pdta)Cl$_2$ (Figure 15.7), which demonstrated antitumoral activity *in vivo* (EAT, L1210, P388, MX-1, M5076) with low general toxicity [27]. The presence of the two chloride ligands in the *cis*-conformation was found to be an important feature of its biological activity; under physiological conditions, the chlorides rapidly hydrolyze [28]. Ru(pdta)Cl$_2$ was found to alter the DNA conformation in pHV14 DNA and inhibit DNA lysis by restriction enzymes [29]. The structurally similar K[Ru^{III}(eddp)Cl$_2$] complex (where eddp = ethylenediamine-*N,N*-di-3-propionate) also displayed cytotoxicity *in vitro* (HeLa, BT-20, HT-29) and was found to induce DNA cleavage [30]. A related compound, K$_2$[Ru^{III}(dmgly)Cl$_4$] (where dmgly = *N,N'*-dimethylglycine), was reported to be cytotoxic toward the murine C6 astrocytoma cell line but not toward primary rat astrocytes, further demonstrating the selective toxicity of ruthenium complexes toward cancer cells [31]. Another class of pac–Ru complexes that has been extensively studied, K[Ru^{III}(pac)Cl], contains ethylenediaminetetraacetatic acid and its derivatives as the pac ligands (Figure 15.7) [26, 32]. Like Ru(pdta)Cl$_2$, these complexes also rapidly hydrolyze to yield Ru(pac)(H$_2$O) species at low pH. *In vitro* cytotoxicity against tumor cell lines (MCF-7, NCI-H460, SF-268) showed K[Ru^{III}(pdta)Cl] and K[Ru^{III}(edta)Cl] to be more efficacious inhibitors of cell growth, presumably because of their higher reactivity toward nucleotides [26, 32, 33]. In addition, several of these complexes were found to be effective NO scavengers and protease inhibitors; thus they could be used to treat various diseases or serve as antiviral agents [34].

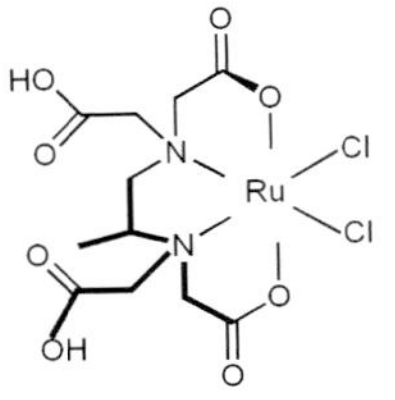

Figure 15.7 Examples of Ru–pac complexes investigated for anticancer activity (pdta = 1,2-propylenediaminetetraacetate; edta = ethylenediaminetetraacetic acid; cdta = cyclohexane-1,2-*trans*-diamine-N,N,N′,N′-tetraacetate).

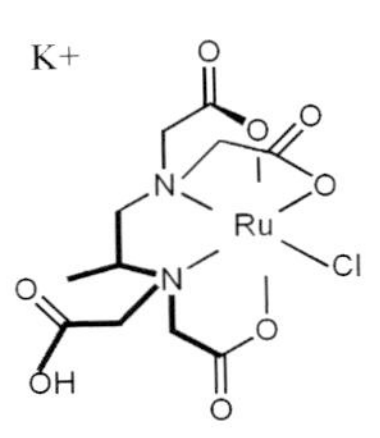

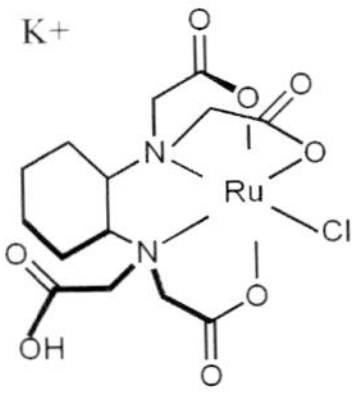

15.2.3
Ru-Dimethyl Sulfoxide Complexes

The *cis*- and *trans*-Ru(DMSO)$_4$X$_2$ (where X = Br, Cl; DMSO = dimethyl sulfoxide) were screened for antitumor activity (C75B1/BD2F1 female mice with lung carcinoma) and *trans*-Ru(DMSO)$_4$Cl$_2$ was 20-fold more active against metastasis than its *cis* counterpart (Figure 15.8) [35, 36]. In leukemic mice, the complexes were able to prolong the lifespan of the host without affecting the number of tumor cells [35, 37]. No further development of *cis*- and *trans*-Ru(DMSO)$_4$Cl$_2$ was reported, presumably in favor of NAMI-A which was found to be a strong antimetastatic agent by the same investigators. However, analogues of *cis*- and *trans*-Ru-(DMSO)$_4$Cl$_2$ containing chelating DMSO ligands such as bis(methylsulfinyl)ethane, bis(ethylsulfinyl)ethane (BESE), bis(propylsulfinyl)ethane and bis(methylsulfinyl)propane have since been reported and studied *in vitro* (CHO) [38]. The compounds were found to be noncytotoxic under aerobic and hypoxic conditions, despite accumulating significantly within the cellular DNA. *cis*-Ru(DMSO)$_4$Cl$_2$, in particular, has been used as a synthon for numerous compounds such as *cis,cis,trans*-RuL$_2$(DMSO)$_2$Cl$_2$ (where L corresponds to cytotoxic nitrofurylsemicarbazone ligands), although the conjugated products did not offer improved cytotoxicity (MCF-7, TK-10, HT-29) [39]. Similarly, complexes with derivatized DMSO, *cis,cis,trans*-RuL$_2$(DMSO)$_2$Cl$_2$ and *cis*-Ru(DMSO)$_2$(mal)$_2$ type compounds (where L = metroidazole and mal = maltol or ethylmaltol), show no advantageous improvement in cytotoxicity (MDA-MB-435S) [40]. *cis*-Ru(DMSO)$_4$Cl$_2$ has been

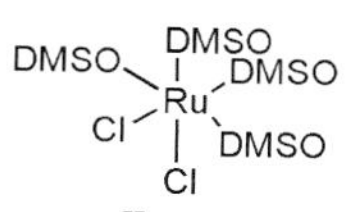

cis-Ru^{II}(DMSO)$_4$Cl$_2$

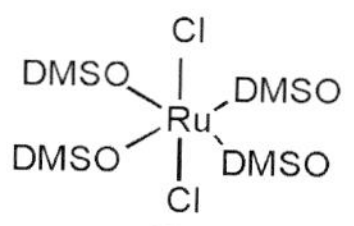

trans-Ru^{II}(DMSO)$_4$Cl$_2$

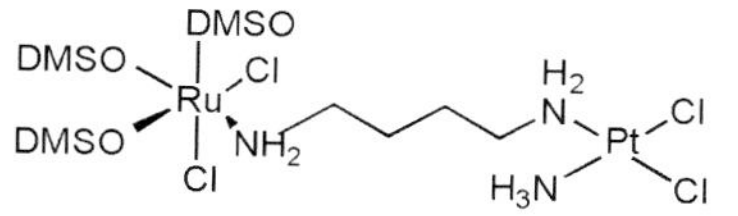

[{cis,fac-RuCl$_2$(DMSO)$_3$}(NH$_2$(CH$_2$)$_4$NH$_2$){cis-Pt(NH$_3$)Cl$_2$}]

Figure 15.8 Examples of DMSO–ruthenium compounds evaluated for antitumor activity.

used to prepare heterodinuclear Pt–Ru compounds crosslinked by 1,4-diaminobutane, that is, [{cis,fac-RuCl$_2$-(DMSO)$_3$}(NH$_2$(CH$_2$)$_4$NH$_2$){cis-Pt(NH$_3$)Cl$_2$}], with a view to develop a compound that could crosslink DNA–DNA or DNA–protein structures (Figure 15.8) [41]. On the basis of DNA-binding studies, it was demonstrated that the heterodinuclear compound could form specific DNA lesions that could crosslink proteins to DNA, but its sensitivity to light and rapid hydrolysis prevented further evaluation [25, 42]. More recently, both cis- and trans-Ru-(DMSO)$_4$Cl$_2$ were investigated for their phototoxicity under UVA illumination [43]. Both complexes exhibited photodependent cytotoxicity against human and murine melanoma cell lines (SK-MEL 188, S91) – the trans isomer was more photocytotoxic.

15.2.4
Ru–Arylazopyridine Complexes

Arylazopyridine ruthenium(II) complexes, Ru(azpy)$_2$Cl$_2$ (where azpy = 2-phenylazopyridine), represent a class of well-characterized anticancer compounds with a strong structural dependency. Although these complexes can exist in up to five different isomeric forms, arising from the lack of a twofold symmetry axis, only three have been reported, namely the α, β and γ isomers, and their structures unequivocally established by x-ray crystallography (Figure 15.9) [44, 45]. The structural characteristics have a significant impact on the efficacy of the compounds as cytotoxic agents. Reedijk et al. reported that the high cytotoxicities of the α and γ isomers in vitro (A498, EVSA-T, H226, IGROV,MCF-7, WIDR, M19) were comparable to those of cisplatin and 5-fluorouracil and about 10-fold higher than that of the corresponding β isomer. This result is surprising since the α and β isomers are structurally similar and differ only in the orientation of the azpy ligand [45, 46]. The addition of methyl groups to either the pyridine or phenyl moiety, as in Ru(tazpy)$_2$Cl$_2$ and Ru(mazpy)$_2$Cl$_2$ (tazpy = o-tolylazopyridine; mazpy = 4-methyl-2-phenylazopyridine), did not alter the trend, validating the structure–activity

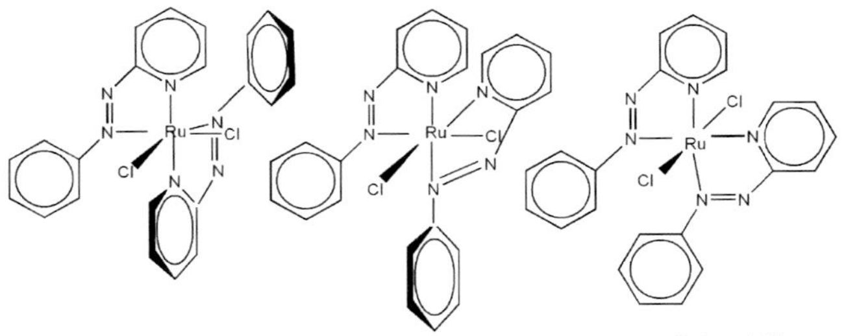

Figure 15.9 Different isomeric forms of Ru(azpy)$_2$Cl$_2$ complexes.

relationship of the isomers on cytotoxicity [46]. Density Functional Theory (DFT) calculations suggest that the ability of the Ru(mazpy)$_2$Cl$_2$ isomers to intercalate to DNA decreases from γ > α > β isoforms on the basis of the geometric and electronic factors, which correlates with the observed cytotoxicity [47]. The γ isomer has the most preferential geometric arrangement of the mazpy ligand for DNA intercalation, as well as the lowest LUMO energy level and smallest HOMO–LUMO (highest occupied molecular orbital–lowest unoccupied molecular orbital) energy gap. Accordingly, it is the most reactive toward DNA. A mixed-ligand analog, *cis*-Ru(bpy)(azpy)Cl$_2$, which is structurally similar to α-Ru(azpy)$_2$Cl$_2$, was found to be 2–10-fold more cytotoxic *in vitro* (A498, EVSA-T, H226, IGROV, MCF-7, WIDR, M19) than *cis*-Ru(bpy)$_2$Cl$_2$ but much less cytotoxic (>50-fold) than either α- or β-Ru(azpy)$_2$Cl$_2$ [48].

The replacement of the chloride ligands with a 2,2′-bipyridine group, as in [Ru(bpy)$_n$(azpy)$_{3-n}$](PF$_6$)$_2$, also did not offer any significant advantage over the parent α-Ru(azpy)$_2$Cl$_2$ [49]. Some water-soluble derivatives of the α isoform, where the chloride ligands are replaced by bridging carboxylate ligands, for example, oxalate, malonate or 1,1-cyclobutanedicarboxylate, overcome one of the main limitations of these complexes [50]. Although the compounds are 5–10-fold less cytotoxic than α-Ru(azpy)$_2$Cl$_2$ (A2780, A2780cisR) and slightly less cytotoxic than cisplatin, the cytotoxicity is comparable to that of carboplatin. Another water-soluble derivative [NEt$_4$]$_2$[Ru(sazpy)$_2$Cl$_2$] (where sazpy = 2-phenylazopyridine-5-sulfonate), which contains a sulfonate functionality on the azpy group [51], was ~100-fold less cytotoxic (A2780, A2780cisR) than α-Ru(azpy)$_2$Cl$_2$. More recently, dinuclear analogues with bridging azpy ligands, comprising two azpy units joined at the *para* position of the phenyl rings by a bridging methylene group, have been reported [52]. The supramolecular complexes each contain two Ru(azpy)$_2$Cl$_2$ moieties arranged in either the α or γ isoforms. Three isomers have been isolated that contain either α/α-, α/γ- or γ/γ- "Ru(azpy)$_2$Cl$_2$" units and their structures have been confirmed by x-ray crystallography (Figure 15.10). The α/γ and γ/γ isomers were tested *in vitro* on cancer and nontumorigenic cell lines (T47D, HBL-100), with the γ/γ isoform exhibiting the highest cytotoxicity, ~30-fold higher than that of cisplatin.

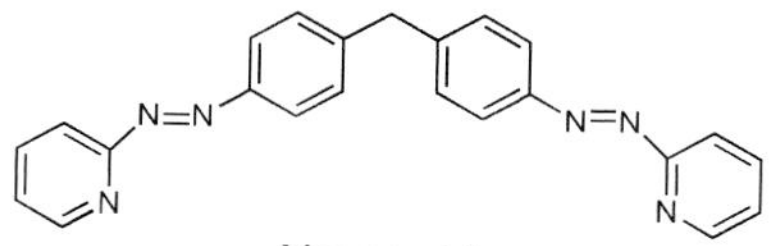

Figure 15.10 The azpy derivative ligand bisazopyridine,

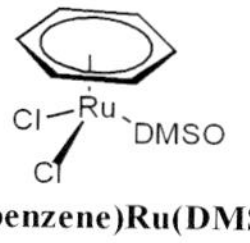

(η^6-benzene)Ru(DMSO)Cl$_2$

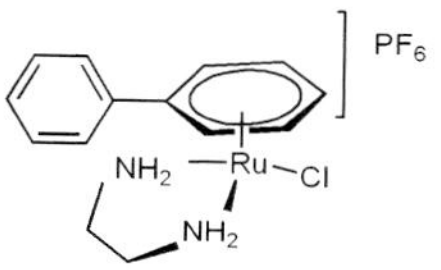

[(η^6-biphenyl)Ru(en)Cl]PF$_6$

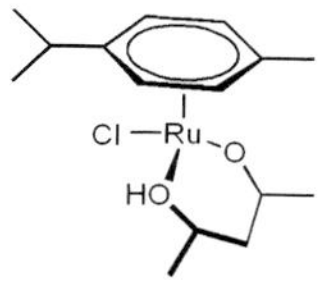

(η^6-cymene)Ru(acac)Cl

Figure 15.11 Examples of organometallic (η^6-arene) ruthenium anticancer drugs.

15.2.5
Ru–Organometallic Arene Complexes

Organometallic ruthenium complexes bearing η^6-arene ligands have been investigated extensively as anticancer drug candidates. One of the earlier examples is (η^6-benzene)Ru(DMSO)Cl$_2$ (Figure 15.11), which has been shown to strongly inhibit topoisomerase II activity by cleavage complex formation [53]. The authors suggested that the ruthenium complex interacts with DNA and forms crosslinks with topoisomerase II. The complex exhibited antiproliferative activity *in vitro* (Crit-2), but it is inconclusive whether there is a direct link to its ability to inhibit topoisomerase II activity. Arene–ruthenium complexes containing BESE, which is essentially a bidentate "DMSO" ligand, have been tested *in vitro* (MDA-MB-435s), but their cytotoxicities were more than 5-fold higher than that of cisplatin, and no further investigation was reported [54]. In contrast, monofunctional Ru(II) complexes of the type [(η^6-arene)Ru(en)X]$^+$ (where en = ethylenediamine or its derivatives and X = halide) exhibit high cytotoxicity *in vitro* (A2780, A2780cisR, A2780adr, HT29, Panc-1, NX02), comparable to that of cisplatin, that is dependent on the arene ligand (Figure 15.11) [55]. Large arene ligands, such as biphenyl and tetrahydroanthracene, improved the cytotoxicity of the drug. Replacement of the en ligand with bpy or tmeda (*N,N,N′,N′*-tetramethylethylenediamine) resulted in complexes with poor cytotoxicity [56]. Extensive oligonucleotide studies have been carried out and $[(\eta^6\text{-cymene})\text{Ru}(\text{en})_X]^+$ have been found to preferentially bind to guanine bases to form monofunctional DNA adducts [57–59]. With large arene rings systems, for example, dihydroanthracene and tetrahydroanthracene, binding of $[(\eta^6\text{-arene})\text{Ru}(\text{en})_X]^+$ complexes to nucleotide bases was promoted by hydrophobic arene–purine base π–π stacking interactions, which could explain the enhanced cytotoxicity of those derivatives [60]. There was also evidence of dynamic chiral recognition of ethylguanine by the diastereomers of $[(\eta^6\text{-biphenyl})\text{Ru}(\text{Et-en})_{Cl}]^+$, realizing the concept of induced-fit recognition of DNA by chiral derivatives of ruthenium complexes [61].

15.2.6
NAMI-A Type Complexes

Imidazolium *trans*-[tetrachloro(DMSO)(imidazole)ruthenate(III)] H$_2$im[*trans*-RuCl$_4$-(DMSO)Him], more commonly known as NAMI-A (Figure 15.3), was the first ruthenium-based anticancer drug to enter clinical trials – as a drug candidate against non-small cell lung cancer (NSCLC) [62]. Compared to other transition-metal-based drugs such as cisplatin, NAMI-A is unique. *In vitro*, it is virtually devoid of cytotoxicity, while *in vivo*, it inhibits lung formation of metastases and reduces the weight of metastases without affecting the primary tumor [63], in contrast to platinum-based drugs which typically exhibit strong growth inhibition on primary tumors. Indeed, both *in vitro* and *in vivo* data appear to exclude DNA as the primary target, in line with the observation that the binding of NAMI-A to DNA is much weaker than that of platinum complexes [63–65]. Instead, strong binding to serum proteins is observed and the drug could potentially exploit receptor-mediated delivery by transferrin for selective delivery to cancer cells [66]. There is strong evidence linking the antimetastatic behavior to the lack of cytotoxicity *in vitro*, cell-cycle changes corresponding to cell arrest in the premitotic G2-M phase, and inhibition of Matrigel invasion, which could constitute an *in vitro* screening strategy for ruthenium complexes with similar properties [67]. Several NAMI-A analogues have been studied for potential antimetastatic activity as a means to ascertain the structure–activity relationships in this class of compounds. NAMI-A derivatives [*mer*-RuCl$_3$(DMSO)(acv)(H$_2$O)] and [*mer*-RuCl$_3$(DMSO)(acv)-(MeOH)] and [*trans*-RuCl$_4$(DMSO)guaH] (where acv = acyclovir and gua = guanine) exhibited low cytotoxicity *in vitro* (TS/A), similar to NAMI-A [68]. Analogues with the dmtp ligand (where dmtp = 5,7-dimethyl[1,2,4]triazolo[1,5-*a*]pyrimidine) were also investigated and found to be slightly cytotoxic *in vitro* (TS/A, KB, B16-F10) [69]. In particular, Hdmtp[*trans*-RuCl$_4$(DMSO)dmtp] exhibited a similar *in vivo* profile to NAMI-A (CBA mice with MCa carcinoma), strongly inhibiting lung metastasis formation without significantly affecting the primary tumors [69]. This trend has also been replicated in NAMI-A analogues with pyrazine, pyrazole, and bidentate N-heterocyclic ligands, for example, 4,4′-bipyridine, 1,2-bis(4,4′-pyridyl)ethane, suggesting that the imidazole fragment is not an essential feature for the antimetastatic property of NAMI-A [70].

15.2.7
The Transferrin Delivery Mechanism

Indazolium trans-[tetrachlorobis(1H-indazole)ruthenate(III)] H_2in[trans-RuCl$_4$(Hin)$_2$], KP1019 (Figure 15.12), the only other ruthenium drug presently undergoing clinical evaluation, contains two indazole ligands in the trans conformation (whereas in NAMI-A the imidazole ligand is trans to a DMSO moiety). Unlike NAMI-A, KP1019 is significantly cytotoxic in vitro against colorectal cell lines (SW480, HT29) by induction of apoptosis [71]. There is also evidence of P-glycoprotein (Pgp)-mediated drug resistance to KP1019, although other multidrug resistance (MDR)-associated proteins (MRP1, BCRP, and LRP) did not affect its activity significantly [72]. The drug was evaluated in vivo against autochthonous colorectal tumors in rats and was found to be highly effective in reducing tumor growth (superior to 5-fluorouracil, the most effective drug in clinical use against colorectal cancer) [11]. In contrast, cisplatin was inactive in the in vivo model. The activity of KP1019 is attributed, at least in part, to transferrin-mediated drug transport, with KP1019 binding strongly to transferrin in the iron-binding pockets [73]. Indeed, its imidazolium analogue, which binds more weakly to transferrin, is taken up less effectively by transferrin and is also less cytotoxic in vitro [71]. Another possible mechanism in play is that the drug is activated by reduction, from Ru(III) to Ru(II), selectively in hypoxic tumor tissue by endogenous bioreductants such as glutathione. It appears that KP1019 induces apoptosis in colorectal cell lines predominantly by the intrinsic mitochondria pathway and that DNA could be a target, although studies have shown that the DNA lesions formed by KP1019 are different from that by cisplatin [11, 74]. In phase 1 clinical trials, the drug was well tolerated, and five out of six patients treated achieved disease stabilization [11, 13]. A large

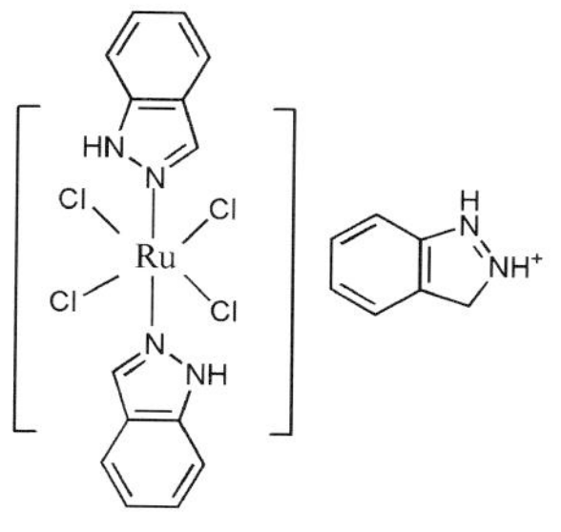

Figure 15.12 The indazolium compound KP1019.

number of KP1019 analogues, comprising different types and numbers of N-heterocyclic rings, have been reported [75, 76]. Notably, it was found that increasing the number of indazole ligands, as in trans-Ru(II)Cl$_2$-(Hin)$_4$ and [trans-Ru(II)Cl$_2$(Hin)$_4$]Cl, improved the *in vitro* cytotoxicity significantly (CH1, SW480) [76]. This was correlated with the increased cellular uptake and a higher reduction potential of the homologues, in line with the "activation by reduction" hypothesis.

15.2.8
Discerning Estrogen Receptor Modulators Based on Ru

Selective estrogen receptor (ER) modulators (SERMs), such as tamoxifen (see Figure 15.13), are a class of drug that have been successfully used to treat hormone-dependent (ER-positive) breast cancer, that is, tumors which express the estrogen receptor ERα. At the same time, there is no satisfactory therapeutic treatment for hormone-independent (ER-negative) breast tumors that contain another estrogen receptor ERβ, which accounts for about one-third of breast cancer cases. Jaouen *et al.* have developed a series of hydroxytamoxifen derivatives comprising ferrocene (hydroxyferrocifen) (Figure 15.13), and other potentially cytotoxic organometallic fragments of rhodium, manganese, titanium, and rhenium have been reported, representing a strategy to defeat breast tumors which contain both ERα and ERβ receptors [77]. Hydroxytamoxifen analogues containing organometallic ruthenocenes and their biological evaluation *in vitro* (MCF7, MDA-MB231) have also been reported [78]. Although the hydroxytamoxifen–Ru complexes were found to bind strongly to both ERα and ERβ, unlike hydroxyferrocifen, hydroxytamoxifen–Ru exhibits no cytotoxic effect toward ER-negative breast cancer cell lines, possibly because of the increased stability of the ruthenium fragment compared to the iron

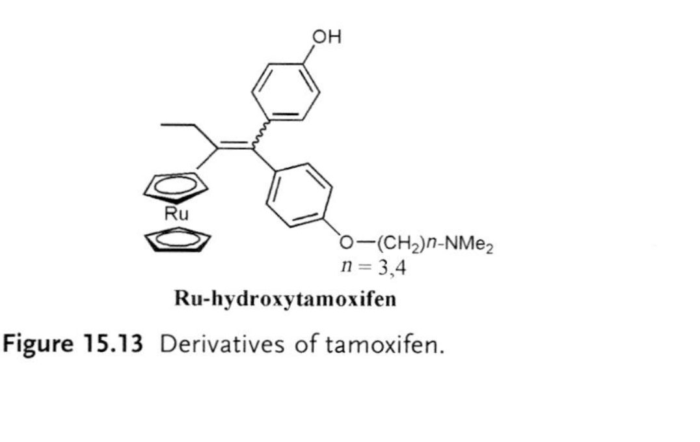

Figure 15.13 Derivatives of tamoxifen.

analogue with respect to oxidation, and thus would not offer any therapeutic advantage over existing treatment options. However, the authors noted that the hydroxytamoxifen–Ru complexes could be suitable for radioimaging of tumor cells using either of two γ-emitting Ru isotopes, ^{97}Ru and ^{103}Ru [78].

15.2.9
Ru–Ketoconazole Complexes

Ketoconazole (KTZ) and clotrimazole (CTZ) are azole compounds that were originally developed as antifungal agents. They also exhibit anticancer properties and ketoconazole in particular is being used in the clinic as a second-line agent for hormone-refractory prostate cancer. Ketoconazole–ruthenium Ru(KTZ)$_2$Cl$_2$ and clotrimazole–ruthenium Ru(CTZ)$_2$Cl$_2$ complexes were originally developed to treat tropical diseases [79], but more recently they have been evaluated for anticancer activity. *In vitro*, Ru(KTZ)$_2$Cl$_2$ and Ru(CTZ)$_2$Cl$_2$ were found to be more effective inhibitors of cell growth proliferation (C8161) than the parent ligands KTZ and CTZ [80]. The effects of Ru(KTZ)$_2$Cl$_2$ on various cell signaling pathways were investigated. In particular, it was observed that A431 spheroids, known to overexpress EGF-R and to be resistant to either Ru(KTZ)$_2$Cl$_2$ or C225 *anti*-hEGF-R monoclonal antibody (MAb), were susceptible to a combination treatment of Ru(KTZ)$_2$Cl$_2$ and the C225 antibody. C225 is an experimental monoclonal antibody presently under clinical evaluation as a targeted therapeutic agent against EGFR-expressing metastatic colorectal cancer. The authors suggested that Ru(KTZ)$_2$Cl$_2$ might be used to enhance other targeted therapeutic treatment methods [80].

15.2.10
Protein Kinase Inhibitors Based on Ru

With a view to developing synthetic compounds with superior biological activity, Meggers *et al.* reported an organometallic ruthenium complex that mimics the shape of a known protein kinase inhibitor, staurosporine (Figure 15.14) [81, 82]. The strategy exploits the relative ease of synthesizing ruthenium moieties of a specific spatial conformation to mimic aspects of an inhibitor not easily accessible using purely organic scaffolds [81, 83, 84]. The authors further demonstrated the

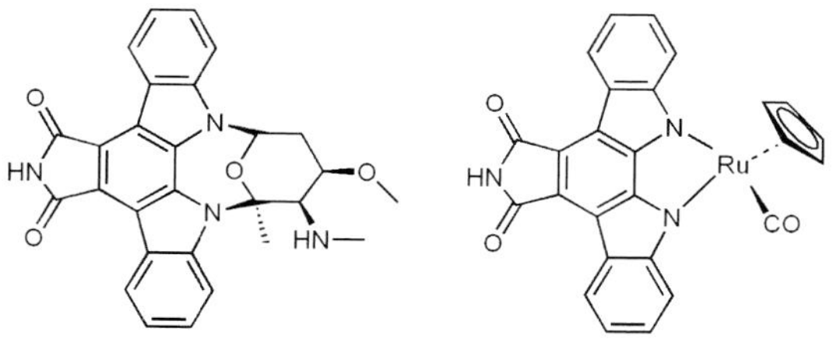

Figure 15.14 Staurosporine and an organometallic ruthenium mimic.

superior binding of the ruthenium mimic to Pim-1, a protein kinase, and reported co-crystallization of the protein with the ruthenium complex [81]. This provides a basis for the development of further transition-metal-based enzyme inhibitors and could lead to new types of organometallic-targeted therapeutic agents.

15.2.11
Ru–RAPTA Complexes

The RAPTA compounds comprise a class of organometallic ruthenium(II) complexes with a monodentate 1,3,5-triaza-7-phosphatricyclo[3.3.1.1]decane (pta) ligand and a η^6-arene ligand [85, 86] RAPTA compounds are generally air-stable complexes with good thermodynamic stability. The prototype [(η^6-cymene)Ru^{II}(pta)Cl$_2$], RAPTA-C, has been central to biological evaluation and is the reference compound from which other RAPTA compounds are developed (Figure 15.15). On the basis of discovering more chemotherapeutic applications for RAPTA complexes, more than 20 RAPTA complexes of the general formula (arene)Ru^{II}(pta)X$_2$, where X = Cl, Br, I, SCN or bridging carboxylate ligands, have been synthesized and studied. In addition, RAPTA analogues containing other organometallic fragments, namely [CpRu], [Cp*Ru], [Cp*Rh] and their cytotoxicities have also been studied [88]. The observation that RAPTA-C **1** induces pH-dependent DNA damage against *E coli* pBR322 DNA plasmids, with a significant retardation of the migration of the supercoiled DNA at pH < 7.0 because of unwinding of the DNA as a result of drug interactions, has prompted the study of RAPTA complexes as potential anticancer drugs [89]. This provided a means of targeting cancer cells, since they generally exhibit lower pH as a result of metabolic changes, partly because of accelerated cell division. However, the RAPTA complexes do not show selective binding to DNA *in vitro*; proteins and RNA appear to be the main intracellular targets [90–92].

The cytotoxicity of the RAPTA complexes is generally very low, at least in both TS/A murine adenocarcinoma and nontumorigenic HBL-100 human mammary cell lines in comparison to cisplatin [90]. However, the lead RAPTA complexes (Figure 15.15) exhibit selective cytotoxicity toward the TS/A cancer cell lines relative to HBL-100, which provided early indication that the complexes are benign toward healthy cells. This is important since high systemic toxicity is associated with the drastic side-effects of cisplatin and related drugs and limits the amount of drug that can be administered. The pta fragment appears to play a significant part in determining this selectivity. When pta was replaced with pta-Me$^+$ (Figure 15.15), selectivity was lost and the compounds were equally toxic in both the cancerous and nontumorigenic cell lines [90]. The presence of the ester group significantly improved the drug uptake in RAPTA-CE, with a corresponding improvement in cytotoxicity against TS/A cells [90]. Addition of functional groups such as benzocrown and an imidazolium moiety appeared to improve cytotoxicity toward TS/A, but not to a sufficient degree to merit further studies [90]. RAPTA complexes with hydrogen-bonding groups attached to the arene ring were developed and found to be generally more reactive than lead RAPTA compounds (Figure 15.15)

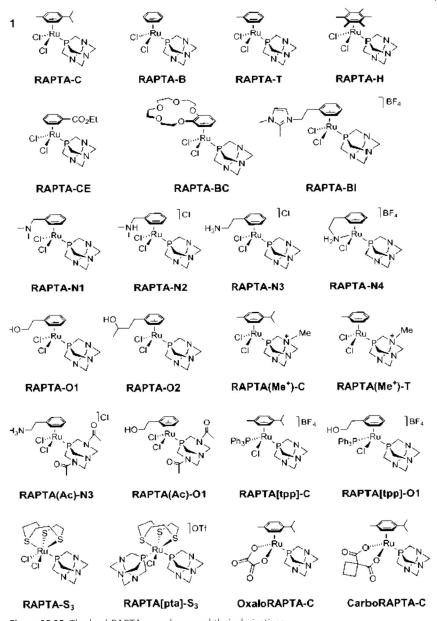

Figure 15.15 The lead RAPTA complexes and their derivatives.

[93, 94]. The replacement of the aromatic fragment with a [9]aneS3 ligand, RAPTA-S3, results in only a slight decrease in selectivity and cytotoxicity, supporting to the idea that the aromatic fragment may not be an essential feature for *in vitro* activity and could be effectively replaced by another face-capping ligand with low steric demand (Figure 15.15) [95]. Interestingly, RAPTA[pta]-S3, with [9]aneS3 and two pta ligands shows good selectivity and cytotoxicity toward TS/A cells.

RAPTA complexes are prone to hydrolysis and would have to be administered in saline to suppress the cleavage of the chloride ligands. With a view to developing drugs that could resist hydrolysis in aqueous media, RAPTA complexes bearing chelating carboxylate ligands instead of the two chloride ligands have been developed (Figure 15.15) [96]. They were found to be kinetically more stable than 1, and essentially, retained the carboxylate ligand in aqueous solution. Preliminary investigations show that these derivatives exhibit the same order of cytotoxicity *in vitro* (A549, T47D, MCF7, HT29) [96]. RAPTA complexes conjugated through the arene ring to ethacrynic acid, a known inhibitor of GST enzymes, were found to be effective GST inhibitors with significantly increased cytotoxic activity, that is, showing comparable cytotoxicity to that of cisplatin in cell lines known to contain elevated levels of GST (A549, HT29, T47D). Other potential enzyme targets are presently being investigated (Figure 15.16) [97].

With a view to rationally developing other ruthenium-based antimetastatic agents, a new class of imidazole (η^6-arene)ruthenium complexes of the general formula [(η^6-arene)Ru^{II}Cl$_2$(imid)], [(η^6-arene)Ru^{II}Cl(imid)$_2$]X, or [(η^6-arene)Ru^{II}(imid)$_3$]X$_2$ (where imid = imidazolium ligand, X = Cl, BF$_4$, BPh$_4$) have been prepared, combining the unique structural aspects of NAMI-A and the "piano-stool" arene ruthenium(II) complexes [98–101]. The derived complexes exhibit similar cytotoxicity to RAPTA complexes *in vitro* (TS/A, HBL-100) and several of the complexes exhibit selectivity toward cancer cells. Specifically, [(η^6-cymene)Ru^{II}Cl(vinylimid)$_2$]Cl and [(η^6-benzene) Ru^{II}(mimid)$_3$](BF$_4$)$_2$ have been identified for further *in vivo* experiments (Figure 15.17) [101–103].

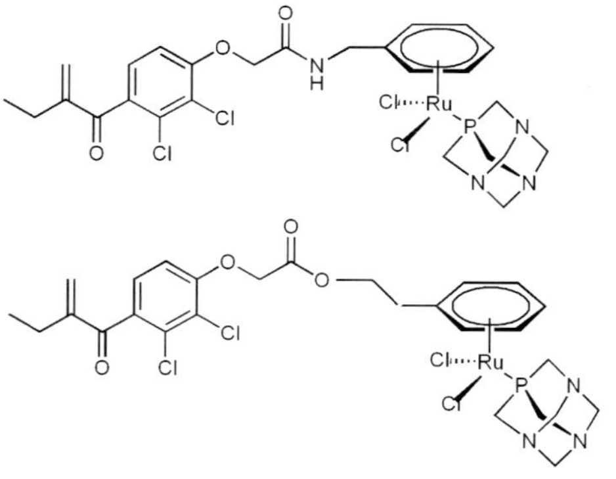

Figure 15.16 RAPTA complexes designed to defeat GST-mediated drug resistance.

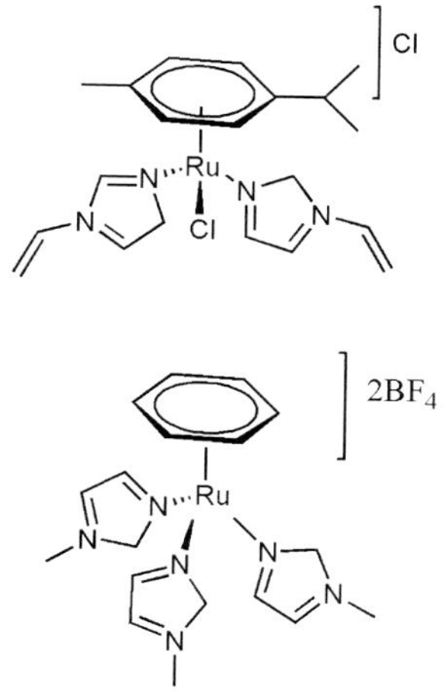

Figure 15.17 New class of RAPTA-NAMI antimetastatic drug candidates.

15.3
Chemical Nucleases as Anticancer Drugs

Model biomimetic compounds reported for the blue copper proteins, involved in molecular oxygen and electron transfer, have been much explored on the frontier of bioinorganic chemistry during the last 30 years [104]. Similar copper complexes have also been reported as chemical nucleases in the presence or absence of reducing agents *in vitro*. These compounds can be explored as chemotherapeutics, such as anticancer agents, metal-mediated antibiotics, antibacterials, and antivirals [105].

Thus, design of DNA and RNA specific agents capable of controlled chemical cleavage are of paramount importance owing to their potential use for the DNA-targeted chemotheraupeutic drugs. Metal complexes are attractive reagents for this nucleolytic activity owing to their inherent diverse structure and redox properties. Examples of such complexes include $[Fe(edta)]^{2-}$ (edta = ethylenediaminetetraacetic acid), $[Cu(phen)_2]^{2+}$ (phen = 1,10-phenanthroline), metalloporphyrins, Ni-azamacrocycles, $[Mn(salen)]^{3+}$ (salen = N,N'-ethylenebis(salicylaldeneaminato), [Cu-desferal], [Co-cyclam], $[Rh(phen)_2(phi)]^{3+}$, and $[Rh(en)_2(phi)]^{3+}$ (en = N,N'-ethylenediamine; phi = 9,10-phenanthrenequinone dimine) [106–113]. However, in most cases the cleavage reaction must be initiated by excess exogenous agents such as H_2O_2, mercaptopropionic acid, dithiothreitol, or light, limiting their *in vitro* applications. As a result, self-activating systems that require no further activation to bring about DNA cleavages are desirable. A review of the literature indicates only few self-activating DNA cleaving systems as follows: (1) [Fe-BLM] (BLM = bleomycin) in the presence of molecular oxygen [114]; (2) hydroxysalen-Cu(II) (hydroxysalen = bis(hydroxysalicylidene)ethylenediamine) complexes spontaneously form the oxidant species Cu^{III} and cleave DNA [115]; (3) the marine natural product tambjamine E and the corresponding pyrrolidine derivatives which induce double-strand DNA cleavage in the presence of Cu(II) and molecular oxygen without addition of any external reducing agent (Figure 15.18) [116]. Bleomycin is broadly known to bind to several metal ions such as Cu, Fe, Ni, Co, etc and to form metal complexes which cleave DNA in the presence of molecular oxygen. Although bleomycin and its metal complex derivatives exhibit good antitumor activity, little attention has been paid to the DNA cleavage activity, *in vivo* [117, 118]. It is demonstrated that the related natural product tambjamine E binds DNA effectively and facilitates single-strand or double-strand DNA cleavage in the presence of Cu(II) and molecular O_2. The prodigiosins (derivatives of tambjamine E) are the red pigments produced by microorganisms such as *Streptomyces* and *Serratia*, which also possess promising anticancer, antimicrobial, and immunosuppressive activities, cleave DNA through double strand scission. The methoxy group is suggested to be critical for their anticancer properties whereby they cause apoptosis and exhibit selective activity against liver cancer cell lines [119, 120].

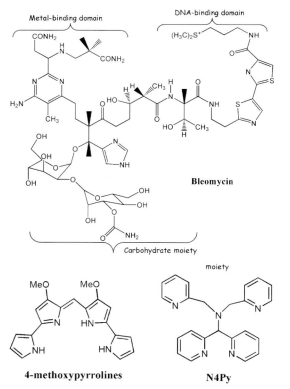

Figure 15.18 The self-activating "chemical nucleases" on complexation.

Recently our group reported the discovery of a copper(II) complex, prepared from Hpyramol [121], that catalytically cleaves target DNA in the absence of reductant by attacking at multiple positions of the nucleotide (Figure 15.19). It was suggested that the DNA cleavage reaction is oxidative, through a non-diffusible radical mechanism, because radical scavengers do not significantly inhibit the DNA cleavage reaction (Figure 15.20). The cytotoxicities of [Cu(pyrimol)Cl] (Figure 15.21) and cisplatin, were studied, under similar conditions, on L1210 murine leukemia cancer cell lines, which are sensitive (Figure 15.20c, lane 1) and resistant (lane 2) to cisplatin, and A2780 human ovarian carcinoma cell lines, which are sensitive (lane 3) and resistant to cisplatin (lane 4). The IC$_{50}$ values for the cell lines 1, 3, and 4 are in the range of those observed with cisplatin. For the L1210 cancer cell lines 2, the copper complex is more active than cisplatin. These results are promising for a potential clinical application since the activity of this labile

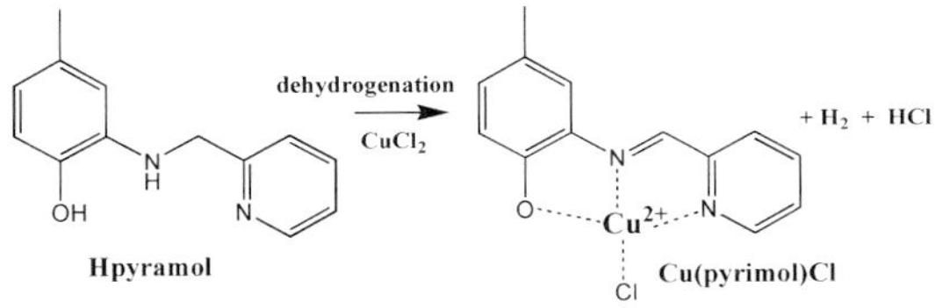

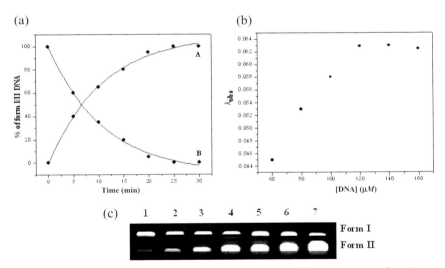

Figure 15.19 The ligand dehydrogenation of Hpyramol on coordination with Cu(II).

Figure 15.20 (a) Time course experiments of DNA cleavage (80 μM of supercoiled DNA (in base pairs) + 20 μM of Cu-pyrimol) over a period of 30 minutes. Plot of the increase in form II with time (A) and plot of the decrease in form I with time (B); (b) Plot of the k_{obs} versus the concentration of DNA from "true-Michaelis–Menten" kinetic studies; the highest k_{cat} value obtained is 0.063 min^{-1}. (c) Saturation kinetics of the cleavage of ΦX174 supercoiled plasmid DNA using 20 μM Cu-pyrimol.

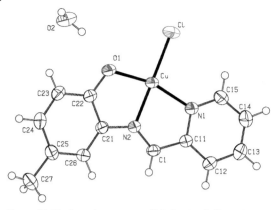

Figure 15.21 Crystal structure of [Cu(pyrimol)Cl]·(H$_2$O).

Figure 15.22 Agarose gel electrophoresis of φX174 phage DNA (20 μM b.p.) after 8 hours incubation with 100 μM of zinc complexes (with acetate and chloride anions), without reductant in phosphate buffer, pH 7.2, 37 °C. The oxidative cleavage from form I to form II.

copper complex is comparable to that of the platinum drug. Interestingly, the copper complex has overcome the cross-resistance of cisplatin that is observed for both the cancer cell lines that are resistant to cisplatin, as expressed by the resistance factors, RF, which are low for the copper complex compared to cisplatin [121]. This copper complex is self-activated by the dehydrogenation (Figure 15.19) of the precursor Hpyramol ligand upon coordination to the metal ion; that is, the dehydrogenation of the ligand Hpyramol to Pyrimol has also been observed by us with transition metals such as Cu^{II}, Fe^{III}, Mn^{II}, and Zn^{II} [122, 123]. The oxidative DNA-cleavage mechanism appears to be purely ligand-based, as the redox-inactive zinc complex also shows DNA cleavage [124].

The DNA cleavage reactions (Figure 15.22) carried out with the zinc complexes are not inhibited by the presence of various radical scavengers, that is, NaN$_3$, superoxide dismutase, DMSO, ethanol, and D$_2$O. The use of distamycin or an excess of NaCl also does not inhibit the cleavage of DNA, and form II is still detected if the digestion is performed under argon, or using dark conditions. These observations corroborate a hydrolytic pathway for the cleavage of DNA mediated by the zinc complexes. Religation experiments on DNA cleavage products obtained from digestions with the zinc complexes have been carried out to determine the nature of the DNA cleavage. Remarkably, these experiments show clearly that DNA cleavage mediated by zinc complexes is not hydrolytic but is purely oxidative, since DNA cleavage products are not religated, while control φX174 DNA, digested by the restriction endonuclease *Pst*I, is religated with >35% efficiency. One can reasonably expect the involvement of radical species in an oxidative cleavage of DNA, which would be inhibited in the presence of radical scavengers during the digestion experiments. In the absence of evidence of any free diffusible radical, a hydrolytic cleavage of the DNA strands may be anticipated. However, this mechanism is excluded because the nicked forms of the DNA digested by the zinc complexes are not religated by the T4 ligase enzyme. The results achieved from the experiments in the presence of various radical scavengers, and from the religation experiments show that a nonhydrolytic pathway not related to OH*, superoxide, or singlet oxygen is involved [126]. To further investigate in detail the nature of the cleaving properties of zinc complexes, pUC19 plasmid DNA treated with the zinc complexes was religated and was used for cell transformation tests. As observed, the religation of *Pst*I-linearized pUC19 DNA is achieved with an efficiency of about 94%. In contrast, the pUC19 DNA cleavage products obtained from partial digestion with the zinc complexes do not show any cell transformations, since the observed colony-forming units are purely due to the presence of residual undigested supercoiled pUC19 DNA.

Traces of organic radical in [Zn_2(pyrimol)$_2$Cl$_2$] (Figure 15.23) at solid state have been detected by EPR at room temperature. The amount of radical species present could not be determined, most likely owing to their decay or delocalized nature (Figure 15.24) [127]. The radical is expected to originate from the phenolate moiety of the ligand coupled with the dehydrogenation of the Hpyramol ligand. Copper complexes with phenolate ligands are known in the literature as structural/

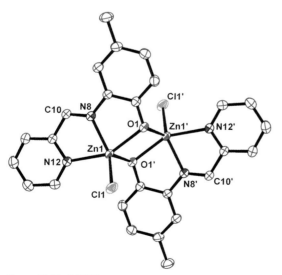

Figure 15.23 ORTEP perspective view of zinc(II) complex [Zn_2(pyrimol)$_2$(Cl)$_2$].

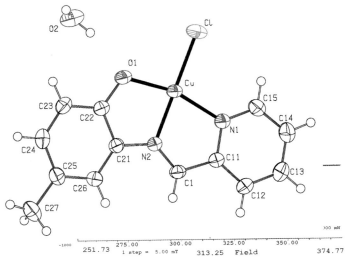

Figure 15.24 Solid-state powder EPR spectrum of [Zn_2(pyrimol)$_2Cl_2$], recorded at room temperature.

functional models for the active site of the enzyme galactose oxidase (Figure 15.25) [128–130]. This enzyme oxidizes primary alcohols selectively to aldehydes by hydrogen abstraction in the presence of dioxygen, which is reduced to dihydrogen peroxide. We found that the Cu–pyrimol complex also oxidizes benzyl alcohol selectively to benzaldehyde [131]. Accordingly, the catalytic DNA cleavage observed with the Cu–pyrimol complex can be explained by the initial hydrogen abstraction from the deoxyribose sugar in the presence of dioxygen, generating dihydrogen peroxide *in situ*. Next, the dihydrogen peroxide couples with Cu^{II} in a Fenton-type reaction to produce reactive oxygen species, which ultimately turns the DNA cleavage oxidative and catalytic. Similar results of oxidative DNA cleavage with Cu–salen type complexes, without any reductant, have been reported [132]. In the case of zinc complexes, the Zn^{II} ions cannot play a redox role like Cu^{II} and form hydrogen peroxide, so the DNA cleavage observed is strictly stoichiometric. Our observations with the zinc complexes [124, 131] and their cleavage activity clearly depict the mechanism as involving phenoxyl radical. Thus the Cu–pyrimol and pyrimol-related complexes are expected to induce cytotoxicities in tumor cell-lines targeting the cellular chromosomal DNA/RNA damage via a non-diffusible radical pathway.

15.4
Inorganic Chemotherapy for Cancer: Outlook

Many metallic elements play a crucial role in living systems. A characteristic of metals is that they easily lose electrons from the familiar elemental or metallic

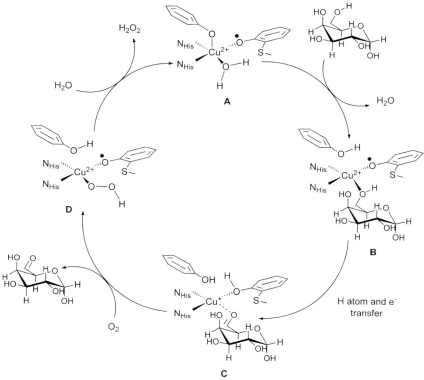

Figure 15.25 Active site of the enzyme galactose oxidase. Primary alcohol oxidation and reduction of molecular oxygen to H_2O_2.

state to form positively charged ions which tend to be soluble in biological fluids. It is in this cationic form that metals play their role in biology. Whereas metal ions are electron-deficient, most biological macromolecules such as proteins and DNA are electron-rich. The attraction of these opposing charges leads to a general tendency for metal ions to bind to and interact with biopolymers. When inorganic complexes (based on any element in the periodic table) are considered for tumor treatment, the first point to be noted is the threshold value for activity in the human body without toxicity. The area of optimum physiological response for the successful metal complexes *in vivo* will vary greatly according to the element, its speciation and oxidation state, and the biochemistry of the specific compound from which it derives (marine/natural products). Therefore, the areas of deficiency, toxicity, and optimum physiological response can be dramatically varied by considering a combination of these variables, as well as design features of the potential ligand which may be altered to tune the delivery of that metal ion into the biological system.

The fact that cisplatin is still unique in cancer treatment is based on its property of inducing p53 – the tumor suppressor phosphoprotein which regulates the cell cycle, DNA repair, and apoptosis. The stabilization of the p53 in the cisplatin-sensitive cell-lines has been proved compared to that in the cisplatin-resistant cell-lines [125]. Hence hydrolysis and DNA binding of cisplatin is not the only features which make for tumor suppression.

Future strategies for drug design should be based on certain principles:

1. Selective targeting of tumor cells rather than normal cells (targeted chemotherapy).
2. Effectiveness in inducing the p53 pathway to apoptosis (biosignaling interference).
3. Low general toxicity.
4. The ligand sphere should also assist the excretion of the heavy metals involved in the drugs during metabolism
5. Metal ions in the 3d series such as Mn, Fe, Cu, and Zn are used by Nature in abundance. Drugs involving these metal ions have positive metabolic cycles. These metal centers inside the folded protein are mainly involved in "activation of oxygen" and electron transfer. In the literature, most of the chemical nucleases are reported to cleave DNA in the presence of molecular oxygen and reductant. These complexes may be a good choice of candidates to target DNA in the tumor cells but in a classical way.
6. The cisplatin analogues and the alternates based on ruthenium metal ions discussed above have drawbacks not only because of their incompatible mechanisms, but essentially due to solvation problem in H_2O. Most of the drugs discussed as alternatives are less soluble in H_2O and a fair amount of DMF or DMSO is essential to at least mix them in the salinated tumor sample cell-lines. Results based on such experiments may become valueless when the compounds go to trials.
7. The DNA binding experiments, including DFT calculations, NMR experiments and so on, *in vitro* give no clue whether they reflect in the *in vivo* trial conditions. More experiments under physiological conditions may help to move toward effective solutions for drug resistance and toxicity problems.
8. Inorganic chemists should be proud that cisplatin has no single carbon atom, while this simplicity has not yet been achieved with any other drug. Most of the effort has been paid to rational drug design without none devoted to structure–activity relationships.

The evolution in medicinal inorganic chemistry of entirely novel strategies to competing with Nature's irregularity in tumor formation is possible with fundamental inorganic chemistry coupled with the understanding of the biochemistry of metals in human body.

15.5
Acknowledgments

The author thanks Prof. Jan Reedijk for his support and encouragement throughout the postdoctoral training period (May 2004–March 2007). Financial support from COST Action D21/003/01, Dutch WFMO (Werkgroep Fundamenteel-Materialen Onderzoek) CW (Foundation for the Chemical Sciences), NWO, and EET is gratefully acknowledged. Prof. Jaap Brouwer, Prof. Gilles P. van Wezel and Dr. Sharief Barends from Leiden Institute of Chemistry, Leiden, are thanked for support in performing the biological experiments.

References

1 Mansour, V.H., Rosenberg, B., Vancamp, L. and Trosko, J.E. (1969) *Nature*, **222**, 385–6.
2 Boulikas, T. and Vougiouka, M. (2003) *Oncology Reports*, **10**, 1663–82.
3 Pasini, A. and Zunino, F. (1987) *Angewandte Chemie (International Ed. in English)*, **26**, 615–24.
4 a. Wong, E. and Giandomenico, C.M. (1999) *Chemical Reviews*, **99**, 2451–66. b. Galanski, M., Jakupec, M.A. and Keppler, B.K. (2005) *Current Medicinal Chemistry*, **12**, 2075–94.
5 Vouillamoz-Lorenz, S., Buclin, T., Lejeune, F., Bauer, J., Leyvraz, S. and Decosterd, L.A. (2003) *Anticancer Research*, **23**, 2757–65.
6 a. Ang, W.H., Khalaila, I., Allardyce, C.S., Juillerat-Jeanneret, L. and Dyson, P.J. (2005) *Journal of the American Chemical Society*, **127**, 1382–3. b. Ang, W.H., Pilet, S., Scopelliti, R., Bussy, F., Juillerat-Jeanneret, L. and Dyson, P.J. (2005) *Journal of Medicinal Chemistry*, **48**, 8060–9. c. Barnes, K.R., Kutikov, A. and Lippard, S.J. (2004) *Chemistry and Biology*, **11**, 557–64. d. Muller, P., Schroder, B., Parkinson, J.A., Kratochwil, N.A., Coxall, R.A., Parkin, A., Parsons, S. and Sadler, P.J. (2003) *Angewandte Chemie (International Ed. in English)*, **42**, 335–9.

7 a. Cerasino, L., Intini, F.P., Kobe, J., de Clercq, E. and Natile, G. (2003) *Inorganica Chimica Acta*, **344**, 174–82. b. Descôteaux, C., Provencher-Mandeville, J., Mathieu, I., Perron, V., Asselin, E., Bérube, G. and Mandal, S.K. (2003) *Bioorganic and Medicinal Chemistry Letters*, **13**, 3927–31. c. Hasinoff, B.B., Wu, X. and Yang, Y.W. (2004) *Journal of Inorganic Biochemistry*, **98**, 616–24. d. Reedijk, J. (1996) *Chemical Communications*, 801–6. e. Farrell, N. (2004) *Metal Ions in Biological Systems*, 251–96.
8 a. Allardyce, C.S. and Dyson, P.J. (2001) *Platinum Metals Review*, **45**, 62–9. b. Kostova, I. (2006) *Current Medicinal Chemistry*, **13**, 1085–107.
9 a. Allardyce, C.S., Dorcier, A., Scolaro, C. and Dyson, P.J. (2005) *Applied Organometallic Chemistry*, **19**, 1–10. b. Galanski, M., Arion, V.B., Jakupec, M.A. and Keppler, B.K. (2003) *Current Pharmaceutical Design*, **9**, 2078–89.
10 Dyson, P.J. and Sava, G. (2006) *Dalton Transactions (Cambridge, England: 2003)*, 1929–33.
11 Hartinger, C.G., Zorbas-Seifried, S., Jakupec, M.A., Kynast, B., Zorbas, H. and Keppler, B.K. (2006) *Journal of Inorganic Biochemistry*, **100**, 891–904.
12 Rademaker-Lakhai, J.M., Van Den Bongard, D., Pluim, D., Beijnen, J.H. and

Schellens, J.H.M. (2004) *Clinical Cancer Research*, **10**, 3717–27.
13 Jakupec, M.A., Arion, V.B., Kapitza, S., Reisner, E., Eichinger, A., Pongratz, M., Marian, B., Graf v. Keyserlingk, N. and Keppler, B.K. (2005) *International Journal of Clinical Pharmacology and Therapeutics*, **43**, 595–6.
14 Varmus, H. (2006) *Science*, **312**, 1162–5.
15 Sebolt-Leopold, J.S. and English, J.M. (2006) *Nature*, **441**, 457–62.
16 Schultz, R.M. (2005) *Advances in Targeted Cancer Therapy* (eds P.L. Herrling and A. Matter), Birkhäuser.
17 a. Goldberg, A.L. and Rock, K. (2002) *Nature Medicine*, **8**, 338–40.
b. Gschwind, A., Fischer, O.M. and Ullrich, A. (2004) *Nature Reviews. Cancer*, **4**, 361–70.
18 Kay, P. (2006) *Seminars in Oncology Nursing*, **22**, 1–4.
19 Novakova, O., Kasparkova, J., Vrana, O., Vanvliet, P.M., Reedijk, J. and Brabec, V. (1995) *Biochemistry*, **34**, 12369–78.
20 Vanvliet, P.M., Toekimin, S.M.S., Haasnoot, J.G., Reedijk, J., Novakova, O., Vrana, O. and Brabec, V. (1995) *Inorganica Chimica Acta*, **231**, 57–64.
21 a. Mazumder, U.K., Gupta, M., Bera, A., Bhattacharya, S., Karki, S., Manikandan, L. and Patra, S. (2003) *Indian Journal of Chemistry. Section A*, **42**, 313–7. b. Mazumder, U.K., Gupta, M., Bhattacharya, S., Karki, S.S., Rathinasamy, S. and Thangavel, S. (2004) *Journal of Enzyme Inhibition and Medicinal Chemistry*, **19**, 185–92.
c. Mazumder, U.K., Gupta, M., Karki, S.S., Bhattacharya, S., Rathinasamy, S. and Sivakumar, T. (2005) *Bioorganic and Medicinal Chemistry*, **13**, 5766–73.
d. Mazumder, U.K., Gupta, M., Karki, S.S., Bhattacharya, S., Rathinasamy, S. and Thangavel, S. (2004) *Chemical and Pharmaceutical Bulletin*, **52**, 178–85.
e. Mishra, L., Sinha, R., Itokawa, H., Bastow, K.F., Tachibana, Y., Nakanishi, Y., Kilgore, N. and Lee, K.H. (2001) *Bioorganic and Medicinal Chemistry*, **9**, 1667–71. f. Mishra, L., Yadaw, A.K., Bhattacharya, S. and Dubey, S.K. (2005) *Journal of Inorganic Biochemistry*, **99**, 1113–8.

22 Chan, H.L., Liu, H.C., Tzeng, B.L.C., You, Y.S.Y., Peng, S.M., Yang, M.S. and Che, C.M. (2002) *Inorganic Chemistry*, **41**, 3161–71.
23 Karidi, K., Garoufis, A., Tsipis, A., Hadjiliadis, N., den Dulk, H. and Reedijk, J. (2005) *Dalton Transactions (Cambridge, England: 2003)*, 1176–87.
24 Anderberg, P.I., Harding, M.M., Luck, I.J. and Turner, P. (2002) *Inorganic Chemistry*, **41**, 1365–71.
25 van der Schilden, K., Garcia, F., Kooijman, H., Spek, A.L., Haasnoot, J.G. and Reedijk, J. (2004) *Angewandte Chemie (International Ed. in English)*, **43**, 5668–70.
26 Chatterjee, D., Mitra, A. and De, G. (2006) *Platinum Metals Review*, **50**, 2–12.
27 Vilaplana, R., Romero, M., Quiros, M., Salas, J. and Vilchez and Gonzalez, F. (1995) *Met-Based Drugs*, **2**, 211–9.
28 Gonzalez Vilchez, F., Vilaplana, R., Blasco, G. and Messori, L. (1998) *Journal of Inorganic Biochemistry*, **71**, 45–51.
29 Gallori, E., Vettori, C., Alessio, E., Vilchez, F.G., Vilaplana, R., Orioli, P., Casini, A. and Messori, L. (2000) *Archives of Biochemistry and Biophysics*, **376**, 156–62.
30 Grguric-Sipka, S.R., Vilaplana, R.A., Perez, J.M., Fuertes, M.A., Alonso, C., Alvarez, Y., Sabo, T.J. and Gonzalez-Vilchez, F. (2003) *Journal of Inorganic Biochemistry*, **97**, 215–20.
31 Djinovic, V., Momcilovic, M., Gruric-Sipka, S., Trajkovic, V., Stojkovic, M.M., Miljkovic, D. and Sabo, T. (2004) *Journal of Inorganic Biochemistry*, **98**, 2168–73.
32 Chatterjee, D., Mitra, A., Hamza, M.S.A. and van Eldik, R. (2002) *Journal of the Chemical Society. Dalton Transactions*, 962–5.
33 Chatterjee, D., Sengupta, A., Mitra, A. and Basak, S. (2005) *Inorganica Chimica Acta*, **358**, 2954–9.
34 a. Chatterjee, D., Hamza, M.S.A., Shoukry, M.M., Mitra, A., Deshmukh, S. and van Eldik, R. (2003) *Dalton Transactions (Cambridge, England: 2003)*, 203–9. b. Chatterjee, D., Mitra, A., Sengupta, A., Saha, P. and Chatterjee, M. (2006) *Inorganica Chimica Acta*, **359**,

2285–90. c. Chatterjee, D., Sengupta, A., Mitra, A., Basak, S., Bhattacharya, R. and Bhattacharyya, D. (2005) *Journal of Coordination Chemistry*, **58**, 1703–11.

35. Alessio, E., Mestroni, G., Nardin, G., Attia, W.M., Calligaris, M., Sava, G. and Zorzet, S. (1988) *Inorganic Chemistry*, **27**, 4099–106.

36. Sava, G., Pacor, S., Zorzet, S., Alessio, E. and Mestroni, G. (1989) *Pharmacological Society*, **21**, 617–28.

37. Coluccia, M., Sava, G., Loseto, F., Nassi, A., Boccarelli, A., Giordano, D., Alessio, E. and Mestroni, G. (1993) *European Journal of Cancer*, **29A**, 1873–9.

38. Yapp, D.T.T., Rettig, S.J., James, B.R. and Skov, K.A. (1997) *Inorganic Chemistry*, **36**, 5635–41.

39. Cabrera, E., Cerecetto, H., Gonzalez, M., Gambino, D., Noblia, P., Otero, L., Parajon-Costa, B., Anzellotti, A., Sanchez-delgado, R., Azqueta, A., de Cerain, A. L. and Monge, A. (2004) *European Journal of Medicinal Chemistry*, **39**, 377–82.

40. Wu, A., Kennedy, D.C., Patrick, B.O. and James, B.R. (2003) *Inorganic Chemistry*, **42**, 7579–86.

41. Qu, Y. and Farrell, N. (1995) *Inorganic Chemistry*, **34**, 3573–6.

42. Van Houten, B., Illenye, S., Qu, Y. and Farrell, N. (1993) *Biochemistry*, **32**, 11794–801.

43. Brindell, M., Kulis, E., Elmroth, S.K.C., Urbanska, K. and Stochel, G. (2005) *Journal of Medicinal Chemistry*, **48**, 7298–304.

44. a. Seal, A. and Ray, S. (1984) *Acta Crystallographica. Section C*, **40**, 929–32. b. Krause, R.A. and Krause, K. (1980) *Inorganic Chemistry*, **19**, 2600–3.

45. Velders, A.H., Kooijman, H., Spek, A.L., Haasnoot, J.G., de Vos, D. and Reedijk, J. (2000) *Inorganic Chemistry*, **39**, 2966–7.

46. Hotze, A.C.G., Caspers, S.E., De Vos, D., Kooijman, H., Spek, A.L., Flamigni, A., Bacac, M., Sava, G., Haasnoot, J.G. and Reedijk, J. (2004) *Journal of Biological Inorganic Chemistry: JBIC: A Publication of the Society of Biological Inorganic Chemistry*, **9**, 354–64.

47. Chen, J.C., Li, J., Qian, L. and Zheng, K.C. (2005) *Journal of Molecular Structure (Theochem)*, **728**, 93–101.

48. Hotze, A.C.G., van der Geer, E.P.L., Caspers, S.E., Kooijman, H., Spek, A.L., Haasnoot, J.G. and Reedijk, J. (2004) *Inorganic Chemistry*, **43**, 4935–43.

49. Hotze, A.C.G., Van Der Geer, E.P.L., Kooijman, H., Spek, A.L., Haasnoot, J.G. and Reedijk, J. (2005) *European Journal of Inorganic Chemistry*, 2648–57.

50. Hotze, A.C.G., Bacac, M., Velders, A.H., Jansen, B.A.J., Kooijman, H., Spek, A.L., Haasnoot, J.G. and Reedijk, J. (2003) *Journal of Medicinal Chemistry*, **46**, 1743–50.

51. Hotze, A.C.G., Kooijman, H., Spek, A.L., Haasnoot, J.G. and Reedijk, J. (2004) *New Journal of Chemistry*, **28**, 565–9.

52. Hotze, A.C.G., Kariuki, B.M. and Hannon, M.J. (2006) *Angewandte Chemie (International Ed. in English)*, **45**, 4839–42.

53. Gopal, Y.N.V., Konuru, N. and Kondapi, A.K. (2002) *Arch BiochemBiophys*, **401**, 53–62.

54. Huxham, L.A., Cheu, E.L.S., Patrick, B.O. and James, B.R. (2003) *Inorganica Chimica Acta*, **352**, 238–46.

55. a. Yan, Y.K., Melchart, M., Habtemariam, A. and Sadler, P.J. (2005) *Chemical Communications*, 4764–76. b. Peacock, A.F.A., Habtemariam, A., Fernandez, R., Walland, V., Fabbiani, F.P.A., Parsons, S., Aird, R.E., Jodrell, D.I. and Sadler, P.J. (2006) *Journal of the American Chemical Society*, **128**, 1739–48.

56. Morris, R.E., Aird, R.E., Murdoch, Del Socorro, P., Chen, H., Cummings, J., Hughes, N.D., Parsons, S., Parkin, A., Boyd, G., Jodrell, D.I. and Sadler, P.J. (2001) *Journal of Medicinal Chemistry*, **44**, 3616–21.

57. Novakova, O., Chen, H., Vrana, O., Rodger, A., Sadler, P.J. and Brabec, V. (2003) *Biochemistry*, **42**, 11544–54.

58. Chen, H., Parkinson, J.A., Morris, R.E. and Sadler, P.J. (2003) *Journal of the American Chemical Society*, **125**, 173–86.

59. Fernandez, R., Melchart, M., Habtemariam, A., Parsons, S. and Sadler, P.J. (2004) *Chemistry – A European Journal*, **10**, 5173–9.

60 Chen, H., Parkinson, J.A., Parsons, S., Coxall, R.A., Gould, R.O. and Sadler, P.J. (2002) *Journal of the American Chemical Society*, **124**, 3064–82.

61 Chen, H., Parkinson, J.A., Novakova, O., Bella, J., Wang, F., Dawson, A., Gould, R., Parsons, S., Brabec, V. and Sadler, P.J. (2003) *Proceedings of the National Academy of Sciences of the United States of America*, **100**, 14623–8.

62 Alessio, E., Mestroni, G., Bergamo, A. and Sava, G. (2004) *Current Topics in Medicinal Chemistry*, **4**, 1525–35.

63 Cocchietto, M., Zorzet, S., Sorc, A. and Sava, G. (2003) *Investigational New Drugs*, **21**, 55–62.

64 Frausin, F., Cocchietto, M., Bergamo, A., Searcia, V., Furlani, A. and Sava, G. (2002) *Cancer Chemotherapy and Pharmacology*, **50**, 405–11.

65 Ravera, M., Baracco, S., Cassino, C., Colangelo, D., Bagni, G., Sava, G. and Osella, D. (2004) *Journal of Inorganic Biochemistry*, **98**, 984–90.

66 Bergamo, A., Messori, L., Piccioli, F., Cocchietto, M. and Sava, G. (2003) *Investigational New Drugs*, **21**, 401–11.

67 Zorzet, S., Bergamo, A., Cocchietto, M., Sorc, A., Gava, B., Alessio, E., Iengo, E. and Sava, G. (2000) *The Journal of Pharmacology and Experimental Therapeutics*, **295**, 927–33.

68 Turel, I., Pecanac, M., Golobic, A., Alessio, E., Serli, B. and Bergamo, A., Sava, G. (2004) *Journal of Inorganic Biochemistry*, **98**, 393–401.

69 Velders, A.H., Bergamo, A., Alessio, E., Zangrando, E., Haasnoot, J.G., Casarsa, C., Cocchietto, M., Zorzet, S. and Sava, G. (2004) *Journal of Medicinal Chemistry*, **47**, 1110–21.

70 a. Alessio, E., Iengo, E., Zorzet, S., Bergamo, A., Coluccia, M., Boccarelli, A. and Sava, G. (2000) *Journal of Inorganic Biochemistry*, **79**, 173–7. b. Bergamo, A., Stocco, G., Casarsa, C., Cocchietto, M., Alessio, E., Serli, B., Zorzet, S. and Sava, G. (2004) *International Journal of Oncology*, **24**, 373–9. c. Bergamo, A., Stocco, G., Gava, B., Cocchietto, M., Alessio, E., Serli, B., Iengo, E. and Sava, G. (2003) *The Journal of Pharmacology and Experimental Therapeutics*, **305**, 725–32.

71 Kapitza, S., Pongratz, M., Jakupec, M.A., Heffeter, P., Berger, W., Lackinger, L., Keppler, B.K. and Marian, B. (2005) *Journal of Cancer Research and Clinical Oncology*, **131**, 101–10.

72 Heffeter, P., Pongratz, M., Steiner, E., Chiba, P., Jakupec, M.A., Elbling, L., Marian, B., Korner, W., Sevelda, P., Micksche, M., Keppler, B.K. and Berger, W. (2005) *The Journal of Pharmacology and Experimental Therapeutics*, **312**, 281–9.

73 a. Hartinger, C.G., Hann, S., Koellensperger, G., Sulyok, M., Groessl, M., Timerbaev, A.R., Rudnev, A.V., Stingeder, G. and Keppler, B.K. (2005) *International Journal of Clinical Pharmacology and Therapeutics*, **43**, 583–5. b. Piccioli, F., Sabatini, S., Messori, L., Orioli, P., Hartinger, C.G. and Keppler, B.K. (2004) *Journal of Inorganic Biochemistry*, **98**, 1135–42. c. Pongratz, M., Schluga, P., Jakupec, M.A., Arion, V.B., Hartinger, C.G., Allmaier, G. and Keppler, B.K. (2004) *J Anal At Spectrom*, **19**, 46–51.

74 Malina, J., Novakova, O., Keppler, B.K., Alessio, E. and Brabec, V. (2001) *Journal of Biological Inorganic Chemistry: JBIC: A Publication of the Society of Biological Inorganic Chemistry*, **6**, 435–45.

75 a. Arion, V., Eichinger, A., Jakupec, M. and Keppler, B.K. (2001) *Journal of Inorganic Biochemistry*, **86**, 129 b. Reisner, E., Arion, V.B., Eichinger, A., Kandler, N., Giester, G., Pombeiro, A.J.L. and Keppler, B.K. (2005) *Inorganic Chemistry*, **44**, 6704–16. c. Reisner, E., Arion, V.B., Fatima, M., Da Silva, C.G., Lichtenecker, R., Eichinger, A., Keppler, B.K., Kukushkin, V.Y. and Pombeiro, A.J.L. (2004) *Inorganic Chemistry*, **43**, 7083–93.

76 Jakupec, M.A., Reisner, E., Eichinger, A., Pongratz, M., Arion, V.B., Galanski, M., Hartinger, C.G. and Keppler, B.K. (2005) *Journal of Medicinal Chemistry*, **48**, 2831–7.

77 a. Vessieres, A., Top, S., Beck, W., Hillard, E. and Jaouen, G. (2006) *Dalton Transactions (Cambridge, England: 2003)*, 529–41. b. Top, S., Vessieres, A., Jaouen,

G. and Fish, R.H. (2006) *Organometallics*, **25**, 3293–6.
78. Pigeon, P., Top, S., Vessieres, A., Huche, M., Hillard, E.A., Salomon, E. and Jaouen, G. (2005) *Journal of Medicinal Chemistry*, **48**, 2814–21.
79. a. Navarro, M., Lehmann, T., Cisneros-Fajardo, E.J., Fuentes, A., Sanchez-Delgado, R.A., Silva, P. and Urbina, J.A. (2000) *Polyhedron*, **19**, 2319–25.
b. Sanchezdelgado, R.A., Lazardi, K., Rincon, L., Urbina, J.A., Hubert, A.J. and Noels, A.N. (1993) *Journal of Medicinal Chemistry*, **36**, 2041–3.
c. Sanchez-Delgado, R.A., Navarro, M., Lazardi, K., Atencio, R., Capparelli, M., Vargas, F., Urbina, J.A., Bouillez, A., Noels, A.F. and Masi, D. (1998) *Inorganica Chimica Acta*, **276**, 528–40.
80. Rieber, M.S., Anzellotti, A., Sanchez-Delgado, R.A. and Rieber, M. (2004) *International Journal of Cancer*, **112**, 376–84.
81. Debreczeni, J.E., Bullock, A.N., Atilla, G.E., Williams, D.S., Bregman, H., Knapp, S. and Meggers, E. (2006) *Angewandte Chemie (International Ed. in English)*, **45**, 1580–5.
82. Zhang, L., Carroll, P. and Meggers, E. (2004) *Organic Letters*, **6**, 521–3.
83. Bregman, H., Carroll, P.J. and Meggers, E. (2006) *Journal of the American Chemical Society*, **128**, 877–84.
84. Williams, D.S., Atilla, G.E., Bregman, H., Arzoumanian, A., Klein, P.S. and Meggers, E. (2005) *Angewandte Chemie (International Ed. in English)*, **44**, 1984–7.
85. Bennett, M.A. and Smith, A.K. (1974) *Journal of the Chemical Society. Dalton Transactions*, 233–41.
86. Geldbach, T.J., Brown, M.R.H., Scopelliti, R. and Dyson, P.J. (2005) *Journal of Organometallic Chemistry*, **690**, 5055–65.
87. Phillips, A.D., Gonsalvi, L., Rornerosa, A., Vizza, F. and Peruzzini, M. (2004) *Coordination Chemistry Reviews*, **248**, 955–93.
88. a. Akbayeva, D.N., Gonsalvi, L., Oberhauser, W., Peruzzini, M., Vizza, F., Bruggeller, P., Romerosa, A., Sava, G. and Bergamo, A. (2003) *Chemical Communications*, 264–5. b. Dorcier, A., Ang, W.H., Bolaño, S., Gonsalvi,L., Juillerat-Jeannerat, L., Laurenczy, G., Peruzzini, M., Phillips, A.D., Zanobini, F. and Dyson, P.J. (2006) *Organometallics*, **25**, 4090–6.
89. Allardyce, C.S., Dyson, P.J., Ellis, D.J. and Heath, S.L. (2001) *Chemical Communications*, 1396–7.
90. Scolaro, C., Bergamo, A., Brescacin, L., Delfino, R., Cocchietto, M., Laurenczy, G., Geldbach, T.J., Sava, G. and Dyson, P.J. (2005) *Journal of Medicinal Chemistry*, **48**, 4161–71.
91. Bergamo, A., Scolaro, C., Sava, G. and Dyson, P.J. (unpublished).
92. Dorcier, A., Dyson, P.J., Gossens, C., Rothlisberger, U., Scopelliti, R. and Tavernelli, I. (2005) *Organometallics*, **24**, 2114–23.
93. Scolaro, C., Geldbach, T.J., Rochat, S., Dorcier, A., Gossens, C., Bergamo, A., Cocchietto, M., Tavernelli, I., Sava, G., Rothlisberger, U. and Dyson, P.J. (2006) *Organometallics*, **25**, 756–65.
94. Chaplin, A.B., Scolaro, C. and Dyson, P.J. (unpublished results).
95. Serli, B., Zangrando, E., Gianferrara, T., Scolaro, C., Dyson, P.J., Bergamo, A. and Alessio, E. (2005) *European Journal of Inorganic Chemistry*, 3423–34.
96. Ang, W.H., Daldini, E., Scolaro, C., Scopelliti, R., Juillerat-Jeannerat, L. and Dyson, P.J. (2006) *Inorganic Chemistry*, **45**, 9006–13.
97. Ang, W.H. and Dyson, P.J. (unpublished results).
98. Khailaila, I., Bergamo, A., Bussy, F., Sava, G. and Dyson, P.J. (2006) *International Journal of Oncology*, **29**, 261–8.
99. Allardyce, C.S., Dyson, P.J., Abou-Shakra, F.R., Birtwistle, H. and Coffey, J. (2001) *Chemical Communications*, 2708–9.
100. Allardyce, C.S., Dyson, P.J., Ellis, D.J., Salter, P.A. and Scopelliti, R. (2003) *Journal of Organometallic Chemistry*, **668**, 35–42.
101. Vock, C., Scolaro, C., Phillips, A.D., Scopelliti, R., Sava, G. and Dyson, P.J. (2006) *Journal of Medicinal Chemistry*, **49**, 5552–61.
102. a. Timerbaev, A.R., Hartinger, C.G., Aleksenko, S.S. and Keppler, B.K. (2006)

Chemical Reviews, **106**, 2224–48. b. Polec-Pawlak, K., Abramski, J.K., Semenova, O., Hartinger, C.G., Timerbaev, A.R., Keppler, B.K. and Jarosz, M. (2006) *Electrophoresis*, **27**, 1128–35. c. Szpunar, J., Makarov, A., Pieper, T., Keppler, B.K. and Lobinski, R. (1999) *Analytica Chimica Acta*, **387**, 135–44. d. Timerbaev, A.R., Rudnev, A. V., Semenova, O., Hartinger, C.G. and Keppler, B.K. (2005) *Analytical Biochemistry*, **341**, 326–33.

103 a. Allardyce, C.S. and Dyson, P.J. (2001) *Journal of Cluster Science*, **12**, 563–9. b. Rosenberg, E., Spada, F., Sugden, K., Martin, B., Milone, L., Gobetto, R., Viale, A. and Fiedler, J. (2003) *Journal of Organometallic Chemistry*, **668**, 51–8.

104 Stubbe, J. and van der DonK, W.A. (1998) *Chemical Reviews*, **98**, 705–62.

105 Wang, Y.D., DuBois, J.L., Hedman, B., Hodgson, K.O. and Stack, T.D.P. (1998) *Science*, **279**, 537.

106 Dervan, P.B. (1986) *Science*, **232**, 464–71.

107 Sigman, D.S., Bruice, T.W., Mazumder, A. and Sutton, C.L. (1992) *Accounts of Chemical Research*, **26**, 98–104.

108 Stubbe, J. and Kozarich, J.W. (1987) *Chemical Reviews*, **87**, 1107–36.

109 Babrowaik, J.C., Ward, B. and Goodisman, G. (1989) *Biochemistry*, **28**, 3314–22.

110 Burrows, C.J. and Rokita, S.E. (1994) *Accounts of Chemical Research*, **27**, 295–301.

111 Gravert, D.J. and Griffin, J.H. (1995) *Inorganic Chemistry*, **35**, 4837–47.

112 Joshi, R.R., Likhite, S.M., Kumar, R.K. and Ganesh, K.N. (1994) *Biochimica Et Biophysica Acta*, **1199**, 285–92.

113 Yam, V.W.-W., Choi, S.W.K., Lo, K.K.W., Dung, W.F. and Kong, R.V.C. (1994) *Journal of the Chemical Society, Chemical Communications*, 2379–80.

114 a. Henner, W.D., Rodriguez, L.O., Hecht, S.M. and Haseltine, W.A. (1983) *The Journal of Biological Chemistry*, **258**, 711. b. Janicek, M.F., Haseltine, W.A. and Henner, W.D. (1985) *Nucleic Acids Research*, **13**, 9011. c. Breen, A.P. and Murphy, J.A. (1995) *Free Radical Biology and Medicine*, **18**, 1033.

115 Lamour, E., Routier, S., Bernier, J.L., Catteau, J.P., Bailly, C. and Vezin, H. (1999) *Journal of the American Chemical Society*, **121**, 1862–9.

116 Borah, S., Melvin, M.S., Lindquist, N. and Manderville, R.A. (1998) *Journal of the American Chemical Society*, **120**, 4557–62.

117 a. Jekunen, A.P., Kaireme, K.J., Ramsay, H.A. and Kajanti, M.J. (1996) *Clin Nucl Med*, **21**, 129. b. Lazo, J.S. and Humphreys, C.J. (1983) *Proceedings of the National Academy of Sciences of the United States of America*, **80**, 3064. c. Claussen, C.A. and Long, E.C. (1999) *Chemical Reviews*, **99**, 1797–2816. d. Burger, R.M. (1998) *Chemical Reviews*, **98**, 1153–69.

118 a. Domenge, C., Orlowski, S., Luboinski, B., De Baere, T., Schwab, G., Belehradek, J. and Mir, L.M., Jr (1996) *Cancer*, **77**, 956. b. Sebti, S.M. and Lazo, J.S. (1988) *Pharmacology and Therapeutics*, **38**, 321.

119 Melvin, M.S., Tomlinson, J.T., Saluta, G.R., Kucera, G.L., Lindquist, N. and Manderville, R.A. (2000) *Journal of the American Chemical Society*, **122**, 6333–4.

120 Melvin, M.S., Wooton, K.E., Rich, C.C., Saluta, G.R., Kucera, G.L., Lindquist, N. and Manderville, R.A. (2001) *Journal of Inorganic Biochemistry*, **87**, 129–35.

121 Maheswari, P.U., Roy, S., den Dulk, H., Barends, S., van Wezel, G., Kozlevcar, B., Gamez, P. and Reedijk, J. (2006) *Journal of the American Chemical Society*, **128**, 710–1.

122 Pachon, L.D., Golobic, A., Kozlevcar, B., Gamez, P., Kooijman, H., Spek, A.L. and Reedijk, J. (2004) *Inorganica Chimica Acta*, **357**, 3697.

123 de Hoog, P., Pachon, L.D., Gamez, P., Lutz, M., Spek, A.L. and Reedijk, J. (2004) *Dalton Transactions (Cambridge, England: 2003)*, 2614.

124 Maheswari, P.U., Barends, S., Özalp-Yaman, ., de Hoog, P., Casellas, H., Teat, S.J., Massera, C., Lutz, M., Spek, A. L., van Wezel, G.P., Gamez, P. and Reedijk, J. (2007) *Chemistry–A European Journal*, **13**, 5213–22.

125 Yazlovitskaya, E.M., Dehaan, R.D. and Persons, D.L. (2001) *Biochemical and Biophysical Research Communications*, **283**, 732–7.

126 Pogozelski, W.K. and Tullius, T.D. (1998) *Chemical Reviews*, **98**, 1089.
127 Sokolowski, A., Muller, J., Weyhermuller, T., Schnepf, R., Hildebrandt, P., Hildenbrand, K., Bothe, E. and Wieghardt, K. (1997) *Journal of the American Chemical Society*, **119**, 8889.
128 Chaudhuri, P., Hess, M., Muller, J., Hildenbrand, K., Bill, E., Weyhermuller, T. and Wieghardt, K. (1999) *Journal of the American Chemical Society*, **121**, 9599.
129 Pratt, R.C. and Stack, T.D.P. (2003) *Journal of the American Chemical Society*, **125**, 8716.
130 Pratt, R.C. and Stack, T.D.P. (2005) *Inorganic Chemistry*, **44**, 2367.
131 Maheswari, P.U., Hartl, F., Quesada, M., Buda, F., Lutz, M., Spek, A.L., Gamez, P. and Reedijk, J. (2007) *Chemistry – A European Journal*, **13**, 5213–22.
132 Lamour, E., Routier, S., Bernier, J.L., Catteau, J.P., Bailly, C. and Vezin, H. (1999) *Journal of the American Chemical Society*, **121**, 1862.

16
Speeding Up Discovery Chemistry: New Perspectives in Medicinal Chemistry

Matteo Colombo and Ilaria Peretto

The drug discovery process is a long and expensive pathway and rich in failure. It starts from the disease selection and the identification of the correlate gene/target and finishes, after the clinical trials, with the launch of the drug on the market (see Figure 16.1).

The principal issue in the drug discovery process is the high failure rate in the clinical trials, mainly due to liabilities related to poor pharmacokinetics (PK), poor efficacy, and high toxicity. The earlier "lead optimization" (LO) phase then represents a crucial step in the drug discovery process, since it involves the preparation and the selection of suitable drug candidates. In view of the increasing need for speed in the preclinical research and development, the determination of activity and selectivity is performed simultaneously with the evaluation of pharmacokinetic and toxicity properties. This multiparametric approach allows the early selection of the compounds with the best overall balanced druglike profile [1].

Early determination of PK properties (*a*bsorption, *d*istribution, *m*etabolism, *e*xcretion and *t*oxicity, ADMET) has become a fundamental resource of medicinal chemistry in the LO phase. New technologies have been developed to perform a great number of *in vitro* and even *in silico* tests. Currently, the most common early-ADME assays evaluate both physicochemical properties (such as the solubility in an opportune medium, the lipophilicity, and the pK_a) and biophysical properties (such as the permeability through cellular monolayers to predict oral absorption and the metabolic stability after treatment with liver or microsomal subcellular fraction that contains oxidative cytochromes).

From an organic chemistry perspective, the main requirement is the preparation of chemical entities having the following features:
- novelty and patentability, to be original and proprietary;
- high purity (usually higher than 95%), to avoid side-effects from contaminating species;
- amount in the 20–100 mg range, to allow simultaneous multiparametric assays;
- high biological activity and selectivity toward the selected target, to be potent and safe;

Tomorrow's Chemistry Today. Concepts in Nanoscience, Organic Materials and Environmental Chemistry.
Edited by Bruno Pignataro
Copyright © 2008 WILEY-VCH Verlag GmbH & Co. KGaA, Weinheim
ISBN: 978-3-527-31918-3

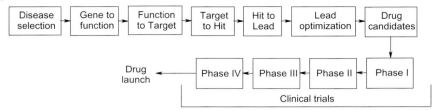

Figure 16.1 The drug discovery process.

- druglikeness and good early pharmacokinetic properties, to give a better chance to advance safely into the drug discovery process.

Chemists must face the challenging task of preparing appropriately complex and decorated scaffolds. New and different synthetic and purification approaches have been developed, often based on the synthesis of small arrays of compounds featuring small chemical variations, in order to explore the physico-chemical and steric properties as well as the structure–activity relationship (SAR). Exploration of the chemical space, both in term of diversity and properties, becomes more important than the mere number of prepared compounds. High-throughput synthesis, which finds wide application for large libraries in the hit identification phase, is not suitable for the LO phase. Although combinatorial chemistry can generate a great number of molecules in a short time, these compounds are usually produced in small amounts, are usually not further developable, and do not present the necessary diversity. Moreover, challenging synthetic routes often cannot be applied to intensive parallel and combinatorial chemistry. Therefore, new synthetic strategies and technologies have become necessary to improve the efficacy and the throughput of organic chemistry in the LO step.

In this chapter we discuss the new speeding-up techniques, optimized during the last decade, such as solid-phase extraction, polymer-assisted solution-phase synthesis, microwave-assisted organic synthesis, and flow chemistry. The improvements obtained with these techniques are not limited to a subset of chemical reactions (e.g., the reported examples), but they are fully applicable to the entire set of chemistry involved in the synthetic drug discovery process.

16.1
Solid-phase Extraction

Solid-phase extraction (SPE) is a purification technique based on the extraction of a compound (or a mixture of compounds) from a solution through absorption on a solid support. The main physical principles involved in the extraction process are ionic interactions between acidic and basic species, and polar and/or nonpolar

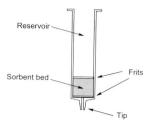

Figure 16.2 General scheme of a cartridge and purification mechanisms.

interactions depending on the chemical peculiarities of the mixture to be purified and on the physical characteristics of the solid support.

The typical frame of the SPE cartridges is depicted in Figure 16.2: a polypropylene syringe (resistant to a great number of solvents and chemicals) is partially filled with an appropriate solid sorbent, contained by a couple of chemically stable frits. The main portion of the reservoir is empty, in order to allow the introduction of the liquid mixture to be purified and of the appropriate washing solutions.

SPE purification is based on two types of mechanism: catch-and-release system and scavenging systems (Figure 16.2). The elution process may be speeded up by a positive or negative (vacuum) pressure, and the eluates are easily collected under the terminal tip. Uniform particle size of the sorbent layer is an essential requirement to ensure maximum and efficacious contact among solvent, sorbent, and compounds to be purified. In addition, a good particle size distribution avoids solvent flow through the path of least resistance (channeling), which would lead to a reduction of the capacity of the sorbent phase. Excellent results are often reached by the use of home-filled and packed SPE cartridges: the experience and sensibility of the chemist are the best bases for the choice of the right sorbent material and suitable amounts. On the other hand, many suppliers (e.g., Biotage AB; Agilent Technologies, Inc.; Varian, Inc.) [2], offer a wide selection of disposable prepacked cartridges with different fillings and sizes.

The most widely used and marketed prepacked SPE cartridges have the following sorbent layers [3]:
- silica gel, for the removal of highly polar impurities from crude reaction products;
- alumina (acidic, neutral, or basic), similar to silica gel but sometimes more suitable for acid-sensitive compounds;
- magnesium oxide silica gel, which adsorbs the polar compounds less strongly;
- C_8- or C_{18}-functionalized silica gel, for the extraction of nonpolar products (reversed-phase technique);
- benzenesulfonic acid (SCX, strong cation exchange), for the scavenging or the catch-and-release purification of amines and basic compounds;

- quaternary ammonium salts (SAX, strong anion exchange), for isolating or removing acidic species;
- sodium sulfate, for a very quick drying of organic mixtures.

The use of SPE cartridges allows a considerable simplification of the work-up procedures, avoiding liquid–liquid extractions and tedious chromatographic columns. Moreover, parallel processing is possible, because prepacked 48- and 96-well format plates with different sorbent fillings are commercially available.

On the other hand, this purification technique is mainly limited to quite clean reaction mixtures or to chemically known impurities: complex mixtures with a lot of by-products need accurate and time-consuming standard purification methodologies. In addition, the SPE strategy is little used in multigram scale, mainly because of the need for careful scaling up and the shortage of suitable disposable kits.

16.2
Polymer-assisted Solution-phase Synthesis

Solid-supported reagents and scavengers have increasingly been employed in organic chemistry during recent years, since they allow the simplification of both synthetic procedures and isolation or purification steps (Scheme 16.1) [4].

The most common solid supports employed for reagents and scavengers in organic synthesis (polymer-assisted solution-phase synthesis (PASP))are polystyrene-based polymers, which are chemically stable to common reaction conditions. Crosslinking with divinylbenzene (from low, 1–5%, to high, 20–30%) or grafting with different moieties (such as Tentagels [5] and JandaGels [6]) allows mechanical stabilization and modulation of such physicochemical properties as swelling, hydrophilicity, and access of the solvent to the inner surface of the beads. More recently, silica-supported reagents and scavengers have been developed [7], characterized by faster kinetics and higher chemical, mechanical, and thermal stability compared with polystyrene supports.

The use of solid-supported reagents and scavengers presents significant advantages and improvements in organic synthesis [8]. As mentioned above, purification and work-up operations are considerably simplified. The use of a large excess of reagents (often necessary to drive reactions to completion) is then possible without requiring additional purification steps. Toxic, noxious, or hazardous reagents and their by-products can be immobilized and so are not released into solution, thereby improving their general acceptability, utility, and safety profile. Due to site isolation of reagents on the resin bead, species that are incompatible in solution may be used together to achieve one-pot transformations that are not possible in classical homogenous solution. Moreover, reaction monitoring can be performed using conventional methods (thin-layer chromatography (TLC), liquid chromatography (LC)–mass spectrometry (MS) (LC-MS), nuclear magnetic resonance (NMR), etc.).

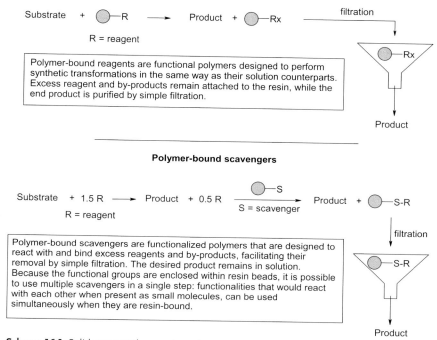

Scheme 16.1 Solid-supported reagents and scavengers.

However, these materials have some limitations. The main drawback is their lower reactivity compared with the homogeneous counterparts, due to the nature of the solid support and to the kinetic limit of diffusion of the solution species in the solid support bearing the reactive functionalities. In order to overcome their poor reactivity, these materials are usually employed in large excess, thus resulting in higher costs, especially if the materials are not recyclable. These limits become particularly significant when considering large-scale PASP synthesis.

Almost every kind of organic reaction described in homogeneous phase can be performed today with the corresponding supported reagents. The use of supported reagents has proved particularly advantageous in multistep synthesis, and particularly for natural products. One of the most challenging targets accomplished using solid-supported reagents and scavengers is the total synthesis of epothilone C (a potent inhibitors of tumor cell proliferation inducing mitotic arrest), described by Storer, Ley and colleagues in 29 steps [9], where almost any type of chemical transformation has been reproduced by PASP. All the conventional total syntheses reported for epothilone derivatives required extensive silica gel chromatography to provide material free from contaminating by-products: such processes are then suitable neither for larger-scale production nor for parallel synthesis of analogues.

In Scheme 16.2, the synthesis of one of the fragments of epothilone C with supported reagents and scavengers is illustrated. The use of supported inorganic bases and acids (such as PS-carbonate, PS-sulfonic acid) avoids all the aqueous work-up steps; supported organic bases (like PS-triazabicyclodecene, PS-pyridine) can be removed by filtration instead of requiring chromatographic purification. Similarly, supported triphenylphosphine is employed to generate the partner for the Wittig reaction, so that supported triphenylphosphine oxide can be easily removed from the reaction mixture. All the inorganic oxidizing agents are also employed in the polymer-supported form (the use of PS-TEMPO and PS-periodate is exemplified in the last step of Scheme 16.2).

The last steps of the synthetic strategy leading to epothilone C are reported in Scheme 16.3. Again, the extensive use of inorganic and organic supported reagents and scavengers avoided all the intermediate purifications. The final macrocycle was purified first by a cation-exchange protocol, which removed most of the non-basic impurities, and finally only a single chromatography procedure on silica over 29 steps was required to obtain the pure product.

The commercial availability of almost every kind of functionality in the supported form (from oxidizing to reducing agents, to acids and bases, catalysts, chiral auxiliaries, and so on) makes the use of PASP particularly versatile and attractive to organic chemists. Moreover, supported reagents can be employed also in combination with other emerging technologies such as microwave heating and flow chemistry, further enhancing the efficiency of the synthetic process.

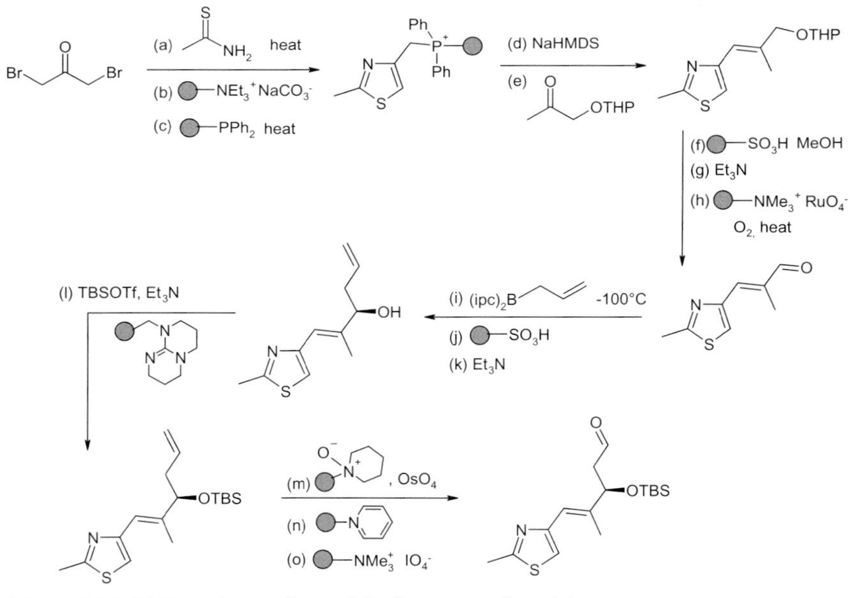

Scheme 16.2 PASP synthesis of one of the fragments of epothilone C.
THP = Tetrahydropyran; TBS = t-butyldimethylsilyl;
ipc = isopinocampheyl; Tf = Triflate.

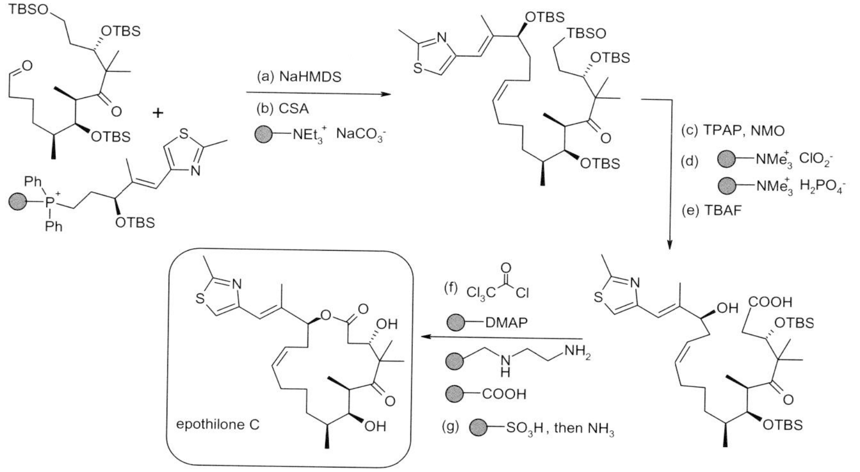

Scheme 16.3 PASP synthesis of epothilone C, final steps.
CSA = camphorsulfonic acid;
TPAP = Tetrapropylammonium perruthenate;
NHO = N-methyl morpholine N-oxide;
TBAF = Tetrabutylammonium fluovide;
DMAP = 4-dimethylaminopyridine.

16.3
Microwave-assisted Organic Synthesis [10, 11]

High-speed synthesis with microwaves has attracted a considerable amount of attention in recent years. In 1986 two different research groups (Gedye *et al.* [12] and Giguere *et al.* [13]) published the first work regarding the use of microwave heating to increase the speed of chemical transformations. After a decade of slow uptake, since the late 1990s an exponentially growing number of reports based on microwaves heating have appeared in the literature, not only from academic research groups but also from industrial ones, indicating that microwave-assisted organic synthesis (MAOS) is becoming forefront support for rapid optimization of reactions, for the efficient synthesis of new chemical entities, and for discovering and probing new chemical reactivity. Following this trend, some companies, such as Biotage AB, CEM Corporation, Milestone srl, and Anton Parr GmbH [14], have started the production and optimization of chemistry-dedicated microwave equipment. These instruments allow accurate control of reaction parameters (temperature, pressure, power), in contrast with the unsafe and barely controlled modified domestic household microwave ovens previously employed. In this way the reproducibility, the safety, and the applicability of microwave-heated chemical reactions have increased remarkably.

MAOS is mainly based on the efficient heating of materials by the microwave dielectric heating effect [15] mediated by dipolar polarization and ionic conduction. When irradiated at microwave frequencies, the dipoles (e.g., the polar solvent molecules) and ions of the sample align to the applied electric field. If the applied field oscillates, the dipoles and ions tend to realign themselves with the alternating field. This continual reorientation produces molecular friction and dielectric loss that generate energy in the form of heat. In this way, microwave irradiation produces efficient internal heating compared with the traditional external heat source, such as an oil bath, which transfers heat by conduction.

The use of microwave irradiation brings significant advantages:
- higher reaction temperatures by combination of microwave heating with sealed vessels;
- reduced reaction times and, sometimes, higher yields and cleaner reaction profiles;
- the possibility of using lower-boiling-point solvents under pressure in sealed vessels;
- faster processes by direct "in core" heating of reaction mixtures;
- specific heating of strongly microwave-absorbing metal catalysts;
- more reproducible experimental conditions by accurate on-line control of temperature and pressure profiles;
- easy adaptation to automated sequential or parallel synthesis.

All the points just described match the requirements of medicinal chemistry: the opportunity to synthesize a large number of new chemical entities in very short times, in high purity, in a reproducible way, and making use of scarcely reactive species. The following examples might better illustrate the advantages of the application of MAOS in organic synthesis.

Leadbeater and Marco described a very rapid synthesis of a small array of substituted biphenyls by a ligand-free palladium-catalyzed Suzuki coupling [16]. The couplings were performed under pressure in sealed vessels using water as solvent (cheap, environmental friendly, not toxic, not flammable, having a good response to microwaves, low solubility for organic compounds, and pseudo-organic behavior at elevated temperature) and tetrabutylammonium bromide (TBAB) as phase-transfer catalyst (Scheme 16.4a). Similar conditions were used in our laboratories to perform the Suzuki coupling between modified phenylpropionic acids and arylboronic acids (Scheme 16.4b) [17].

Another example of reaction-rate enhancement was reported for the microwave-assisted Paal–Knorr synthesis of a series of tetrasubstituted pyrroles [18]. Following the standard procedure, 1,4-dicarbonyl compounds were converted to pyrrole rings via acid-mediated dehydrative cyclization in presence of primary amines. The main limitation of the standard protocol is the harsh reaction conditions (reflux in acetic acid for extended times). The use of microwaves slashes the reaction times to few minutes, giving good isolated yields of the desired products (Scheme 16.5).

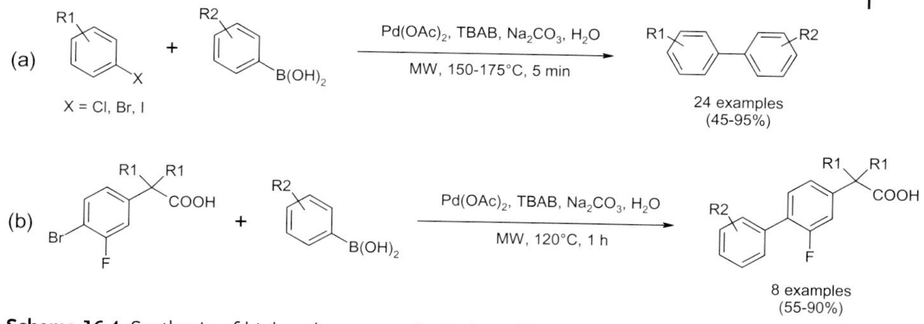

Scheme 16.4 Synthesis of biphenyl compounds via ligand-free Suzuki coupling.

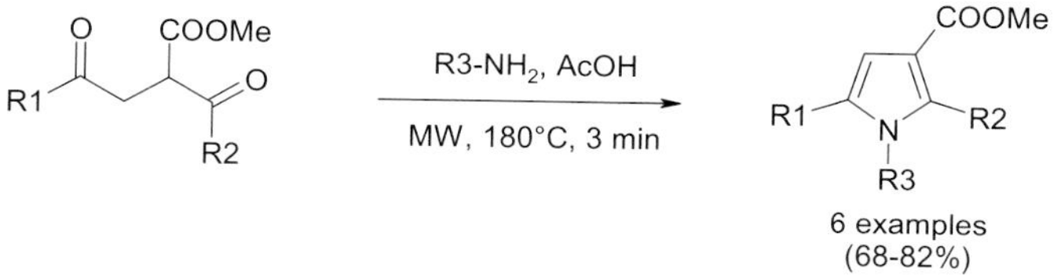

Scheme 16.5 Synthesis of tetrasubstituted pyrroles by microwave-assisted Paal–Knorr reaction.

Microwave heating may promote reactions of poorly reactive moieties as well, based on the possibility of reaching high temperatures in short times without the typical overheating phenomena of traditional oil baths. In our laboratories, a series of condensation between carboxylic acids and substituted anilines was performed (Colombo H. and Bossoto S., unpublished results). The aniline ring was substituted with electron-withdrawing groups (such as halogens, sulfonamides, nitro groups, and carboxylates), which lower the electron density and cause a considerable reduction of the nitrogen nucleophilicity. Classical coupling agents were tested, but, because of the low nucleophilicity of nitrogen atom, only the reaction with acyl chlorides afforded the expected amides in acceptable yields. A better one-pot protocol (based on the elimination of the time-consuming preactivation step of the carboxylic acids) was set up, entailing the application of the microwave heating to a mixture of aniline, carboxylic acid, and phosphorus trichloride in tetrahydrofuran (THF) (Scheme 16.6). A small collection of products was quickly prepared thanks to the microwave instrument automation, which allows the use of the microwave oven in a sequential way, processing up to 60 different vials. In addition, the process is acceptable clean: a very short SPE silica cartridge was sufficient to eliminate the by-products, obtaining the high purities required for the biological assays.

Microwave-assisted organic chemistry may be also coupled to inorganic-supported solvent-free conditions, thus allowing the simplification of work-up procedures (in many cases the pure expected products can be obtained directly by

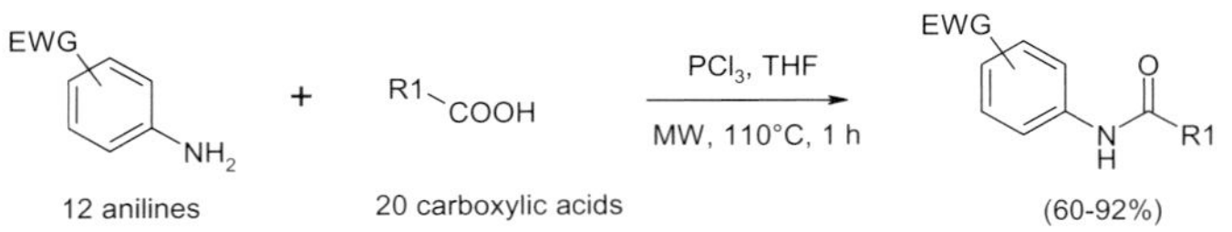

Scheme 16.6 Condensation of electron-withdrawing substituted anilines with carboxylic acids.

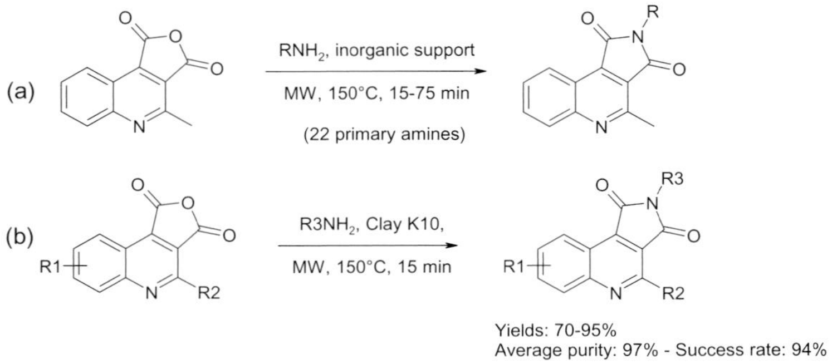

Scheme 16.7 (a) Condensation of 2-methylquinoline-3,4-dicarboxylic anhydride with primary amines using inorganic solid support; (b) application of the methodology to the synthesis of a small library of imide derivatives.

simple extraction, distillation, or sublimation), the elimination of solvent wastes, hazards, and toxicity, the opportunity to work with open vessels, and the ability to scale up the reaction to multigram amounts. Microwave-assisted formation of imide derivatives was reported as a result of a preliminary investigation of the synthesis on inorganic solid supports achieved in our laboratories [19]. The imide group was obtained by condensation between substituted quinoline-3,4-dicarboxylic anhydride and a set of primary amines (Scheme 16.7a). Silica, alumina (neutral, acidic, and basic) and K10 clay (montmorillonite) were tested as solid supports in comparison with the classical condensation in toluene. The best results were obtained with the use of an equimolar mixture of substrate and primary amine absorbed on wet K10 clay. After 15–75 minutes of microwave irradiation, the conversion was complete, whereas the same results were achieved in longer times using toluene as solvent. Afterwards, the optimized conditions were used to expand a library of substituted imido-quinolines: 10 substituted quinoline-3,4-dicarboxylic anhydrides were reacted with 52 commercially available primary amines (Raucati F., unpublished results). After microwave heating, the products were recovered in high yields and in excellent purity by simple basic alumina SPE (Scheme 16.7b).

Another simple and useful application of microwave heating is the combination of solid supports and organic synthesis. Usually, the synthetic steps involving polymeric supports require repeated runs and longer reaction times than the corresponding solution-phase protocol, to reach high conversions. Microwave heating again allows reduction of the times and improvement of the loading of the functionalized solid support, employing both traditional polystyrene support and soluble polymers (poly(ethylene glycol), PEG) and fluorous phase synthesis. The following simple examples demonstrate the advantages of this combination.

In the first example, Stadler and Kappe published the substitution of chlorinated Wang resin (polystyrene matrix) with a series of cesium salts of carboxylic acids [20]. Within only 3–15 minutes, the synthetic conversions were at least 85% (Scheme 16.8a). In comparison, the same reactions performed with the conventional heating method required longer times (up to 48 hours), higher amounts of carboxylic acid and base and the presence of potassium iodide as additive. Moreover, high loadings of the resin-bound esters could be obtained by microwave heating even with sterically demanding acids.

The second example involves the transformation of iodoaryl derivatives into the corresponding tetrazoles (Alterman and Hallberg) [21]. The 4-iodobenzoic acid residue, bound to a Tentagel Rink support, was converted in very short times to the cyano derivative by reaction with zinc cyanide, and subsequently was transformed into a tetrazole ring by treatment with sodium azide (Scheme 16.8b). Both synthetic steps were improved by microwave heating: the reaction times were very short and, despite the very high temperatures, only negligible decomposition of the solid support was observed. Another advantage of this procedure was the possibility of preparing the expected functionalized tetrazole in a one-pot reaction, avoiding the use of the tetrakis(triphenylphosphine)palladium as catalyst.

MAOS can also be coupled to polymer-assisted solution phase synthesis (PASP, see above). In 2005 a short review of the combination of these two techniques was published [22], gathering some representative examples of the recent literature.

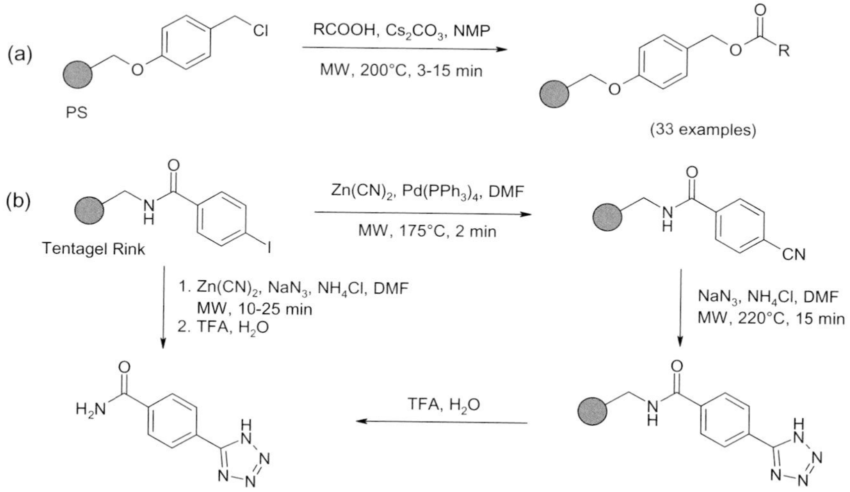

Scheme 16.8 (a) Polystyrene resin functionalization with carboxylic acids; (b) substituted tetrazole synthesis on solid phase.

As previously discussed, the most used support is variously crosslinked polystyrene, because it is chemically stable in the usual solvents and conditions used with microwave heating.

Room temperature ionic liquids (RTILs), belonging to an interesting class of nonclassical solvents, are useful tools that are also utilized in MAOS. RTILs consist of organic cations (e.g., 1-alkyl-3-methylimidazolium, 1-alkylpyridinium) and appropriate inorganic anions (mainly hexafluorophosphate, tetrafluoroborate, chloride). They have negligible vapor pressure and are immiscible with a range of organic solvents. Thus, the possibility of removing the organic product by solvent washing, the nonflammability, the wide accessible temperature range, the ease of use and recycling, and the optimal interaction with the microwaves through the ionic conduction mechanism have made the use of ionic liquids a practical and safe synthetic technique. A complete treatment of the combination of ionic liquids and MAOS, with an exhaustive list of examples, is reported in two recent reviews [23]. However, ionic liquids should not be considered only as alternative "green" solvents, particularly suitable for microwave heating. The extremely wide class of RTILs (evaluated in more than 10^{18} combinations!) allows their application on large scale and in different areas. The modulation of physicochemical and electronic properties, based on the opportune choice of anions and cations, has been only partially studied and understood, but should increase the research and industrial potential of salts of this kind [24].

MAOS presents some limitations and drawbacks, such as the costs of the apparatus and above all the scalability of the processes. In fact, most of the published MAOS examples were carried out on a small scale (usually less than 1 gram scale and up to 10–20 ml reaction volume) due to the instrumental restriction on cavity-volume. Thus, microwave synthesis has become a useful ally in the preparation of lead compounds and for lead optimization, but it cannot yet be used in the scale-up process. Currently, two different approaches are being developed to resolve these issues. The first approach entails the experimental use of large batch-type reactors, such as the prototype multimode apparatus used by Kappe and co-workers [25], in order to scale up several different organic reactions from 1 to 100 mmol, by partitioning the reaction mixture in eight 60 ml volume vessels, which were heated simultaneously and in parallel. A second possibility employs the continuous-flow technique, such as the continuous-flow microwave reactor fabricated by Laporterie, Dubac and colleagues for the Friedel–Crafts acylation of aromatics [26], which allowed them to obtain the expected products in high yields with a 1.2l h^{-1} flow rate. However, the scalability of microwave reactions still requires more development, especially in the technology and engineering areas.

16.4
Flow Chemistry

Despite the noteworthy developments in new synthetic methodologies during the last ten years (achieved in catalysis, asymmetric synthesis, combinatorial chemistry, and other fields), organic synthesis is generally still being carried out in a very traditional way. Flow-through processes can represent a significant breakthrough toward more efficient syntheses, in view of the overall need for new chemical technologies.

In flow chemistry, a chemical reaction is run in a continuously flowing stream: pumps move solutions containing the reactants into a network of interconnecting tubes and, where tubes join one another, the fluids come into contact and the reaction takes place. In the realization of a flow process, then, technical and engineering aspects have the same importance as chemical considerations, even in laboratory scale.

The main features, advantages, and disadvantages of continuous-flow processes are summarized in Table 16.1 [27, 28].

Table 16.1 Features of continuous-flow processes.

Continuous synthesis

Features	Performance benefits	Chemistry benefits	Disadvantages
Elimination of dosing time: constant mixture composition	No accumulation of unreacted reagents: enhanced safety	Side-reaction suppression; handling of unstable batch processes	Flow rates must ensure full conversion of the starting material
Pressuring the system: working above solvent boiling point	Enhanced safety	Expanded temperature range	The presence of bubbles can affect the homogeneity of the system. Issue: incompatible with precipitating products
Possible coupling with microwave heating	Optimized energy transfer and temperature control	Expanded temperature range	
Possible use of supported reagents–scavenger	Simultaneous reaction and filtration	Byproducts are retained on the solid phase	
Assembling a line of reactors	Linear, divergent as well as convergent multistep syntheses are also feasible		Issue: solvent switching. If reaction times are not similar, switching and recycling valves are required

Continuous synthesis in microstructured systems

Features	Performance benefits	Chemistry benefits	Disadvantages
Efficient mixing. High surface-volume ratio	Optimized heating transfer: precise temperature control	Side-reaction suppression	Only soluble substrates-reagents are compatible. Limited reaction-time range; small scale

Microreactors (flow reactors with micrometer scale) were first employed in organic synthesis to perform chemical reactions in flow processes. The small dimensions of microreactors allow the use of minimal amounts of reagent under precisely controlled conditions, and the rapid screening of reaction conditions with improved overall safety of the process. To obtain synthetically useful amounts of material, either the microreactors are simply allowed to run for a longer period of time ("scale-out"), or several reactors are placed in parallel ("numbering up") [29].

Challenging applications, such as the synthesis of β-peptides [30] or examples of flash chemistry (like the reaction of electrochemically generated reactive cation pools [31]) have been successfully realized using microreactors.

Due to the possibility of producing a large number of compounds in high purity and in short time, microreactors have also been employed in combinatorial chemistry [32]. Flow chemistry in microreactors allows the rapid synthesis of combinatorial arrays [33] and also their on-line purification by chromatography [34]. The concept can be extended to include high-throughput assays in the flow system, in order to speed up the whole hit identification and lead optimization process [35]. In one example reported by Hawkes *et al.* [36] an array of six sulfonamides was synthesized in a hyphenated platform combining synthesis and screening microfluidic equipment (Scheme 16.9).

Examples of laboratory flow systems operating on a larger scale than microreactors have been described recently and are commonly used in organic synthesis. In particular, the use of flow systems for catalytic heterogeneous hydrogenation has found wide application, since this technology has become commercially available. Both literature reports [37] and some applications conducted in our laboratories (Mazzacani A., unpublished results) confirmed significant advantages compared with batch hydrogenation reactions, as illustrated in Table 16.2.

In addition to hydrogenation reactions, many examples in the field of organic synthesis and multigram-scale preparations have been also reported [38, 39]. In

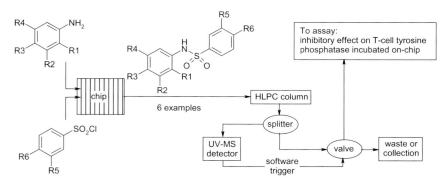

Scheme 16.9 Combination of synthesis and screening in microfluidic apparatus.

Table 16.2 Examples of heterogeneous flow-through hydrogenations.

Reaction	Flow-through conditions	Batch conditions (r.t.)
Ref. 39a X = NO$_2$ → NH$_2$	Catalyst: Pd/C cartridge Flow: 1 ml min^{-1}, 30 °C H$_2$: 1 bar Conversion: 100% (3 minutes)	Catalyst: Pd/C H$_2$: 1 bar Conversion: 2% (3 min), 100% (20 minutes)
Ref. 39c X = CH$_2$Ph → H	Catalyst: Pd/C cartridge Flow: 1 ml min^{-1}, 40 °C H$_2$: 1–2 bar Conversion: 81–95%	Catalyst: Pd/C H$_2$: 3 bar Moderate to high yields (8–24 hours)
Ref. 40 X = NO$_2$ → NH$_2$	Catalyst: Ni-Raney cartridge Flow: 1 ml min^{-1}, 60 °C H$_2$: 1 bar Conversion: 100%	Catalyst: Ni-Raney H$_2$: 2.1 bar Conversion: 100% (4 hours)
Ref. 40 X = NO$_2$ → NH$_2$	Catalyst: Ni-Raney cartridge Flow: 1 ml min^{-1}, 60 °C H$_2$: 1 bar Conversion: 100%	Catalyst: Pd/C H$_2$: 2.8 bar Conversion: 100% (24 hours)
Ref. 40 X = CN → CH$_2$NH$_2$	Catalyst: Pd/C cartridge Flow: 1 ml min^{-1}, 100 °C H$_2$: 80 bar Conversion: 75%	Catalyst: Pd/C H$_2$: 3.1 bar Conversion: 50% (16 hours)

this respect, the use of flow-through processes can be advantageously coupled with polymer-supported reagents and scavengers [40] and also with enzymes [41] to achieve better and more efficient synthesis protocols.

The total synthesis of the natural alkaloid oxomaritidine [42] represents the first multistep flow-through preparation of a natural product utilizing flow methods and techniques. In this study, multiple synthetic transformations were achieved by directly coupling glass reaction columns packed with polymer-supported reagents, avoiding any product isolation, distillation, crystallization or column chromatography: the natural product can be produced in automated sequence from readily available starting materials in less than one day (Scheme 16.10).

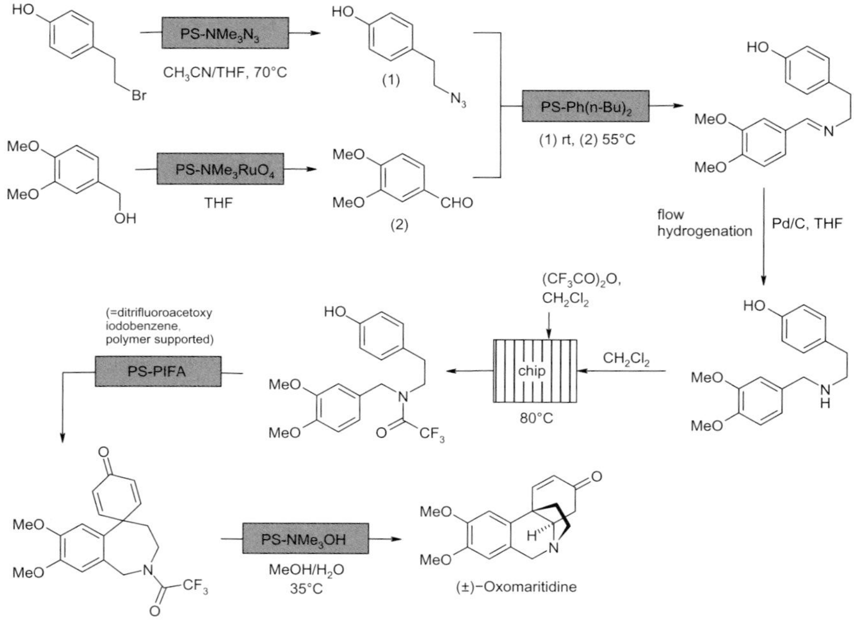

Scheme 16.10 Flow-through synthesis of oxomaritidine.

Continuous-flow processes display significant advantages during scale-up and development in process chemistry, for example, for fast reactions with reactive intermediates or products that can degrade under extended batch processing, for careful control of reaction parameters, and for processes in which the stream may be recycled through the reactor to increase process yields. The problem of inefficient mixing, the most common cause of scale-up difficulties, can be successfully overcome with continuous-flow technologies. However, some limits of these techniques must be considered in large-scale development, such as incompatibility with the presence of solids and bubbles, safety issues, and longer set-up and optimization of conditions. Many examples of implementation of industrial flow processes have been reported, including exothermic reactions, oxidations, enzymatic processes, recycling of immobilized catalysts, and electrochemical and photochemical reactions [43].

16.5
Analytical Instrumentation

The analytical technologies have followed the improvements of synthetic apparatus in order to speed up both the in-process evaluation of the reactions and the conclusive structural analyses of the potential leads. Usually in the LO phase the main structural information about a molecule is determined by ^{1}H- and/or ^{13}C-

NMR (functional groups and atom connectivity) and LC-MS (purity, molecular ion mass and fragmentation). Even if a hyphenated instrument is commercially available [44], it is not simple to conduct this kind of analysis in a parallel on a large number of samples. It is more advantageous to speed up the serial processes, mainly in two ways: reducing the time of each single analysis, and automating the analytical sequences. To do this, the NMR technologies offer new instruments with stronger magnetic fields, automation accessories (for example, automatic handling of NMR tubes), and user-friendly software that manages both the automation and the registration and the transformation of spectra.

Regarding the LC-MS improvements, a new chromatographic solution, called UPLC (Ultra Performance Liquid Chromatography, Waters) [45], is available and can easily be coupled with a MS system. This new technology is based on the use of analytical columns featuring 1.7 mm filling particles that ensure more selective and efficacious separations, drastically reducing the analysis times compared with the traditional high-performance liquid chromatography (HPLC) runs. Moreover, all the apparatus can be automated in order to further reduce the analysis times.

Another emerging technology is supercritical fluid chromatography (SFC) that uses supercritical carbon dioxide as the apolar eluent [46]. The main advantage of SFC, which can be applied both in the analytical and in the purification area, is the higher resolution than the traditional HPLC, allowing time reduction and more efficient separations. Also this technique can be advantageously coupled with a MS detector, to further improve the full analytical process.

16.6
Conclusions

In the preclinical phases of the long drug discovery process, some improvements, in terms of time and quality, may be achieved by the use of new techniques optimized in recent years that are reported in this chapter. Besides the necessity for ever more sophisticated technologies, in future the organic chemist will increasingly be called to further improve the existing chemical reactions and to develop new ones, in order to meet the urgent and constant needs for new drug candidates.

References

1 Dondio, G.M. (2003) *Seminars in Organic Synthesis*, Società Chimica Italiana, 71–90 [and references cited therein].

2 Biotage, A.B.:www.biotage.com; Agilent Technologies Inc.: www.agilent.com; Varian Inc.: www.varianinc.com (accessed date: 10-23-2007).

3 For a more complete list of cartridges and for an accurate selection and operation guide, see the catalogues of the above cited provider companies.

4 For a definition of polymer supported reagents and scavengers, see http://www.biotage.com/DynPage.aspx?id=24303 (accessed date: 10-23-2007).

5 http://www.rapp-polymere.com/preise/tent_s_d.htm (accessed date: 10-23-2007).
6 Roice, M. and Rajasekharan Pillai, V.N. (2005) *Journal of Polymer Science: Part A*, **43**, 4382–92.
7 Thompson, L.A., Combs, A.P., Trainor, G.L., Wang, Q., Langlois, T.J. and Kirkland, J.J. (2000) *Combinatorial Chemistry and High Throughput Screening*, **3**, 107–15. On Web, see: http://www.sigmaaldrich.com/Area_of_Interest/Chemistry/Drug_Discovery/Resin_Explorer/Functionalized_Silica_Gels.html.
8 For reviews on the use of solid-supported reagents and scavengers in organic synthesis see: a. Ley, S.L., Baxendale, I.R., Bream, R.N., Jackson, P.S., Leach, A.G., Longbottom, D.A., Nesi, M., Scott, J.S., Storer, R.I. and Taylor, S.J. (2000) *Journal of the Chemical Society. Perkin Transactions*, **1** (23), 3815–4195. b. Battacharyya, S. (2004) *Current Opinion in Drug Discovery and Development*, **7**, 752–64. c. Sherrington, D.C. (2001) *Journal of Polymer Science Part A: Polymer Preprints*, **39**, 2364–77.
9 Storer, R.I., Takemoto, T., Jackson, P.S., Brown, D.S., Baxendale, I.R. and Ley, S.V. (2004) *Chemistry – A European Journal*, **10**, 2529–47.
10 Books covering the subject: a. Loupy, A. (2002) *Microwaves in Organic Synthesis*, Wiley-VCH Verlag GmbH, Weinhheim. b. Hayes, B.L. (2002) *Microwave Synthesis: Chemistry at the Speed of Light*, CEM publishing, Matthews NC. c. Tierney, J.P. and Lidström, P. (2005) *Microwave-Assisted Organic Synthesis*, Blackwell, Oxford. d. Kappe, C.O. and Stadler, A. (2005) *Microwaves in Organic and Medicinal Chemistry*, Wiley-VCH Verlag GmbH, Weinhheim.
11 Recent reviews: a. Kappe, C.O. (2004) *Angewandte Chemie (International Ed. in English)*, **43**, 6250–84. b. Hayes, B.L. (2004) *Aldrichim Acta*, **37**, 66–77. c. Kappe, C.O. and Dallinger, D. (2006) *Nature Reviews. Drug Discovery*, **5**, 51–64.
12 Gedye, R., Smith, F., Westaway, K., Ali, H., Baldisera, L., Laberge, L. and Rousell, J. (1986) *Tetrahedron Letters*, **27**, 279–82.
13 Giguere, R.J., Bray, T.L., Duncan, S.M. and Majetich, G. (1986) *Tetrahedron Letters*, **27**, 4945–58.
14 Biotage, A.B.: www.biotage.com; CEM corporation: www.cem.com; Milestone srl: www.milestonesci.com; Anton Parr GmbH: www.anton-parr.com (accessed date: 10-23-2007).
15 Gabriel, C., Gabriel, S., Grant, E.H., Halstead, B.S. and Mingos, D.M.P. (1998) *Chemical Society Reviews*, **27**, 213–23.
16 Leadbeater, N.E. and Marco, M. (2002) *Organic Letters*, **4**, 2973–6.
17 Peretto, I., Radaelli, S., Parini, C., Zandi, M., Raveglia, L.F., Dondio, G., Fontanella, L., Misiano, P., Bigogno, C., Rizzi, A., Riccardi, B., Boscaioli, M., Marchetti, S., Puccini, P., Catinella, S., Rondelli, I., Cenacchi, V., Bolzoni, P.T., Caruso, P., Villetti, G., Facchinetti, F., Del Giudice, E., Moretto, N. and Imbimbo, B.P. (2005) *Journal of Medicinal Chemistry*, **48**, 5705–20.
18 Minetto, G., Raveglia, L.F. and Taddei, M. (2004) *Organic Letters*, **6**, 389–92.
19 Mortoni, A., Martinelli, M., Piarulli, U., Regalia, N. and Gagliardi, S. (2004) *Tetrahedron Letters*, **45**, 6623–7.
20 Stadler, A. and Kappe, C.O. (2001) *European Journal of Organic Chemistry*, 919–25.
21 Alterman, M. and Hallberg, A. (2000) *The Journal of Organic Chemistry*, **65**, 7984–9.
22 Bhattacharyya, S. (2005) *Molecular Diversity*, **9**, 253–7.
23 Leadbeater, N.E., Torenius, H.M. and Tye, H. (2004) *Combinatorial Chemistry and High Throughput Screening*, **7**, 511–28. Habermann, J., Ponzi, S. and Ley, S.V. (2005) *Mini-Review in Organic Chemistry*, **2**, 125–37.
24 For a complete treatment of ionic liquids see: a. Wasserscheid, P. and Keim, W. (2000) *Angewandte Chemie (International Ed. in English)*, **39**, 3772–89. b. Wasserscheid P. and Welton, T. (2003) *Ionic Liquids in Synthesis*, Wiley-VCH Verlag GmbH, Weinhheim. c. Rogers, R.D. and Seddon, K.R. (2005) *Ionic Liquids IIIB: Fundamentals, Challenges and Opportunities*, ACS, Washington DC.

25 Stadler, A., Yousefi, B.H., Dallinger, D., Walla, P., Van der Eycken, E., Kaval, N. and Kappe, C.O. (2003) *Organic Process Research and Development*, **7**, 707–16.
26 Marquié, J., Salmoria, G., Poux, M., Laporterie, A., Dubac, J. and Roques, N. (2001) *Industrial and Engineering Chemistry Research*, **40**, 4485–90.
27 Wille, G. (0000) Microreactor Technology: Alternative Tool for Chemical Challenges. Symposium on Medicinal Chemistry and Emerging Technologies in Drug Discovery, November 7, 2006, Baranzate, Milan, Italy.
28 Jas, G. and Kirschning, A. (2003) *Chemistry – A European Journal*, **9**, 5708–23.
29 Geyer, K., Codeè, J.D.C. and Seeberger, P.H. (2006) *Chemistry – A European Journal*, **12**, 8434–42.
30 Flogel, O., Codeé, J.D.C., Seebach, D. and Seeberger, P.H. (2006) *Angewandte Chemie (International Ed. in English)*, **45**, 7000–3.
31 Yoshida, J. (2005) *Chemical Communications*, 4509–16.
32 Fletcher, P.D.I., Haswell, S.J., Pombo-Villar, E., Warrington, B.H., Watts, P., Wong, S.Y.F. and Zhang, X. (2002) *Tetrahedron*, **58**, 4735–57.
33 a. Watts, P. and Haswell, S.J. (2003) *Current Opinion in Chemical Biology*, **7**, 380–7. b. Schwalbe, T., Autze, V. and Wille, G. (2002) *Chimia*, **56**, 636–46. c. Watts, P. (2005) *QSAR and Combinatorial Science*, **24**, 701–11. d. Watts, P. (2005) *Chemical Engineering and Technology*, **28**, 290–301.
34 a.Watts, P. (2004) *Current Opinion in Drug Discovery and Development*, **7**, 807–12. b. Watts, P. (2005) *Analytical and Bioanalytical Chemistry*, **382**, 865–7.
35 a.Dittrich, P.S. and Manz, A. (2006) *Nature Reviews*, **5**, 210–8. b. Hughes, I. (2006) *Application of flow technologies to enhance the Lead Optimization process.* First European Chemistry Congress, August 2008, Budapest, pp. 27–31.
36 Wong Hawkes, S.Y.F., Chapela, M.J.V. and Montembault, M. (2005) *QSAR and Combinatorial Science*, **24**, 712–21.
37 a. Jones, R., Gödörházy, L., Szalay, D., Gerencsér, J., Dormán, G., Ürge, L. and Darvas, F. (2005) *QSAR and Combinatorial Science*, **24**, 722–7. b. Jones, R., Gödörházy, L., Varga, N., Szalay, D., Ürge, L. and Darvas, F. (2006) *Journal of Combinatorial Chemistry*, **8**, 110–6. c. Desai, B. and Kappe, C.O. (2005) *Journal of Combinatorial Chemistry*, **7**, 641–3. d. Saaby, S., Knudsen, K.R., Ladlow, M. and Ley, S.V. (2005) *Chemical Communications*, 2909–11.
38 Liu, Z., Zhang, J., Chen, C. and Wang, P.G. (2002) *A European Journal of Chemical Biology*, **3**, 348–55. a Nahalka, J., Liu, Z., Chen, X. and Wang, P.G. (2003) *Chemistry – A European Journal*, **9**, 373–7.
39 Baumann, M., Baxendale, I.R., Ley, S.V., Smith, C.D. and Tranmer, G.K. (2006) *Organic Letters*, **8**, 5231–4.
40 Hodge, P. (2005) *Industrial and Engineering Chemistry Research*, **44**, 8542–53.
41 a. Chen, P., Han, S., Lin, G. and Li, Z. (2002) *The Journal of Organic Chemistry*, **67**, 8251–3. b. Sandy, A.J., Petra, D.G.I., Reek, J.N.H., Kramer, P.C.J. and van Leewen, P.W.N.M. (2001) *Chemistry – A European Journal*, **7**, 1202–8.
42 Baxendale, I.R., Deeley, J., Griffiths-Jones, C.M., Ley, S.V., Saaby, S. and Tranmer, G. K. (2006) *Chemical Communications*, 2566–8.
43 Anderson, N.G. (2001) *Organic Process Research and Development*, **5**, 613–21.
44 Biospin, B. Corporation: www.bruker-biospin.com (accessed date: 10-23-2007).
45 Waters Corporation: www.waters.com (accessed date: 10-23-2007).
46 Mettler-Toledo International Inc.: www.us.mt.com (accessed date: 10-23-2007).

17
Overview of Protein-Tannin Interactions

Elisabete Barros de Carvalho, Victor Armando Pereira de Freitas and Nuno Filipe da Cruz Batista Mateus

17.1
Phenolic Compounds

Phenolic compounds are one of the most common groups of plant secondary metabolites, with functions that range from pigmentation to growth and include defense against microbiological attack and predators [1].

There are thousands of phenolic compounds in plants, divided into several classes according to their structures. The simplest structures are C_6, C_6-C_1 benzenic compounds such as benzoic acids (Figure 17.1a), and C_6-C_3-C_6 compounds composed of two benzenic rings (A and B) and a heterocyclic pyranic ring C, characteristic of flavonoids (Figure 17.1c). The more complex are in general polymers of these structures.

The flavonoid group is very diverse and contains several compounds including flavanones, flavones, flavonols, dihydroflavonols, isoflavonoids, anthocyanins, flavan-3,4-diols, flavan-4-ols, and flavan-3-ols. Flavan-3-ols are the structural units of the polymeric compounds termed condensed tannins abundant in plants.

This basic structure of flavan-3-ols can vary according to the degree of hydroxylation of ring B, which can have one (afzlechin/epiafzlechin), two (catechin/epicatechin), or three (gallocatechin/epigallocatechin) hydroxyl groups and to the stereochemistry of the asymmetric carbons in the heterocyclic ring C, yielding (+)-afzelechin/(+)-catechin/(+)-gallocatechin and (−)-epiafzlechin/(−)-epicatechin/(−)-epigallocatechin (Figure 17.1c). The enantiomers with a 2S configuration in carbon-2, such as ent-catechin and ent-epicatechin are relatively rare in nature [2].

Flavan-3-ols can also contain acyl or glycosyl groups [3], the most common being a galloyl group linked to the hydroxyl group in position C3 through an ester linkage.

Tomorrow's Chemistry Today. Concepts in Nanoscience, Organic Materials and Environmental Chemistry.
Edited by Bruno Pignataro
Copyright © 2008 WILEY-VCH Verlag GmbH & Co. KGaA, Weinheim
ISBN: 978-3-527-31918-3

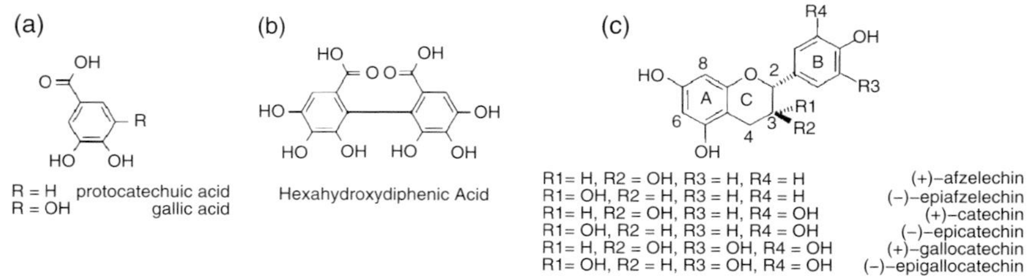

Figure 17.1 Structure of some representative phenolic compounds: (a) benzoic acids, (b) hexahydroxydiphenic acid, (c) flavan-3-ols (flavonoids).

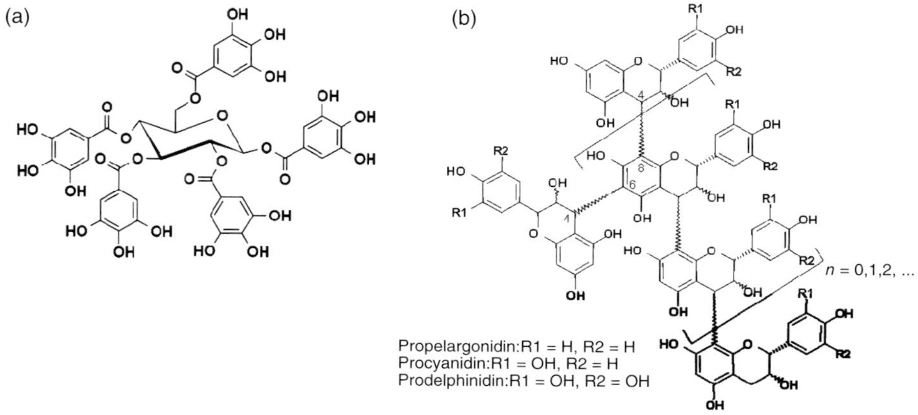

Figure 17.2 Structure of tannins: (a) hydrolyzable tannin (pentagalloylglucose); (b) general structure of condensed tannins.

17.2 Tannin Structures

Tannins are phenolic compounds that have the ability to complex proteins. They are ubiquitous compounds widespread in the plant kingdom, and are present in a wide variety of foodstuffs of plant origin such as fruits and grains, and also in several beverages such as wine, beer, and tea.

Tannins are classically divided into hydrolyzable tannins and condensed tannins according to their chemical structure (Figure 17.2). Hydrolyzable tannins are composed of a polyol, such as glucose, connected by ester linkages to at least one gallic acid (gallotannins) or hexahydroxydiphenic acid (Figure 17.1b) (ellagic tannins). In Figure 17.2a pentagalloylglucose (PGG) is represented as an example of a gallotannin.

Condensed tannins are polymers of flavan-3-ols (Figure 17.2c). Usually flavan-3-ol units are linked through C–C interflavanol bonds established between the C4 of one flavan-3-ol unit and the C8 or C6 of another. There are also tannins with an additional ether linkage between the C2 of the upper unit and the oxygen-bearing C7 or C5 of the lower unit, in addition to the usual C4–C8 or C4–C6 interflavanol bond. Condensed tannins can be found with different degrees of polymerization with two or more units of flavan-3-ols, reaching as high as 80 units of flavanol [4].

When heated in acidic medium, condensed tannins produce anthocyanidins; hence they are also called proanthocyanidins. Depending on the degree of hydroxylation of their ring B extension units, proanthocyanidins can be classified as propelargonidins (monohydroxylated), procyanidins (dihydroxylated), or prodelphinidins (trihydroxylated) since they yield pelargonidin, cyaniding, or delphinidin, respectively, upon acid treatment.

17.2.1
Dietary Burden and Properties of Phenolic Compounds

Phenolic compounds can be found in almost all foodstuffs of vegetable origin. The daily intake of phenolic compounds is highly dependent on the diet. Drinks such as tea and wine are widely consumed tannin-rich beverages. Phenolic compounds constitute from 35% to 56% of solids in green tea and 10% of solids in black tea [5]. Red wines can contain up to $2\,g\,l^{-1}$ of tannins [6]. Other foods with high contents in phenolic compounds include apples, persimmon, banana, dark chocolate, berries, barley, cider, beans and beer. Therefore, depending on the diet, the consumption of phenolic compounds can reach a few grams on a daily basis, particularly from a diet rich in fruits and vegetables [7]. Proanthocyanidins are more common in the diet than hydrolyzable tannins [3].

Phenolic compounds contribute to the organoleptic characteristics of foodstuff, namely color and flavor. In fact, anthocyanins (phenolic pigments) contribute importantly to the majority of red and blue colors observed in plants. In wines, these pigments can interact with proanthocyanidins, resulting in the formation of new anthocyanin-derived pigments (such as the orange-red anthocyanin pyruvic acid adducts and the blue portisins) and leading to changes in their color characteristics [8].

Phenolic compounds also have important antioxidant properties, protecting food from oxidation [9]. The antioxidant properties of phenolic compounds can have an impact on human health and they are regarded as having a protective effect against low-density lipoprotein (LDL) oxidation and cardiovascular diseases [3]. Recently phenolic compounds have been widely used in cosmetic preparations to delay aging [10].

Phenolic compounds also contribute directly to the flavor due to their astringency and bitter taste characteristics [11]. In fact, astringency is believed to be due to the interaction between tannins and salivary proteins, resulting in the formation of protein–tannin aggregates in the mouth, as discussed in more detail below [12–15].

17.3
Interactions between Proteins and Tannins

One of the most important properties of tannins is precisely their ability to interact with proteins. In fact, the term "tannin" is typically employed to designate the substances of vegetable origin able of transform fresh skin into leather, thereby interacting and precipitating proteins of the animal skin.

It is believed that proteins and tannins interact via hydrogen bonding and hydrophobic effects. Hydrogen bonds can be formed between the hydroxyl groups of phenolic compounds and carbonyl and amide groups of proteins. Hydrophobic interaction can occur between the benzenic nuclei of phenolic compounds and the apolar side-chains of amino acids such as leucine, lysine, or proline in proteins. Several authors have observed the occurrence of hydrophobic interactions between proteins and tannins [16–20].

The proline residue is generally considered to be a good binding site as it provides a flat, rigid, hydrophobic surface that is favorable to interactions with other planar hydrophobic surfaces such as benzenic rings. Effectively, NMR studies have shown the existence of hydrophobic stacking between the benzenic ring of several phenolic compounds and the apolar face of the proline ring of peptides rich in this amino acid [21–23]. These studies have indicated that the principal interactions sites are proline residues, and that a particularly favored site is the first proline residue of a Pro-Pro sequence.

However, other studies seem to suggest that the driving force of protein–tannin interactions is mainly hydrogen bonding [24, 25]. Simon et al. [25], using a dimeric procyanidin and a proline-rich peptide, have observed by NMR and molecular modeling studies that procyanidins bind to the hydrophilic side of the proline-rich peptide, suggesting the prevalence of hydrogen bonding. These authors have observed no hydrophobic stacking by NMR, in contrast to several studies using similar systems of procyanidins and proline-rich peptides [19–22]. However, they attribute the apparently contradictory results to the use of different solvent systems.

In fact, it seems that both type of interactions are present, strengthening each other, depending on the protein and tannin structure and medium conditions [17, 26, 27].

The affinity of tannins to bind proteins is favored by their ability to work as multidentate ligands in which one tannin is able to bind to more than one point in the same protein structure or bind to more than one protein at a time [22]. Proteins can also wrap around tannins [16], and the tannins have the ability to autoassociate with other tannins, forming stacks, even when they are bound to proteins [21].

17.4
Experimental Studies of the Interactions between Proteins and Tannins

Protein–tannin interactions have been studied using several techniques, such as NMR [18, 22, 23, 25], microcalorimetry [28, 29], enzyme inhibition [30–32], microscopy techniques [16, 33, 34], radioactivity measurements [35], high-performance liquid chromatography (HPLC) [15, 36], fluorescence [37], electrophoresis [38–40], electrospray ionization mass spectrometry (ESI-MS) [25, 41], infrared spectroscopy [42], marker utilization [43] or molecular modeling [25].

However, the most intuitive way of measuring aggregate formation in solution is the measurement of the light scattered by "protein–tannin" particles. In fact nephelometry and other light-scattering measurement techniques have been widely used to study protein–tannin interactions [33, 34, 44–49].

Some studies have been made relating specifically to astringency. Some of these studies have focused directly on interactions between tannins and salivary proteins [21, 23, 25, 46, 50–53], and on the changes in saliva protein composition after interaction with tannins [15, 38, 54, 55]. Other studies have correlated the sensorial astringency with protein–tannin interactions [45, 56, 57]. In fact, the astringency felt when sampling different tannin solutions can be correlated with the ability of these tannins to precipitate proteins. This effect has been observed with several proteins such as mucin [45], ovalbumin and gelatin [57], bovine serum albumin (BSA), and salivary proteins [58].

An electronic tongue based on protein–tannin interactions has been developed to measure astringency [42].

Apart from salivary proteins, other proteins have been used in studies of kind: mucin [56], gliadin [49], β-glycosidase [32, 59], hemoglobin [60], and BSA [24, 35–37, 48, 61–64]. Other proteins have been used because of certain characteristics that make them similar to proline-rich proteins (PRPs), like casein [33, 65], gelatin [42, 49, 65–67], and polyproline [34, 66, 68]. Although it is not a protein, the polymer polyvinylpolypyrrolidone has also been used in these studies [68, 69].

Concerning the phenolic compounds used in these studies, procyanidins have been widely used, either as pure compounds [19, 25, 26, 41, 54, 66], or as a mixture of several procyanidins [38, 42, 45, 50, 68, 70] Other studies used simple flavanols [24, 27, 34, 37, 43, 49], particularly (−)-epigallocatechin gallate (EGCG), a major component of tea [16, 18, 51, 61, 71]. Hydrolyzable tannins have also been used in these studies [29, 48, 49, 52, 61], particularly PGG [21, 23, 27, 47, 65, 71–74].

17.4.1
Nephelometric Studies of BSA and Condensed Tannin Aggregation

The interaction between condensed tannins and BSA, a model protein widely used in these studies, can be determined directly in solution by measuring the amount of tannin/BSA aggregates, using the nephelometric technique. When increasing concentrations of BSA are added to a condensed tannin solution, a typical behavior can be observed: the tannin/BSA aggregates dispersed in solution increase with the addition of protein up to a maximum, from which no further precipitation occurs. From this point, the progressive addition of BSA causes a decrease in aggregation (Figure 17.3).

This behavior has been observed by other authors with gelatin–PGG and gelatin–tannic acid systems [49, 65]. A conceptual model has been proposed by

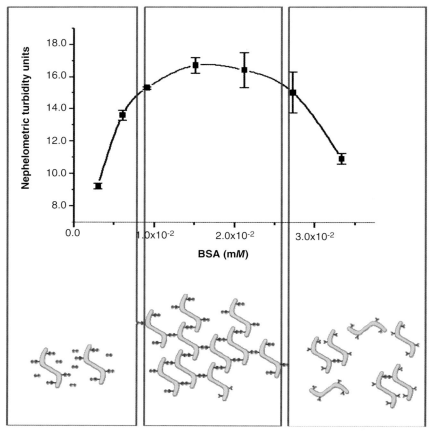

Figure 17.3 Influence of protein concentration on protein–tannin aggregation. Top: Influence of BSA on the aggregation with oligomeric grape seed procyanidins 0.10 mM (in 12% aqueous ethanol, 0.1 M acetate buffer, pH 5.0) [78]. Bottom: Conceptual model to explain the influence of protein concentration on aggregation, based on Siebert et al. [49, 75]. The gray "S" shapes represents proteins with a number of tannin binding places, and the black arrows represent tannins with protein binding sites.

Siebert et al. [49, 75] to explain this behavior (Figure 17.3). Proteins are supposed to have a number of sites to which tannins can bind; tannins are supposed to have a number of ends that can bind to proteins. As the concentration of protein increases, the tannin ends start to occupy the protein binding sites, forming aggregates. When the concentration of tannin ends is equal to the number of protein binding sites, a large crosslinked network is formed, corresponding to the largest complexes and resulting in maximum light scattering. When a large excess of protein is present, each tannin molecule should be able to bridge some protein molecules, but it is unlikely that there would be sufficient excess tannins to crosslink between these complexes. Therefore, an excess of protein would result in small aggregates and less light scattering.

This does not seem to occur in all cases. Some authors have observed that with an increase in tannin/protein ratio there is the formation of large particles that eventually precipitate. A three-phase model has been proposed to explain this phenomenon [33, 51]. The simultaneous binding of the multidentate tannin to several places in the protein leads to the protein wrapping around the tannins. As the concentration of tannin increases, several tannin molecules bind to the protein surface and crosslink with other proteins, leading to protein association. As more tannin is added, more protein–tannin complexes aggregate, forming larger particles that precipitate.

17.5
Factors That Influence the Interactions between Proteins and Tannins

Apart from the relative concentrations of protein and tannin, protein–tannin interactions are influenced by several other factors.

17.5.1
Structural Features

Protein–tannin interactions are affected by the type of tannin involved and their structure [18, 27]. In general, the tannins' affinity to complex proteins increases with the molecular weight and the degree of galloylation, probably because they have more interaction sites [21, 34, 36, 43, 50, 76]. For condensed tannins, the type of interflavanol bond (C4–C6 or C4–C8) also influences their interaction with proteins, with dimers C4–C8 showing higher affinities toward proteins than the corresponding dimers C4–C6, possibly due to conformational constraints imposed by C4–C6 linkages [50]. It has been suggested that more flexible tannins have better ability to bind proteins because they are more efficient crosslinkers [29, 50]. For the bigger structures of condensed tannins (>3400 Da), despite the high number of potential interaction sites there is a reduction in protein affinity, which could be explained by the decrease in their flexibility [46, 59].

Protein structure also influences the interaction. Apparently the binding between proteins and tannins is made in a selective and specific manner [68]. The interaction can be affected by the size of the protein [68, 70], its charge [68], the presence of side-chains [70], and its conformation [18].

As already indicated, the presence of proline is apparently a common characteristic of proteins with high affinities toward tannins [68]. Proline residues, apart from being binding sites, are also useful in maintaining the peptide in extended conformation, providing more surface of the protein for binding [21].

17.5.2
pH and Ionic Strength

Another factor that influences protein–tannin interactions is pH. Apparently there is higher protein precipitation at pH values close to the protein isoelectric point (pI) [27, 64, 77] where protein–protein electrostatic repulsions are minimized [68]: proteins with acidic pI have higher affinities for complexation with tannins at lower pH, whereas basic proteins aggregate preferentially at higher pH [40].

An interesting study by Charlton *et al.*, with EGCG and a basic PRP fragment of 19 amino acids, has shown that pH greatly affects precipitation of aggregates but does not change the protein–tannin binding affinities [51]. These authors suggested that protein–tannin particles are in a colloidal state, stabilized by repulsions of proteins of the same charge. When this charge is diminished (at pH = pI), there is no more repulsion and particles aggregate together and precipitate. So, although the initial binding is due to specific protein–tannin interactions, the aggregation process seems to be due mainly to superficial charge effects [51].

This could also explain the influence of ionic strength in protein–tannin interactions. Figure 17.4 shows the influence of ionic strength on the interactions between BSA and procyanidins [78]. The precipitation of BSA by tannins has been shown to be very sensitive to ionic strength, decreasing regularly with the increase in ionic strength. Effectively, aggregation decreased about 80% when the ionic strength increased from 70 to 500 mM. These results suggested an important contribution of hydrogen bonding to the interaction between BSA and procyanidins under the conditions studied.

This decrease in aggregation could be due to ion adsorption on the tannin or protein surface restraining the interaction, or to adsorption of ions onto the surface of the protein or protein–tannin complex leading to higher solvation.

In fact Hagerman *et al.* have proposed that while the interaction between BSA and hydrolyzable tannins (more hydrophobic than procyanidins) is governed by

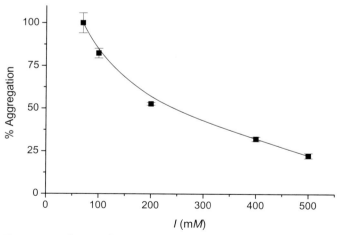

Figure 17.4 Influence of ionic strength on the aggregates formation between BSA (6.1×10^{-3} mM) and oligomeric procyanidins from grape seed (0.10 mM) (in 12% aqueous ethanol, 0.1 M acetate buffer, pH 5.0) [78].

hydrophobic effects, the interaction with procyanidins seems to be due to hydrogen bonding [27].

Rawel et al. [79] have observed a similar effect, as the increase in ionic strength led to a decrease in the interaction between BSA and quercetin. These authors have suggested that the presence of salt in solution leads to alterations in the BSA structure, exposing different amino acid residues that have different abilities to interact with quercetin. This outcome seems to contradict some results by other authors who reported that the increase of the ionic character of the medium decreases the protein–tannin solubility and improves precipitation. However, these studies were focused on the interaction between BSA and more hydrophobic tannins such as PGG [65, 77]. The authors suggested that the hydrophobic interactions could be enhanced in the presence of salt due to a salting-out effect improving precipitation [65, 77, 80].

Protein–tannin interactions also seem to be affected by the presence of ethanol [2, 62] and other compounds in solution [65] and by temperature [2, 77].

17.5.3
Influence of Polysaccharide on the Interactions between Protein and Tannin

Protein–tannin interactions appear to be inhibited by the presence of some carbohydrates in solution [32, 65, 78, 81, 82]. This is particularly evident in the graph of Figure 17.5.

Two mechanisms have been proposed to explain the influence of polysaccharides on protein–tannin interactions [2, 65] (Figure 17.6):
(1) Polysaccharides, being polyelectrolytes, could form ternary complexes with the protein–tannin aggregate, enhancing its solubility in aqueous medium.
(2) Polysaccharides could encapsulate tannins, interfering with their ability to bind and precipitate proteins.

The neutral carbohydrates, such as glucose, β-cyclodextrin, or arabinogalactan needed to be present in much higher amounts than the other carbohydrates to inhibit the aggregation. This low affinity of neutral carbohydrates for complexation of tannins has been reported by other authors [65, 83].

The ionic character of pectin, xanthan, polygalacturonic acid, and gum arabic is probably at the origin of their higher effectiveness in preventing protein–tannin aggregation. The highest decrease in the percentage of aggregates occurs with xanthan and polygalacturonic acid, which have a higher ionic character. In fact, xanthan is a heteroglycan composed of glucose, mannose, and glucuronic acid, but many of the mannose residues are pynivated contributing to the molecular charge. Polygalacturonic acid consists of a polymer of galacturonic acid molecules. In the case of pectin, the galacturonic structural units have the carboxyl groups partially esterified with methanol, which reduces its polarity.

However, another explanation for the differences in behavior between carbohydrates could be their ability to develop a gel-like structure in solution able to

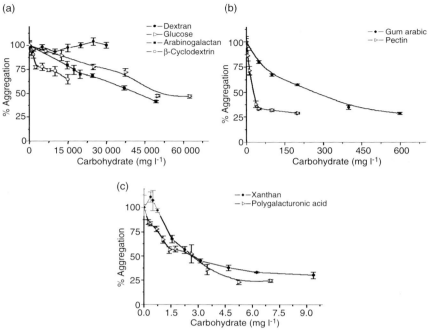

Figure 17.5 Influence of carbohydrate concentration on the formation of aggregates between BSA (6.1×10^{-3} mM) and oligomeric procyanidins from grape seed (0.10 mM) (in 12% aqueous ethanol, 0.1 M acetate buffer, pH 5.0). [78].

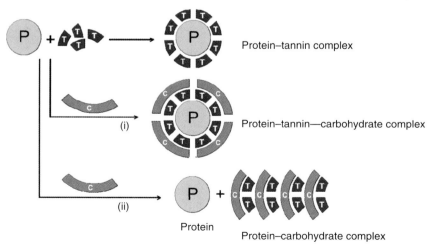

Figure 17.6 Possible mechanisms (i and ii) involved in the inhibition of the aggregation of tannins (T) and proteins (P) by carbohydrates (C) [82].

encapsulate tannins and prevent their interaction with proteins, as was suggested for xanthan and polygalacturonic acid, according to mechanism (ii) of Figure 17.6 [2, 32].

17.6
Flow Nephelometric Analysis of Protein–Tannin Interactions

The flow nephelometric technique is a very fast and reliable method for studying the interaction between tannins and proteins (Figure 17.7) [84]. The possibility of changing the flow and changing the gradient of the different substrates involved is very convenient and versatile and the technique can be applied to complex samples such as different phenolic extracts and beverages (red and white wines, fruit juices, beers).

The tannin specific activity (TSA) of phenolic extracts and beverages can be directly determined from different sample flows without previous dilutions or any other treatment.

Figure 17.8 shows a typical flow nephelometric graphic involving the interactions between three substrates: protein, tannin, and carbohydrate. In an initial phase there is an increase in turbidity corresponding to the stabilization of protein and tannin concentrations. The maximum value of turbidity corresponds to the TSA of the tannin solution in relation to the protein (BSA). In a second phase a gradient of increasing carbohydrate concentration is started, leading to a decrease

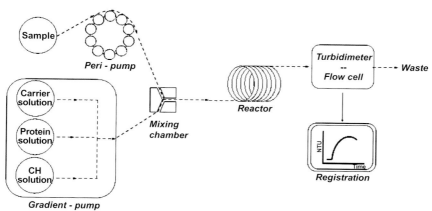

Figure 17.7 Schematic of the flow-nephelometric apparatus. The apparatus is composed of a liquid pump system that has the ability to form a gradient of carbohydrate concentration at a constant flow rate using three solutions: A, eluent; B, protein in A; C, carbohydrate in A. This solution is brought into contact in a mixing valve with the tannin solution (sample) pumped by a peristaltic pump at a constant flow rate. The mixture passes through a reactor and a flow cell, where the online detection of protein–tannin aggregates is carried out in a laboratory turbidimeter detector coupled to a PC equipped with software for data acquisition [81].

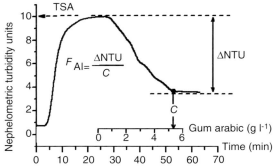

Figure 17.8 Influence of an increasing gradient of gum arabic concentration on the interactions between BSA (5×10^{-3} mM) and a fraction of procyanidins from grape seed (gallate tetramers and pentamers, 0.034 g l^{-1}) (in 12% aqueous ethanol, 0.1 M acetate buffer, pH 5.0) [81].

Table 17.1 Aggregation inhibitor factor (F_{Al}) of the interaction between BSA (5×10–3 mM) and procyanidins of different molecular weights (in 12% aqueous ethanol, 0.1 M acetate buffer, pH 5.0), at the concentrations indicated under [Fraction], chosen to obtain similar NTU values (around 12 NTU) [81].

Procyanidin fraction	MW range	[Fraction] (g l^{-1})	F_{Al} (NTU g l^{-1})
I	300–600	0.73	129 ± 8
II	600–900	0.37	127 ± 5
III	1000–1500	0.13	123 ± 9
IV	1500–1600	0.034	86 ± 13
V	1600–3500	0.050	60 ± 12

in protein–tannin aggregation. A factor of inhibition of aggregation (F_{Al}) can be calculated by dividing the change of aggregation (ΔNTU) by the carbohydrate concentration at the point of maximum inhibition.

As an example of the application of this technique, the effect of xanthan on the interaction between several procyanidin fractions (with different molecular weights) and BSA was studied (Table 17.1). Different concentrations of the different condensed tannin fractions were chosen to obtain similar turbidity values (around 12 NTU) [44]. The concentration of procyanidin decreased with the increase in molecular weight as a result of the stronger reactivity of the higher-molecular-weight condensed tannins vis-à-vis BSA. Fraction V required a slightly higher concentration than fraction IV probably because the structure is so large that it may lose some conformational mobility, which would hinder the interaction with BSA [46, 81]. Regarding the effect of the carbohydrate F_{Al}, procyanidin fractions could be divided in two groups: xanthan affects the binding of fractions I–III toward BSA more effectively than that of fractions IV and V. These two groups showed significant differences between them ($P < 0.01$). These results point to a higher efficiency of xanthan in the inhibition of tannin–BSA aggregation for condensed tannin fractions that have lower molecular weights. This is in agreement with similar nephelometric studies in batch [82]. It seems that the smaller phenolic compounds are more easily encapsulated in xanthan pores than larger ones, reducing their ability to complex BSA.

17.7
Interactions of Tannins with Salivary Proteins – Astringency

Astringency is a sensation that can be perceived in a variety of foods and beverages such as unripe fruits, red wines, teas, and beers. The sensation is usually defined as an array of perceptions such as dryness (lack of lubrication or moisture resulting in friction between oral surfaces), roughness (harsh texture of the oral cavity marked by edges and projections that are felt when oral surfaces contact each other), and constriction (feeling of puckering and contraction felt in the mouth lips and cheeks) resulting from the exposure to substances such as tannins [85, 86]. It is widely accepted that astringency results from the interaction between tannins and salivary proteins, resulting in the formation of protein–tannin aggregates in the mouth [12–15].

Astringency is often perceived as a negative attribute of tannin-rich products. However, several factors influence the preference for these products, particularly social factors, because astringent products such as tea, wine, and coffee have a "social" context of consumption associated with them, which may eventually lead to a preference for these products [11].

Astringency is one of the most important attributes of a red wine. High-quality red wines have balanced levels of astringency: they should not have an excess that could mask other wine characteristics and make it harsh and dry, and they should not have a level so low that could lead the wine to be flat and uninteresting [87].

The sensation of astringency is felt differently by tasters [86] probably due to differences in individuals' saliva in terms of its protein composition [11, 47]. Astringency is also affected by the structure of the phenolic compounds [88], pH [89], the presence of other substances [90–92], and viscosity [11]. In fact it is believed that complex beverages such as wine and beer have subtle sub-qualities of sensorial descriptors related to astringency (soft, grainy, harsh, green, chalky, etc.), that are not perceived in tannin model solutions, which could be due to the presence of other molecules [87].

Saliva is produced by two groups of glands: parotid glands and submandibular/sublingual glands. Submandibular/sublingual saliva is produced in nonstimulated conditions (between meals) and its major organic component is mucins, proteins that have lubricatory properties. Parotid saliva is secreted in stimulated conditions and is mainly composed of 30% α-amylase and 70% PRPs with high levels of proline (35–40%) [93]. PRPs can be divided into three major groups: acidic, basic, and glycosylated proteins [94].

Some studies have shown that human α-amylase is inhibited [31, 95] or precipitated by tannins [46]. However, this enzyme can be protected from inhibition by tannins in the presence of other salivary proteins, such as histatins or PRPs, which is generally the case [40].

Salivary histatins have a high ability to precipitate tannins [18, 40, 71]. Histatins are histidine-rich proteins that can be found in both sublingal/submandibular saliva and parotid saliva and represent about 2.6% of the total salivary proteins. It has been proposed that histatins could be important in neutralizing tannins in situations where they come not from the diet but from other situations when the saliva is not stimulated, such as when cotton workers breathe tannin-rich dust [18].

All salivary PRPs have shown affinity for dietary tannins [43] and are easily precipitated [15, 40]. Basic PRPs seem to be more effective in precipitating tannins when compared to acidic and glycosylated PRPs [96].

A group of 11 proteins (IB1-IB7, IB8a, b and c and IB9) can be found among the family of basic PRPs having very similar sequences and displaying repetitive patterns, in which some sequences of amino acids can be found several times in the same protein and in different proteins [97] (Figure 17.9). It has been proposed that the presence of multiple repeat regions rich in proline could provide sites favorable for tannin binding, and flexible hinges on the protein allow it to fold and "wrap around" the tannin, thereby increasing the association by cooperative intermolecular interactions [98]. PRPs have been suggested to work as tannin sponges [25] and it is believed that PRPs could act as a primary line of defense against possible deleterious effects of tannin consumption [99]. Effectively, it was observed that basic PRPs have the ability to decrease the transport of phenolic compounds across intestinal cells by forming insoluble complexes [100], specially with the bigger structures. This suggested that salivary proteins prevent the transport of large phenolic compounds that could be harmful, but allow the passage of small phenolics that could have antioxidant effects [101].

Some recent evidences have shown that a basic PRP binds condensed tannins much more effectively than α-amylase. In Figure 17.10, it can be seen that the light-scattering intensity caused by aggregation increases with the concentration of procyanidins until a maximum after which the light scattered intensity remains constant even with a further increase in tannin concentration.

It can be observed that, for similar amounts of α-amylase and IB8c (0.4 and 0.5 µM, respectively), the amount of tannin bound by IB8c at the point of maximum

^{1}SPPGK**PQGPPP**QGGNQ**PQGPPP**
PPGK**PQGPPP**QGGNK**PQGPPP**
PGK**PQGPPP**QGGSKSRSA^{61}

Figure 17.9 Complete amino acid sequence of the salivary basic PRP IB8c, aligned to indicate the recurring motifs.

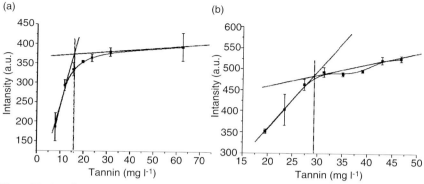

Figure 17.10 Influence of the concentration of procyanidins (gallate tetramers and pentamers) on the aggregation with two salivary proteins: (a) α-amylase (0.4 µM) and (b) IB8c (0.5 µM) in 12% aqueous ethanol, 3.1 mM acetate buffer, pH 5.0 [104].

aggregation (corresponding to the interception point of the two curves) was almost double, even though this protein is about 10 times smaller than α-amylase. This difference can be explained by the 3D structure of the proteins: α-amylase is a globular protein, and IB8c is likely to be adopt an extended random coil conformation with some degree of type II helix, like the congener IB7 [25], which would allow the protein to offer more contact sites to interact with proteins. This finding corroborates the theory of basic PRPs working as tannin sponges. However, α-amylase seems to be more specific and selective than PRPs in the aggregation with samples containing different amounts of procyanidins [53].

17.8
Polysaccharides and Astringency

As already mentioned, polysaccharides inhibit protein–tannin interactions. This inhibition has been proposed to contribute to the loss of astringency during ripening of some fruits. In fact, as the cellular structure softens during fruit ripening there is an increase in soluble pectin fragments that inhibit salivary protein–tannin interactions in the mouth, leading to a decreased astringent response [32, 65, 90].

In tannin-rich beverages like wine, polysaccharides are also present and are expected to affect the mouthfeel properties and mainly astringency. Major wine polysaccharides are arabinogalactan proteins (AGP) and rhamnogalacturonan II (RGII), pectic polysaccharides which originate from grape cell walls, and mannoproteins (MP) produced by yeast during wine fermentation [102].

Vidal *et al.* have demonstrated that procyanidin astringency decreases in the presence of RGII, while MPs and AGPs decreases bitterness [103].

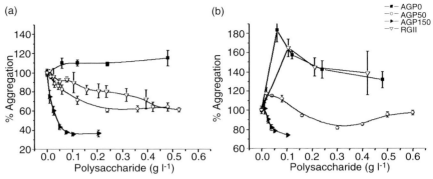

Figure 17.11 Influence of polysaccharide concentration on aggregate formation between procyanidins 15.6 mg l^{-1} and α-amylase 0.4 μM (a) or procyanidins 31.2 mg l^{-1} and IB8c 0.5 μM (b) in 12% aqueous ethanol, 3.1 mM acetate buffer, pH 5.0, (where 100% refers to intensity at zero polysaccharide concentration) [104]. The procyanidins used correspond to a mixture of gallate tetramers and pentamers.

Carvalho et al. [104] have studied the influence of wine polysaccharides on salivary protein–tannin interactions (Figure 17.11). Polysaccharides (AGP, RGII, and MP) were isolated from wine, α-amylase was isolated from saliva, and the peptide IB8c was chemically synthesized. The results showed that the most acidic fractions of AGPs have the ability to inhibit the formation of aggregates between condensed tannins and the two different salivary proteins. RGII has the same ability toward α-amylase, but not for IB8c under the conditions studied. The most neutral fraction of AGP seemed to have the reverse effect on the aggregation between protein and tannin, showing an increase in the light scattered by the protein–tannin aggregates. RGII also favored the formation of aggregates between IB8c and tannin. This could be due to a co-aggregation of polysaccharide with protein and tannin complexes, forming larger aggregates.

Some MP fractions obtained from wine also had the ability to inhibit protein–tannin aggregation (unpublished results). Polysaccharides showed effects at concentrations at which they are present in wine, which means that they could have an influence on wine astringency.

As described above, astringency is a very complex sensation that occurs in several foodstuff and beverages, involving different kinds of compounds (phenolic compounds, proteins), and that is affected by medium conditions (pH, alcohol, presence of other molecules in solution, etc.). This sensation constitutes one of the main organoleptic properties of red wine and has been a matter of research over the years for wine chemists. Winemakers believe that polysaccharides contribute to the "mouthfeel volume" of a wine, but this is a rather empirical observation.

The results obtained directly from compounds isolated from wine and directly involved in the astringency sensation (tannins, polysaccharides, and human salivary proteins) highlight the importance of wine polysaccharides in reducing astringency, even in wines with a high tannin content.

Studies performed over the years have contributed to better understanding of the interactions between proteins and tannins, which are important not only due to their astringency but also because of their impact on food nutritional characteristics, on human health, and on plant metabolism. It is clear that protein–tannin interactions are influenced by several factors, among which polysaccharides could be important because they are also present in tannin-rich vegetables. Much remains to be studied in this field, particularly the specific phenomenon that occurs between proteins, tannins, and polysaccharides that leads to a decrease in aggregation, and further studies are needed involving other salivary proteins and digestive enzymes.

17.9
Acknowledgments

Project funding by PTDC/AGR-ALI/67579/2006. E.C. was supported by a grant from Fundação para a Ciência e a Tecnologia (SFRH/BD/9325/2002).

References

1 Naczk, M. and Shahidi, F. (2006) *Journal of Pharmaceutical and Biomedical*, **41**, 1523–42.
2 Haslam, E. (1998) *Pratical Polyphenolics: From Structure to Molecular Recognition and Physiological Action*, Cambridge University Press.
3 Santos-Buelga, C. and Scalbert, A. (2000) *Journal of the Science of Food and Agriculture*, **80**, 1094–117.
4 Souquet, J.M., Cheynier, V., Brossaud, F. and Moutounet, M. (1996) *Phytochemistry*, **43**, 509–12.
5 Bennick, A. (2002) *Critical Reviews in Oral Biology and Medicine: An Official Publication of the American Association of Oral Biologists*, **13**, 184–96.
6 Sun, B.S. and Spranger, M.I. (2005) *European Food Research and Technology*, **221**, 305–12.
7 Scalbert, A. and Williamson, G. (2000) *The Journal of Nutrition*, **130**, 2073S–2085S.
8 de Freitas, V. and Mateus, N. (2006) *Environmental Chemistry Letters*, **4**, 175–83.
9 Moure, A., Cruz, J.M., Franco, D., Dominguez, J.M., Sineiro, J., Dominguez, H., Nunez, M.J. and Parajo, J.C. (2001) *Food Chemistry*, **72**, 145–71.
10 Macheix, J., Fleuriet, A. and Jay-Allemand, C. (2005) *Les Composés Phénoliques des Végétaux*, PPUR.
11 Lesschaeve, I. and Noble, A.C. (2005) *The American Journal of Clinical Nutrition*, **81**, 330S–335S.
12 Green, B.G. (1993) *Acta Psychologica*, **84**, 119–25.
13 de Wijk, R.A. and Prinz, J.F. (2005) *Food Quality and Preference*, **16**, 121–9.
14 De Wijk, R.A. and Prinz, J.F. (2006) *Journal of Texture Studies*, **37**, 413–27.
15 Kallithraka, S., Bakker, J. and Clifford, M.N. (1998) *Journal of Sensory Studies*, **13**, 29–43.
16 Jobstl, E., Howse, J.R., Fairclough, J.P.A. and Williamson, M.P. (2006) *Journal of Agricultural and Food Chemistry*, **54**, 4077–81.
17 Oh, H.I., Hoff, J.E., Armstrong, G.S. and Haff, L.A. (1980) *Journal of Agricultural and Food Chemistry*, **28**, 394–8.

18 Wroblewski, K., Muhandiram, R., Chakrabartty, A. and Bennick, A. (2001) *European Journal of Biochemistry / FEBS*, **268**, 4384–97.
19 Hatano, T. and Hemingway, R.W. (1996) *Chemical Communications*, 2537–8.
20 Hatano, T., Yoshida, T. and R.W. (1999) Hemingway, in *Plant Polyphenols 2: Chemistry, Biology, Pharmacology, Ecology*, Plenum Publishers.
21 Baxter, N.J., Lilley, T.H., Haslam, E. and Williamson, M.P. (1997) *Biochemistry*, **36**, 5566–77.
22 Charlton, A.J., Haslam, E. and Williamson, M.P. (2002) *Journal of the American Chemical Society*, **124**, 9899–905.
23 Murray, N.J., Williamson, M.P., Lilley, T.H. and Haslam, E. (1994) *European Journal of Biochemistry / FEBS*, **219**, 923–35.
24 Frazier, R.A., Papadopoulou, A. and Green, R.J. (2006) *Journal of pharmaceutical and Biomedical*, **41**, 1602–5.
25 Simon, C., Barathieu, K., Laguerre, M., Schmitter, J.M., Fouquet, E., Pianet, I. and Dufourc, E.J. (2003) *Biochemistry*, **42**, 10385–95.
26 Artz, W.E., Bishop, P.D., Dunker, A.K., Schanus, E.G. and Swanson, B.G. (1987) *Journal of Agricultural and Food Chemistry*, **35**, 417–21.
27 Hagerman, A.E., Rice, M.E. and Ritchard, N.T. (1998) *Journal of Agricultural and Food Chemistry*, **46**, 2590–5.
28 Beart, J.E., Lilley, T.H. and Haslam, E. (1985) *Phytochemistry*, **24**, 33–8.
29 Frazier, R.A., Papadopoulou, A., Mueller-Harvey, I., Kissoon, D. and Green, R.J. (2003) *Journal of Agricultural and Food Chemistry*, **51**, 5189–95.
30 Fickel, J., Pitra, C., Joest, B.A. and Hofmann, R.R. (1999) *Comparative Biochemistry and Physiology. C*, **122**, 225–9.
31 Kandra, L., Gyemant, G., Zajacz, A. and Batta, G. (2004) *Biochemical and Biophysical Research Communications*, **319**, 1265–71.
32 Ozawa, T., Lilley, T.H. and Haslam, E. (1987) *Phytochemistry*, **26**, 2937–42.
33 Jobstl, E., O'Connell, J., Fairclough, J.P.A. and Williamson, M.P. (2004) *Biomacromolecules*, **5**, 942–9.
34 Poncet-Legrand, C., Edelmann, A., Putaux, J.L., Cartalade, D., Sarni-Manchado, P. and Vernhet, A. (2006) *Food Hydrocoll*, **20**, 687–97.
35 Hagerman, A.E. and Butler, L.G. (1980) *Journal of Agricultural and Food Chemistry*, **28**, 944–7.
36 Kawamoto, H., Nakatsubo, F. and Murakami, K. (1996) *Phytochemistry*, **41**, 1427–31.
37 Papadopoulou, A., Green, R.J. and Frazier, R.A. (2005) *Journal of Agricultural and Food Chemistry*, **53**, 158–63.
38 Gambuti, A., Rinaldi, A., Pessina, R. and Moio, L. (2006) *Food Chemistry*, **97**, 614–20.
39 Papadopoulou, A. and Frazier, R.A. (2004) *Trends in Food Science and Technology*, **15**, 186–90.
40 Yan, Q.Y. and Bennick, A. (1995) *The Biochemical Journal*, **311**, 341–7.
41 Sarni-Manchado, P. and Cheynier, V. (2002) *Journal of Mass Spectrometry*, **37**, 609–16.
42 Edelmann, A. and Lendl, B. (2002) *Journal of The American Chemical Society*, **124**, 14741–7.
43 Bacon, J.R. and Rhodes, M.J.C. (1998) *Journal of Agricultural and Food Chemistry*, **46**, 5083–8.
44 Chapon, L. (1993) *Journal of the Institute of Brewing*, **99**, 49–56.
45 Condelli, N., Dinnella, C., Cerone, A., Monteleone, E. and Bertuccioli, M. (2005) *Food Quality and Preference*.
46 de Freitas, V. and Mateus, N. (2002) *Journal of the Science of Food and Agriculture*, **82**, 113–9.
47 Horne, J., Hayes, J. and Lawless, H.T. (2002) *Chemical Senses*, **27**, 653–9.
48 Lin, H.C., Chen, P.C., Cheng, T.J. and Chen, R.L.C. (2004) *Analytical Biochemistry*, **325**, 117–20.
49 Siebert, K.J., Troukhanova, N.V. and Lynn, P.Y. (1996) *Journal of Agricultural and Food Chemistry*, **44**, 80–5.

50 de Freitas, V. and Mateus, N. (2001) *Journal of Agricultural and Food Chemistry*, **49**, 940–5.
51 Charlton, A.J., Baxter, N.J., Khan, M.L., Moir, A.J.G., Haslam, E., Davies, A.P. and Williamson, M.P. (2002) *Journal of Agricultural and Food Chemistry*, **50**, 1593–601.
52 Bacon, J.R. and Rhodes, M.J.C. (2000) *Journal of Agricultural and Food Chemistry*, **48**, 838–43.
53 Mateus, N., Pinto, R., Ruao, P. and de Freitas, V. (2004) *Food Chemistry*, **84**, 195–200.
54 Kallithraka, S., Bakker, J. and Clifford, M.N. (2001) *Journal of the Science of Food and Agriculture*, **81**, 261–8.
55 Sarni-Manchado, P., Cheynier, V. and Moutounet, M. (1999) *Journal of Agricultural and Food Chemistry*, **47**, 42–7.
56 Monteleone, E., Condelli, N., Dinnella, C. and Bertuccioli, M. (2004) *Food Quality and Preference*, **15**, 761–9.
57 Llaudy, M.C., Canals, R., Canals, J.M., Rozes, N., Arola, L. and Zamora, F. (2004) *Journal of Agricultural and Food Chemistry*, **52**, 742–6.
58 Troszynska, A., Amarowicz, R., Lamparski, G., Wolejszo, A. and Barylko-Pikielna, N. (2006) *Food Quality and Preference*, **17**, 31–5.
59 Haslam, E. (1974) *The Biochemical Journal*, **139**, 285–8.
60 Bate-Smith, E.C. (1973) *Phytochemistry*, **12**, 907–12.
61 Hatano, T., Hori, M., Hemingway, R.W. and Yoshida, T. (2003) *Phytochemistry*, **63**, 817–23.
62 Serafini, M., Maiani, G. and FerroLuzzi, A. (1997) *Journal of Agricultural and Food Chemistry*, **45**, 3148–51.
63 McManus, J.P., Davis, K.G., Beart, J.E., Gaffney, S.H., Lilley, T.H. and Haslam, E. (1985) *J Chem Soc Perkin Trans*, **2**, 1429–38.
64 Naczk, M., Oickle, D., Pink, D. and Shahidi, F. (1996) *Journal of Agricultural and Food Chemistry*, **44**, 2144–8.
65 Luck, G., Liao, H., Murray, N.J., Grimmer, H.R., Warminski, E.E., Williamson, M.P., Lilley, T.H. and Haslam, E. (1994) *Phytochemistry*, **37**, 357–71.
66 Ricardo da silva, J.M., Cheynier, V., Souquet, J.M., Moutounet, M., Cabanis, J.C. and Bourzeix, M. (1991) *Journal of the Science of Food and Agriculture*, **57**, 111–25.
67 Calderon, P., Vanburen, J. and Robinson, W.B. (1968) *Journal of Agricultural and Food Chemistry*, **16**, 479–482.
68 Hagerman, A.E. and Butler, L.G. (1981) *The Journal of Biological Chemistry*, **256**, 4494–7.
69 Laborde, B., Moine-Ledoux, V., Richard, T., Saucier, C., Dubourdieu, D. and Monti, J.P. (2006) *Journal of Agricultural and Food Chemistry*, **54**, 4383–9.
70 Maury, C., Sarni-Manchado, P., Lefebvre, S., Cheynier, V. and Moutounet, M. (2003) *American Journal of Enology and Viticulture*, **54**, 105–11.
71 Naurato, N., Wong, P., Wroblewski, Y., Lu, K. and Bennick, A. (1999) *Journal of Agricultural and Food Chemistry*, **47**, 2229–34.
72 Kawamoto, H., Mizutani, K. and Nakatsubo, F. (1997) *Phytochemistry*, **46**, 473–8.
73 Verge, S., Richard, T., Moreau, S., Richelme-David, S., Vercauteren, J., Prome, J.C. and Monti, J.P. (2002) *Tetrahedron Letters*, **43**, 2363–6.
74 Chen, Y.M. and Hagerman, A.E. (2004) *Journal of Agricultural and Food Chemistry*, **52**, 4008–11.
75 Siebert, K.J. (1999) *Journal of Agricultural and Food Chemistry*, **47**, 353–62.
76 Kawamoto, H., Nakatsubo, F. and Murakami, K. (1995) *Phytochemistry*, **40**, 1503–5.
77 Kawamoto, H. and Nakatsubo, F. (1997) *Phytochemistry*, **46**, 479–83.
78 de Freitas, V., Carvalho, E. and Mateus, N. (2003) *Food Chemistry* **81**, 503–9.
79 Rawel, H.A., Meidtner, K. and Kroll, J. (2005) *Journal of Agricultural and Food Chemistry*, **53**, 4228–35.
80 Poncet-Legrand, C., Cartalade, D., Putaux, J.L., Cheynier, W. and Vernhet, A. (2003) *Langmuir: The ACS Journal of Surfaces and Colloids*, **19**, 10563–72.

81. Carvalho, E., Póvoas, M., Mateus, N. and De Freitas, V. (2006) *Journal of the Science of Food and Agriculture*, **86**, 891–6.
82. Mateus, N., Carvalho, E., Luis, C. and de Freitas, V. (2004) *Analytica Chimica Acta*, **513**, 135–40.
83. Haslam, E. (1996) *Journal of Natural Products*, **59**, 205–15.
84. Carvalho, E., Mateus, N. and de Freitas, V. (2004) *Analytica Chimica Acta*, **513**, 97–101.
85. Clifford, M.N. (1997) *Astringency in Phytochemistry of Fruits and Vegetables: Proceedings of the Phytochemical Society of Europe*, (eds. Tomás-Barberán, F.A., Robins, R.J.) Oxford University Press, Oxford.
86. Gawel, R., Iland, P.G. and Francis, I.L. (2001) *Food Quality and Preference*, **12**, 83–94.
87. Gawel, R. (1998) *Australian Journal of Grape and Wine Research*, **4**, 74–95.
88. Peleg, H., Gacon, K., Schlich, P. and Noble, A.C. (1999) *Journal of the Science of Food and Agriculture*, **79**, 1123–8.
89. Kallithraka, S., Bakker, J. and Clifford, M.N. (1997) *Journal of Agricultural and Food Chemistry*, **45**, 2211–6.
90. Taira, S., Ono, M. and Matsumoto, N. (1997) *Postharvest Biology and Technology*, **12**, 265–71.
91. Valentova, H., Skrovankova, S., Panovska, Z. and Pokorny, J. (2002) *Food Chemistry*, **78**, 29–37.
92. Vidal, S., Francis, L., Noble, A., Kwiatkowski, M., Cheynier, V. and Waters, E. (2004) *Analytica Chimica Acta*, **513**, 57–65.
93. Dodds, M.W.J., Johnson, D.A. and Yeh, C.K. (2005) *Journal of Dentistry*, **33**, 223–33.
94. Kauffman, D.L. and Keller, P.J. (1979) *Archives of Oral Biology*, **24**, 249–56.
95. McDougall, G.J., Shpiro, F., Dobson, P., Smith, P., Blake, A. and Stewart, D. (2005) *Journal of Agricultural and Food Chemistry*, **53**, 2760–6.
96. Lu, Y. and Bennick, A. (1998) *Archives of Oral Biology*, **43**, 717–28.
97. Kauffman, D.L., Bennick, A., Blum, M. and Keller, P.J. (1991) *Biochemistry*, **30**, 3351–6.
98. Charlton, A.J., Baxter, N.J., Lilley, T.H., Haslam, E., McDonald, C.J. and Williamson, M.P. (1996) *FEBS Letters*, **382**, 289–92.
99. Shimada, T. (2006) *Journal of Chemical Ecology*, **32**, 1149–63.
100. Cai, K.H., Hagerman, A.E., Minto, R.E. and Bennick, A. (2006) *Biochemical Pharmacology*, **71**, 1570–80.
101. Cai, K.H. and Bennick, A. (2006) *Biochemical Pharmacology*, **72**, 974–80.
102. Doco, T., Williams, P., Moutounet, M. and Pellerin, P. (2000) *Bulletin de l'O IV*, **73**, 785–92.
103. Vidal, S., Courcoux, P., Francis, L., Kwiatkowski, M., Gawel, R., Williams, P., Waters, E. and Cheynier, V. (2004) *Food Quality and Preference*, **15**, 209–17.
104. Carvalho, E., Mateus, N., Plet, B., Pianet, I., Dufourc, E. and De Freitas, V. (2006) *Journal of Agricultural and Food Chemistry*, **54**, 8936–44.

18
Photochemical Transformation Processes of Environmental Significance

Davide Vione

18.1
Introduction and Overview of Environmental Photochemistry

Photochemical processes are important pathways for the transformation of organic compounds of both natural and anthropic origin in environmental compartments, including the atmospheric gas phase, atmospheric water and airborne particulate matter, surfaces exposed to the atmosphere such as snow and ice layers, and surface waters and soil [1–5]. While the driving force in all such processes is represented by sunlight, and in particular the UV-Vis component that reaches the ground ($\lambda > 290$ nm, [6]), the reaction pathways and the reactive species involved depend strongly on the phase under consideration. The transformation of organic compounds, and in particular of organic pollutants, will often lead to their degradation and therefore to an overall effect of decreasing pollution, with reduced impact on human health and the environment, but in some cases induces the formation of more harmful compounds than the parent ones [1, 2].

18.1.1
Photochemical Processes in the Atmosphere

In the atmospheric gas phase the main reactive species are $^{\bullet}OH$, $^{\bullet}NO_3$, O_3, and sunlight itself which can be involved in direct photolysis processes. In the latter case a sunlight-absorbing molecule reaches an electronically and vibrationally excited state after absorption of a photon of appropriate wavelength. The surplus energy can be dissipated by vibrational relaxation (i.e., thermally lost), fluorescence, phosphorescence, or chemical reactivity. The latter is often in the form of bond breaking (photolysis), induced by the excess of vibrational energy that can sometimes increase vibration amplitude beyond the threshold where the atoms involved in the bond (B and C in Equation 18.1) are permanently separated [7].

$$A-B-C-D + h\nu \rightarrow A-B^{\bullet} + C-D^{\bullet} (\text{or } A-B^{+} + C-D^{-}) \qquad (18.1)$$

Tomorrow's Chemistry Today. Concepts in Nanoscience, Organic Materials and Environmental Chemistry.
Edited by Bruno Pignataro
Copyright © 2008 WILEY-VCH Verlag GmbH & Co. KGaA, Weinheim
ISBN: 978-3-527-31918-3

Note that bond breaking can be homolytic or heterolytic, leading in the former case to the generation of neutral radical species and in the latter to charged ones. Homolytic bond breaking is understandably favored in apolar media and in the gas phase, where the vast majority if not all of the relevant processes are radical-based, while aqueous solution stabilizes charged species and therefore enhances charge separation. In spite of this, many photochemical processes occurring in aqueous solution are radical processes as well [8]. As an alternative, energy connected with photon absorption can be higher than the ionization energy of the molecule and photoionization can take place (Equation 18.2) [9]. In many a case such a process is favored when the molecule is adsorbed on a polar solid surface [10].

$$A-B-C-D + h\nu \rightarrow A-B-C-D^+ + e^- \tag{18.2}$$

The direct photolysis of compounds such as HONO, O_3, HCHO, and $^\bullet NO_2$ in the tropospheric gas phase is a very important source of reactive species, which are then involved in the transformation of organic compounds. Additionally, some organic molecules including organic pollutants undergo photolysis as a significant or even the main process of removal from the atmosphere. It is for instance the case for nitronaphthalenes, the atmospheric lifetime of which can be as low as a couple of hours because of direct photolysis [11, 12].

As far as $^\bullet OH$, $^\bullet NO_3$, and O_3 in the tropospheric gas phase are concerned, the occurrence of these species is directly or indirectly linked to, and in some cases strictly controlled by, photochemistry. Ozone is a very significant source of the other two species as well as being an important reactive species itself. The hydroxyl radical $^\bullet OH$ is formed by photolysis of O_3, HONO, and H_2O_2 and is by far the main sink of natural and anthropogenic organic compounds in the troposphere, including many aliphatic and aromatic species. Due to formation by sunlight and elevated reactivity, it is exclusively present in the troposphere during daytime [3, 11, 13–18].

$$O_3 + h\nu \rightarrow O_2 + O^{1D} \tag{18.3}$$

$$O^{1D} + H_2O \rightarrow 2\,^\bullet OH \tag{18.4}$$

$$HONO + h\nu \rightarrow\, ^\bullet NO +\, ^\bullet OH \tag{18.5}$$

$$H_2O_2 + h\nu \rightarrow 2\,^\bullet OH \tag{18.6}$$

The nitrate radical $^\bullet NO_3$ is formed upon reaction between O_3 and $^\bullet NO_2$ (see below for the processes that lead to the atmospheric generation of the two reactants). As O_3 and $^\bullet NO_2$ concentrations are higher during daytime, the same applies to the formation rate of $^\bullet NO_3$, but the nitrate radical is highly unstable under sunlight because of very fast photolysis. Accordingly, only the $^\bullet NO_3$ that is formed after sunset, albeit at a somewhat lower rate, can escape photolysis and undergo atmospheric reactions. $^\bullet NO_3$ thus plays a role in nighttime atmospheric chemistry. Although less reactive than $^\bullet OH$ it can still take part to the transformation, including nitration [19–21], of a large number of compounds [12–17]. For our recent reviews on the subject see [22, 23].

$$O_3 + {}^\bullet NO_2 \rightarrow O_2 + {}^\bullet NO_3 \tag{18.7}$$

$$^\bullet NO_3 + h\nu \rightarrow {}^\bullet NO_2 + O^{3P} \tag{18.8}$$

Finally, O_3 is a major photochemical smog component that is formed via a complex reaction chain involving sunlight, nitrogen oxides, volatile organic compounds (in particular those containing double C=C bonds) [13–17], and $^\bullet OH$. In the early morning, when O_3 is lacking, the main atmospheric source of $^\bullet OH$ is the photolysis of HONO (Equation 18.5) [24, 25].

$$>\!C\!=\!C\!<\; +\; {}^\bullet OH \;\rightarrow\; >\!\overset{|}{C}\!-\!\overset{}{C}\!-\!OH \tag{18.9}$$

$$>\!\overset{|}{C}\!-\!\overset{}{C}\!-\!OH\; +\; O_2 \;\rightarrow\; {}^\bullet O\!-\!O\!-\!\overset{|}{C}\!-\!\overset{|}{C}\!-\!OH \tag{18.10}$$

$$^\bullet NO\; +\; {}^\bullet O\!-\!O\!-\!\overset{|}{C}\!-\!\overset{|}{C}\!-\!OH \;\rightarrow\; {}^\bullet NO_2\; +\; {}^\bullet O\!-\!\overset{|}{C}\!-\!\overset{|}{C}\!-\!OH \tag{18.11}$$

$$^\bullet NO_2 + h\nu \rightarrow {}^\bullet NO + O^{3P} \tag{18.12}$$

$$O_2 + O^{3P} \rightarrow O_3 \tag{18.13}$$

O_3 is mainly reactive toward alkenes, of which a nonnegligible fraction is of biogenic origin even in urban atmospheres, and for which it can be an important atmospheric sink in addition to $^\bullet OH$ and $^\bullet NO_3$ [16, 26].

Photochemical reactivity in the atmosphere can also involve compounds that are present on particulate matter or inside suspended water droplets in the upper layer of fog and cloud. Interestingly, because of radiation diffusion and reflection, the irradiation intensity on top of cloud can be double than at the ground [27].

Direct photolysis processes on the surface of airborne particulate matter can be important sinks of sunlight-absorbing compounds (see [28] for a recent review by our group on this subject), and in particular of polycyclic aromatic hydrocarbons (PAHs) [11]. The particles can protect adsorbed substrates against reaction with species such as $^\bullet OH$ and $^\bullet NO_3$ from the gas phase, and enhance the relative role of direct photolysis. However, it should be considered that black carbonaceous particles can also shield sunlight and protect as a consequence adsorbed substrates against direct photolysis [29–32], which could therefore be slower on fine particles than in the gas phase or inside droplets.

Apart from direct photolysis that is highly substrate-dependent, indirect photoreactions sensitized by quinones, furans, aromatic carbonyls (e.g., benzaldehyde derivatives), all important constituents of atmospheric aerosols, can lead to the transformation of many organic compounds [33–35]. In such processes the photosensitizer P absorbs radiation and is then able to cause transformation of a substrate S because of energy transfer, electron or atom abstraction.

$$P + h\nu \rightarrow P^* \tag{18.14}$$

$$P^* + S \rightarrow \text{Products} \tag{18.15}$$

The process of sensitized photolysis causes some energy to be dissipated in the transfer reaction of Equation 18.15 between P* and S, but given the substantial absorption of sunlight by some sensitizers the reactions they induce can be faster than direct photolysis [34].

Other important processes can be induced by components of particulate matter able to absorb sunlight, such as nitrate. As found recently by our group, NH_4NO_3 can account for around 50% of the fine particle mass (PM_{10}, particle diameter <10 μm) in polluted areas during winter [36]. We have also recently shown that the photolysis of solid nitrate in the presence of water vapor produces $^{\bullet}OH$ and $^{\bullet}NO_2$ [37], which can be involved in aromatic oxidation and nitration processes in the gas phase [18], and $HOONO/ONOO^-$ (peroxynitrous acid/peroxynitrite). Photoproduction of HOONO, an oxidizing and nitrating agent [38, 39], can be enhanced by NH_4^+, a stronger H^+ donor than water vapor.

$$NO_3^- + h\nu + H^+ (H_2O) \rightarrow {}^{\bullet}OH + {}^{\bullet}NO_2 \tag{18.16}$$

$$NO_3^- + h\nu + H^+ (NH_4^+) \rightarrow HOONO \tag{18.17}$$

Oxidation/hydroxylation of aromatic compounds by $^{\bullet}OH$ and HOONO is expected to enhance their degradation rate and hence decrease their lifetime on particulate matter, which in the case of pollutants is beneficial from the point of view of human health. Oxidation of PAHs could also lead to the production of photosensitizers such as quinones and aromatic carbonyls [10, 40, 41]. These compounds, if present in the gas phase, are also able to form aggregates and are therefore involved in the formation of secondary organic aerosol [42]. In contrast, nitration induced by $^{\bullet}OH + {}^{\bullet}NO_2$ or HOONO could lead to highly mutagenic nitro-PAHs [43] or phytotoxic nitrophenols [44, 45], in which case the health and environmental impact of the reaction intermediates is not negligible and is sometimes higher than that of the parent molecules.

A number of photochemical reactions can also take place in clouds and fog [27]. Direct and sensitized photolysis processes can be operational, the latter (see

Equations 18.14 and 18.15) induced by the same classes of compounds already seen for particulate matter [46]. Transformation of organic compounds dissolved in water droplets can also take place because of reaction with the powerful oxidizing agent ˙OH. This species can be produced on photolysis of H_2O_2 (see Equation 18.6, [12, 27]), nitrate (Reaction 18.16, which is also valid for aqueous solutions), and nitrite [47].

$$NO_2^- + h\nu + H^+ \rightarrow {}^\bullet OH + {}^\bullet NO \tag{18.18}$$

$$NO_2^- + {}^\bullet OH \rightarrow {}^\bullet NO_2 + OH^- \tag{18.19}$$

Nitrite can be oxidized by ˙OH to the nitrating agent nitrogen dioxide. In a number of recent studies we have demonstrated that ˙NO_2 formation on nitrate photolysis or nitrite photooxidation can be involved into the nitration of aromatic and in particular phenolic compounds in aqueous solution [48, 49]. There is, for instance, evidence that the powerful phytotoxic agent 2,4-dinitrophenol arises on mononitrophenol nitration in cloud [50], a process that is likely to involve ˙NO_2 and sunlight. Our proposed nitration pathways for the formation of 2,4-dinitrophenol under photochemical conditions are shown in Figure 18.1a,b [51].

Differently from the case of 2,4-dinitrophenol, there is still debate concerning the mononitrophenol sources to the atmosphere (direct traffic emissions, phenol nitration in the gas phase or in aqueous solution [23]). However, a recent study of nitrophenol partitioning between gas and aqueous phases suggests that a significant fraction of mononitrophenols that are found in the atmospheric gas phase could actually have been formed in solution and then undergone partitioning to the gas phase [52]. Considering that nitrophenols are among the most abundant semivolatile organic compounds in the atmospheric aqueous phase [53], these data

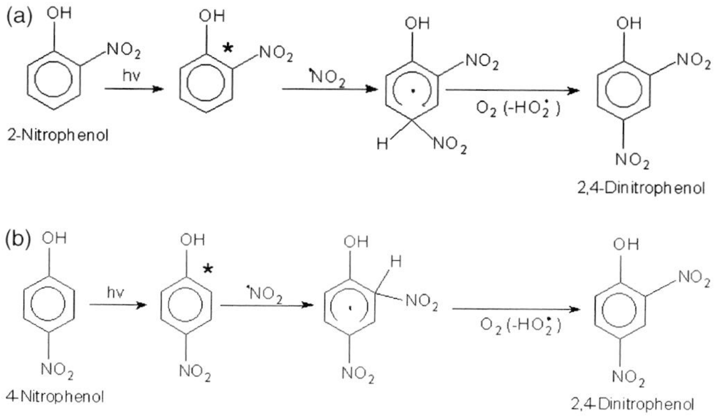

Figure 18.1 (a) Proposed pathway for the nitration of 2-nitrophenol to 2,4-dinitrophenol. (b) Corresponding pathway for the nitration of 4-nitrophenol. The reaction schemes are based on the experimental data of reference [51].

give an indication of the significance of photochemical processes in atmospheric water droplets.

Another interesting feature of disperse droplets is their small diameter (1–10 μm), which affords a high surface-to-volume ratio and causes surface processes to be very important. Accordingly, surface accumulation of organic molecules [54] and polarizable anions [55] can have a significant impact on chemical reactions, including photochemical ones, as demonstrated in a recent study by our group: 15% of the reaction transforming benzene into phenol with photolysis of nitrate would occur in just 0.1% of the volume of a droplet of 1 μm radius [56].

Further description of photochemical processes in the atmospheric aqueous phase and on particulate matter is included in our recent review [28].

18.1.2
Photochemical Reactions in Ice and Snow

The ice and snow layers present in polar regions and in high-mountain glaciers throughout the world contain many photoactive organic and inorganic compounds. These cold surfaces are also able to condense volatile molecules that are mainly present in the gas phase at temperate latitudes and, if stable enough, can persist in the atmosphere until they undergo deposition by condensation. This way, following the so-called "global distillation effect," volatile and persistent organic pollutants such as organochlorine compounds can be accumulated in ice and snow in the cold regions [57].

Much attention has understandably been devoted to the photochemistry of chlorinated pollutants, including chlorophenols, polychlorobiphenyls, and dioxins included in an ice matrix [58, 59]. There has been found an interesting difference between reaction pathways and photodegradation products in water ice matrix and in aqueous solution, but also a similarity when irradiation in ice was carried out at relatively high subzero temperatures (>−5 °C). Such a finding suggests the presence of a quasi-liquid layer on the surface of ice and snow grains at relatively high subzero temperatures, where reactivity is expected to be similar to that in water [59].

Another interesting research topic in this context arose from the discovery that sunlight-irradiated snow layers in polar regions are important sources of $^\bullet NO_2$ and HONO apart from the limited role of anthropic emissions in such remote environments. The additional observation that photogenerated nitrogen species can induce photochemical smog cycles in the Arctic [60] has prompted research on their sources within the snow components. It is very likely that $^\bullet NO_2$ and HONO arise from the photolysis of nitrate in ice [61]. In the case of gaseous HONO, acidification processes linked with snow aging [62] would play a role because nitrate photolysis yields nitrite. We have demonstrated that the photochemical generation of nitrite from nitrate is strongly enhanced in the presence of organic snow components such as formic acid and formaldehyde [63]. Ice photochemistry is a very interesting research topic in rapid development, given that the impact of

18.1.3
Photochemical Reactions in Surface Waters

As already seen for the atmospheric aqueous phase, photochemical reactions in surface freshwater (e.g., rivers and lakes) can consist of direct and indirect (sensitized) photolysis, and of reaction with transient species. The rate of direct photolysis of a given molecule will depend on its absorption spectrum in the environmentally significant wavelength range, on the photolysis quantum yield, and on the features of the water body, most notably radiation absorption and scattering by the water column. Water column effects would be common to all the photoactive compounds in a water body, while the absorption spectrum and photolysis quantum yield are strictly dependent on the substrate and may vary significantly. Accordingly, the importance of direct photolysis should be studied separately for each compound and it is difficult to generalize the data, even within homogeneous classes [64, 65].

Sensitized photolysis mainly involves the photochemistry of dissolved organic matter (DOM), consisting of water-dissolved organic compounds that usually derive from the microbiological transformation of animal and plant spoils. A very important role is played in this context by humic and fulvic acids and in particular by their aromatic/quinonoid moieties, which mainly arise from the biodegradation of lignin and are considerably photoactive [8].

Sensitized photodegradation can be caused by energy, electron, or atom transfer involving photoexcited DOM components and ground-state substrates, or can take place via the intermediate role of singlet oxygen ($^1O_2$) [64, 66, 67]. Assume DOM to be the photosensitizer and S the substrate, ^3X representing a triplet state and ^1X a singlet state. Excited states at higher energy are indicated by *, while the ground states are represented without asterisks. Ground states are ^1DOM, ^1S and $^3O_2$ for sensitizers, organic substrates, and oxygen, respectively.

$$^1DOM + h\nu \rightarrow {}^1DOM^* \tag{18.20}$$

$$^1DOM^* \rightarrow (\text{Inter-system crossing}) \rightarrow {}^3DOM^* \tag{18.21}$$

$$^3DOM^* + {}^1S \rightarrow \text{Products} \tag{18.22}$$

$$^3DOM^* + {}^3O_2 \rightarrow {}^1DOM + {}^1O_2^* \tag{18.23}$$

$$^1S + {}^1O_2^* \rightarrow \text{Products} \tag{18.24}$$

A number of studies, also by our group, have been devoted to the comparison of the kinetics of direct and sensitized photolysis in real or simulated surface waters (see e.g. [68–74]), but it is difficult to draw a definite conclusion because the relative role of the two processes depends strongly on the chosen substrate and the irradiation conditions. For instance, in some cases direct photolysis of the substrate is quite slow and DOM can enhance photodegradation; in other cases DOM has an inhibitory effect because it competes with the substrate for radiation absorption and subtracts energy to a very fast photolysis process.

While the relative importance of direct and sensitized photolysis has to be assessed case by case, it is very likely that these two processes play a substantial role in phototransformation of most organic solutes in surface waters, together with the photolysis of species that can act as ligands for Fe(III) (e.g. carboxylates). Indeed, the photolysis of Fe(III)–carboxylate complexes plays a major role in natural DOM photomineralization [75–78].

$$[Fe^{III}-OOC-R]^{2+} + h\nu \rightarrow Fe^{2+} + R-COO^{\bullet} \tag{18.25}$$

$$R-COO^{\bullet} \rightarrow R^{\bullet} + CO_2 \tag{18.26}$$

Equation 18.26 is directly involved in DOM photomineralization, and Equation 18.25 yields Fe^{2+}. Complexation of Fe(III) by organic ligands is in competition with the precipitation of ferric oxide colloids [79], and the formation of ferrous iron on photolysis of Fe(III)–carboxylate complexes is an important factor in defining the bioavailability of iron in aquatic systems. Iron bioavailability, minimal for the oxides and maximal for Fe^{2+}, is considerably enhanced by the formation of Fe(III)–organic complexes and their subsequent photolysis. Iron bioavailability plays a key role in phytoplankton productivity in oceans [80–82], while that of freshwater is mainly controlled by nitrogen and phosphorus.

The reaction between organic compounds and reactive radical species such as $^{\bullet}OH$, $CO_3^{\bullet-}$, $^{\bullet}NO_2$, $Cl_2^{\bullet-}$ and so on would often play a secondary role in the photodegradation kinetics in surface waters. However, a number of organic pollutants of high concern nowadays (among which are some pesticides and pharmaceuticals) are refractory to degradation by direct or sensitized photolysis, and for them reaction with $^{\bullet}OH$ plays a major role in their transformation kinetics [83–86]. The reactivity toward these substrates of the radical $CO_3^{\bullet-}$, formed on reaction between $^{\bullet}OH$ and carbonate or bicarbonate, could also be important in the phototransformation process, and compensate for the scavenging of $^{\bullet}OH$ by inorganic carbon species [87].

Finally, our recent research work has demonstrated that hazardous compounds can be formed in the presence of the nitrating agent $^{\bullet}NO_2$, arising from nitrate photolysis or nitrite oxidation [88–91], and the chlorinating $Cl_2^{\bullet-}$ [92]. Formation of the latter from $^{\bullet}OH$ and Cl^- can only take place in acidic solution, but chloride oxidation is, for instance, possible upon charge-transfer processes in the presence of irradiated Fe(III) oxide colloids (represented as $=Fe^{3+}-OH^-$ in Equation 18.32) [93].

$$HCO_3^- + {}^{\bullet}OH \rightarrow CO_3^{\bullet-} + H_2O \tag{18.27}$$

$$CO_3^{2-} + {}^{\bullet}OH \rightarrow CO_3^{\bullet-} + OH^- \tag{18.28}$$

$$NO_2^- + {}^\bullet OH \rightarrow {}^\bullet NO_2 + OH^- \tag{18.29}$$

$$Cl^- + {}^\bullet OH \rightleftarrows HOCl^{-\bullet} \tag{18.30}$$

$$HOCl^{-\bullet} + H^+ \rightleftarrows H_2O + Cl^\bullet \tag{18.31}$$

$$=Fe^{3+}-OH^- + Cl^- + h\nu \rightarrow Fe^{2+} + OH^- + Cl^\bullet \tag{18.32}$$

$$Cl^\bullet + Cl^- \rightleftarrows Cl_2^{\bullet -} \tag{18.33}$$

This research topic is quite recent and could lead to a better understanding of how natural freshwater systems work and hazardous compounds are formed in photochemical reactions. The remaining part of the chapter will be devoted to the reactions involved in the photochemical generation of some radical species in surface waters, and to their reactivity toward organic compounds of both natural and anthropic origin.

18.2
Transformation Reactions Induced by ${}^\bullet OH$, ${}^\bullet NO_2$ and $Cl_2^{\bullet -}$ in Surface Waters

18.2.1.
Reactions Induced by ${}^\bullet OH$

Hydroxyl radicals in surface waters can be formed on photolysis of nitrate (Equation 18.16), nitrite (Equation 18.18), DOM, and probably Fe(III) involved in a photo-Fenton process [94–98]. The role of Fe(III) is the most difficult to assess because the study of surface water photochemistry requires use of filtered samples for laboratory operation and to achieve biological stabilization, avoiding possible interference by microbiological processes. Filtration removes Fe(III) colloids and the complexes with coarse DOM, thereby decreasing the iron loading of water. Nevertheless, an important role of the photo-Fenton chemistry (Equations 18.34–18.36) has been evidenced in iron-rich natural water samples (iron content above 10 μM) [99]. In Equation 18.34, L is often an organic ligand of Fe(III); in Equation 18.35, $=Fe^{III}-OH$ represents the surface hydroxyl groups of Fe(III) oxide colloids.

$$[Fe^{III}-L]^{2+} + h\nu \rightarrow Fe^{2+} + L^\bullet \tag{18.34}$$

$$=Fe^{III}-OH + h\nu \rightarrow Fe^{2+} + {}^\bullet OH \tag{18.35}$$

$$Fe^{2+} + H_2O_2 \rightarrow Fe^{3+} + {}^\bullet OH + OH^- \tag{18.36}$$

Photo-Fenton systems, enhanced by Fe(III) complexation by humic and fulvic acids [100–102], could be a very important ${}^\bullet OH$ source in surface waters. However, for the moment an exact quantification of their role is still lacking due to the difficulty of modeling natural DOM and iron speciation under laboratory conditions.

The photogeneration of ·OH by DOM is most likely a consequence of water oxidation by photoexcited triplet states (Equations 18.20 and 18.37) [8].

$$^3DOM^* + H_2O \rightarrow {}^1DOM\text{-}H^\bullet + {}^\bullet OH \tag{18.37}$$

While an exact assessment of the Fe(III) contribution is still lacking, we were able to quantify the relative role of nitrate, nitrite and DOM as ·OH sources in filtered water samples under simulated sunlight irradiation [103, 104]. We adopted a UV intensity of $22\,W\,m^{-2}$ to be scaled under field conditions for actual sunlight intensity as far as the absolute generation rates of ·OH are concerned. Note that a sunny summer day (15 July at 45°N latitude) would correspond to 9.5 hours steady irradiation under $22\,W\,m^{-2}$ UV intensity of simulated sunlight. Absolute rates are proportional to the irradiation intensity, and the relative role of the different sources to ·OH photoproduction would be independent of intensity as a first approximation. The relative role could still vary with the irradiation spectrum, which would change with the water column depth because the shorter wavelengths undergo more extensive absorption [103, 104]. Given these premises, in the surface layer of natural waters in the absence of water column absorption and under summertime sunlight irradiation, nitrate and DOM would be comparable ·OH sources for a $NO_3^-/NPOC$ ratio of $3.3 \times 10^{-5}\,mol\,NO_3^-\,(mg\,C)^{-1}$. Higher values of the ratio would favor nitrate and lower ones DOM as source. Note that NPOC (nonpurgeable organic carbon), expressed in $mg\,C\,l^{-1}$, is the usual way to quantify DOM in surface waters, in the presence of elevated amounts of bicarbonate and carbonate [103]. As far as the relative role of nitrate and nitrite is concerned, on an equimolar basis nitrite would be 160 times more effective than nitrate for ·OH photoproduction because of higher photolysis quantum yield and more extensive sunlight absorption [104]. This fact often compensates for the lower concentration values of nitrite compared to nitrate in surface waters [104, 105]. The rates of ·OH photoproduction are also proportional to the concentration values of nitrate and nitrite; thus their relative role in a surface water sample under summertime irradiation conditions, given the actual concentrations of the two species, can be expressed as follows:

$$\frac{r_{\bullet OH}^{NO_2^-}}{r_{\bullet OH}^{NO_3^-}} = 160 \cdot \frac{[NO_2^-]}{[NO_3^-]} \tag{18.38}$$

Nitrite photochemistry in surface waters has often been underestimated because ion chromatography, the standard technique adopted for anion (and therefore nitrate) quantification in such matrices is poorly sensitive for nitrite, which would often be below detection limit although its photochemistry could still be very significant. Accordingly, the assessment of nitrite photochemistry in surface water samples requires more sensitive, dedicated analytical techniques [104, 106, 107].

Water column absorption would modify the sunlight intensity and spectrum with depth, and UV radiation would undergo higher depletion than visible. Nitrate under sunlight undergoes the most effective photolysis at 315 nm, nitrite at 340 nm, and DOM shows significant absorption also in the visible region. Figure 18.2a,b shows the summertime sunlight spectrum reaching the ground at 50°N ($I_{sun}(\lambda)$, [6]), the absorption spectra of nitrate and nitrite ($\varepsilon_{NO_3^-}(\lambda)$, $\varepsilon_{NO_2^-}(\lambda)$), their photolysis quantum yields ($\phi_{NO_3^-,\bullet OH}(\lambda)$ and $\phi_{NO_2^-,\bullet OH}(\lambda)$, the former being constant in the wavelength interval under consideration), and the product

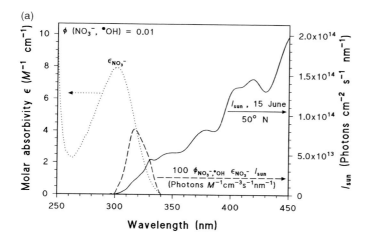

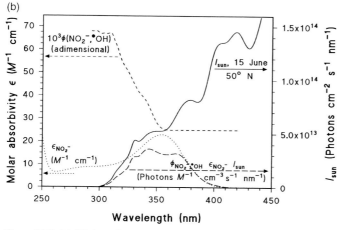

Figure 18.2 (a) (Right ordinate) Intensity spectrum $I_{sun}(\lambda)$ of sunlight at the ground (50°N, based on data from reference [6]); (left ordinate) absorption spectrum of nitrate ($\varepsilon_{NO_3^-}(\lambda)$), and the product $\phi_{NO_3^-}(\lambda)\varepsilon_{NO_3^-}(\lambda)I_{sun}(\lambda)/I_{sun}$. (b) (Right ordinate) Intensity spectrum $I_{sun}(\lambda)$ and (left ordinate) absorption spectrum of nitrite ($\varepsilon_{NO_2^-}(\lambda)$), and the product $\phi_{NO_2^-}(\lambda)\varepsilon_{NO_2^-}(\lambda)I_{sun}(\lambda)$.

$\phi_X(\lambda)\ \varepsilon_X(\lambda)\ I_{sun}(\lambda)$, where X = NO_3^- or NO_2^-. The integral on λ of the cited product is proportional to the rate of photolysis. Radiation absorption by natural water increases with decreasing wavelength; thus the species that require shorter-wavelength radiation to undergo photolysis will be more strongly inhibited with increasing depth. Given the photochemical properties of the different species, the extent of water column inhibition of ·OH photoproduction would be nitrate > nitrite > DOM. We have quantitatively assessed the effect of water column absorption on the rates of photolysis of nitrate and nitrite [108]. Our conclusion was that, because of the combination of the relative role of the different species as ·OH sources in the surface layer of natural waters, and of water column absorption, DOM would play on average the main role toward ·OH photoproduction in surface water bodies, followed by nitrite and by nitrate [103, 104].

The radical ·OH in surface waters is quickly consumed by organic compounds, bicarbonate, carbonate, and nitrite. It has a typically low steady-state concentration of around 10^{-16} M. For this reason it cannot be directly detected, and quantification in laboratory experiments is usually carried out by means of reactions of known kinetics. The formation of phenol from benzene, of 4-hydroxybenzoic from benzoic acid, and the disappearance of nitrobenzene are suitable systems if intermediate monitoring is carried out by liquid chromatography, while the disappearance kinetics of butyl chloride is suitable for headspace sampling and gas-chromatographic analysis [64].

Based on irradiation experiments in which ·OH was quantified with the reaction transforming benzene into phenol, we have been able to model the steady-state [·OH] in the surface layer of natural waters under summertime irradiation conditions (22 W m^{-2} intensity in the UV). Source data were the direct assessment of ·OH photoproduction by nitrate and nitrite, the literature rate constants for the reactions between ·OH and bicarbonate, carbonate and nitrite [109], and the correlation found between DOM (NPOC) and the ·OH sources and sinks [103]. The equation can be expressed as follows as a function of water composition [110]:

$$[^\bullet OH] = \frac{1.7 \times 10^{-7}[NO_3^-] + 2.6 \times 10^{-5}[NO_2^-] + 5.7 \times 10^{-12}\ NPOC}{8.5 \times 10^6[HCO_3^-] + 3.9 \times 10^8[CO_3^{2-}] + 1.0 \times 10^{10}[NO_2^-] + 5.0 \times 10^4\ NPOC} \quad (18.39)$$

The quantities in square brackets are expressed in mol l^{-1}, NPOC in (mg C) l^{-1}. Note that in the vast majority of actual cases DOM is by far the main ·OH sink, followed by inorganic carbon (contribution of 10% or lower) and nitrite (around 1%). Our model can be simplified, albeit with some loss of accuracy, considering that carbonate and bicarbonate are correlated with pH and inorganic carbon (IC, sum of H_2CO_3, HCO_3^- and CO_3^{2-}, expressed in (mg C) l^{-1}) of the water samples, and on average in lakewater it is $[NO_3^-] \approx 190[NO_2^-]$ [103, 104, 110]. Given these premises one obtains:

$$\begin{cases} [^{\bullet}OH] = \dfrac{3.1 \times 10^{-7} \times [NO_3^-] + 5.7 \times 10^{-12} \times NPOC}{\xi(IC) + 5.0 \times 10^4 \times NPOC + 5.2 \times 10^7 \times [NO_3^-]} \\ \xi(IC) = \dfrac{IC}{12000 \times (10^{-10.2} + 10^{-pH})} \times (8.5 \times 10^6 \times 10^{-pH} + 0.025) \\ pH = 1.95 \times (1 - 10^{-0.075 \times IC}) + 6.32 \end{cases} \quad (18.40)$$

In this way [$^{\bullet}$OH] can be expressed as a function of nitrate, DOM (as NPOC) and inorganic carbon (IC) only. Note that inorganic carbon is more easily measured than the values of [HCO_3^-] and [CO_3^{2-}].

A traditional paradigm concerning [$^{\bullet}$OH] in surface water states that it is directly proportional to the concentration of nitrate/nitrite and inversely to DOM (NPOC) [95–97]. This is valid for the cases where nitrate/nitrite are the main $^{\bullet}$OH sources and DOM the main sink, but we have found that in many instances (and in particular in lakewater) DOM is both the main source and sink of hydroxyl radicals, and [$^{\bullet}$OH] would therefore be independent of or very little dependent on NPOC. In this context it should be emphasized that DOM composition is not constant at all in surface water bodies and that NPOC as a global measure is only partially representative of the potential of DOM as $^{\bullet}$OH source and sink. Accordingly, data scattering in [$^{\bullet}$OH] versus NPOC plots is unavoidable [103, 105].

Figure 18.3a,b reports the 3D plots of [$^{\bullet}$OH] vs. nitrate and NPOC at constant IC (a) and of [$^{\bullet}$OH] versus NPOC and IC at constant nitrate (b), based on Equation 18.40. The intervals and values chosen for NPOC, IC, and nitrate are representative of those found in surface waters [103].

Based on our Equation 18.39 or Equation 18.40, four possible scenarios can be envisaged for the trend of [$^{\bullet}$OH] versus NPOC: (i) If nitrate/nitrite were the main $^{\bullet}$OH sources and DOM the main sink, [$^{\bullet}$OH] would be inversely proportional to NPOC (traditional paradigm, curves **a** of Figure 18.3a,b). (ii) If DOM were both the main source and sink, [$^{\bullet}$OH] would be independent of NPOC (plateau **b** of Figure 18.3a,b). (iii) Independence between [$^{\bullet}$OH] and NPOC would also be expected in the (presumably very rare) case that nitrate/nitrite were the main sources and inorganic carbon the main sink. (iv) If DOM were the main $^{\bullet}$OH source and inorganic carbon the main sink, one would expect direct proportionality between [$^{\bullet}$OH] and NPOC, that is, the opposite of the traditional paradigm (curves **c** of Figure 18.3a,b). Such a situation would, however, require DOM-poor, carbonate-rich waters (for inorganic carbon to prevail as a sink) deprived of nitrate and nitrite, for instance because of biological consumption (for the limited amount of DOM to be the main source). While it is not impossible for such conditions to be reached, they are probably very rare in the environment and would possibly be associated with low [$^{\bullet}$OH] [103].

The half-life time of an organic compound S for reaction with $^{\bullet}$OH would depend on the steady-state [$^{\bullet}$OH] and the second-order rate constant for the reaction, $k_{^{\bullet}OH,S}$. Additionally, in the environment [$^{\bullet}$OH] would not be constant because of varying solar irradiation, starting from the diurnal cycle. From the [$^{\bullet}$OH] value

(a)

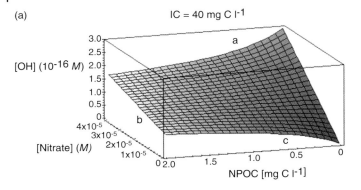

(b)

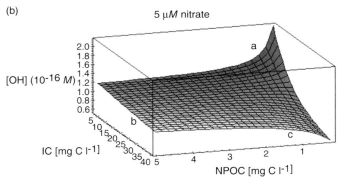

Figure 18.3 (a) Plot of [•OH] vs. NPOC and nitrate at constant IC = 40 mg C l⁻¹. (b) Plot of [•OH] vs. NPOC and IC at constant [NO₃⁻] = 5 × 10⁻⁶ M. For the meaning of curves **a, b, c**, see the text. The plots are based on Equation 18.40.

obtained with our model (22 W m^{-2} intensity in the UV) it should be considered the equivalence between the adopted UV sunlight intensity and the summertime one. Given these premises one obtains:

$$t_{1/2} = \frac{2.03 \times 10^{-5}}{k_{\bullet OH,S} \times [^\bullet OH]} \tag{18.41}$$

where $t_{1/2}$ is in outdoor summer sunny days, $k_{\bullet OH,S}$ in $M^{-1} s^{-1}$, and [•OH] is derived from Equation 18.39 or Equation 18.40 (22 W m^{-2} sunlight intensity in the UV). Figure 18.4 shows the values of $t_{1/2}$ versus [•OH] for compounds having different $k_{\bullet OH,S}$, the latter reported in [109]. Steady-state [•OH] values measured under 22 W m^{-2} sunlight UV in different water samples are also reported [103].

Finally, note that the rate of photochemical reactions, including those involving photoformed •OH, is maximum in the water surface layer and decreases with the water column depth because of decreasing sunlight intensity. Accordingly, the

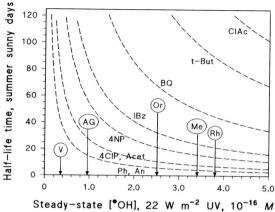

Figure 18.4 Half-life time in outdoor summer sunny days equivalent to 15 July at 45°N, of selected organic compounds, as a function of the steady-state [•OH] measured under 22 W m^{-2} sunlight irradiation. Ph, An = phenol, aniline ($k_{•OH,S} = 1.4 \times 10^{10}\, M^{-1}\, s^{-1}$, [109]); 4ClP, Acet = 4-chlorophenol, acetochlor ($k_{•OH,S} = 7.5 \times 10^9\, M^{-1}\, s^{-1}$); 4NP = 4-nitrophenol ($k_{•OH,S} = 3.8 \times 10^9\, M^{-1}\, s^{-1}$); IBz = iodobenzoate anion ($k_{•OH,S} = 2.5 \times 10^9\, M^{-1}\, s^{-1}$); BQ = 1,4-benzoquinone ($k_{•OH,S} = 1.2 \times 10^9\, M^{-1}\, s^{-1}$); t-But = tert-butyl alcohol ($k_{•OH,S} = 6.0 \times 10^8\, M^{-1}\, s^{-1}$); ClAcet = chloroacetate anion ($k_{•OH,S} = 4.0 \times 10^8\, M^{-1}\, s^{-1}$). V = Lake Viverone (N.W. Italy); AG = Lake Grande in Avigliana (N.W. Italy); Or = Lake Orta (N.W. Italy); Me = Lake Meugliano (N.W. Italy); Rh = Rhône River delta (S France).

deeper is the water column, the less important are the photochemical processes in the whole water body. For this reason, as far as water column effects are concerned, photochemical reactions would be most important in shallow and clear water bodies. As shown in our recent work, good candidates from this point of view are high-mountain shallow lakes [110], where also the incident UV intensity is elevated. Figure 18.5a shows the sunlight intensity spectrum and the absorption spectra ($A(\lambda)$, 1 cm optical pathlength) of water samples from Lake Piccolo in Avigliana and Lake Rouen. They are both located in the Province of Torino (Piedmont, N.W. Italy), the former at 356 m above sea level (a.s.l.) (average depth 7.7 m), the latter at 2391 m a.s.l. With an average depth around 1.5 m Lake Rouen is a clear, shallow high-mountain lake. In the Lambert–Beer approximation the sunlight intensity at the depth b (in cm) in a lake water column can be expressed as follows:

$$I_{sun}(\lambda, b) = I_{sun}(\lambda)\exp[-2.3 A(\lambda) b] \qquad (18.42)$$

where $I_{sun}(\lambda)$ is the sunlight intensity at the ground (see Figure 18.5a). It is possible to calculate the rate of photolysis of nitrate or nitrite at the depth b in the lake, relative to that at the surface, given $I_{sun}(\lambda, b)$ and nitrate and nitrite absorption spectra and quantum yields:

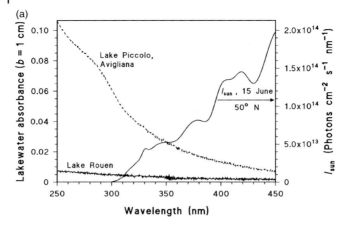

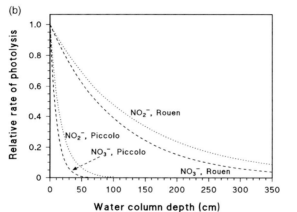

Figure 18.5 (a) Intensity spectrum of sunlight (right ordinate) and absorption spectra of water samples from Lake Piccolo in Avigliana and Lake Rouen (lowland and high-mountain lake, respectively, in N.W. Italy) (left ordinate). The lakewater spectra were measured with an optical path length of 1 cm. (b) Relative rates of photolysis of nitrite and nitrate, referred to the photolysis rate at the surface, as a function of the water column depth in Lake Piccolo in Avigliana and Lake Rouen.

$$\frac{\text{Rate}_{X,b}^{\bullet OH}}{\text{Rate}_{X,\text{surf}}^{\bullet OH}} = \frac{\int_\lambda \phi_X(\lambda) \times \varepsilon_X(\lambda) \times I_{\text{sun}}(\lambda, b) d\lambda}{\int_\lambda \phi_X(\lambda) \times \varepsilon_X(\lambda) \times I_{\text{sun}}(\lambda) d\lambda} \tag{18.43}$$

where $X = NO_3^-$ or NO_2^-, and $I_{\text{sun}}(\lambda, b)$ is from Equation 18.42. Figure 18.5b displays Equation 18.43 as a function of the water column depth b for nitrate and nitrite in Lake Piccolo and Lake Rouen. It is evident that nitrite photochemistry is less influenced than that of nitrate by water column effects, but also that lakewater absorption has a major role in defining the impact of photochemistry. In the case of Lake Piccolo, ·OH photoproduction by nitrate and nitrite would be completely inhibited within the first meter of the water column, while for Lake Rouen it would be active in the whole lake, which is not deeper than 3.5 m at any point.

18.2.2
Reactions Induced by ·NO$_2$

The nitrating agent nitrogen dioxide can be produced in surface waters by nitrate photolysis (Equation 18.16) and nitrite oxidation (Equation 18.19). It is unstable in aqueous solution and can either undergo dimerization and hydrolysis (Equations 18.44 and 18.45), or react with dissolved organic compounds:

$$2\,^{\bullet}NO_2 \rightleftarrows N_2O_4 \tag{18.44}$$

$$N_2O_4 + H_2O \rightarrow NO_2^- + NO_3^- + 2H^+ \tag{18.45}$$

Based on our calculations, performed on literature data, most likely candidates among DOM components for reaction with ·NO$_2$ are the phenolic moieties. However, given the usual DOM loading in surface waters and the expected second-order rate constants for reaction with ·NO$_2$, we have shown that hydrolysis would usually be the most important sink of nitrogen dioxide in aqueous solution [104, 111–113].

Considering the kinetic system made up of Equations 18.16, 18.19, 18.44, 18.45, the relevant rate constants [109], and applying the steady-state approximation to [·OH] and [·NO$_2$], we have obtained the following result for [·NO$_2$]:

$$[^{\bullet}NO_2] = \sqrt{88.8 \times [^{\bullet}OH] \times [NO_2^-] + 1.49 \times 10^{-15} \times [NO_3^-]} \tag{18.46}$$

Our equation would yield steady-state [·NO$_2$] values in the range 10^{-11}–10^{-9} M in the surface layer of natural waters [104, 112, 113]. It is valid for constant 22 W m^{-2} sunlight UV irradiation, and its application to the outdoor environment requires the equivalence between steady and actual irradiation conditions, considering for instance the day 15 July at 45°N latitude as reference (see page 438). Equation 18.46 was able to predict quite well the rate of nitrophenol formation from phenol on irradiation of nitrate-rich groundwater [103, 104], considering that the nitration kinetics of phenol by ·NO$_2$ in the initial rate approximation is d[NP]/dt = 3.2 × 10^3 [Phenol][·NO$_2$], where [NP] = [2-nitrophenol] + [4-nitrophenol], d[NP]/dt is in mol l^{-1} s^{-1}, and the (initial) concentration values are all in molarity. These data are very interesting because nitrogen dioxide could cause nitration of phenolic compounds in a few days in surface waters, and the reaction could therefore be very significant in shallow water bodies rich in nitrate and nitrite [112].

Phenol nitration is also an interesting process for the measurement of the steady-state [·NO$_2$] under laboratory conditions and we have recently used it for the assessment of the rate constants of other photonitration reactions. Assume S to be an aromatic substrate and S—NO$_2$ its nitro derivative. Because at sufficiently low substrate concentration [S] the steady-state [$^{\bullet}$NO$_2$] would be independent of [S], and also of the kind of substrate under otherwise identical irradiation conditions, the rate constant $k_{S,^{\bullet}NO_2}$ for the nitration of S by $^{\bullet}$NO$_2$ would be

$$k_{S,^{\bullet}NO_2}[M^{-1}S^{-1}] = 3.2 \times 10^3 \frac{\lim_{[S]\to 0}\{(d[S-NO_2]/dt)/[S]\}}{\lim_{[Phenol]\to 0}\{(d[NP]/dt)/[Phenol]\}} \qquad (18.47)$$

Following the procedure described we could calculate a second-order rate constant of 650 $M^{-1}s^{-1}$ for the nitration of 2,4-dichlorophenol into 2,4-dichloro-6-nitrophenol [113].

Field data collected in the Rhône river delta (southern France) showed the occurrence of quite elevated levels (around $10^{-8}M$) of 2,4-dichlorophenol and 2,4-dichloro-6-nitrophenol in late spring–early summer. The former compound most likely arises from the hydrolysis/photolysis of the herbicide dichlorprop, heavily used in the flooded rice farming that is a major activity in the delta. The nitro derivative would likely be formed on aqueous-phase nitration as suggested by the time evolution and spatial distribution data of the two compounds. From water composition data and the application of Equation 18.47 for the assessment of 2,4-dichlorophenol nitration kinetics and of Equations 18.39 and 18.46 for the assessment of [$^{\bullet}$OH] and of [$^{\bullet}$NO$_2$], we have concluded that photonitration of 2,4-dichlorophenol to 2,4-dichloro-6-nitrophenol could take place in a couple of weeks in the shallow water (10 cm depth) of the rice fields. These results are compatible with the field data and are reasonable because it is in rice fields that herbicide application primarily occurs, and transformation into 2,4-dichlorophenol and subsequent nitration are most likely. Additionally, shallow water favors photochemical reactions [113].

18.2.3
Reactions Induced by Cl$_2^{\bullet-}$

As shown by various researchers, and also by our group, the oxidizing [109, 111] and chlorinating [92, 93] agent Cl$_2^{\bullet-}$ can be formed on chloride oxidation by $^{\bullet}$OH in acidic solution (Equations 18.30, 18.31, 18.33) or by irradiated Fe(III) oxide colloids under a wide variety of pH conditions (Equations 18.32 and 18.33). The latter process is understandably more relevant to surface waters [93], while the former can be significant in atmospheric aerosols. The radical Cl$_2^{\bullet-}$, although capable of oxidizing/chlorinating many aromatic compounds, is considerably less reactive than $^{\bullet}$OH [109, 111]. The obvious consequence is that, when formation of Cl$_2^{\bullet-}$ takes place because of chloride oxidation by $^{\bullet}$OH in acidic solution, the overall transformation rate of dissolved organic compounds decreases with increasing chloride concentration. Furthermore, potentially harmful chlorinated intermediates can be formed in the process [92, 93].

The photoinduced oxidation of chloride by Fe(III) oxide colloids is relevant to surface waters, and in particular to estuarine areas because of the chloride build-up [93]. Fe(III) oxides are usually poorly active in the photodegradation of organic compounds in neutral to basic solution, because (i) the efficiency of ˙OH photogeneration by Fe(III) colloids (Equation 18.35) is much lower than the corresponding reactions induced by the monomeric hydroxocomplexes (most notably FeOH^{2+}) that are found under acidic conditions. Additionally, (ii) Fe(III) oxide colloids under irradiation (but also the much smaller oligomeric and polymeric species that can be found under mildly acidic conditions) can induce transformation via charge-transfer processes, but these are usually slow when involving organic compounds [114]. Indeed, most organic molecules are more likely to undergo abstraction of hydrogen atoms than of electrons [109, 115]. In contrast, inorganic anions can undergo relatively fast electron abstraction [115].

As an interesting consequence we have shown that nitrite, an ˙OH scavenger that inhibits transformation processes mediated by the hydroxyl radical, is able to enhance phenol degradation by Fe(III) oxide colloids such as α-Fe$_2$O$_3$ and β-FeOOH. This would happen because direct phenol photodegradation by α-Fe$_2$O$_3$ and β-FeOOH is quite slow, while nitrite can be more readily photooxidized to ˙NO$_2$. While less reactive than ˙OH, and mainly a nitrating agent, ˙NO$_2$ is able to transform phenol faster than the charge-transfer processes induced by Fe(III) colloids [48].

As far as chloride is concerned, we have found that it enhances the photodegradation of carbamazepine (an antiepileptic drug that is found at elevated concentration in surface waters) by Fe(III) oxide colloids at pH >5 [93]. Under more acidic conditions the elevated efficiency of ˙OH photoproduction by Fe(III) monomeric species, stable at acidic pH, and the scavenging of ˙OH by chloride would cause Cl$^-$ to inhibit photodegradation. In contrast, at neutral to basic pH the Fe(III) oxide colloids would photooxidize chloride to Cl$_2$˙$^-$ faster than they directly degrade carbamazepine, and further reaction between Cl$_2$˙$^-$ and carbamazepine would account for the enhancement of the degradation rate by chloride. Additionally, chloride is not able to scavenge ˙OH in neutral to basic solution.

The described enhancement of carbamazepine photodegradation by Fe(III) oxide colloids could take place in deltas and estuaries, and our laboratory data suggest that the photochemical consequences of the chloride build-up could more than compensate the decrease of iron, due to colloid coagulation and sedimentation, which is usually observed in these environments. Interestingly, carbamazepine photodegradation on interaction between Fe(III) oxide colloids and chloride was more important than DOM-sensitized photolysis or direct photolysis, which are usually major photodegradation pathways for organic pollutants in surface waters [93].

Should the data concerning carbamazepine be of more general validity, estuarine areas, and even more the deltas because of their shallower water, could be important locations for the photochemical transformation of pollutants transported by river water. Also note that carbamazepine underwent oxidation and dimerization in the presence of Fe(III) + chloride, and chlorination only to a very limited extent [93], confirming that $Cl_2^{\cdot-}$ is mainly an oxidizing agent [92]. Finally, carbamazepine is an interesting example of the photochemical transformation of a compound to yield a more hazardous intermediate: the main product of carbamazepine's direct photolysis is in fact the toxic and mutagenic acridine [93].

18.3
Conclusions

Very significant photochemical processes can take place in various environmental compartments and account for the transformation of organic and inorganic compounds, including pollutants released by human activities. In many cases pollutant transformation is beneficial to the environment and to human health because it decreases the lifetime and hence the possible impact of harmful compounds. However, in some cases the environmental transformation of pollutants and of some otherwise harmless xenobiotics can yield compounds having much higher impact than the parent compounds (e.g., the case of carbamazepine transformation into acridine). It is therefore very important to assess the transformation pathways, including the photochemical ones, of compounds naturally present in the environment or released by human activities.

Also note that climate change could have an important impact on the photochemistry of surface water bodies. The expected decrease of precipitation in the Mediterranean region could lead to desertification of some areas, and in particular to the transformation of permanent water bodies (rivers and lakes) into ephemeral ones. Present ephemeral (temporary) lakes, common in some dry areas of the southern Mediterranean and of South West and Central Asia (Anatolia, Caspian region) undergo extensive evaporative concentration during the dry season. This phenomenon causes lake water to become considerably more saline than sea water, with a transition from freshwater to brackish to brine [116–118] that, according to our recent studies [93, 103, 104, 110, 113], would deeply impact the photochemical processes in the surface water layer. Also, the water level decrease due to evaporation could enhance the importance of photochemical reactions by reducing the lake volume that is not reached by sunlight. These processes, including the possible formation of harmful and environmentally persistent (nitrated, chlorinated) compounds, could influence the biodiversity of ephemeral ecosystems that host a peculiar endemic fauna and microflora [119, 120], and that constitute an important water reservoir for arid regions. Studies into the different features of temporary lakes, including an extension of the current knowledge of the impact of water components, and inorganic anions in particular, on surface water photochemistry are important for the management of water resources. Major aims are to avoid repeating the damage to the environment and human communities caused by the wrong interventions made in the past, ironically with the aim of preserving the water supplies, as was the case of the Lake of Aral and of Kara-Bogaz Gol in Central Asia [117], and to get information on the correct management of temporary water resources that will become precious when many European water bodies become ephemeral due to climate change.

18.4
Acknowledgments

Financial support by PNRA – Progetto Antartide, INCA consortium (GLOB CHE. M. Working group) and CIPE – Regione Piemonte (Project A142) is gratefully acknowledged.

References

1 Ravishankara, A.R. and Longfellow, C.A. (1999) *Physical Chemistry Chemical Physics: PCCP*, **1**, 5433–41.
2 Ravishankara, A.R. (2003) *Chemical Reviews*, **103**, 4505–7.
3 Monks, P.S. (2005) *Chemical Society Reviews*, **34**, 376–95.
4 Rossi, M.J. (2003) *Chemical Reviews*, **103**, 4823–82.
5 Spicer, C.W., Plastridge, R.A., Foster, K.L., Finlayson-Pitts, B.J., Bottenheim, J.W., Grannas, A.M. and Shepson, P.B. (2002) *Atmospheric Environment*, **36**, 2721–31.
6 Frank, R. and Klöpffer, W. (1988) *Chemosphere*, **17**, 985–94.
7 Calvert, J.G. and Pitts, J.N. (1966) *Photochemistry*, John Wiley & Sons, Inc., New York.
8 Boule, P., Bahnemann, D.W. and Robertson, P.K.J. (eds) (2005) *Environmental Photochemistry Part II*, (*The Handbook of Environmental Chemistry*, Vol. 2, Part M), Springer.
9 Balzani, V., Beleskaya, I.P., Bolton, J.R., Chanon, M., Lewis, N.A., Marcus, R.A., Mataga, N. and Zchariasse, K.A. (1996) *Pure and Applied Chemistry. Chimie Pure Et Appliquee*, **68**, 2223–88.
10 Reyes, C.A., Medina, M., Crespo-Hernandez, C., Cedeno, M.Z., Arce, R., Rosario, O., Steffenson, D.M., Ivanov, I.N., Sigman, M.E. and Dabestani, R. (2000) *Environmental Science and Technology*, **34**, 415–21.
11 Neilson, A.H. (ed.) (1998) *PAHs and Related Compounds*, (*The Handbook of Environmental Chemistry*, Vol. 3, Part I), Springer.
12 Vione, D., Maurino, V., Minero, C. and Pelizzetti, E. (2003) *Annali di chimica (Rome)*, **93**, 477–88.
13 Finlayson-Pitts, B.J. and Pitts, J.N., Jr (1986), *Atmospheric Chemistry*, John Wiley & Sons, Inc., New York.
14 Finlayson-Pitts, B.J. and Pitts, J.N., Jr (1997) *Science*, **276**, 1045–52.
15 Atkinson, R. (1985) *Chemical Reviews*, **85**, 69–201.
16 Atkinson, R. (1994) *Journal of Physical and Chemical Reference Data*, Monograph, 2.
17 Atkinson, R. and Arey, J. (1994) *Environmental Health Perspectives*, **102**, 117–26.
18 Olariu, R.I., Klotz, B., Barnes, I., Becker, K.H. and Mocanu, R. (2002) *Atmospheric Environment*, **36**, 3685–97.
19 Atkinson, R., Aschmann, S.M. and Arey, J. (1992) *Environmental Science and Technology*, **26**, 1397–403.
20 Barletta, B., Bolzacchini, E., Meinardi, S., Orlandi, M. and Rindone, B. (2000) *Environmental Science and Technology*, **34**, 2224–30.
21 Bolzacchini, E., Bruschi, M., Hjorth, J., Meinardi, S., Orlandi, M., Rindone, B. and Rosenbohm, E. (2001) *Environmental Science and Technology*, **35**, 1791–7.
22 Vione, D., Barra, S., De Gennaro, G., De Rienzo, M., Gilardoni, S., Perrone, M.G. and Pozzoli, L. (2004) *Annali di chimica (Rome)*, **94**, 257–68.

23 Harrison, M.A.J., Barra, S., Borghesi, D., Vione, D., Arsene, C. and Olariu, R.I. (2005) *Atmospheric Environment*, **39**, 231–48.
24 Kurthenbach, R., Becker, K.H., Gomes, J.A.G., Kleffmann, J., Lörzer, J.C., Spittler, M., Wiesen, P., Ackermann, R., Geyer, A. and Platt, U. (2001) *Atmospheric Environment*, **35**, 3385–94.
25 Stemmler, K., Ammann, M., Donders, C., Kleffmann, J. and George, C. (2006) *Nature*, **440**, 195–8.
26 Atkinson, R. and Arey, J. (1998) *Accounts of Chemical Research*, **31**, 574–83.
27 Warneck, P. (1999) *Physical Chemistry Chemical Physics: PCCP*, **1**, 5471–83.
28 Vione, D., Maurino, V., Minero, C., Pelizzetti, E., Harrison, M.A.J., Olariu, R.I. and Arsene, C. (2006) *Chemical Society Reviews*, **35**, 441–53.
29 Yokley, R.A., Garrison, A.A., Wehry, E.L. and Mamantov, G. (1986) *Environmental Science and Technology*, **20**, 86–90.
30 Dunstan, T.D.J., Mauldin, R.F., Jinxian, Z., Hipps, A.D., Wehry, E.L. and Mamantov, G. (1989) *Environmental Science and Technology*, **23**, 303–8.
31 Behymer, T.D. and Hites, R.A. (1985) *Environmental Science and Technology*, **19**, 1004–6.
32 Behymer, T.D. and Hites, R.A. (1988) *Environmental Science and Technology*, **22**, 1311–9.
33 McDow, S.R., Sun, Q., Vartiainen, M., Hong, Y., Yao, Y., Fister, T., Yao, R. and Kamens, R.M. (1994) *Environmental Science and Technology*, **28**, 2147–53.
34 Jang, M. and McDow, S.R. (1995) *Environmental Science and Technology*, **29**, 2654–60.
35 Jang, M. and McDow, S.R. (1997) *Environmental Science and Technology*, **31**, 1046–53.
36 Minero, C., Maurino, V. and Gianotti, E. (2004) Final Report to the Province of Torino Government, Program "Evaluation of the Impact of Different Emission Sources on Urban Air Quality," Technical Report, University of Torino, Italy.
37 Borghesi, D., Vione, D., Maurino, V. and Minero, C. (2005) *Journal of Atmospheric Chemistry*, **52**, 259–81.
38 Vione, D., Maurino, V., Minero, C., Lucchiari, M. and Pelizzetti, E. (2004) *Chemosphere*, **56**, 1049–59.
39 Vione, D., Maurino, V., Minero, C. and Pelizzetti, E. (2005) *Environmental Science and Technology*, **39**, 1101–10.
40 Barbas, J.T., Sigman, M.E. and Dabestani, R. (1996) *Environmental Science and Technology*, **30**, 1776–80.
41 Mallakin, A., Dixon, D.G. and Greenberg, B.M. (2000) *Chemosphere*, **40**, 1435–41.
42 Forstner, H.I.L., Flagan, R.C. and Seinfeld, J.H. (1997) *Environmental Science and Technology*, **31**, 1345–58.
43 Enya, T., Suzuki, H., Watanabe, T., Hirayama, T. and Himasatsu, Y. (1997) *Environmental Science and Technology*, **31**, 2772–6.
44 Natangelo, M., Mangiapan, S., Bagnati, R., Benfenati, E. and Fanelli, R. (1999) *Chemosphere*, **38**, 1495–503.
45 Schüssler, W. and Nitschke, L. (2001) *Chemosphere*, **42**, 277–83.
46 Anastasio, C., Faust, B.C. and Rao, C.J. (1997) *Environmental Science and Technology*, **31**, 218–32.
47 Anastasio, C. and McGregor, K.G. (2001) *Atmospheric Environment*, **35**, 1079–89.
48 Vione, D., Maurino, V., Minero, C. and Pelizzetti, E. (2002) *Environmental Science and Technology*, **36**, 669–76.
49 Vione, D., Maurino, V., Minero, C., Borghesi, D., Lucchiari, M. and Pelizzetti, E. (2003) *Environmental Science and Technology*, **37**, 4635–41.
50 Lüttke, J., Scheer, V., Levsen, K., Wünsch, G., Cape, J.N., Hargreaves, K.J., Storeton-West, R.L., Acker, K., Wieprecht, W. and Jones, B. (1997) *Atmospheric Environment*, **16**, 2637–48.
51 Vione, D., Maurino, V., Minero, C. and Pelizzetti, E. (2005) *Environmental Science and Technology*, **39**, 7921–31.
52 Harrison, M.A.J., Heal, M.R. and Cape, J.N. (2005) *Atmospheric Chemistry and Physics*, **5**, 1679–95.
53 Lüttke, J., Levsen, K., Acker, K., Wieprecht, W. and Möller, D. (1999)

International Journal of Environmental Analytical Chemistry, **74**, 69–89.
54. Sadiki, M., Quentel, F., Elléouet, C., Huruguen, J.-P., Jestin, J., Andrieux, D., Olier, R. and Privat, M. (2003) *Atmospheric Environment*, **37**, 3551–9.
55. Knipping, E.M., Lakin, M.J., Foster, K.L., Jungwirth, P., Tobias, D.J., Gerber, R.B., Dadub, D. and Finlayson-Pitts, B.J. (2000) *Science*, **288**, 301–6.
56. Vione, D., Minero, C., Hamraoui, A. and Privat, M. (2007) *Atmospheric Environment*, **41**, 3303–14.
57. Simonich, S.L. and Hites, R.A. (1995) *Science*, **269**, 1851–4.
58. Klán, P. and Holoubek, I. (2002) *Chemosphere*, **46**, 1201–10.
59. Klánova, J., Klán, P., Nosek, J. and Holoubek, I. (2003) *Environmental Science and Technology*, **37**, 1568–74.
60. Beine, J.H., Allegrini, I., Sparapani, R., Ianniello, A. and Valentini, F. (2001) *Atmospheric Environment*, **35**, 3645–58.
61. Dubowski, Y., Colussi, A.J., Boxe, C. and Hoffmann, M.R. (2002) *The Journal of Physical Chemistry A*, **106**, 6967–71.
62. Rothlisberger, R., Mulvaney, R., Wolff, E.W., Hutterli, M.A., Bigler, M., De Angelis, M., Hansson, M.E., Steffensen, J.P. and Udisti, R. (2003) *Journal of Geophysical Research–Atmospheric*, **108**, art No. 4526.
63. Minero, C., Maurino, V., Bono, F., Pelizzetti, E., Marinoni, A., Mailhot, G., Carlotti, M.E. and Vione, D. (2007) *Chemosphere*, **68**, 2111–7. DOI: 10.1016/j.chemosphere.2007.02.011
64. Stumm, W. (ed.) (1990) *Aquatic Chemical Kinetics*, Wiley, NY.
65. Czaplicka, M. (2006) *Journal of Hazardous Materials*, **B134**, 45–59.
66. Canonica, S. and Freiburghaus, M. (2001) *Environmental Science and Technology*, **35**, 690–5.
67. Canonica, S., Hellrung, B., Muller, P. and Wirz, J. (2006) *Environmental Science and Technology*, **40**, 6636–41.
68. Vialaton, D., Richard, C., Baglio, D. and Paya-Perez, A.B. (1998) *Journal of Photochemistry and Photobiology A: Chemistry*, **119**, 39–45.
69. Vialaton, D., Baglio, D., Paya-Perez, A. and Richard, C. (2001) *Pest Management Science*, **57**, 372–9.
70. Vialaton, D., Pilchowski, J.F., Baglio, D., Paya-Perez, A., Larsen, B. and Richard, C. (2001) *Journal of Agricultural and Food Chemistry*, **49**, 5377–82.
71. Vialaton, D. and Richard, C. (2002) *Aquatic Sciences*, **64**, 207–15.
72. Zamy, C., Mazellier, P. and Legube, B. (2004) *Water Research*, **38**, 2305–14.
73. Walse, S.S., Morgan, S.L., Kong, L. and Ferry, J.L. (2004) *Environmental Science and Technology*, **38**, 3908–15.
74. Chiron, S., Minero, C. and Vione, D. (2007) *Annali di chimica (Rome)*, **97**, 135–9.
75. Zuo, Y. and Jones, R.D. (1997) *Water Research* **31**, 850–8.
76. Brinkmann, T., Sartorius, D. and Frimmel, F.H. (2003) *Aquatic Sciences*, **65**, 415–24.
77. Brinkmann, T., Hörsch, P., Sartorius, D. and Frimmel, F.H. (2003) *Environmental Science and Technology*, **37**, 4190–8.
78. Meunier, L., Laubscher, H., Hug, S.J. and Sulzberger, B. (2005) *Aquatic Sciences*, **67**, 292–307.
79. Cullen, J.T., Bergquist, B.A. and Moffett, J.W. (2006) *Marine Chemistry*, **98**, 295–303.
80. Borer, P.M., Sulzberger, B., Reichard, P. and Kraemer, S.M. (2005) *Marine Chemistry*, **93**, 179–93.
81. Hiemstra, T. and van Riemsdijk, W.H. (2006) *Marine Chemistry*, **102**, 181–97.
82. Gerringa, L.J.A., Rijkenberg, M.J.A., Wolterbeek, H.Th., Verburg, T.G., Boye, M. and de Baar, H.J.W. (2007) *Marine Chemistry*, **103**, 30–45.
83. Chin, Y.P., Miller, P.L., Zeng, L.K., Cawley, K. and Weavers, L.K. (2004) *Environmental Science and Technology*, **38**, 5888–94.
84. Lam, M.W., Tantuco, K. and Mabury, S.A. (2003) *Environmental Science and Technology*, **37**, 899–907.
85. Miller, P.L. and Chin, Y.P. (2002) *Journal of Agricultural and Food Chemistry*, **50**, 6758–65.

86 Stangroom, S.J., Macleod, C.L. and Lester, J.N. (1998) *Water Research*, **32**, 623–32.
87 Canonica, S., Kohn, T., Mac, M., Real, F.J., Wirz, J. and von Gunten, U. (2005) *Environmental Science and Technology*, **39**, 9182–8.
88 Vione, D., Maurino, V., Minero, C., Vincenti, M. and Pelizzetti, E. (2001) *Chemosphere*, **44**, 237–48.
89 Vione, D., Maurino, V., Minero, C. and Pelizzetti, E. (2001) *Chemosphere*, **45**, 893–902.
90 Vione, D., Maurino, V., Minero, C. and Pelizzetti, E. (2001) *Chemosphere*, **45**, 903–10.
91 Vione, D., Maurino, V., Pelizzetti, E. and Minero, C. (2004) *International Journal of Environmental Analytical Chemistry*, **84**, 493–504.
92 Vione, D., Maurino, V., Minero, C., Calza, P. and Pelizzetti, E. (2005) *Environmental Science and Technology*, **39**, 5066–75.
93 Chiron, S., Minero, C. and Vione, D. (2006) *Environmental Science and Technology*, **40**, 5977–83.
94 Russi, H., Kotzias, D. and Korte, F. (1982) *Chemosphere*, **11**, 1041–8.
95 Hoigné, J. and Haag, W.R. (1985) *Chemosphere*, **14**, 1659–71.
96 Zepp, R.G., Hoigné, J. and Bader, H. (1987) *Environmental Science and Technology*, **21**, 443–50.
97 Brezonik, P.L. and Fulkerson-Brekken, J. (1998) *Environmental Science and Technology*, **32**, 3004–10.
98 Vaughan, P.P. and Blough, N.V. (1998) *Environmental Science and Technology*, **32**, 2947–53.
99 White, E.M., Vaughan, P.P. and Zepp, R.G. (2003) *Aquatic Sciences*, **65**, 402–14.
100 Voelker, B.M., Morel, F.M.M. and Sulzberger, B. (1997) *Environmental Science and Technology*, **31**, 1004–11.
101 Kwan, W.P. and Voelker, B.M. (2003) *Environmental Science and Technology*, **37**, 1150–8.
102 Southworth, B.A. and Voelker, B.M. (2003) *Environmental Science and Technology*, **37**, 1130–6.
103 Vione, D., Falletti, G., Maurino, V., Minero, C., Pelizzetti, E., Malandrino, M., Ajassa, R., Olariu, R.I. and Arsene, C. (2006) *Environmental Science and Technology*, **40**, 3775–81.
104 Minero, C., Chiron, S., Falletti, G., Maurino, V., Pelizzetti, E., Ajassa, R., Carlotti, M.E. and Vione, D. (2007) *Aquatic Sciences*, **69**, 71–85.
105 Takeda, K., Takedoi, H., Yamaji, S., Ohta, K. and Sakugawa, H. (2004) *Analytical Sciences: The International Journal of the Japan Society For Analytical Chemistry*, **20**, 153–8.
106 Kieber, R.J. and Seaton, P.J. (1996) *Analytical Chemistry*, **67**, 3261–4.
107 Kieber, R.J., Li, A. and Seaton, P.J. (1999) *Environmental Science and Technology*, **33**, 993–8.
108 Vione, D., Minero, C., Maurino, V. and Pelizzetti, E. (2007) *Annali di chimica (Rome)*, **97**, 699–711.
109 Buxton, G.V., Greenstock, C.L., Helman, W.P. and Ross, A.B. (1988) *Journal of Physical and Chemical Reference Data*, **17**, 1027–284.
110 Minero, C., Lauri, V., Maurino, V., Pelizzetti, E. and Vione, D. (2007) *Annali di chimica (Rome)*, **97**, 685–98.
111 Minero, C., Maurino, V., Pelizzetti, E. and Vione, D. (2006) *Environmental Science and Pollution Research International*, **13**, 212–4.
112 Minero, C., Maurino, V., Pelizzetti, E. and Vione, D. (2007) *Environmental Science and Pollution Research International*, **14**, 241–3. DOI: 10.1065/espr2007.01.382.
113 Chiron, S., Minero, C. and Vione, D. (2007) *Environmental Science and Technology*, **41**, 3127–33.
114 Mazellier, P., Mailhot, G. and Bolte, M. (1997) *New Journal of Chemistry*, **21**, 389–97.
115 Neta, P., Huie, R.E. and Ross, A.B. (1988) *Journal of Physical and Chemical Reference Data*, **17**, 1027–228.
116 Giralt, S., Julia, R., Leory, S. and Gasse, F. (2003) *Earth and Planetary Science Letters*, **212**, 225–39.
117 Leroy, S., Marret, F., Giralt, S. and Bulatov, S.A. (2006) *Geophysical Journal International*, **150**, 52–70.

118 Kazanci, N., Toprak, O., Leroy, S., Onvel, S., Ileri, O., Emre, O., Costa, P., Erturac, K. and McGee, E. (2006) *Applied Geochemistry*, **21**, 134–51.

119 Baxevanis, A.D. and Abatzopoulos, T.J. (2004) *Journal of Biological Research*, **1**, 107–14.

120 Baxevanis, A.D., Kappas, I. and Abatzopoulos, T.J. (2006) *Molecular Phylogenetics and Evolution*, **40**, 724–38.

Index

a

α-alkoxyalkyl radical 347
α-amino acids
   – α-aminoamides 349
α-aminoalkyl radicals
   – amides 340
   – Fe(II) 340
   – Fenton 340
   – hydrogen peroxide 340
α-amylase 421
accelerated aging 251, 255
acetonitrile 10
acidity
   – basification 24
   – pK_a 7, 20–21
   – sulfanilic acid 7, 8
acridine 448
active site
   – type 1 active site 102
   – type 2 active site 103
   – type 3 active site 104, 113
   – type 4 active site 104
acyl radical
   – acylation 339
   – aldehyde 339
   – decarbonylation 343
   – decarboxylation 339
   – Fe(II) 339
   – hydroperoxide 339
   – persulfate 339
addressable groups 222
adsorption 47
aedamers 272
airborne particulate matter 429
alcohol
   – allylic 326
   – benzylic 326
   – dehydrogenation 325
   – donor alcohol 325
   – secondary 327
amicyanin 102
amphiphilic rotaxane 147
Anderson-type cluster 37
anthocyanins 411
antiparallel β-sheets 85
arabinogalactan 417
   – proteins 423
arctic, photochemical smog cycles 434
arrays 216
artificial molecular muscles 143
ascorbate oxidase 104
astringency 411
atomic force microscopy (AFM) 189
axial ligation 285
azurin 102

b

β-amino-α-hydroxyesters 347
β-cyclodextrin 417
β-sheet 80
   – stacking 89
barrel-stave ion channel 273
basic PRPs 422
bicarbonate 440
bioconjugation 220
bioelectronics
   – construction between GO$_x$ and gold electrode 148
   – membrane transport rotaxane 149
bioinspired models 101ff.
biomedical applications
   – bioimaging 226
   – biomedicine 222
biomimetic chemistry 101
biomimetic compounds
   – copper proteins 373
biomimetic methods, *see* biomimetics

biomimetic synthesis, *see* biomimetics
biomimetics
 – amino acids 66
 – bacteria *Bacillus* subtilis 66
 – biostructures 67
 – enzymes 67
 – exoskeletons 67
 – phospholipids 67
 – polyamines 67
 – polypeptides 67
 – proteins 67
 – silica 64
biomimetism 101
biosensors 53, 228
biquinoline 20
blue copper proteins 102
blue oxidases 104
bond dissociation energy (BDE) 341
bottom-up techniques
 – building blocks 47
 – nanoblocks 47
 – secondary building units (SBUs) 49
 – self-assembly 47
bottom-up 73
bovine serum albumin 413
brine 448
building block
 – inorganic clusters 31
 – nanometer-scale 129
 – p6P 186
 – polyoxometalates (POMs) 31

c

calorimetric techniques 237
cancer cells, demand for iron 356
cancer therapy 358
capping agent 57–58
carbamazepine 447
carbonate 440
carvone 329
cascade 20
catalysis
 – acid-catalyzed 48
 – alkylation 49
 – biocatalysts 67
 – condensation reactions 53
 – cracking 49
 – esterification 53
 – for asymmetric catalysis 66
 – isomerization 49
 – olefin cracking 61
 – selective catalysts 54
catalyst 322

catechin 409
catechol oxidase
 – catalytic reaction mechanism 107
 – crystal structure 105
 – *deoxy* state 105f, 116
 – function 105
 – *met* state 105, 116
 – model systems 108
 – *oxy* state 105
catechol oxidase, tyrosinase 114
catechol 102, 105, 107–111, 113–114, 117, 121, 123–124
catecholase activity 105, 111
cellular materials 58
ceruloplasmin 104
chain extension 89
chain scission 255–257
charge separation 286
chelate
 – entropic driving force 18, 23
 – effect 23
chemoselective reduction 323
chimeric virus technology 219
chirality
 – diastereomers 7
 – diastereoselectivity 8
 – Kamlet–Taft 8
 – stereocenters 7
chromophore, NDIs 269, 287
chromophoric ligand 163, 168
cisplatin, antitumor activity 355
clean protocols
 – fine chemicals 321
 – organic synthesis 321
 – oxidation 321
 – reduction 321
climate change 448
combinatorial chemistry 390
complexes bearing bridging phosphane ligands 305
conducting polymers, electron transfer 147
confined liquids 238
confinement effect 239, 243
conjugated polymers 74
conproportionation 16
controlled porosity of cationic surfactants 51
conventional chemotherapy
 – drugs interfering with tumor cells 358
cooperative selection by iron and copper 17ff

coordination chemistry of 2-(2-
    Pyridyl)phosphole Derivatives
    – Cu(I) bimetallic complexes 307
    – π–π interactions 313
    – isomerization into 2-phospholene
        ring 301
    – nanosized rectangles 313
    – nonlinear optical activities 304
    – Pd(I) and Pt(I) Bimetallic
        Complexes 306
    – Pd(II) centers 300, 303
    – Ruthenium complexes 304
    – two coordination centers with
        different stereoelectronic
        properties 300
coordination
    – type 170, 178
    – geometry 175ff.
copper
    – active sites 102, 114,
    – containing proteins 102
    – coordination geometries 102
Cowpea mosaic virus (CPVC) 216
CO_x 105
CPMV, see *Cowpea mosaic virus* 216
crosslinking 237, 244–245, 251, 253, 414
crown ether hosts
    – binding properties 130
    – cationic guests 130
crystallinity 251
crystallizability 253
customized DNA-targeting agents 360
cyclic voltammetry 227
cyclin-dependent kinases (CDKs) 358
cytochrome *c* 104
    – oxidase 105

d

1D aggregates 80
3D organization 237
Dawson structures
    – short S···S interaction 42
    – sulfite anions 42
Dawson-like clusters, W-based 41
Dawson-type clusters {Mo_{18}} 41
denaturing gel electrophoresis 221
density functional (DFT)
    – calculations 26f.
    – delocalized 26
    – HOMO 26
deposition rates
    – organic molecules 193
    – temperature window 193, 196

desertification 448
DFT, see density function
1,5-dialkoxynaphthalene (DAN) 272
2,4-dichloro-6-nitrophenol 446
2,4-dichlorophenol 446
dichlorprop 446
dicopper-dioxygen
    – "end-on" 107, 118
    – helicates 8
    – "side-on" 107, 118
    – *trans* 118
dihydrogen peroxide 106, 110, 113,
    121–122, 124
dihydroxynaphthalene 278
2,4-dinitrophenol 433
dimethy solfoxide (DMSO) 7, 22
direct photolysis 429
disproportionation 5
dissolved organic matter 435
3,5-di-*tert*-butylcatechol ($DTBCH_2$) 108, 117
3,5-di-*tert*-butyl-*o*-benzoquinone (DTBQ)
    108
3,5-di-*tert*-butylquinone 121
ditopic bis-zinc(II) porphyrin tweezers 285
DMSO, see dimethy solfoxide
DNA cleavage reactions
    – zinc complexes 376
    – Cu(II) and molecular O_2 374
    – time course experiments 375
DNA
    – charge separation 279
    – sensing 278
donor–π-acceptor system 170–171
dopamine-β-hydroxylase 104
(3D) organization of polymers and gels 237
double-helices 87
drug delivery 67
drug discovery 389
DSC, see photo-differential scanning
    calorimetry
dynamic combinatorial chemistry 5f.
dynamic libraries 13
dynamic memory storage 144

e

elastomers 244
electrochemical studies, see cyclic
    voltammetry
electron paramagnetic resonance (EPR)
    spectra 270
electron transfer (ET)
    – charge-separation 281, 283, 286
    – covalent models 281

– NDIs 281
– noncovalent models 284
electronic effects 22
emission 161–163, 170–171, 173–175, 168–169, 180
encapsulation 35
– cations 36
energy conversion systems 284
energy production and storage 67
energy transfer
– directionality 283
– NDI 282
environmental photochemistry 429
environmental pollution 101
enzyme mimics 151
ephemeral ecosystems 448
epidermal growth factor receptor (EGFR) 358
(–)-epigallocatechin gallate (EGCG) 413
epitaxy 188
epothilone C 393
estrogen receptor (ER), tamoxifen 368
estuarine areas 447
Eu 166, 170, 173, 178, 180

f

Fe(III) oxide colloids 447
ferrocenylnaphthalene diimide 278
f–f emission 180
f–f transition 164, 168–170, 180
fiberlength
– MOP4 aggregates 194
– nanofibers from CLP4 194
– nanostructures from NMeP4 194
– p6P fibers 194
fibers 81, 87
flash chemistry 402
flavan-3-ols 409
flavonoids 409
flavors and fragrances 325
flow nephelometric technique 419
flow processes 401
fluorescence microscopy 230
– nanofibers emit polarized blue fluorescence 188
– symmetrically functionalized p4Ps' nanoaggregates 201
fluorescence
– NDIs 266, 269
– quenching 284
fluorophores, NDIs 269, 286, 287
foams, Freon 58
fog 431

foldamer 74, 272
folding of the polymer 79
free-radical processes
– enthalpic effect 337
– polar effect 337
functional POM cluster 37
functionalization
– heterogenize homogeneous catalysts 53
– immobilize biomolecules 53
functionalized *para*-quaterphenylenes 185

g

galactose oxidase 104
gallocatechin 409
gas adsorption 67
gels 237, 244
Gibbs–Thomson equation 239f.
global distillation effect 434
glucose 417
green chemistry 101
green gap 288, 290
grid, tetracopper 12

h

half-life time 441
halide-selective sequence 274
Hammett effect 22
harmonic generation 161
hazardous compounds 437
heating, microwave 396
helical bundles 87
helicate 23f.
hemocyanin 104, 108
herringbone packing 203
– lattice constants 190
heterogeneous catalysts 321f.
heterogeneous hydrogenation 402
hierarchical materials 57–62
hierarchical self-organization
– parallel orientation 85
– oligopeptide-polymer conjugate 85
hierarchical structure 73
high-mountain shallow lakes 443
histatins 422
hole transport 282
homolytic bond breaking 430
host–guest chemistry 38
hydrogels 246
hydrogen acceptor 326
– polyunsaturated compounds 326
hydrogen bonding 3, 85, 89, 412
– NDIs 284, 287

hydrogen transfer
– hydrogenation 321
– oxidation 321
– selective transformations 321
hydrophobic effect 89, 412
hydroxyl radical 430
hydroxymethyl radical
– Hydroxymethylation 343
hyperpolarizability 176–178

i

imine
– aminal 6
– hemiaminal 6
immobilization 229
immunosuppressive activities
– *Serratia* 374
– *Streptomyces* 374
indazolium compound KP1019 367
indirect photoreactions 432
infrared spectroscopy (IR) 85, 166, 168, 174
inorganic carbon 440
inorganic chemotherapy for cancer 378
inorganic precursor
– organometallic 58
– trialkoxysilane-functionalized metal nanoparticles 55
interflavanol bonds 410
inverse microemulsion 57–58
ion channels 273ff.
ion-exchange 48–49, 52
ionic radius 177–178
ionic strength 415
IR, *see* infrared
iron 17, 436
irradiation intensity 438
isoelectric point 415–416
isomeric forms 364
isomerization 331

k

Keplerates 38
kinetics
– photoDSC 247, 249,
– chain scissions 255, 258
knowledge of interactions 130

l

laccase 104
lakewater spectra 444
Lambert–Beer approximation 443
Langmuir-Blodbett (LB) films 79

lanthanide complexes 181
lanthanide complexes 161, 165, 169, 173, 176, 180
LC-MS (liquid chromatography – mass spectrometry) 405
lead optimization 389
light scattered 413
light-emitting organic nanoaggregates 185
light-harvesting 288
linear free energy relationship
– α (hydrogen bond donor strength) 8
– β parameter 8
linear optics
– *para*-phenylenes 201
– shift of the peak emission 201
linker groups 37
living anionic polymerization 83
luminescence 163, 170–171
luminescent 161

m

macro- and mesoporous, *see* meso-macroporous materials
macrocycle
– Borromean 4
– catenate 11
macrocyclic ligand 116, 119–120, 124
macromonomer 82
manipulation with the SFM tip 93
Mannich reaction
– radical-type 348
manno-proteins (MP) 423
MCM-41 51, 53, 55–56, 64–68
MCM-48 51, 59
MCM-50 51
medicinal chemistry 389
melanin 104–105
menthol 330
mesh size distribution 244
meso-macroporous materials 62, 66
mesoporous material
– anodic alumina 61
– chiral mesoporous silica 64–65
– circular ordered mesoporous 67f.
– crystallinity 54
– helical structures 61, 63
– periodic metal-incorporated 55
metal-based anticancer drugs
– oxidation stage 355
metallo-organic 3
metallomacrocycle 23
mica phlogopite 188
Michaelis–Menten behavior 117, 121

microemulsion   57–58
micro-mesoporous materials   47, 56
micromolding   47, 61–62
microporous materials
   –zincophosphates   57
microreactors   402
   –enzymes   403
   –polymer-supported reagents and scavengers   403
   –processesdevelopment   404
microwave
   –condensation   397
   –formation of imide derivatives   398
   –heating   396
   –inorganic supported reagent   397
   –ionic liquids   400
   –MAOS (microwave assisted organic synthesis)   395
   –Paal–Knorr synthesis   396
   –PASP (polymer assisted solution phase synthesis)   399
   –reactors   267
   –scale-up   400
   –solid supports   398
   –solvent-free conditions   397
   –Suzuki coupling   396
mimics   369
mint oil   330
Mo-based pentagonal building blocks   34
Mo-based POM cluster   33
mold   48, 57, 59, 66
molecular electronics
   –rotaxanes   145
molecular lifts   142
molecular machines
   –at interfaces   152
   –extension–contraction motions   143
   –next generation   152
   –operation in solution   152
molecular motors   152
molecular muscles   143
molecular oncology   358
molecular shuttle
   –acid–base-controlled   139, 142
   –light-driven   140
   –rotaxanes   137
molecular wires   26
monomer preorganization   78
multicomponent   346
multicopper oxidases   104
multidentate chelate   415
multiphoton absorption   164f., 168, 173, 175

multiphoton excitation   173
muscovite mica
   –growth direction   187

n
NAMI-A type complexes   366
nanobiotechnology   215
nanobranches   198
nanobuilding blocks   215
nanocasting   59
nanofiber growth
   –frequency doublers   203
   –muscovite mica   187
   –p6P   187
nanomaterials   57
nanoneedles   185
nanoparticles   47, 66
nanoscale supercluster   33
nanoscopic   73
nanoshaping   200
nanostructured carbon   56
nanotechnology 215
naphthalene derivatives   277
naphthalene diimides, see NDIs
native gel electrophoresis   221
natural rubber   245
Nd complex   174
NDI (naphthalene diimides) 265
   –absorptions   269
   –catenanes synthesis   276
   –chromatographic behavior   275
   –core substituents   266. 268, 270, 287
   –electron transfer   281
   –host–guest chemistry   272
   –I-DAN foldamers   272
   –nanotubes   279
   –optical properties   272
   –redox properties   271
   –sensing of aromatic molecules   277
   –solubility   270
   –synthesis   266
   –unsubstituted   266, 268
nephelometry   413
neutral carbohydrates   417
N-heteroaromatic bases
   –electron-rich   338
   –Friedel–Crafts   338
N-hydroxyphthalimide
   –bond dissociation energy (BDE)   341
nitrate photolysis   445

nitrate radical   430
nitrate   432
nitrate-rich groundwater   445
nitrite oxidation   445
nitrite   433
nitrogen dioxide   433
nitro-PAHs   432
nitrophenols   432
nitrous oxide reductase   104–105
NLO, *see* nonlinear optics
NMR Cryoporometry   241–242
NMR   405–412
nonlinear activity   177
nonlinear harmonic generation   163, 169, 176
nonlinear optics (NLO)
 – second-harmonic signal   204
 – frequency doublers   203
 – nonsymmetric substitution   203
nonlinear processes
 – second harmonic generation (SHG)   163, 169,173
 – third harmonic generation (THG)   163
nonlinear properties   180f.
nonpurgeable organic carbon   438
n-type semiconductor   265, 287
nucleophilic carbon-centered radical   337
nucleophilic radical addition   338

o
oligomer   24
oligopeptide polymer conjugates   78, 89
oligopeptides   80
optical properties
 – nanofibers from p6P   186
optoelectronically active polymers   73
organic light-emitting diode
 – electroluminescence (EL) quantum yield   299
 – low turn-on voltage   299
 – maximum brightness   299
organic nanofibers   185
organic pollutants   430
organic templates   49f. 59
organic–inorganic nanostructures, *see* periodic mesoporous organosilicas
organic–lanthanide complexes   165, 176
organometallic ruthenium mimic   369
outdoor environment   445
oxidase
 – *oxy* state   106
oxido-reductive process   337

oxomaritidine   403
oxygen   342
ozone   430

p
π–π interaction   54, 60
π-conjugated derivatives incorporating phosphole ring
 – cyclic voltammetry (CV)   298
 – Fagan–Nugent method   296
 – physical properties   296
 – Sonogashira coupling reactions   296
 – synthesis   296
 – theoretical studies   298
 – UV-vis spectrum   298
 – x-ray diffraction studies   297
palladium nanoparticles   55
para-hexaphenylene (p6P)
 – deposition on substrates   185
para-phenylene nanofibers   187, 190
para-quaterphenylenes
 – organic nanoaggregates   193
 – synthesis   189
parotid saliva   421
PASP: *see* polymer assisted solution phase synthesis
PDIs, see perylene diimides
pectin   417
pentagalloylglucose   410
pentylenetetrazole (PTZ)   279
peptide coupling   84
perfluoroalkyl radical
 – electrophilic character   344
 – perfluoroakylation   344
periodic mesoporous organosilicas   53
permanent water bodies   448
peroxide   254
perylene diimides (PDIs)   265
phenol nitration   445
phenolic compounds
 – antioxidant properties   411
 – dietary burden   411
 – properties   411
phenylalanine hydroxylase   104
phenylthiourea   106, 107
phosphole ligands 300, 303
photo-aging   254
photocatalysis
 – anatase   341
 – rutile   341
 – sunlight   343
 – T_iO_2 341

photochemical reactions   434
  – indirect   432
  – in ice and snow   434
photochemical smog   431
photochemical transformation
  processes   442, 445
photoconductive polymers   79
photocuring   247
photoDSC   237, 247ff.
photo-differential scanning calorimetry,
  see photoDSC
photoexcited triplet states   438
photo-Fenton process   437
photoinduced oxidation of chloride   447
photoinduced shuttling movement   141
photoionization   430
photolysis quantum yield   435
photomineralization   436
photonic crystals
  – inverse opal   64
  – natural opals   64
  – silica microspheres   64
photonitration   446
photooxidation   433
photophysical properties   161, 171
photopolymerization   247–249
photosensitizer   432
photosynthesis, modeling   281288
plant viral capsids   215
plastocyanin   102
platinum-based drugs   355
polar regions   434
pollutant transformation   448
poly(acetylene)   77
poly(diacetylene)s   79
poly(ethylene oxide)   255
poly(isocyanide)   77
polyamidoamine (PAMAM) dendrimers   280
polycyclooctene   251
polycyclic aromatic hydrocarbons   431
polygalacturonic acid   417
polymer conjugates   80
polymer-assisted solution-phase synthesis
  (PASP)   392
polymeric membranes   246
polymeric catenanes   27
polymerization   24
  – by UV   92
polyolefins, see also polyunsaturated
  compounds   246
polyoxometalate-based functional
  nanosystems   42

polyoxometalates, see POMs
polyphenols   104
polysaccharide   417
polysisoprene   245
polyunsaturated compounds   323
POM building block   33
POM clusters
  – conductivities of   40
  – design   43
  – Fe^{III} "linker"   39
  – heteropolyanions   32
  – isopolyanion   32
  – Mo-blue   32
  – Mo-brown   32
  – $\{Mo_2\}$ "linker"   39
  – paramagnetic centers   39
  – physical propertie   31
  – superclusters   38
  – thermally switchable   40
  – thermochromic   40
pore size distribution   242
porosity   240
porous materials   242
potassium carbonate   267
power dependence   163, 166, 173f.
p-quaterphenylenes
  – brightly blue fluorescent nanofibers
    from MOP4   194
  – monofunctionalized   199
  – nonsymmetrically disubstituted   192
  – nonsymmetrically functionalized   197, 204
  – symmetrically disubstituted   192
  – symmetrically functionalized   194
proanthocyanidins   411
programming
  – assembly instructions   13
  – subroutines   27
proline   412
proline-rich proteins (PRP)   417, 421f.
properties of phosphole rings
  – optical and electrochemical   299
  – versatile reactivity   298
protein concentration   414(f)
protein kinase inhibitors   369
proteins   64, 272
protein–tannin aggregates   411
PRPs, see proline-rich proteins
pseudorotaxane   131
pyrazole based ligand   114
pyromellitic diimides (PMIs)   265, 270

q

quantum dots
- CdSe 58
quantum efficiency 170
quaternary structure 93
quercetin 417
quinone 105, 107, 117, 123, 344

r

racemization
- alcohols 331
radiation absorption 435
radical Cl_2
- chloride 446
- α-aminoalkyl 340
- acyl 339, 343
Randles–Sevcik equation 227
RAPTA complexes
- GST-mediated drug resistance 372
RAPTA-NAMI
- antimetastatic drug candidates 373
rate of photolysis 440
reactive radical species 436
redox-activated switches 144
redox activity 53
- ferrocene 227
- viologen 227
reductive radical addition
- aldimine 345
- aminyl radical 345
- amminium radical 346
- imine 345
reticulation 244
rhamnogalacturonan II 423
Rhône river delta 446
ribbons 80, 81, 87, 89
rotaxane
- catalytically active 151
- hydrogen bonding 134
- rotaxane 9^{6+} 141
Ru–arylazopyridine complexes 363
Ru-dimethyl sulfoxide complexes 362
Ru–ketoconazole complexes 369
Ru–organometallic arene complexes 365
Ru–polyaminocarboxylate complexes
- antiviral agents 361
Ru–RAPTA complexes, cytotoxicity 357, 365, 370
Russell–Saunders coupling 164
ruthenium-based drug
- lower toxicity 355
- mimic iron in the binding to biological molecules 355
- range of oxidation states 355
- Ru–polypyridyl complexes 359
- tetradentate cyclam rings 360

s

salivary proteins 411
satraplatin 355
SBA-15 52, 59
$Sc(OTf)_3$ 283
scanning force microscopy images 87
scattering 435
scavengers 392
second- and third-harmonic generation 168
second-harmonic generation (SHG) 163, 169, 171, 173, 178f.
secondary building units (SBUs) 49
selection rules
- even-to-even parity 164
- even-to-odd parity 164
- odd-to-odd parity 164
- parity selection rule 164
- parity-allowed 164
selection rules 165, 168
selective estrogen receptor modulators (SERMs) 368
self-assembly
- cooperative assembly 50
- geometry 51–53, 61, 63, 67
- inorganic precursors 52
- isomorphic substitution 52–53
- liquid crystal 50, 56
- micelle 50, 52
- multiscale self-assembly 59
- surfactant 50, 61, 64, 68
- swelling agents 52
- three-block copolymers 51
self-organization 3
semiquinone 110, 113, 121, 124
sensitized photolysis 432
sensors
- NDIs 277
separation 47, 53, 67
serinol 10
Serratia 374
SFC, see supercritical fluid chromatography
shallow and clear water bodies 443
SHG, see second harmonic generation
shuttling movement 141
soft materials 243
soft matter 237
solar energy conversion 284

solid-phase extraction (SPE)
  – catch-and-release system 391
  – purification 390
  – scavenging systems 391
solid-state REDOR and DOQSY NMR experiments 85
solid-supported reagents 392
sorting
  – differential selectivity 14
SPE, *see* solid phase extraction 390
speeding-up techniques 390
spin crossover 20
stability 52, 56, 61
staurosporine 369
steady-state concentration 440
steric repulsion 10, 20
strain 13
Strecker synthesis
  – radical version 349
Streptomyces 374
structural features 415
structure-directing moieties 36
subcomponent self-assembly 3
submandibular/sublingual saliva 421
substitution
  – heteroanion 40
substrate temperature 196
summertime sunlight spectrum 439
superamphiphiles 77
supercapacitor 56
supercritical fluid chromatography 405
superoxide dismutase 104
supported
  – reagents 392
  – seavengers 392
supramolecular arrays 130
supramolecular assemblies 296, 310f.
supramolecular chemistry
  – NDIs 265
supramolecular 5–6, 13
surface force apparatus 241
surface layer 438
surface waters 435, 441
  surfactant
    – bio-inspired surfactants 64
    – cationic surfactant 51
    – cationic surfactant 65
    – CTAB 64, 68
    – lecithin 68
    – neutral surfactant 61–62
    – quaternary amine 64
    – three-block copolymer 51

Suzuki cross-coupling reactions
  – nonsymmetrically disubstituted derivatives 191
  – symmetrically substituted compounds 191
swelling 244
  – swelling ratio 246
switching mechanism, counterions 146
synthesis of rotaxanes
  – donor–acceptor interactions 135
  – hydrophobic interactions 133
  – synthesis 131
  – transition-metal coordination 136
  – Van der Waals interactions 132

t

tannin specific activity (TSA) 419
tannin 410
tannin/protein ratio 415
tapes 80–81, 87, 89
targeted therapy 358
technologies in medicinal chemistry 389
temperatures range
  – generating nanofibers 193
template
  – key functions 49
  – molecular templates 48
  – porogen 49
  – requirements 48
template-assisted synthesis 47
temporary water resources 449
thermodynamic 3, 6, 18, 20
thermoporosimetry 237, 241–243
THG (third harmonic generation) 163
Ti(III), reduction power 345
Ti(IV), Lewis acid 345
top-down 73
topochemical polymerization 78
  – diacetylene 79
topological 11, 27
topology
  – Borromean link 4
  – catenanes 4, 11, 27
  – helicates 4
  – macrocycles 4
  – rotaxanes 4
  – Solomon link 4
  – template 4
transferrin delivery mechanism
  – indazolium compound KP1019 367
transferrin-mediated drug transport 367

transitions in confined geometries 238
transmission electron microscopy 227
triazines 285
tropospheric gas phase 430
turnover rate
— FAD to PQQ 148
— gold nanoparticles 148
tweezers, molecular 285
tyrosinase activity 124
tyrosinase 104, 108, 113

u

umbrella-rotaxane 150
unsatisfied valences 13
up-conversion 162f.
UPLC (ultra performance liquid chromatography) 405
UV and Raman spectra 92
UV-visible Spectroscopy
— π–π^* transitions 9
— metal-to-ligand charge transfer 9
vacuum sublimation processes 185

v

valence satisfaction 13–14
— coordinative saturation 15
— hetero-ligand 15
— homo-ligand 15
vascular endothelial growth factor (VEGF) 358
viruses
— animal viruses 216
— bacteriophages 216
— plant viruses 216
— viral capsids 215
— viral particles 216
volatile organic compounds 431

w

walking sticks
— MOCNP4 198
water column 435
W-based POM clusters 35
Wells-Dawson structure 40
wetting layer 189, 203
wine polysaccharides 423
wine 411
wires 59f., 80, 95, 146, 266, 280

x

xanthan 417
xenobiotics 448
x-ray diffraction (XRD) 53, 189

z

zeolite
— Brönsted acidity 48
— fluoride route 49
— hydrothermal treatment 49
— isomorphic substitution 48
— ITQ-29 60
— LTA zeolite 60
— mesoporous zeolites 55
— quaternary amines 48–49
— Si/Al 49
— zeotypes 49
— ZSM-5 49
zinc complexes 377
zinc(II)porphyrin (ZnP) 281, 285f.